TEXTBOOK OF THEORETICAL BOTANY

Hortus Sanitatis Mainz 1491

(*Frontispiece*). A Botany Class in the Middle Ages. The teacher might be Albertus Magnus of the University of Paris, and, judging by the ferocity of his expression, the younger figure on the right could be Roger Bacon, who, as a student, incurred the enmity of Albertus by getting the better of him in a debate.

TEXTBOOK OF
THEORETICAL
BOTANY

BY

R. C. McLEAN
M.A., D.Sc., F.L.S.

AND THE LATE

W. R. IVIMEY-COOK
B.Sc., Ph.D. F.L.S.

VOLUME 4

With 85 illustrations in the text.

LONGMAN

580

LONGMAN GROUP LIMITED
London
*Associated companies, branches and representatives
throughout the world*

© Longman Group Limited 1973

First published 1973

ISBN 0 582 44114 5 ✓

PRINTED IN HONG KONG BY
DAI NIPPON PRINTING CO. (H.K.) LTD

ACKNOWLEDGMENTS

I have to acknowledge with gratitude the essential help given by Mrs. Ivimey Cook in undertaking the whole of the typing of this Volume as of the previous Volumes. Her technical knowledge of the subject matter was a great asset in this laborious task.

I wish also to acknowledge the kindness of Professor G. F. Asprey in allowing me the continued use of a room in the Department of Botany at University College, Cardiff.

Thanks are also due to the following for permission to use illustrations: Academic Press Inc. for a figure from Williams, *Pattern in the Balance of Nature*; Aldine-Atherton Inc. for a figure from Bunting, *The Geography of Soils*; Cambridge University Press for a figure from Richards, *The Tropical Rain Forest*; The Clarendon Press for a figure from Raunkiaer, *The Life Forms of Plants and Plant Geography*: Ecological Society of America for a figure from *Ecology* 1951, Vol. 32; W. H. Freeman & Co. for two figures from *The Scientific American*; Gustav Fischer Verlag for figures from Lundegardh, *Klima und Boden* 4th Edn. and Walter and Lieth, *Weltatlas der Klimatologie*; Harper and Row for a figure from Gate, *Energy Exchange in the Biosphere*; Hohenlohe'sche Buchhandlung Öhringen for a figure from *Der Wald als Lebensgemeinschaft*; H.M.S.O. for a photograph of Wistmans Wood; Hutchinson Publishing Group Ltd. for a figure from Bunting, *Geography of Soils*; Dr. W. Junk N. V. for a figure from Croizat, *Manual of Phytogeography*; Longman Group Ltd. for two figures from Good, *The Geography of the Flowering Plants*; McGraw-Hill Book Company for a figure from Duggar, *Biological Effects of Radiation* Vol. 2; Springer-Verlag for a figure from Stocker and Hodlheide, *Handbuch der Pflanzlichen Physiologie* Vol. 3 and figures from Braun-Blanquet, *Planzensoziologie*; Tropical Science Center for a figure from Holdridge, *Life Zone Ecology*; United States Government Printing Office for a figure from *U.S. Yearbook of Agriculture 'Water'* 1955; John Wiley & Sons Inc. for a figure from Daubenmire, *Plants and Environments* and Hutchinson, *Limnology* Vol. 1.

PREFACE TO VOLUME 4

It was originally intended to conclude the Text Book with this Volume, including in it Plant Pathology and Economic Botany, but this proved to be too much for one Volume, so it was decided to reserve the latter two subjects for a fifth and final volume, which, it is hoped, may follow without undue delay.

Ecology has lately attracted a good deal of popular attention and the word has been mis-applied in a variety of ways, particularly in its ostensible relationship to Sociology. Most of the extensive literature of Plant Ecology deals with special aspects of the subject and attempts to survey its general principles have been few. Many people who meet this unfamiliar word for the first time are moved to ask, 'What is Ecology?' and 'What does it do?' The present sketch may be useful in answering such questions, so far as plant life is concerned.

CONTENTS

PART TWO

PRINCIPLES OF PLANT GEOGRAPHY

PART ONE

PRINCIPLES OF PLANT ECOLOGY

CHAPTER XXXIX

A RETROSPECT OF ECOLOGY

ECOLOGY is a comparatively junior branch of science and its territory has not yet been defined with the exactitude and general acceptance that we expect from the older disciplines. There are considerable differences of approach between different writers of equal experience. To some it is a limited field of study which they define and sometimes subdivide in a systematic fashion according to their own creed, which, it may be added, often reflects conditions in the part of the world in which they happen to live. To others it represents more vaguely a summation or synthesis of all biological knowledge, metaphorically a sort of cream which rises to the top when biological knowledge has sufficiently matured.

Observations of the kind we would now call ecological were made by Theophrastus and other ancient writers, but the ideas involved were never clearly grasped or formulated until recent times. The name was given by Ernst Haeckel in 1869, based on the Greek word *oikos*, which means a dwelling place or home,* and it was intended to signify, in the most general terms, the study of the relationships between living organisms and their natural environments.

This concept grew out of a study which is older and in many respects different, namely, that of the geography of plants, which was in the first place purely floristic, that is to say it was concerned with the geographical distribution of plant species. This study, which may be said to have begun with Christian Menzel in the mid-eighteenth century, was at first purely systematic in its outlook. The idea of plant communities was not present, and when, for example, Linnaeus uses terms like *pinetum* or *ericetum* they are simply descriptive of localities. In 1790, however, Willdenov began to recognize that there were natural communities of plants, most obviously those characterized by one prominent species. Willdenov passed on his ideas to Alexander von Humboldt, who held them in mind during his extensive travels in Tropical and South America. As a result he developed, in a series of publications, his views on 'the phenomenon of social plants', treating at first only of groupings of single species, which he realized were more characteristic of temperate than of tropical climates, but later extending the concept of 'social plants' to include recurrent groupings of plants of distinct species, which marked the birth of the concept of 'plant associations' (1805), since developed through many hands into a veritable science of Plant Sociology. This we may regard as the earliest aspect of what has now become Ecology.

* More correctly, the word signifies a homestead and, hence, embodies the idea of a self-contained community.

Plant Sociology is distinct from, though closely related to Plant Geography. The former is bound to enter into any treatment of the latter subject, but pure sociology need not have any geographical content, as it concerns itself with the relationships existing within social groups of plants. In other words its outlook is intensive while that of Plant Geography is extensive.

Although we have indicated Plant Sociology as the groundwork of Ecology, it is only fair to say that it has been strenuously claimed by some of its exponents that it is a distinct branch of science. If, however, we call Ecology the study of the plant in relation to its natural environment, the conclusion is inescapable, that the other plants in whose company it grows are an integral part, indeed are often a determining part, of its environment and few if any ecologists at the present day would exclude sociological analysis from the scope of their interests. Indeed, in applying ecological ideas to any area for the first time, the primary endeavour has usually been to recognize and define the natural plant 'associations' occurring in the area and thereafter to determine their dependence on the circumstances of their environment; thus repeating, in effect, the historical sequence in the development of the science.

An early successor to von Humboldt was F. J. Mayen, whose *Grundriss der Pflanzengeographie* was published in 1836 and in English translation in 1846. A traveller himself, he wrote in a lively and descriptive style, but his interest in plant communities as such was very limited and he speaks of the social growth of plants as if it were only a phenomenon of single species, favoured by local soil conditions He contests from his experience the received idea that the social growth of plants did not occur in the tropics. He uses twenty physiognomic forms of the same kind as von Humboldt used, and these he employs as the basis for his delimitation of vegetational regions. He criticizes previous attempts, such as that of Schouw, to found such regions on floristic grounds, for the good reason that too little was yet known of systematic distribution, whereas physiognomy is obvious and recognizable at sight, and also that the causes of systematic distribution were quite unknown. He has some interesting remarks or the benefit it would be to the study of distribution if the mutability of species could be admitted and he says that 'many naturalists would be very much inclined to admit it', if only there were good evidence in its favour. However, he relies on the principle that 'a species is that which is constant in nature' (Link) and if it be not constant then it is not a true species; an excellent example of begging the question.

During the greater part of the nineteenth century there was great activity in exploration, with a consequent immense extension of knowledge about the world's vegetation. Much of the new information was purely floristic, but Grisebach, in a series of annual reports between 1840 and 1853, succeeded in establishing the concept of plant 'formations'. He also conceived in 1866 the idea of a deeper Plant Geography in which physiological, morphological and systematic considerations would all play a part, under the name of 'Geobotany', an idea which grew until it found full expression in Schimper's classic *Plant Geography on a Physiological Basis* which appeared in 1898 (English translation, 1903). (See also p. 3325.)

Morphology entered early into the discussion of plant societies under the title of 'physiognomy', that is the discrimination of plant societies according to the predominant growth form, which determines their appearance. Humboldt used this characteristic widely, and he distinguished altogether nineteen forms, explicitly only as a beginning. However, he used criteria of form which had little or no ecological significance, but were based upon the characteristic growth forms of certain systematic affinities; for example, the palm form, the cactus form, the aroid form, the grass form, the banana form. These forms have for the most part an hereditary basis and are sometimes called 'phyads'. Their employment as ecological indices proved a blind alley, a defect recognized by Heer who, in 1835, substituted a much simpler division into trees, shrubs, semi-shrubs and herbs, physiognomical classes which centuries before had misled the early systematists, but which now found a true ecological meaning. This system was elaborated by Kerner in 1863, who defined twelve physiognomical classes which he called '*Grundformen*'.* These included some additions to the four primary forms, such as: leaf plants (mostly ferns), felted plants (mostly mosses), climbing plants and, finally, encrusting plants (mostly lichens). A social plant-mass of the same ground-form he called a '*Bestand*'. which may be translated as a *stock* or simply a *stand*, which might be either 'open' or 'closed' according to its density. He used the term *plant formation*, now often called a 'phytocoenosis', for a linked complex in which various *Bestände* occurred as distinct layers. This term 'formation', which has haunted Ecology ever since, is thus seen to have had a morphological origin in the ground-forms, but these forms had a living, ecological, not a purely systematic importance. These were fertile ideas which later developed in a variety of ways, as we shall see.

While comprehensive ideas of plant communities were being created, other workers directed their interest towards detailed analysis of the specific constitution of communities. Here we must mention Lecoq, who surveyed French heathlands in 1844. He introduced the term 'association', which has since developed into a most important concept of Plant Sociology, but which he did not use in a sociological sense at all but simply to designate a particular sample of heath. He did, however, analyse, with the use of a numerical scale of 1–10, the relative importance of species in the vegetation and used terms of classification which have retained a value. Thus species in grade 10 he called dominant species, 9–6 essential species, 5–3 accessory species, and 2–1 accidental species. In thus arranging a hierarchy of components it is difficult to believe that he had not already grasped instinctively the idea of an ordered community among plants, although the concept of the permanent individuality of plant communities had not yet arrived.

About the same time a notable step was taken by Colniess in Russia, who first used the *quadrat* system of analysis on the vegetation of the South Russian Steppes. This involved the use of squares as test areas in the various plant

* Although this term seems straightforward enough, there does not seem to be an English equivalent which expresses the sense of the German. Perhaps 'archetypes' would be the most suitable translation.

communities of the grass steppes, with the exact marking in each area of the individual plants of each species found there. His maps, which were published by Köppen in 1845, were thus the first to show the details of the grouping of species and of their distribution within a plant community.

This trend towards closer analysis of vegetation found a notable exponent in Hampus von Post who worked in middle Sweden from 1842 onwards. In 1851 he published his conclusions on the methods of vegetation analysis, which were notable for their clear and modern outlook. He advocated that all conclusions about the character of vegetation must be founded on the examination of the actual vegetation on the ground and not on more or less hypothetical views about the environmental conditions. This scheme, which seems obvious to a worker at the present time, was in fact a revolt against what was called the 'deductive school' represented by Sendtner, which based the concept of the plant community on the physical and chemical features of the particular area and the knowledge, or supposed knowledge, of the requirements of the species comprising the local flora. Given such knowledge, it was argued, the plant community was bound to consist of the selection of species whose requirements coincided with the characteristics of any given area. Such a theoretical approach produced ideas of plant communities which were purely subjective constructions, seldom with any relation to observed facts. Even presupposing perfect physiological information, which still lies in the future, the deductive scheme left out of account too many other factors, biological or historical, to stand much chance of accurately reflecting nature.

Hampus von Post called each individual example of any type of vegetation a 'locality', but we still lack a universally acceptable name for such entities, that is for a given area of bog, woodland or whatever it may be, as a thing in itself. We shall come back to this later in this chapter (p. 3404), but von Post first recognized the need for a term. His analysis was directed to finding four facts: which species lived in the area, what was their relative abundance, what was their relative coverage of the area, and to what physiognomical life-form they each belonged. His life-forms were very simple, being trees and shrubs, herbs and grasses respectively, cryptogams being lumped together under the last heading. The result was that his localities were for the most part natural plant communities, and localities with closely similar communities were brought together as 'vegetation groups', corresponding more or less to formations.

In 1862 he published a survey of the vegetation of middle Sweden based on his principles and united his Vegetation Groups into Vegetation Provinces and Vegetation Kingdoms, insisting that a study of localities was the true basis of Plant Geography. He was also the first to draw attention to the parellelisms between plant communities and those of animals and men, which later led to remarkable developments of Ecology.

In his later writing von Post seems to extend this term 'locality' to the plant community itself, giving it the meaning of a group of species which occur together as a natural community depending on certain environmental conditions, which brings it close to what we would now call an 'association'.

Among the workers of this period we must mention Ragnar Hult, who laid down in the clearest form the opposition between the deductive school, which we have mentioned, and the inductive methodology, which he claimed to be the only truly scientific approach to plant sociology. His declared aim was the search for the general ground principles in the structure of plant formations, without regard to local conditions or other extraneous factors. This may be regarded as a declaration of 'pure' sociology which it may be granted here has, *ex hypothesi*, little to do with ecology. It still has followers, and Hult was one of the founders of the Uppsala school of sociologists which has had some distinguished adherents, such as Sernander and du Rietz.

Hult's method consisted of the listing of species in homogeneous samples of vegetation, arranged in quantitative order on a scale of five grades, the highest grade being in fact those species with the greatest coverage. Species were also grouped in seven layers and in each layer the (ten) *Grundformen* were arranged in the same manner as the species into quantitative grades. This information was then arranged in diagrammatic form as a diagnosis of the community. The diagram was later modified by Sernander, as shown in Fig. 4.1.

In spite of its claim to use inductive principles there is an air of artificiality about all this which recalls the efforts of early systematists to found systems of classification on the logical application of selected criteria. It should be pointed out that although the method is inductive the criteria employed are not, but are a group of subjective assessments resting on theoretical assumptions. Furthermore, such a view of vegetation is open to the criticism that it is static and treats each vegetative type, or formation, as if it were a permanent entity, much as the pre-Darwinian systematists regarded species.

Two developments militated against this formality. One was the rise of physiological plant anatomy which gave an impetus to the interpretation of plant structures on a functional basis, thus leading away from the formal concept of *Grundformen*. The second was, of course, the rise of Darwinism with the ideas of 'fitness' and 'adaptation', which laid so much stress upon the influence of the environment that it seemed useless to consider plant communities in a purely sociological framework without reference to the character of the environment with which they were associated. Moreover, sociological analysis had previously tended to ignore the fact that for each species involved the community itself was a biotic environment. Hence there arose a school for whom physiological considerations assumed the prime importance in questions of plant geography and in the study of vegetation. Among such workers a truly ecological outlook was born. To them, the *Grundformen* became simply an expression of convergence of structure among unrelated species due directly to adaptation to a similar environment, the 'epharmonic convergence' of Vesque (1882).

One of the first exponents of these views was Oscar Drude, the pupil and follower of Grisebach, who published between 1876 and 1918. He is particularly known for his *Handbuch der Pflanzengeographie*, published in 1890, in which he presented an outline of the world's vegetation in which he sought to substitute biological factors for the physiognomical *Grundformen* of Humboldt and Grisebach. To Drude the unit of classification is still the formation, and in fact his formations are mostly distinguished from each other on a physiognomical basis, though he seeks to relate them to corresponding environmental conditions. This cannot be regarded as simply an extension of Grise-

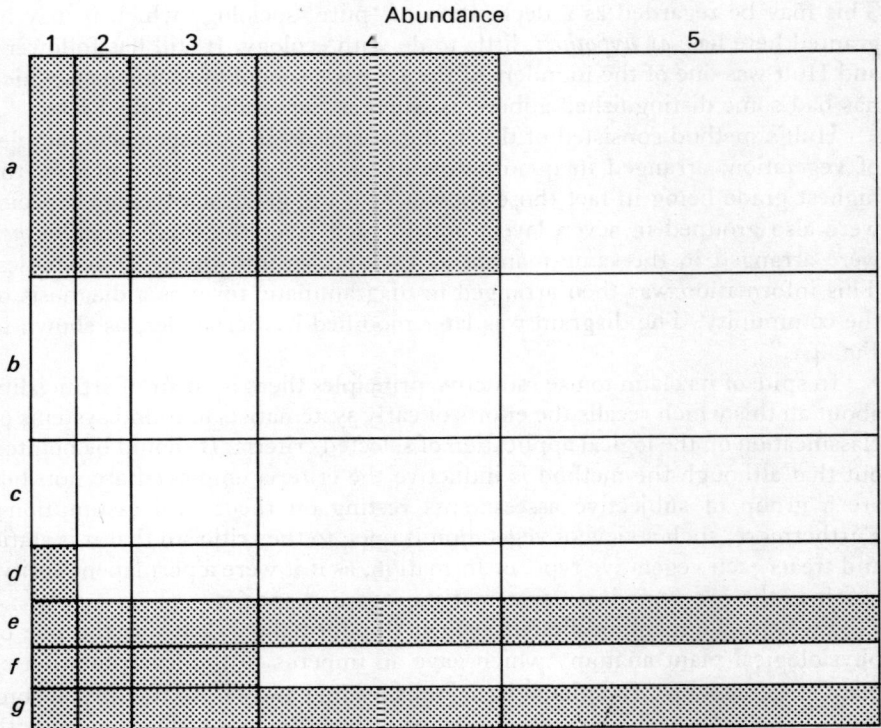

Abundance

1 2 3 4 5

a
b
c
d
e
f
g

FIG. 4.1. The Hult-Sernander analysis diagram, from an example by Lagerberg. A moss-rich pine wood with a dense *Vaccinium* layer. Abundance classes: 1. Isolated. 2. Scanty. 3. Distributed. 4. Abundant. 5. Covering. *a.* High-tree layer (*Pinus*). *b.* Lower-tree layer. *c.* Shrub layer. *d.* High-field layer (*Picea*). *e.* Middle-field layer (*Vacc. myrtillus*). *f.* Lower-field layer (*Vacc. vitis-idaea* and *Calluna*). *g.* Ground layer (*Hylocomium parietinum*).

bach's views. In reality it is a compromise between the purely physiognomical standpoint of Grisebach and the purely physiological standpoint of Sendtner and his followers, who founded their classification of vegetation types on the environmental conditions obtaining in each locality. Although the present-day ecologist believes that the vegetation type cannot be considered apart from the conditions in which it exists, he would not accept the dogma that similar conditions must provide similar vegetation types. That was going too far, and the earlier plant geographers were justly criticized for accepting this dogma too unreservedly and also for applying the argument in reverse, that similar vegetation types implied similar conditions of life, when, in fact, information under both headings was necessarily imperfect over most of the world's surface. Such actions imply too close an adherence to the concept of adaptation and ignore what we now know of the range of tolerance shown by species.

Drude's survey of world vegetation led him to make a clear distinction between floristic provinces and vegetational regions which he expressed in his *Oekologie der Pflanzen* in 1913. This is an important distinction. It means, for example, that if a vegetational region extends into two or more different floristic provinces, the dominant species of a given plant community will differ, according to the floristic province in which it lies, though they may belong to the same physiognomical *Grundform*.

Drude's new viewpoint enunciates that the *Grundformen* are no longer to be regarded as merely natural data, subjectively apprehended, but that they have a discoverable ecological meaning, a viewpoint which is essentially a true ecological approach. The acceptance of such a viewpoint does not invalidate sociological analysis; it becomes no longer an end in itself but, on the contrary, states a question which must have, in the widest sense, an ecological answer. The distribution and grouping of species which are revealed by analysis are not explicable, or predictable, by reference to the physical and chemical environment alone, for the biotic environment has its independent influence. The interactions of species on one another and their effects on the environment, or the actions of animals, including man, may often determine the variation of vegetation over areas which are essentially homogeneous in physical respects.

The formation was for Drude an area, of any size, consisting of a vegetation of more or less uniform physiognomical type, though he did not apply this consistently and sometimes used an ecological definition, as in the Salt Marsh Formation. The formation was divisible into types, according to the prevailing local conditions and the types into facies, distinguished by their predominant species.

Drude's ideas had a very widespread influence since they formed the basis of the great work on the 'Vegetation of the Earth' which he initiated in 1910 with Adolf Engler, a work never completed, but consisting of many volumes of detailed monographs on different countries by specialist authors, among which may be mentioned Engler on Tropical Africa, Willkomm on Spain, Reiche on Chile and Diels on Western Australia.

Following in Drude's footsteps there came Eugene Warming in Denmark, who published in 1896 his *Plantesamfund* or *Foundations of Ecological Plant Geography*. This important work was eventually rewritten for publication in English in 1907 as *Oecology of Plants*, the first textbook to use such a title, though not of course the first use of the name. Warming uses an ecological rather than a physiognomical basis for his classification of plant communities and established thirteen 'classes' of vegetation, influenced by the fundamental division of plants into 'xerophilous' and 'hygrophilous' which was introduced by Thurmann in 1849. These classes, which generally correspond to ideas still in use, were the following:

1. Hydrophytes or aquatic plants.
2. Helophytes or marsh plants.
3. Oxylophytes or plants of acid soils.
4. Psychrophytes or plants of cold soils.
5. Halophytes or plants of saline soils.
6. Lithophytes or plants growing on rocks.
7. Psammophytes or plants of sandy soils.
8. Chersophytes or plants of wastelands.
9. Eremophytes or plants of deserts and steppes.
10. Psilophytes or savannah plants.
11. Sclerophytes or plants of bush and forest with hard evergreen leaves.
12. Coniferous forests.
13. Mesophytes or plants of moderate humidities.

These are intended as groups or unions of formations and they are sub-divided into formations, sometimes on a physiognomical basis, e.g. 'grass steppe', sometimes on an ecological basis, e.g. 'psammophilous halophytes'. It will be noticed also that in classes 11 and 12, above, he uses a physiognomical rather than an ecological criterion of similarity. Nevertheless, Warming lays it down that the decisive factor in the separation of his classes is the amount of water in the soil. Thus, in classes 1 and 2 the soil is *very wet*. In classes 3–5 the soil is *physiologically dry*, a concept which we shall refer to later (p. 3630). In classes 6–8 the soil is *physically dry*. In classes 9–11 both soil and climate are *physically dry*. In class 12, coniferous forest, the soil is either *physiologically* or *physically dry*. Finally, in class 13 we have the large group on soils that are *moderately moist*.

Warming endeavoured to refine the concept of physiognomy by the defini-tion of 'growth-forms', more closely related to environment than were the *Grundformen* and more closely connected to their mode of life than were the types of vegetative form used by Grisebach and Drude. He arranged his growth-forms in six groups; heterotrophic plants (parasites and saprophytes); aquatic plants; muscoid, lichencid and lianoid (woody climbers); and, lastly, all other autonomous land plants. The latter big group is subdivided into monocarpic herbs (annuals and biennials), and polycarpic plants or perennials. The latter he grouped according to their vegetative habit of growth, with special regard to duration of shoots, length and direction of internodes and the structure and position of buds. Thus he made four groups: renascent herbs, with temporary shoots seasonally produced; rosette plants; creeping plants; and plants with erect perennial shoots. Each group is to some extent heterogeneous and includes very various types. The rosette habit, for example, covers such diverse types as *Taraxacum*, *Musa* and *Yucca*.

Physiognomy, to Warming, included more than the growth-form of the prominent species in any given area of vegetation. To build up the picture of the physiognomy of vegetation, which is also to a large extent the physiognomy of landscape, he utilized other criteria such as density, height of plants, dura-tion of life, the number of species involved and their seasonal phases.

Warming made it quite clear that plant communities were not merely theoretical groupings of plants on an ecological basis nor just lists of species, but were natural entities which required to be recognized and examined on the ground in relation to their environment. He has had so many successors that we are apt nowadays to regard such a proposition as a self-evident axiom of Ecology, forgetting that before Warming's time this was not always so.

One of the most prominent and influential contemporaries of Warming was A. F. W. Schimper, who was led through studies on tropical vegetation to an interest in plant geography, which he saw as an ecological science. His great book, published in 1898 and in English in 1903 as *Plant Geography on a Physiological Basis*, gave descriptions of world vegetation types that were so richly factual and so splendidly illustrated that it became universally popular and has remained a classic.

Schimper's units are Formations, which he describes in the following way. The vegetational covering of the earth is determined by three factors: temperature; rainfall (including the influence of wind); and soil conditions. Among these three, the soil conditions modify the material provided by the influence of the other two, leading to the appearance of larger or smaller divisions of unitary ecological and floristic type, which are exactly repeated in similar soil conditions under a uniform climate. On the other hand, similar soil conditions under two different climates will produce different vegetation. He then defines his formations as plant communities determined by the soil conditions, each one being characterized by one or more definitive species; the more abundant subsidiary species can only cause variation in the facies of the formation.

This seems to tie down the formation unambiguously to the soil conditions of which it is the expression, yet Schimper proceeds thereafter to divide them into two groups; climatic formations and edaphic formations. The term 'edaphic' he introduced to designate the sum total of factors acting in or through the soil. This is a classification which has been repeatedly criticized and, indeed, it cannot be maintained in a literal sense: in view of his own definition of a formation it is paradoxical and should never be regarded as a dogma. Nevertheless, it served Schimper himself quite well as a means for discriminating between the broad aspects of vegetation in a geographical sense and the more or less local variations.

The climatically determined vegetation he divided into three physiognomical classes: woodland, grassland and desert, and these, in turn, into units which correspond more nearly to what we now regard as formations, e.g. rain forest, monsoon forest, deciduous summer forest, etc., Schimper's climatology followed broad lines of interrelationship between temperature and rain precipitation, ranging from hot districts constantly moist to cold districts predominantly dry. In each area he deals first with the determining effects of the climate on the overall aspects of the vegetation and then with the edaphic formations which serve to diversify the general picture.

We are now on the threshold of modern Ecology. We have seen how the interest in Plant Geography gave rise to an interest in plant communities and hence to Plant Sociology, and this, in turn, led to the investigation of the environments of the different communities which acquired the name of formations. The natural development of such investigations was towards experiment, which is the keynote of present-day Ecology. The recent history of the subject involves the growth of a number of distinct schools of ecological thinking in widely separated centres, whose diverse views were not unnaturally influenced by the conditions in the areas with which they were most familiar. This has instilled an almost nationalistic flavour into the literature, such that the views a writer holds can be deduced in most cases from the country to which he belongs, a situation peculiar to Ecology. These schools generally owe their origin to the influence of some outstanding teacher or a small group of teachers at a particular university. It is a situation which is doubtless a concomitant of the immaturity of the subject and it reproduces a phase which older sciences left behind them when individualities became merged in universal tenets. Some tenets of Ecology are already practically universal and the differences of schools are not fundamental oppositions, but are due rather to differences of approach and the application of different scales of value in face of facts.

To enumerate all the centres in which Ecology has been studied actively would be almost impossible and to attempt to assign seniority among them would be equally impossible, since historical development has been, in most cases, gradual and slow. The following are therefore arranged alphabetically by countries and the list is limited to those centres which have achieved some celebrity for constructive thinking in advancement of the Science. The names given are those of the men who may be regarded as the founders or chief exponents of the particular schools.

Belgium: Massart.
Denmark: Warming, Raunkiaer.
Finland: Hult, Cajander.
France (Montpellier): Flahault, Pavillard Braun-Blanquet.
Germany: Grisebach, Drude, Kerner.
Great Britain: R. Smith, Tansley, Moss.
Sweden (Uppsala): Sernander, du Rietz.
Switzerland (Zürich): Schröter, Rübel, Gams.
U.S.A.: Cowles (Chicago), Clements (Minnesota).

The differences between these schools arise not only from the use of different terminologies, but also from deeper causes. Tansley* expresses it thus: 'The study of vegetation has suffered, and is still suffering, from the wide and deep differences of training and consequent angle of approach, and from the geographical isolation of individual workers, with the resulting differences in the data they have to work upon.' International appeals by Flahault and Schröter, by Rübel and Gams and by Tansley himself, failed to bring about unification, and lapse of time has, if anything, intensified rather

* TANSLEY, A. G. (1920). 'The classification of vegetation'. *J. Ecol.*, 8, p. 118. A thoughtful survey, with an appeal which unfortunately remains as cogent today as when it was written.

than diminished the cleavage. The human being, like the plant, cannot escape from the influences of his environment and the longer a particular environment or intellectual climate persists the more compelling becomes its impression upon the individual. It follows that no attempt to present Ecology as a science is likely to attract more than a limited acceptance and if it endeavours to make a synthetic approach it may only incur unanimous rejection by all schools.

The incursion of experimental methods into Ecology, which is the most marked feature of the modern period, arose from the increasing appreciation of the determining effect of the environment on the structure of vegetational communities, coupled with the realization of our ignorance of what environments really consisted. The growing concentration of interest on environments, on the *oikos*, produced a drift away from Plant Sociology which allowed a generation to grow up to whom Ecology meant experimental investigation and little else. This justified plant sociologists in declaring that their study was not part of Ecology. Fortunately this particular cleavage has been largely healed. Schröter contested it by coining the term *Synecology* for the study of communities, and the development of modern statistical methods of vegetation analysis and the concept of 'pattern' in vegetation have all helped to preserve the study of communities as an integral part of Ecology.

Indeed, the recognition and designation of distinctive plant communities is part of the foundations of Ecology. It is in no sense a thing apart, but is an essential preliminary and a necessary condition for the study of the internal economy and the external relationships of the plant community, which is the most important task of Ecology. We have here, in application to the community a parallel to the complementary studies of morphology and physiology as applied to the individual organism. The morphology of the community involves its systematic composition and the proportions and relationships of its component species to one another, together with its integrity as a community and its developmental life history. The physiology of the community involves the experimental study of the life histories of the component species and their separate interactions with their environment, as well as the relationships of the community as a whole to determining conditions in climate, soil and in the competition of species, on which the existence of the community as a community depends. This is a very large programme and we are far from having such a full picture of the life of any single community. Ecology is only beginning.

It is difficult to date the introduction of the experimental method into the study of vegetation. The reason is that it grew gradually out of physiological experiments and that a great deal of purely physiological data is relevant to environmental relationships and is a necessary part of their investigation. Clements, in America, refused to draw any distinction between the two fields and in 1907 was teaching them in Minnesota as one subject. A definite landmark, however, was the publication by Clements in 1905 of his *Research Methods in Ecology*, which was the first comprehensive scheme for the application of experimental methods in the field. Though some of it now seems archaic, it undoubtedly opened a door, at least among English-speaking peoples.

Clements was in several ways an innovator, and like most innovators he was the target of much criticism, but time levelled away many rugosities and Clements' valuable contributions can now be better seen and judged. Perhaps his most important effect on Ecology was his insistence on the dynamic nature of vegetation, expressed in the doctrine of *succession*, that is to say, the idea that every plant community is part of a series of communities and is either changing or has changed continuously towards an optimum which he called the *climax*. This, in his view, was fundamentally determined by climate and was uniform and stable so long as the climate was uniform and stable.

This concept, which was first expressed by Cowles, has been criticized as an attempt to import into plant sociology the outworn idea of basing a natural system of classification of species on theoretical phylogeny. The concept was not, however, founded on a suppositious phylogeny of plant communities, but on their ontogeny, which is an observable fact. This will be discussed more fully when we come to consider the nature of plant communities (p. 3380). Clements was also responsible for the theory that a plant community was literally a super-organism, with a life history and a close correlation of parts parallel to the phenomena of the individual organism. This was an extreme view, in which he has not been followed. The comparison is no more than an analogy and it is not difficult to see that on analysis the analogy is defective in important respects. Two other contributions due to Clements may be mentioned. One was the use of 'phytometers' in ecological experiment, that is the use of standard plants of a species whose reactions had been fully investigated in the laboratory, as a means of evaluating the action of the environmental factors in the field, thus replacing, or at least supplementing, instrumental measurements by a living agent, the relationship of which to external influences was direct and immediate. Theoretically this sounds ideal, but in practice it has proved difficult to disentangle the effects on the plant of different factors, some of which may be contradictory of each other. Practically therefore the method is of limited application. Clements also devised an extensive technical terminology for ecological phenomena, some of which has passed into general use, a useful step towards systematizing the science as he conceived it. His work gave a great impetus to the study of Ecology in the United States and has had an enduring influence both there and in Britain, for the ideas he developed were particularly applicable to a large country, where extensive areas of natural vegetation display the effects of climate in a striking manner. One of his associates, Weaver, was responsible in 1919 for developing to a fine art the study of the Ecology of root systems introduced by Shantz in 1906, and facilitated by the deep soils of the prairies, which directed general attention to a neglected, but very important, aspect of the vegetational scene.

Although Clements stands out prominently in the history of Ecology in North America he was not its pioneer. Vegetation studies began before his time and as early as 1891 Ganong made an ecological study of some Canadian peat bogs and MacMillan began describing vegetation in Minnesota. There was also a parallel development in Chicago and the second volume of the *Chicago Textbook of Botany*, written by Cowles in 1911, was devoted to Ecology. It is entirely different from Warming's text. Plant communities are relegated to a few pages at the end of the book, rather the subject matter is what is sometimes called the 'bionomics' or 'natural history' of plants, that is to say, their variety, in form and structure, in relation to the manifold conditions of their lives or, in other words, their adaptations. Cowles was, however, very critical about the use of such a term, and argues at some length against the idea of direct adaptation in response to environmental influences and against teleological ideas generally, by which is implied all interpretations which involve the notion of purposiveness.

Limited conditions in European countries have led naturally to an intensive development of local and regional studies dating back to Wahlenberg's survey of the vegetation of northern Switzerland in 1813, incidentally rather a remarkable date for peaceful survey work. The earliest vegetation map of a locality appears to have been that issued by H. von Post in 1851 of the Äs estate in Södermanland, and that of an entire country was the map of France by Flahault in 1897. So far as Britain was concerned such surveys began in Scotland with the work of R. and W. G. Smith in 1898, which led to the formation in 1904 of the British Vegetation Committee, and this, in turn, led to the production by Tansley in 1911 of *Types of British Vegetation*, with sectional contributions from Moss, Rankin and other active members of the Committee. The foundation of the British Ecological Society in 1913 extended the interest to a national level, while the Society's *Journal of Ecology* provided the first outlet for international discussion of ecological questions.

It is natural that in a small country like Britain with a complex and variable geological structure, the marked local effects of soil factors on vegetation should influence the outlook of ecologists and create resistance to too much reliance upon climate as the sole determining factor in defining the units for the classification of vegetation. With this went the belief that too little was known about the vegetation of the world as a whole to encourage the formulation of worldwide syntheses, and that a period of intensive local investigation must precede attempts to formulate principles of universal application. We are still living in that period.

The cooperative attack upon British vegetation is in contrast with contemporary surveys in the first decade of the century. Graebner's (1909) *Pflanzenwelt Deutschlands* and the volumes of *Die Vegetation der Erde*, beginning in 1910, were all the work of individuals and perpetuated to a relatively late date the early tradition of science, in which personalities were all-important, before teams and ideas became the propelling forces. Such a state of affairs is expressive of the sparse development of ecology at the time and the degree of isolation which its exponents endured.

Since the beginning of the century, and indeed before that, local investigations have been carried on and have extended all over the world, so that the requisite body of knowledge is gradually being built up. The remarkable degree to which specialist studies of particular examples of vegetation have been developed is shown by the sixteen pages of references quoted in Richard's book *The Tropical Rain Forest* (1952).

The variety of interpretations and of nomenclature of plant communities remains with us yet and we intend to discuss various views later (p. 3401). One marked feature of modern development, however, has been the virtual disuse of the old term 'formation'. As an example of this change, we see in Tansley's *Types of British Vegetation* (1911) that formations were treated as fundamental primary units of vegetation, whereas in the same author's *The British Islands and their Vegetation* (1949) the term has almost disappeared and the vegetation described is classified partly on predominant growth-forms and partly by habitats in the Warming tradition, although the divisions recognized are empirical facts, however they may be classified. Warming himself advocated that the term should be dropped, for, as he pointed out, since Grisebach it had been used in so many different senses that it was merely a cause of confusion. We cannot discuss the matter further here (but see p. 3428) and will confine ourselves to noting the change.

The essential differences which separate the various schools of thought with regard to the descriptive technique in plant sociology are differences of approach, as Schröter emphasized as long ago as 1910, and arise from different opinions regarding the degree of relationship subsisting between the vegetation and the habitat. First, there is the physiognomical-floristic approach, which essays to treat plant communities as entities, floristically defined, the relationship of which to habitat factors is a matter for secondary investigation. This is the view of the 'pure' Plant Sociologists. Second, there is the physiognomical-topographical approach, in which the locality, the single stand or example of a community becomes the unit of classification. Lastly, there is the physiognomical-ecological approach, in which the vegetation is regarded as the dependent outcome of the habitat conditions and classification is based on the latter. Compromise seems unattainable in the present state of knowledge, and it can only be hoped that fuller information will show, as has happened often in the history of science, either that one view is right and the others wrong, or, more probably, that none are wholly right and that a synthesis or syncresis is required. The latter outcome is rendered probable by the fact that authors of whatever school are seldom entirely consistent in the application of their principles to nature.

The development in recent years of the application of statistical methods to the analysis of the occurrence and distribution of species in communities has increased the precision with which communities can be distinguished and characterized, and the extension of such methods to tropical countries may very well change our outlook on their plant communities considerably.

It cannot be denied that the theories of Plant Sociology arise out of experiences in the north temperate zone. When a European or North American student of vegetation walks about, he sees more or less clear-cut plant communities, woodland, heath, swamp, grassland, which, on the large scale, have been so long familiar that they have received popular names in all countries. The tropical worker, on the other hand, too often is confronted with what can only be described as a bewildering mess, in face of which the neat systems of the northerner seem quite irrelevant. Subjective judgements are here of little value and only by careful analysis can order, if there is any, be discovered, which means long and arduous application, impossible for the ordinary traveller. This is where tropical research stations have an essential value.

Although professed Ecologists always maintain that the composition and physiognomy of a plant community must be, in some degree, an expression of environmental conditions, they will also agree that it would be a mistake to be dogmatic about this when we know so little about those conditions. Thus is indicated the line of work which has become popular in recent times. Leaving on one side the analysis of populations for specialists of a mathematical bent, more and more attention is being concentrated on experimental study of the physiology and bionomics of particular species, especially those which are sociologically important, in the sense that they are prominent and regular components of well-known types of vegetation. This is what Schröter called *autoecology*. Above all, soil and moisture relationships have attracted investigators, fields in which facts not theories are wanted. British workers have led the way in this respect by the publication, now in course of issue by the British Ecological Society, of a *Biological Flora of Britain*, in which each species is treated by a specialist investigator, and every relevant fact known about the life history, reproductive methods, soil, water and community preferences of the species is recorded.

Invaluable as such a corpus of information may be, it must not be over-looked that a community is an integrated structure which is more than just the sum of its constituent species. However detailed is our knowledge of the physiology and behaviour of individual species in isolation, we have also to reckon with them as community members and to attempt to grasp the mode of life of the community as a natural whole, both by observation and where possible by experiment. Even for the experimental ecologist there can be no divorce from plant sociology if he is to fulfil the requirements of the science.

It should be noticed that the term 'community' has been retained (and its equivalents in other languages) by general consent as a term which may be used for any social grouping of plants, without prejudice to any opinion which may be formed about its status or its place in a classification of communities. In view of the diversity of views which obtain about classification such a non-committal term is very useful.

An alternative to the uncertainties of classification has been sought lately in some quarters by abandoning classification in favour of treating vegetation as a continuum viewed on a principle of 'ordination', a somewhat difficult concept which we shall endeavour to explain later (p. 3480). This, in effect, challenges the traditional concept of the plant association and supports the individualistic views put forward by Gleason in 1926 in the United States, views which are also implicit in contemporaneous studies of seriation in wood-land communities by Sukachev in Russia and Cajander and Ilressalo in Finland.

Whether we accept whole-heartedly or not the formal concept of 'succession' as laid down by Clements, and widely adopted in Britain and America, it is at least possible to accept that vegetation cannot be treated as static. Even forests are mutable and in dealing with vegetation the ubiquity of change must always be in the forefront of our minds. Not change itself, but the direction and rate of change are the subjects of inquiry.

The sorting out of the species which make up the flora of any region into social groupings may be regarded as an aspect of natural selection. Direct adaptation in response to existing conditions can be ruled out as genetically untenable. The situation is not that the plant adapts itself, but that it is genetically pre-adapted to grow best in certain conditions, though with a definite range of tolerance, sometimes wide and sometimes, as gardeners know, narrowly restricted. The interplay of the factors of reproductive capacity, means of dispersal, range of tolerance, life-form, rate of growth, and what may be summarized as competitive power, decide which species will be able to settle permanently in any given locality and the assortment of species thus selected forms the community. On the equilibrium of so many forces depends the stability, always relative, often precarious, of the community so formed.

Ecology began among plants, whose communities and reactions are more obvious than those of animals, but once its ideas were formulated they were very soon adapted to animal study, while eventually both sides were united in the concept of the 'biome', proposed by Vestal in 1915, the complex system in which both plants and animals play a part. Latterly the basic ideas have also been applied to man and have penetrated deeply into anthropology. It is perhaps to human ecology that we may look with most hope for a better understanding of our own communities and the principles which govern their development into improved and perhaps more stable forms.

CHAPTER XL

SCOPE AND OBJECTIVES OF ECOLOGY

In his textbook Warming gives the following as the scope of Ecology from the point of view of his own day.

1. To find out which species are commonly associated together upon similar habitats (stations).
2. To sketch the physiognomy of the vegetation and the landscape.
3. To answer the questions:
   Why each species has its own special habit and habitat?
   Why the species congregate to form definite communities?
   Why these have a characteristic physiognomy?
4. To investigate the problems concerning the economy of plants, the demands that they make on their environment and the means that they employ to utilize the surrounding conditions and to adapt their external and internal structure and general form for this purpose.

Much of this is still acceptable, though in the last item the fallacy of direct adaptation still survives.

Warming understood the 'physiognomy of vegetation', as distinct from that of individual species, to include the following data:

1. Dominant growth forms.
2. Density of the vegetation (number of individuals).
3. Height of vegetation.
4. Colour of vegetation.
5. Seasonal relationships (i.e. phases of vegetation).
6. Duration of life of species.
7. The number of species.

Under the last heading he goes on to point out that the number of species depends largely on the means of competition possessed by the several species. This is not the whole truth, however, for a large number of species in a community may also mean that the locality is so favourable that a large number of species can accommodate themselves to it.

He goes on to say: 'Many species can occur in several kinds of formations because the demands they make are bounded by wide limits.' Or again: 'The more peculiar and extreme a habitat is, the more uniform does its vegetation tend to be, because, as a rule, only a few species are so specialized in their adaptation as to be capable of existing in such a place.' These statements, which involve the idea of a range of tolerance, are scarcely compatible with the doctrine of direct adaptation which he had previously enunciated.

We have quoted Warming because he was the first to formulate general principles. He had a worldwide influence, which moulded the thought of a whole generation of ecologists, while many later developments may truly be said to have been founded on him.

Tansley, on the other hand, writing in 1904, took a much narrower view of the problems of Ecology. He insists on the topographical factor as the essential element in the science and translates it as 'topographical physiology'. It is this factor, he maintains, which distinguishes Ecology from Natural History in the sense in which Kerner used that title for his classic book, *The Natural History of Plants*. As an example he cites pollination mechanisms, the study of which in general is part of Natural History, but a particular pollination mechanism, which is associated with a particular environment, becomes part of the subject matter of Ecology.

He went on to state that 'plant associations' are formed according to habitat, which has always been the keynote of the British and largely also of the American schools. The main subject matter of Ecology, according to Tansley is 'the study of such plant associations, the species and individuals composing them, in their relations to one another and to their common environment'.

He laid down that the first stage of ecological work must be descriptive; the plant associations must first be characterized, ennumerated and described. The species making up an association must be catalogued, their obvious relations described and the physical conditions under which they exist noted. This is to be followed by an ecological survey with the object of producing a map to show the topographical distribution of the associations recognized. Mapping is as necessary to the ecologist as to the geologist, but with this difference, that the distribution of rocks over a district does not change within the human scale of time and revision is only called for regarding minor details. Plant associations, however, do change with time, quite apart from changes in our own understanding of them, which may lead to divisions hitherto overlooked. The act of mapping is an integral part of the survey. Just as, it has been said, we have not truly seen an object until we have drawn it, so we do not know an area until we have mapped it.

All this was regarded as no more than a preliminary to the intensive and difficult study of the causes of the facts we have observed. A difficulty, however, arises over the question of characterizing the associations to be recognized. For this is no mere matter of inspection or 'reconnaisance', as it is sometimes called, which may be all that is possible for a traveller in some out of the way region, but is quite insufficient as a preliminary to intensive study. Characterizing associations is, in fact, the whole subject matter of Plant Sociology and it involves techniques of sampling and analysis which have to be learned and applied before we can have any sound ideas about the vegetation complex with which we are concerned. We may be ready to assume, for example, that an area of forest which appears to inspection to be reasonably uniform, is a true association, but the work of Curtis and McIntosh has shown that we may in such a case be dealing with a continuum of 'stands' characterized by different proportions of the constituent species, and passing from one extreme to another in a manner inperceptible except to analysis. What in such a case is the association? Furthermore, as Ilressalo showed in Finland, such a continuum may be associated with a continuously varying proportion between Calcium and soluble Nitrogen in the soil, only revealed by soil analysis.

It does not seem that the two-stage investigation, of first deciding your associations and then studying them, will work. We can indeed make a map based on physiognomy, on the predominant life-forms, and such a map may be a very useful guide and can be compiled by inspection, but it will only be accidentally a map of plant associations. The uncovering of a 'true' association will often be the concomitant of strictly ecological study rather than its preliminary.

It may well be asked, however, whether the delimiting of associations is the final objective of Ecology, and here opinions may well differ, according to different interests. To the pure sociologist it is so and habitat studies are largely irrelevant to his conclusions or at most provide supporting evidence. The experimentally-minded worker, on the other hand, may be indifferent to nice questions of vegetation analysis and nomenclature; his interests centre on the species and its habitat. He takes his plants as he finds them, not a bad thing to do, and only gradually will the idea of community build itself up in the course of his studies, not as a goal in itself but simply as part of his picture.

These are natural differences and as they are inevitable we must learn to live with them. If we wish to preserve a concept of Ecology as a whole we must give some weight to each point of view and endeavour to integrate them into the understanding of the life of wild plants, which involves many approaches.

Two roads towards the synthesis of such an understanding offer themselves. By one road we start with the acceptance of the theoretical existence of definable plant communities and we seek, by the complex procedures of analysis, to establish their real existence and their composition. Having done so we may follow this up by testing their relationships to habitat conditions, the double scheme advocated by Tansley. By the other road we come first of all to habitats and by exploring their characteristics, physical, chemical and biotic, we build up our picture of the vegetation associated with them. The second view is the more laborious and its results are more limited. They give us no general panoramas of vegetation, but they have the unshakeable advantage of being based upon pure scientific induction, without any admixture of hypothesis.

The strict dogma of the dependence of plant communities upon habitats has been often and justly criticized. The range of adjustment is in many cases wider than used to be supposed. In a broad sense, and it is broad, there is certainly a connection, but the extremes do not meet. We would be very surprised to find the vegetation of a chalk grassland growing on a soil of wet peat, though a few species can even stretch tolerance so far. Nevertheless, having examined as fully as possible the characteristics of a given habitat, we cannot assume that it will be occupied by a predictable plant community. Too many other factors enter into the outcome. In particular the history of the locality may be an important influence upon the constitution of the existing vegetation. This may include not only past human interference, of which no other trace may remain, for example, areas which were once used for grazing or, as in the north of England, areas which were completely de-forested many centuries ago. There may be other natural factors in the history which are also influential, periodic fires, for instance. The establishment of any kind of community depends ultimately on the arrival in the area of seeds, spores or other propagules of the constituent species, and we do not have to go as far back as the Glacial Period to realize how this process may have been influenced by all sorts of occurrences more or less remote from the actual area.

Such considerations seem to open up another line of ecological inquiry, namely, the historical one. Clements based his whole system of Ecology on the doctrine of the succession of communities on the same site or, in other words, the developmental ontogeny of communities from an origin on ground originally bare, through a series of temporary phases towards a stable maturity. Unfortunately, in many cases, one would have to go back to the Late Glacial to find the original bare site and a reconstruction of the ontogeny becomes highly theoretical. Only in deep peat soils has any actual record of this secular development been preserved, and it shows itself in these remains to have been anything but a simple primary succession.

There are a few cases in which succession can be directly observed, cases in which the mobility of the ground lays bare new areas to colonization by plants, the history of which can be recorded. Among such are coastal areas, new land built up by silting in lakes and rivers, or land denuded by landslips or volcanic action. These instances are too few and too specialized to yield material for a general theory of succession, but much valuable information can be obtained by the observation of trends of change in existing communities and by the observation of related communities in the same region, which may give pointers, albeit deductive, to the prevailing mode of succession.

It has been humorously remarked that the objects of Ecology may be summed up by repeating the question: 'Why is this plant?' with the accent on each word in turn. Perhaps the best comment is a proverb: 'There is many a true word spoken in jest.'

CHAPTER XLI

ASPECTS AND SUBDIVISIONS OF ECOLOGY

A SUBJECT as wide and varied as Ecology can obviously be approached in a variety of ways. We have already indicated this in a general way and we should now examine the possibilities more particularly.

In the first place we are faced with several different levels of organization which necessitate differences in treatment. The *individual organism* is our starting point, for below this level we pass out of the scope of Ecology into the realms of cytological, anatomical, morphological and physiological analysis of the individual. The next level is that of the *population*, that is to say a local assemblage of individuals of the same kind. Here Genetics also enters the field alongside Ecology, and each may contribute information to the other.

Above the population in the hierarchy stands the *species* as a whole, while various species are integrated into the *community*. In their turn communities may form part of a community-complex or grouping of communities, with certain factors of life in common, which is called an *ecosystem*. Thus far we have only been considering plants, but, if we enlarge our view to include the animal kingdom, then ecosystems will be seen as aspects of the general body of life, the *biome*.

Schröter endeavoured to distinguish two main aspects of Ecology as *Autecology* and *Synecology*, terms which have passed into general use. The first term implies the physiological, experimental study, in relation to environment, of either individuals or more commonly of populations or of species. Synecology takes over where autecology leaves off, and deals with the sociological characteristics of all the higher levels of organization. The latter was the line pursued by Schröter himself and by the school of Zürich which he founded. He did not consider that autecological studies could ever provide a scientific basis for sociology, and considered their value as limited to a knowledge of single species. It is indeed true that the facts ascertained at one level of organization cannot fully account for the characteristics of higher levels, each of which has characteristics which belong to it peculiarly, as an entity in itself and not simply as the arithmetical sum of its constituent parts. In this respect the analogy with organic individuals holds good. Schröter's position is the antithesis of that later taken up by Gleason in 1917, who denied that plant communities were valid entities and wished to resolve them into assemblages of individuals, whose reactions collectively determined those of the community. He did not deny what is undeniable, that characteristic and recognizable assemblages of species exist, but he maintained that they were no more than assemblages of species without a unitary character of their own, and that the attack upon the reactions of the individual species was therefore the only proper ecological approach. This amounts to an outright rejection of plant sociology as a pseudo-science, an extreme position in which he seems to be isolated.

Recent tendencies have led to a blurring of the original contrast between autecological and synecological approaches and place the level of organization concerned in the forefront. Thus we may be personally concerned either with a population, a species or a community, but each of these severally calls for both autecological and synecological investigation, since in all these categories the plants have not only a relationship of their own to the physical environment, but also relationships to each other, and neither aspect can be adequately considered without at the same time involving the other.

As the recognition of communities, and their primary delimitation, depends on their floristic composition, especially with regard to the smaller categories, accurate floristic study is essential. This must be carried down to the identification of sub-species and varieties since some of these, which appear to be separated on small morphological differences, are found to have markedly different ecological preferences.

Whatever line of approach we adopt, we have to remember that Ecology is a distinct field with its own limits, and that within that field it is the ecological point of view that predominates. Subdivisions of the field are useful only so far as they help to simplify work or clarify our ideas, but they are divisions in our understanding and not in nature. They may serve as a system of reference lines, a lattice that reduces a pattern of immense complexity to elements that our minds can grasp, but they have no absolute validity and may be changed to suit ourselves.

The method of study adopted will naturally depend on how we draw our subdivisions. Communities may be treated from the various angles: floristics, physiognomy; topography; chorology (distributional areas); chronology (developmental history); climatology; pedology (soil relationships); hydrology (water relationships); or in relation to other allied communities. Futhermore, their floristic composition may be analysed quantitatively in a variety of ways. These we shall consider in a later section (p. 3434) when we have examined the nature of plant communities more closely.

The individual plant and any population of individuals, exist in a threefold ambience, *climatic*, *biotic* and *edaphic*. The latter was Schimper's term for all factors acting in or through the substratum, whether soil or water, rock or tree-bark. These divisions are not exclusive. Rainfall, for example, is primarily climatic, but it becomes secondarily an edaphic factor, and the two aspects of its influence may be quite different in different cases. Plainly its edaphic influence on an aquatic plant may be negligible, while in a sandy soil it may be vital. The methods of observation and experiment employed will depend on circumstances. Theoretically every possible aspect of the environmental conditions should be explored, but life is short and working time is shorter still, and it will usually be necessary for the worker to decide on *prima facie* evidence which group of factors he considers, in a particular case, to be the most important. Happily nature often reveals this to the eye of common sense and much useless labour may be avoided by following her indications. If such truncated study does not yield the information desired, then, but only then, it may be necessary to probe further into unsuspected influences.

We will endeavour in later chapters to describe some of the more obvious factors of ecological importance.

CHAPTER XLII

CONCEPTS USED IN ECOLOGY

IT is now time to examine the meaning of a number of technical terms which have become current in ecological literature. In doing this we shall reserve the terminology of plant communities for Chapter XLIII, as it is too large a subject to be included here. A large number of descriptive terms have been proposed by various writers from time to time, but only a small proportion have been generally adopted. Some have been considered redundant, others were held to be only of local significance, while yet others have been superseded by terms thought to be more concordant with advancing knowledge. New terms are an occasional necessity, and when the occasion arises some consideration should be given to their appositeness and their euphony as well as their scientific applicability. It is particularly undesirable to use, in a restricted technical sense, words which already have an entirely different meaning in popular usage. When any term has obtained a certain degree of currency, in the interests of uniformity it is better to employ it, rather than to substitute something new which the individual may consider more appropriate. This applies particularly to terms used in describing communities.

The following are general terms, which are constantly to be found in the literature.

Environment

This is the all-embracing term which includes the sum total of all the external influences on the life of either the individual plant or of the community. We may thus use the term in a double sense. Whatever influences the community as a whole will also affect its individual members and be part of their environment, but there may be incidentals in the life of an individual which do not concern the community. Such influences may be remote or immediate. For example, the geographical latitude is a very important, though remote influence in the environment of the community, while the presence of a large stone in the soil may be a very immediate influence on the life of an individual.

Soil drainage is an immediate factor in the community environment, but an accident of topography may give it quite a special effect on the life of some individuals. While every facet of the community environment must be considered in assessing the individual environment, they do not comprise the whole of it. Each individual has its own local environment as well, which may be distinct in many ways from that of the community as a whole. (See further, p. 3489 *et seq.*)

Habitat

The word in Latin means that 'he, she or it inhabits' somewhere. It came into use through the practice of older botanists of writing on the labels of herbarium specimens such remarks as, '*Habitat in Sylvis*', which means that the species 'lives in the woods'. From this by a transition of meaning the verb became a noun of place (*verbo substantivo*) and this was its second meaning, simply that of the position or site of growth, corresponding to the German term *Standort*. However, each site has its own peculiar conditions, and it was not long before the application was widened to cover those conditions which had an obvious significance for the plant. Still later the emphasis passed from the place to the conditions of the place and the word is de-localized and applied to such conditions in general. Thus we find such experessions as, 'the woodland habitat' or 'the aquatic habitat'.

The term remained in a more or less undefined condition, although in constant use, and it was this very vagueness which appeared to justify the plant sociologists in rejecting considerations of habitat as necessary to the delimitation of plant communities.

It will be seen that the word in its latter sense corresponds fairly closely to environment, but with this difference. Environment is an abstract term, habitat is concrete and it connotes an area, large or small, within which certain conditions obtain.

As is the case with 'environment', the term 'habitat' has both a communal and an individual application, but the former cannot be too strictly applied. Identity of extent between community and habitat was insisted on by an earlier generation. Flahault and Schröter demanded uniform habitat conditions for an association. Nichols speaks of conditions being 'essentially uniform throughout'. When practical experience refuted this idea it was modified. Tansley said: 'The existence of a plant association in more than one kind of "habitat" merely shows that the habitat has been conceived too narrowly', and he emphasizes the range of conditions under which a community can exist. Autecological studies confirm this point of view and have revealed marked diversities in the habitat conditions affecting even neighbouring individuals in the same community. Habitat has thus undergone a further expansion of meaning. No longer does it apply to a given set of uniform conditions, it now signifies a range of conditions. If this range is within the *ecological amplitude* or *range of tolerance* of the constituent species, or of the majority of them, then the habitat area may bear the same plant association throughout, despite its diversities. Similarly, if the two ranges arise from two different mean levels, the ranges of conditions in two neighbouring habitats may overlap without making them one. It must be made clear here that when we speak of 'overlap' we are referring to range of conditions not to areas. Such overlaps are quite compatible with two different plant associations occupying neighbouring areas, since it is axiomatic that plants living in conditions which are near to the mean of their range of tolerance are more vigorous and more competitive than those which are living on the fringe.

The more habitat conditions are investigated the more complex do they appear. On any given site there are gradients of factors in all directions, not only horizontally but vertically as well, in such things as temperature, humidity and, sometimes, CO_2 concentration upwards, or acidity, water content and aeration downwards. Moreover, these gradients are changeable, sometimes cyclically with the seasons of the year, sometimes with a permanent drift in one direction or another.

What then becomes of the 'essentially uniform conditions' formerly postulated for the habitat of a plant association? Is there any truth at all in such an idea? Only a very partial and limited truth. It may happen in exceptional cases that one or more factors of the habitat have such an extreme influence that they override all minor irregularities. If only a limited number of species can tolerate such conditions, then, and only then, we may be faced with an essentially uniform community occupying an *essentially* uniform habitat. One may think of the seaward face of mobile dunes, the margin of an alkali lake or a salt desert as examples. Otherwise the idea of a communal habitat should not be allowed to intrude into the delineation of a plant community except in the most general terms, since 'communal habitat' is an abstraction arising from a remarkably complex reality.

Ganong expressed the situation in 1907 in words which can hardly be bettered. He said:

Any plant stands where it does for the reason that the physical demands made by the structure and habit it happens to possess overlap in some degree the physical conditions prevailing at that place, and the better they match the more nearly does the plant find its optimum, and the worse they match the more slender is the hold of the plant upon that place.

As we have said before, the floristic composition of a community depends on more considerations than those of suitability of habitat and the most precise knowledge of the habitat would not enable us to reconstruct the vegetation with any completeness. On the other hand, if we knew the ranges of tolerance of the constituent species we could predicate the prevailing ranges of conditions which would make up the communal habitat. That is why we called the communal habitat an 'abstraction', because on any given site the communal habitat is a composite of a very large number of overlapping factor ranges, which may never be exactly duplicated on any other site, but which may nevertheless provide on any two sites an overall situation sufficiently similar to suit the tolerances of the same, or nearly the same, assemblage of species. That is the extent to which coincidence of community and communal habitat can exist.

Unfortunately we have a very limited knowledge of specific tolerances, which in any case, if we allow for habitat compensation (see below), may be somewhat elastic. That is information which the British Ecological Society and the Ecological Society of America are trying by slow degrees to amass, and it is certainly the key to a deeper understanding of vegetation.

Another aspect of habitat is that of the action of the vegetation itself in producing changes, so that time and development enter into the concept. Some of the changes are fairly obvious, i.e. change of climate by growth of trees, addition of humus and of wind-blown material to the soil, often raising its level, or changes of base ratios in the soils and suppression or encouragement of species both of plants or animals. They form part of the mechanism of succession whereby one community replaces another, and we may say that there is a habitat succession concordantly. Yapp has called such a sequence of habitat changes associated with vegetational changes, the *successional habitat*. To what extent this can be considered a continuum depends on local circumstances. In some cases it should rather be viewed as a succession of different habitats on the same site, especially where the time element is considerable as is the case in some primary successions.

Yapp has added yet another aspect to the concept of habitat, the '*partial habitat*', which he defined as the habitat of an individual plant during any given period or stage of existence. There are, of course, the seasonal cycles, which have an individual as well as a collective effect, imposing very different conditions at different times. So much field work is limited to the summer months that even this very obvious change may be overlooked. How much do we know of the condition of any plant in the depth of winter? Apart from that, the mere growth in size of a plant, from seed to fruit may bring it into contact with successively different environments. This is plainly true of a tree seedling growing to maturity in a forest, but it is also true of a herb growing among other herbs in a meadow, as Yapp himself showed in detail for *Ulmaria*.

Another effect of advancing age upon the partial habitat lies in the competitive energy of a plant. Every plant passes through phases of (1) building, (2) maturity, (3) degeneration, with corresponding changes in its relationship to its biotic environment. Competing plants will have different relationships to each other according to their respective phases of development. During the first two phases a plant, especially one with active vegetative spread, may solidly occupy its site so that competition is only marginal; but in the third phase, when its productivity and hence its competitive potential has fallen off, it may be crowded by other species which are in their building phase and this is effectively a change in its habitat. Pests and parasites also find in the declining vigour of the post-mature plant a lessened barrier to their attacks.

All study of the environment must be 'functional' in the sense that it is a study of processes, dynamic not static. All the forces of a habitat are at work and there is a continual flux of action and reaction of a degree of complexity which may be beyond the reach of our attempts at formulation as a whole, though we may hope to perceive some of its more decisive aspects. It is not necessary, for instance, to know the dynamics of every ripple to tell which way the river is flowing. Our leading question is, 'What is going on here?' And if we can get a partial answer to that we may be able to divine why it is going on, for every habitat is a microcosm of the universe, involved in the universal cycle of *Werden und Vergehen*, coming into being and going out of being, in which philosophers have summed up existence.

Niche

This is a valuable term introduced by Elton, which is distinguishable from habitat in that it has a purely biological significance and is in no way topographical. It implies the way of life of a particular organism, the way in which it is 'slotted in' to an ecosystem. Every organism must find some way in which it can successfully keep itself alive, reproduce, and maintain some status, if it is to survive, and that is its particular 'niche'. Odum has summed it up very neatly by saying that if the habitat is the organism's 'address', the niche is its 'profession'.

The word has also a communal application in that, just as in a human community so in the community of plants, the niche will determine the plant's social status. The more profitable the niche of any species, the more independent it will be and the greater the importance it will tend to assume in the community.

It has been said that no two species can occupy the same niche in the same community, but this is not universally true, for it gives a shade of localization to the term which is foreign to its meaning. It is possible for several members of the same profession to co-exist in a human community and it is equally possible for two related plant species to make the same demands and follow the same way of life and yet to find an equilibrium which allows both to continue existence. What is true, and equally true of their human counterparts, is that this can only happen if the resources of the community are sufficient to support them both. It is generally true that competition between the two *may* be closer than the competition of either with other species, though this is not always a decisive factor between them, since both may stand in danger from the aggression of another species utilizing a different niche.

Nevertheless, the rule of one species to one niche is so generally true that it has been called 'Gause's Principle', for he showed experimentally that when there is a close overlap of niches between two species one will tend to eliminate the other.

Factor

We have been obliged to use this term many times already in the hope that its meaning would be readily understood. It is, of course, borrowed from Mathematics and it signifies one of the fractions into which we try to break down the complexity of the environment so that it may be isolated and studied by itself. In this sense it is common to both Ecology and Physiology. This is confessedly an artificial, though perhaps inevitable procedure. No factor in nature actually operates in isolation, but if our factor has been soundly conceived we may nevertheless find, by studying the effects of its independent variation, the way in which it locks into the complex of other factors.

Factors vary greatly in kind and are not all of equal significance. They can be classified by their physical or chemical nature or by the source from which they arise. The latter gives a broad classification which may then be subdivided on the former basis.

The four main groups of factors according to their sources are: (1) climatic; (2) edaphic; (3) topographic; (4) biotic. The second and third often go together. Schimper first used the term 'edaphic' as the opposite of 'climatic'. In other words climatic factors were those of the sub-aerial environment, while edaphic covered all those operating in or through the substratum, normally the soil. Topographic factors, while differing in their source from others, i.e. through the effects arising from the general ground relief, are not usually separable in their nature; they are really modifying factors. They may modify the action of climatic factors, e.g. by differences of exposure, or they may modify edaphic factors, e.g. by differences in the rate of soil drainage. Biotic factors are the truly sociological factors, including all influences traceable to the company of other plants (including Fungi) or to the activities of animals, from soil protozoa upwards.

There are certain types of factors which may have a special significance. Tansley uses the concept of *master factors* which he applies to those factors which are so powerful in their action that they may be regarded as the disposing or determining factors for a given plant community, in the sense that they preclude the development of all or nearly all other possible communities in the area of their operation. It is an attractive idea, but it may be deceptive if applied, as it often is, only on the basis of subjective impression. Examples of master factors readily present themselves: extreme drought, extreme exposure, salt soils, waterlogged soils, or highly calcareous soils. It will be noted that these are all exceptional influences and where they occur it will be allowed that they do restrict the possibilities of community formation. But that is a negative effect. The positive factor, which in fact determines the actual constitution of the community, may lie in some other influence altogether, which is operative within the area of the master factor. To take one case as an example: exposure is a master factor in the flora of mountain tops, but the community developing on a given mountain area may be determined by a variety of subsidiary factors, such as the nature of the underlying rock, especially its lime content, or by flushing from a spring, or by persistent snow cover.

The idea of master factors is closely bound up with that of *limiting factors*, which arises from the application of Liebig's Law of the Minimum. This implies that if one factor falls below a certain minimum then that factor *by itself* will be limiting and may involve the disappearance of one community or of some of its species. It should be added that the rise of a factor above a certain maximum will have the same effect.

Although it is true that physiological processes are conditioned by the whole complex of operative factors and that a community depends on the balance of environmental influences, yet it is certain that if one factor departs very markedly from its norm that factor alone upsets the balance and becomes 'limiting', for no change in any other factor can redress the lack of balance. An obvious case is water shortage; nothing but an increase in the water supply can make up for the shortage. Water shortage may destroy a community, but a permanent surplus of water may destroy it as effectively by opening the door to the competition of other more vigorous species. Irrigation of a desert changes the character of the vegetation completely.

Such considerations led Billings to the idea of *trigger factors*. These are factors which start a train of changes, the trend of which may be quite unpredictable. A factor of this kind is the disappearance of a grazing animal, such as the rabbit, which can have quite dramatic effects; or, on the other hand, the introduction of an aggressive species into a new area where it may rapidly become invasive; or again, the effect of fire on a community which is unstable, for whatever reason. Anything, in fact, which initiates change, whether it be physical or biotic would come into this category.

One important deduction from the above considerations is that extremes of a factor have more determining effect than the means. It is true that mean temperatures may sometimes mark limits of distribution, but the possibility of a low winter extreme, even if it is seldom reached, may be absolutely exclusive for many species.

Tolerance

We have already made use of this term, but the idea is sufficiently clear. Here again, we have tolerance of species and tolerance of communities. The range of tolerance expresses the limits of environmental factors within which there can be successful life and reproduction. Tolerance has rather a negative sound, but the concept is expressed in this way because the emphasis is on the limits of the range. The middle of the range will usually be the optimum for a species, and here the word tolerance scarcely applies, but as the limits are approached it becomes a question of how much the species can endure, i.e. tolerate. The range of tolerance is a most important characteristic of a species and very influential in determining its geographical distribution as well as its localized occurrence. It is not, however, a simple factor. The range may differ for the different factors of the environment and the factor for which the range is narrowest will be the limiting factor for the species if it is widely variable in the habitat; if not, if it is fairly constant, then it will not be effective in limiting the distribution of the species. Other things being equal, the species with a wide range of tolerance for all factors, will tend to be the most widely distributed.

Good, who first emphasized the importance of the idea of tolerance, regarded it principally as a factor in geographical distribution. Accordingly, he sought to define the tolerance of species in the following terms:

1. Each and every plant species is able to exist and to reproduce successfully only within a definite range of climatic and edaphic conditions. This range represents the tolerance of the species for external conditions.
2. The tolerance of any species is a specific character subject to the laws and processes of organic evolution in the same way as morphological characters, but the two are not necessarily linked.
3. Change in tolerance may or may not be accompanied by morphological change, and morphological change may or may not be accompanied by change of tolerance.
4. Morphologically similar species may show wide differences in tolerance, and species of similar tolerance may show very little morphological similarity.

5. The relative distribution of species with similar ranges of tolerance is finally determined by the result of competition between them.
6. The tolerance of any larger taxonomic group is the sum of the tolerances of its constituent species.

These propositions embody some points which may be considered debatable. Specific tolerances only define the limits within which individual tolerances vary according to environment, nutritional status, age, ontogenetic phase, competition, etc. Species tolerances are determined by all the separate functional tolerances. These are disparate quantities and the specific tolerance is not their simple arithmetical sum.

The species is perhaps too large a unit to consider in this respect. The tolerance of individuals may vary considerably in accordance with their previous history. A hot, dry summer will increase the tolerance for subsequent winter cold. Many Australian species which endure considerable winter cold in their own climate show little or no tolerance when they are removed to our cooler climate. There may be geographical strains which differ. If a species has an extensive north to south area, individuals from the north end of the area will generally display wider tolerances than those from the south. No doubt the ultimate limits of tolerance are genetically fixed, but within those limits there is a good deal of room for variety. The familiar hardening off process by gardeners shows that the range of tolerance can be changed, even in the individual. Plants can be 'educated' to withstand conditions which would have been fatal to them at the outset. This does not imply genetic change, only some degree of physiological adjustment. There is no inheritance of the acquired character.

Additionally, there are natural differences in the range of tolerance at different periods of the plant's life history, or perhaps, more strictly, different ranges of tolerance associated with different processes in the life history. Particularly are there differences between vegetative and reproductive phases. Vegetative growth generally has a wider range of tolerance than reproduction. Cultivators are familiar with the fact that many plants will grow and flower successfully in conditions in which they never produce seed, a fact which is true of wild as well as cultivated plants. If we examine plants of a species in an area near its limit of distribution, it will be found that the limit is set by the ability to reproduce rather than to grow. Beyond the limits isolated individuals may be found growing successfully, but failure to produce seed prevents any permanent extension of the limit; such individuals remain only temporary pioneers. The limits may sometimes be set by the absence of pollinating insects rather than by an intrinsic quality in the plant itself, but this is not always so, since the same phenomenon is shown by wind-pollinated trees as well as by many animals. Under competitive conditions the effects of tolerance limits are naturally sharper than under cultivation and this may often be observed in the sharp delimitation of a community under conditions when the habitat shows only a continuous slope of change.

Descriptively the prefixes *steno-* and *eury-* are applied to express narrow and wide ranges respectively. Thus, *stenothermal* and *eurythermal* or *stenohaline* and *euryhaline* for ranges in regard to temperature or salinity respectively.

When a leading factor like temperature or soil-moisture is optimal the tolerance for other factors may be increased and vice versa. In other words the plant under such conditions is more vital and better able to cope.

There is no such thing as an absolute optimum for any factor since each factor is influenced by others. The idea of an omni-optimum or perfect environment is thus a chimaera. All we are entitled to say of a factor is that under specified conditions it has an optimum related to those conditions.

Ecologically the concept of tolerance warns us against the naïve idea that species or communities occur in certain habitats because they find there the optimum conditions, or, in other words, that there is a close and inevitable concordance between community and habitat. In extreme habitats particularly, it is the case that the community of species exists there because it can tolerate conditions which are outside the range of the chief competitors, not because the species have an intrinsic preference for those conditions. ('Not where they would but where they can.' Salisbury.) There are many known examples of plants and animals growing in conditions well away from their optima, because under optimum conditions they fall victims to some other influence which they cannot withstand, i.e. plant competitors, parasites or predators. These are, of course, biotic factors, whereas we have been considering tolerance to physical factors. It would be well therefore to enlarge our idea of the range of tolerance to include biotic factors, and if we can do this we will find a closer linkage with habitat than when we consider physical range only. It is, however, somewhat difficult to express biotic limits quantitatively, to express, for example, the range of tolerance of species A for the competition of species B. Even if we can manage to do this, how do we express the range of tolerance for the attacks of a fungal parasite? There is often here an all-or-nothing relationship.

While bearing in mind the biotic relationships, we are probably on safer ground if we confine our ideas of range of tolerance to physico-chemical factors, for that is a key concept both ecologically and geographically and has produced a considerable volume of detailed information since it was formulated by the zoologist Shelford in 1913. It may be a one-sided idea, but it gives us some solid ground on which to consider the biotic relationships.

We have stated above that the ultimate limits of a range of tolerance are genetically determined, and if this be so they are liable to mutation and evolution like morphological characteristics. We may expect therefore to find examples of closely related species which differ in being more widely or more narrowly tolerant. The latter is probably the commoner case and gives rise to the phenomenon of *vicarism*, which implies the occurrence of two related species replacing each other on different soils, either one or both having a restricted range of tolerance. There are many interesting examples of this. Serpentine soils, which are derived from rocks rich in Magnesium-Iron silicate, are low in major nutrient elements but high in Magnesium and often in heavy metals such as Chromium and Nickel (see also p. 3662). On these rocks in Central Europe two species or sub-species of *Asplenium*, *A. adulterinum* and *A. serpentini* occur abundantly and vigorously, whereas *A. trichomanes* and *A. adiantum nigrum*, to which the first two species are closely related, are entirely absent. In the Coast Ranges in California one species of *Emmenanthe* (Hydrophyllaceae), *E. rosea*, is confined to serpentine soils, whereas the related *E. penduliflora* is apparently excluded from them. Calamine soils, containing soluble salts of Zinc, support a special sub-species of *Viola lutea*, known as *V. calaminaria* and also *Thlaspi calaminare*, replacing *T. alpestre*. Here again the specialized forms grow abundantly on the calamine soils but do not overstep it. Other vicarious pairs are: *Galium hercynicum* on acid soils and *G. sylvestre* on calcareous soils; *Picris echioides* on pure limestone soils and *P. hieracioides* on dolomitized limestones. Sometimes the difference is apparently geographical and has not been traced to soil preferences, though these may exist. For example, the extensive montane association dominated by *Carex curvula* contains an exclusive associate, *Phyteuma pedemontanum*, in the Swiss Alps, but in the Austrian Alps the latter species does not appear and its place in the association is taken by the equally exclusive *P. pauciflorum*.

The above instances of vicarist species are also examples of a wider class of plants called *indicator plants*, with peculiar or restricted tolerances which confine them to certain habitats. Their presence is thus an indication of these habitat conditions. The term in its widest sense would include aquatic or bog plants, where the conditions are sufficiently obvious to need no other indication, but even among these there are certain differentials. *Phragmites communis*, for example, is an indicator of shallow water and *Cladium mariscus* grows only on peats which are neutral or alkaline not on acid peats. *Clematis vitalba* is limited to highly calcareous soils, and the distribution of limestone or chalk can be mapped by the indication of its presence or absence. *Urtica dioica* is nitrophilous and occurs chiefly on cultivated or well-manured soils, such as those around sheepfolds or where litter has been deposited. *Chamaenerium angustifolium* follows recent burning so closely that it is called 'Fire Weed' in America. Many other examples could be cited. Sometimes, as in the case of the serpentine and calamine plants, the occurrence of a species may be associated with the presence of some unusual metal in the soil. *Amorpha* (Leguminosae) is said to be an indicator of Lead in the soil and *Astragalus* of Selenium. The California miners of 1849 firmly believed that certain plants were indicators of gold-bearing rocks, but their secret, if it was one, has died with them.

The association of a species with the occurrence of certain elements in the soil is often reflected in differential absorption of that element which can be readily detected by the spectroscopic analysis of ash from the leaves, a method which has been utilized in prospecting for Uranium deposits.

Quite apart from these special cases is the study of natural vegetation as an indication of habitat conditions, especially with regard to water and nutrient reserves, as a means of judging the potentialities of the soil for grazing or agriculture, which has been developed in the western United States. The book by Clements, *Plant Indicators,** gives full information about these practices.

In agricultural surveying it is not so much indicator species which are important as the nature of the plant community. The dominant species of the community are those of most value as indicators because these species are most probably those which are enjoying conditions near their optimum. Even a group of dominants should not be relied upon as evidence without an experimental check in the field on the actual conditions obtaining in the area. Better still is an analysis of the numerical relationships between species in the community, since quantitative variations can provide the most sensitive indicators of variations in habitat conditions.

Micro-organisms have proved valuable as indicators of organic pollution in water supplies. Indol-producing strains of *Escherichia coli* are generally accepted as indicators of recent sewage pollution. The growth of *Oscillatoria* also indicates a high organic content (not necessarily sewage) in the water.

Reaction Level

By this term is described the actual frontier of contact between a plant and any external factor. It also represents the level of intensity at which any factor is directly operative upon the plant, which may not be the same as the general level of that factor in the environment. Thus, it is not the average pH of a soil which is important, it is that of the soil with which the roots are in actual contact that matters. It is not the general rainfall on the site which influences the plant so much as the water content of the soil in its root zone. That is its reaction level. With this in mind, we deduce that meteorological data for an area, though valuable as indications, must not be received as defining the true habitat conditions, which are those at the reaction level of the plants. This was first brought prominently forward by Kraus in his book *Boden und Klima auf kleinstem Raum* in 1911, in which he showed what remarkable differences might exist between the conditions affecting plants which were growing near together, in the reaction level of such important factors as water supply and soil reaction or illumination and exposure. This gave rise to the concept of *microclimates*, which implies the local environment as actually experienced by the individual plant. The word is not a very happy term, and 'microenvironment' is more inclusive and correct. The recognition of the individuality of the environment means that data obtained at one point cannot be extrapolated to apply to a whole population or to a community, but by multiplying such observations we can build up a much more accurate picture of the range of conditions as they affect a whole community.

Not only do conditions vary over short horizontal distances, but there are also comparatively steep gradients of change vertically, as was shown by Geiger in 1957 in his book *Climate near the Ground*, which drew attention to the changes in factors (e.g. humidity) which vertical distances of only a few inches could produce. These vertical gradients can be very important, even

* Carnegie Institute of Washington, 1920. See also SHANTZ, J. L. (1911), *Bulletin 201*, Bureau of Plant Industry, Washington, D.C.

determining, in the mainly cryptogamic ground-layer, of a dwarf community like a heath, but they reach their most impressive development in complex forests, where each layer is subject to a different climate, depending on the height above the ground. Since no layer is quite uniform in density, the climate of each level also varies horizontally. As there is little air movement in dense forests, the vertical gradients in the air tend to be less changeable than those outside the forest, but, except as regards illumination, in general they are probably less pronounced than those in the open. More information is needed on the subject.

Vertical gradients also operate downwards and affect the root systems. The chemical and physical properties of the soil, water supply, temperature and aeration all show vertical gradients, which may be much more permanent than those in the air. In arctic climates snow cover may also produce very marked downward gradients. Observations made by McClure and Johnson in Alaska showed that the surface of the soil under 61 cm of snow was more than 27°C warmer than the air immediately above the snow surface. The ground flora and small animals in the soil thus enjoyed what was effectively an entirely different winter climate from that of the tall trees and larger animals.

Exposure

In the ecological sense this means the degree to which the plants are subjected to the prevailing climate. Differences in exposure may arise from biotic causes, i.e. big plants sheltering small plants, but they also arise from topographical causes. These irregularities may be on a large scale, a familiar example being the difference between the north and south slopes of the same mountain, differences sufficient in many cases to engender two quite different types of vegetational cover. On the other hand they may operate on the micro-level, as on the north and south sides of a single tree-trunk, which may be inhabited by two different assemblages of cryptogams.

Exposure must often be considered under its negative aspect of shelter; shelter from full sunlight, from prevailing winds, from frost or any other detrimental influence which may exclude species or so weaken their vigour that they are unable to compete successfully with others and so tend to be replaced. Increasing exposure has several ecological consequences: increased solar radiation, lower average winter temperatures (except in sites where there is prolonged snow cover), and, perhaps most important, increased transpiration as a result of both increased radiation and increased wind-exposure. When the latter factor is intense it may result in the death by desiccation or cold of all young shoots on the windward side of the plant with resulting deformation in growth, a familiar phenomenon in trees near the sea coast.

Physiognomy

This is one of the oldest terms in use in Ecology, and it is employed in much the same sense as it is in popular language, that is to say, it refers to the characteristic and recognizable appearance of a plant or a community. Communities have been distinguished by their physiognomy from the earliest times. Grassland, heath and forest all owe their early recognition to differences in their physiognomy. Appraisal of their other differences came much later.

It was von Humboldt who was the first to point out that differences of physiognomy in plants meant differences in their habit of growth, what is now called the *life-form* of the plants. He tried to establish a classification of life-forms, but, as we have previously mentioned (p. 3319), he chose a systematic basis for his classes, i.e. they were founded on characters which are hereditary and genetical in nature and not directly related to ecological factors. Thus, in 1806 he established his nineteen *Hauptformen*, which he claimed were selected from a survey of the phanerogamic species of the world, as those to which all others could be referred. As we pointed out on p. 3319, these were all based upon systematic classes, Araceae, Malvaceae, Myrtaceae, Cactaceae, etc., and only in the last-named and in the class of the lianes (woody climbers) do these bear any relation to ecological considerations.

The early travellers in Brazil, von Spix and von Martius (1817–21) and Lund (1835) used illustrations on a lavish scale to depict what they called 'vegetation forms' and von Martius published between 1840 and 1869 his *Tabulae physiognomicae Florae Brasiliensis*, in which physiognomy is equated with vegetation forms, which were in effect types or formations of vegetation of an ecological character, such as the primary forest, the 'caatinga' forest and the various types of 'campos' (mostly savannah). Grisebach similarly considered that a distinctive physiognomy, whether due to the predominance of a single species or to the association of a group of species, was the mark of a distinct formation. He established a physiognomic system comprising eventually sixty vegetative forms which he considered to be related to climatic conditions. This is a complete breakaway from Humboldt's views, since here physiognomy is associated with ecological character, not systematic relationship. Physiognomy alone, however, is no safe guide to the designation of plant formations and it led Grisebach to very varied applications of the term, from altitude zones in the Norwegian mountains on the one hand, to the separation of ericaceous hummocks and *Sphagnum* carpets on a peaty moor (low moor) as different formations on account of their different physiognomy.

The influence of von Humboldt waned as more information about world vegetation became available, and Heer in 1835 used a simple physiognomical classification by height, into herbs, half-shrubs, shrubs and trees. A big step was taken by Kerner von Marilaun in 1863 in establishing classes of *Grundformen*, in which the physiognomy was explicitly related to ecological characters, that is to say they were regarded as life-forms.

Once the idea of life-forms was introduced, there followed several interpretations of the idea by different authors, accepting but trying to improve upon Kerner's list of *Grundformen*.

In 1908 Warming carried matters further by endeavouring to find rational principles for the classification of life-forms. His ideas take us onto a new plane, because he views physiognomy as the expression of the ecological characteristics of the plant, not simply, as in the older authors, a matter of form and appearance. He explicitly rejects the difference between mono-podial and sympodial branching as of no physiognomic importance because it is of only slight ecological significance, although the difference has a marked influence on the general build and appearance of trees.

Because of its ecological basis, it is worth while to consider Warming's system in some detail. He uses six main classes:

1. Heterotrophic.
2. Aquatic.
3. Muscoid.
4. Lichenoid.
5. Lianoid.
6. All other autonomous land plants.

Heterotrophic plants include holoparasites and holosaprophytes. It is unde-niably a valid ecological division, but from the point of view of life-forms it is very mixed and unsatisfactory.

The sixth class obviously calls for subdivision, accordingly it is divided between monocarpic (hapaxanthic) plants which only flower and fruit once and polycarpic (pollakanthic) plants, i.e. longer-lived plants which flower and fruit repeatedly. These groups are then further subdivided.

Monocarpic plants: Summer annuals, winter annuals (germinating in autumn) and biennials.

Polycarpic plants: Warming lays down eight factors for their classifica-tion.
    1. Duration of the vegetative shoot.
    2. Length and direction of internodes.
    3. Position of the renewal buds during the dormant season.
    4. Structure of buds, including renewal buds.
    5. Size of the plant.
    6. Duration of leaves
    7. Relationships between the structure of the green shoot and the conditions of transpiration.
    8. The capacity for social life, which affects the physiognomy of the community.

He distinguished four classes of polycarpic plants.

1. Renascent Herbs. Plants with permanent structures underground sending up annual flowering shoots.
2. Rosette Plants. Plants with short internodes and closely set leaves, which may be large or small.
3. Creeping Plants. This group does not distinguish between herbaceous species like *Hydrocotyle vulgaris* and woody types such as *Aretosta-phylos uva-ursi* with quite different physiognomies.
4. Land Plants with long, erect, long-lived shoots. This again is a miscel-lany, including not only trees and shrubs, but sub-shrubs like *Lavan-dula* and also Cacti, Aroids and cushion plants.

This classification deserves recognition as the first essay at a truly ecological classification of life-forms, but it was weak in two respects; first, it is not sufficiently discriminatory between different types of form, due to using too many various criteria, and, secondly, it does not provide distinctive names for all the classes, which makes it unhandy for practical use. The historical endeavour, from Grisebach through Kerner and Drude to Warming, was to reduce the apparently endless variety of plant forms to the simplest possible, but yet all-inclusive classification. This aim was most nearly reached in the system of classification proposed and developed by another Danish botanist, Raunkiaer from 1905 onwards. Its simplicity is attractive, its terminology is clear and, although its basis on a single character, the position of the perennating (renewal or winter) buds, may be criticized as artificial, it does in fact fit remarkably well with habits of growth. Furthermore, it does not rely on systematic affinities, but is wholly ecological and applicable to any type of vegetation.

In its developed form, as set out in *The Life Forms of Plants* (Oxford U. P., 1934), it comprises five classes in descending order of size but ascending order of specialization. This is due to Raunkiaer's assumption that a hot, moist climate stimulates growth but involves less specialization for the renewal of growth, while on the contrary the species of cold or dry climates are smaller but more specialized.

Class 1. Phanerophytes (PH). Perennating buds borne on aerial shoots.

      (*a*) Evergreen Phanerophytes, without bud scales, i.e. with no resting period.

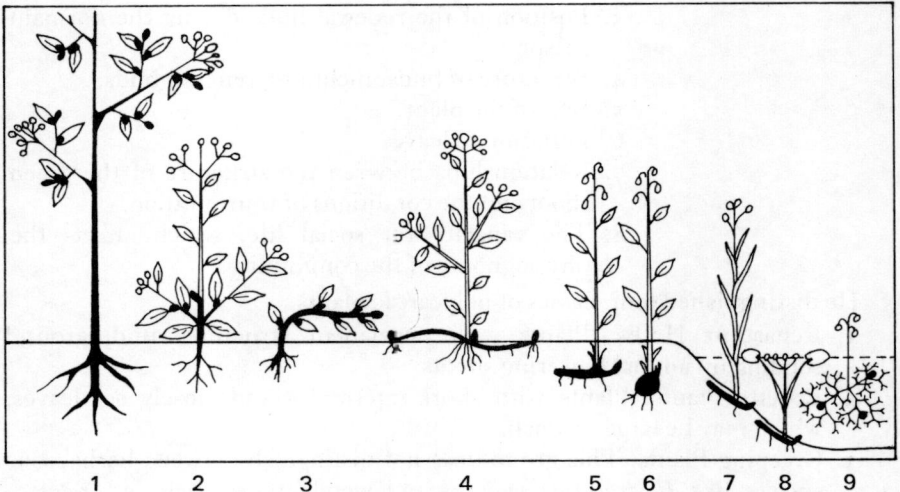

Fig. 4.2. Some of the principal life-forms of Angiosperms, diagrammatically represented following Raunkiaer. The perennating organs are shown in black. 1. Phanerophytes. 2, 3. Chamaephytes. 4. Hemicryptophytes. 5, 6. Geophytes. 7. Helophytes. 8, 9. Hydrophytes.

(b) Evergreen Phanerophytes, with buds protected by bud scales.

(c) Deciduous Phanerophytes, with bud scales.

Each group is subdivided by size, as follows:

Megaphanerophytes (Mg), above 30 metres in height.

Mesophanerophytes (Mo), 8–30 m.

Microphanerophytes (Mi), 2–8 m.

Nanophanerophytes (N), below 2 m.

Class 2. Chamaephytes (CH). Perennating buds borne close to the ground.

(a) Suffruticose Chamaephytes. The aerial shoots die back to near the base in the unfavourable season and perennating buds are formed on the basal parts.

(b) Passive Chamaephytes. Aerial shoots fall prostrate in unfavourable season, so perennating buds are borne close to ground level.

(c) Active Chamaephytes. Creeping plants. Aerial shoots always prostrate.

(d) Cushion Chamaephytes. Aerial shoots all low and compact.

Class 3. Hemicryptophytes (H). Aerial shoots die back completely in unfavourable season ('herbaceous plants'). Perennating buds formed at or in the surface of the soil.

(a) Protohemicryptophytes. Lower leaves less developed than upper leaves.

(b) Partial rosette plants. Lower leaves more developed than those on aerial shoots.

(c) Rosette plants. Leaves all restricted to a rosette at the base.

Class 4. Cryptophytes (CR). Perennating buds borne below ground or below water.

(a) Geophytes (G). Plants with underground storage stems from which the renewal buds arise.

(b) Helophytes (HE). Marsh plants. The perennating buds arise from stems submerged in mud or peat below water level.

(c) Hydrophytes (HY). True aquatic plants. Perennating buds arise from submerged rhizomes or else are detached from the plant (turions) and sink to the bottom.

Class 5. Therophytes (Th). Annual plants with no perennating buds. The unfavourable season is passed in the form of seeds. They may be ephemerals with more than one generation in a season.

To the above should be added two special classes, Stem Succulents (S) and Epiphytes (E), which are of only local occurrence.

Raunkiaer utilized his classification in the analysis by life-forms of floras and vegetation types in various parts of the world, showing how the percentages of species falling into the different classes was expressive of the prevailing ecological conditions. A tabulation of species in this way he called a *Biological Spectrum*. The examples from Raunkiaer given in Table 1 illustrate his contention.

TABLE 1. BIOLOGICAL SPECTRA (after Raunkiaer)

	S	E	MM	M	N	CH	H	G	HY	TH
Labrador coast			2	1	8	17	52	9	5	6
Aden	1			7	26	27	19	3		17
Seychelles	1	3	10	23	24	6	12	3	2	16
Normal spectrum	1	3	6	17	20	9	27	3	1	13

The 'normal spectrum' proposed as a standard of judgement was not an analysis of the world's flora, which would be impracticable but was based on a random sample of 1000 species from a taxonomic catalogue.

Raunkiaer himself did not claim that his system was in any way final, and several authors have attempted to improve or enlarge it. No Cryptogams are included in the original system, but Feldmann in 1937 applied it to perennial Algae with suitably modified classes. Braun-Blanquet also added three extra classes: Planktophytes (plant plankton), Edaphophytes (soil Cryptogams) and Endophytes, which covers both Endolithophytes (plants boring in rocks) and also internal parasites of plants and animals. It is difficult to see that this adds anything to the usefulness of Raunkiaer's classification.

Valuable as the latter undoubtedly is as an index of climatic relationships, it cannot by itself provide a means of classifying plant communities. No more than any other physiognomic-ecological classification can it give a complete picture of communities or their relationships, without taking into account the all-important floristic composition of the communities, to which it must be considered as ancillary.

The criticisms of Raunkiaer's system are directed to three points. First, the limits of the classes are too indefinite and there is some overlapping. The distinction between Hemicryptophytes and Helophytes and between the former and Chamaephytes is often dubious and individual cases difficult to allocate to one class or the other. Second, many plants show marked changes of life-form in different climates or in different areas. Third, that the biological spectrum is not entirely governed by climate but historical causes are also important, so that similar climates in different parts of the world may show different spectra because of their differing floras.

These criticisms are serious enough to modify our views of the utility of life-form classification for world comparisons of remote areas. It still remains valuable for comparisons within a limited geographical compass.

Du Rietz summed up the various ways of regarding life-forms under six headings:

1. *Grundformen*. The original classification, based simply on the appearance of the plant at the season of its greatest development.
2. Growth-forms as used by Warming, based chiefly on shoot formation.
3. Periodicity forms, based on differences of the plant form in the active and passive seasons respectively.
4. Bud-height forms. The Raunkiaer system, based on the height of the perennating buds above the soil.
5. Bud-type forms, based on the characteristics of the perennating buds.
6. Leaf-type forms, based on the shape, size, longevity and formation of leaves.

The last method also appealed to Raunkiaer, for leaves are the organs most likely to show the effects of climate. It is a matter of common observation that the largest leaves are found in moist warm climates, while in cold, dry, unfavourable climates they tend to be small.

After extensive comparisons of leaves from different regions Raunkiaer decided that the following were natural groups in order of size:

1. Leptophyll, up to 25 mm².
2. Nanophyll, 25–225 mm². (25 × 9).
3. Microphyll, 225–2025 mm².
4. Mesophyll, 2025–18,225 mm².
5. Macrophyll, 18,225–164,025 mm².
6. Megaphyll, above 164,025 mm².

The maximum size in each class is nine times that of the previous class. In addition there is a group of Aphylls which have no laminate leaves.

The method has not been widely used, for unless a planimeter is available the measurement of areas is tedious. It has chiefly been employed in tropical forests, in which there is a marked degree of convergence in leaf-shape. Cain has shown that in this vegetation the approximation that the area is two-thirds of the length multiplied by the breadth can be used without vitiating the classification by size. Actual check measurements showed that the average area was 67·4 per cent of the length × breadth product.

Compound leaves are generally treated as made up of simple leaves and in general the individual leaflets tend to be smaller than single leaves. Shade leaves in tropical forest have generally a much greater length–breadth ratio than have sun leaves, and they are often larger. Richards, Cain, and others, have established that approximately 80 per cent of the species and individuals in rain forest have leaves in the Mesophyll class. Taken by forest layers, however, the leaves in the highest ,'emergent', class of trees seldom have Macrophylls, but they bear more Microphylls than do the lower layers. As most of these measures were taken in the American tropics it is possible that a predominance of pinnate-leaved Leguminosae in the canopy may have weighted these measurements towards the smaller leaf-classes.

A striking comparison between lowland rain forest at Mucambo in the Amazon basin and sub-tropical rain forest at Pelotas in southern Brazil, has been made by Cain, Castro, Pires and da Silva, in which the leaf-size classes were arranged according to Raunkiaer life-forms. This showed clearly a climatic effect in the change of prevalent leaf-size in the class of Phanerophytes.

Leaf-size class	Mucambo	Pelotas
Mesophyll	75·9 per cent	31·9 per cent
Microphyll	12·35	48·6

Cain has also shown that when the percentage numbers in the leaf-size classes in an area of Brazilian rain forest were plotted on a log. base the result is an approximately normal curve of distribution with the mode in the Mesophyll class.

The analysis of vegetation into life-forms is particularly useful in descriptions of vegetation in little-known countries, where floristic lists involve considerable difficulty. The life-form can be determined in most cases by inspection and the biological spectrum built up from them is ecologically informative.

Ecosystem

This term was introduced by Tansley in 1935 to replace the idea, which he criticized adversely, of the 'biotic community', which was applied to the whole complex of living organisms, both plant and animal, living in a given habitat. His objection to the idea of a biotic *community* arose from the difficulty of regarding animals as forming part of a community with plants, or vice versa. No doubt both groups or organisms influence one another, often profoundly and the existence of the animal group may indeed depend upon the nature of the plant community, but it is surely a misuse of language to call both one 'community'. Adverting to the human community as a model, since it was from that that the term arose, we see that there must be some similarity of nature and status between the members of a true community. Our sheep and cattle or our crop plants may be of fundamental importance to the human community, but they are not members of it, nor are our dogs and horses or our garden plants, however close to our interest they may be. They do, however, participate in the human biological system, that complex balance of factors within which the human community exists.

If we need a term for the combined forces of plant and animal life in a habitat we can use Clements's neutral term *biome*. There is also the term *biocoenosis* and the corresponding term *phytocoenosis*, which apply to the totality of living organisms or the totality of plants in a habitat. They carry only the connotation of 'living together' and are free from the overtones of the word 'community' and they beg no questions, but it may be doubted if they are really necessary.

The idea of the ecosystem goes much further than any of these, for it involves the concept of an equilibrium, more or less stable, between all the factors, plant, animal and non-living factors, operative in the system. Tansley viewed it as parallel with the physical systems of the universe, which all tend towards equilibrium, whether stability is ever achieved or not. If, in the development of an ecosystem, equilibrium is not achieved, the system will disintegrate and disappear, which happens to many accidental and temporary plant groupings with little or no integration. The more highly integrated the system becomes, the more likely is it to achieve functional stability and the longer will it last.

Anything which has an ecological identity may be a complete ecosystem, there is no condition of size. It may be no more than a crevice in a rock, or it may be a whole forest, but to be a complete ecosystem it should include at least four components. First, the non-living environment complex. Second, the producers, the autotrophic plants. Third, the consumers, living on formed organic substances; and fourth, and not sharply distinguishable from the consumers, the decomposers, mostly Saprophytes, which break down dead organic matter and release simple substances utilizable by the producers.

Of course it is possible to imagine simpler ecosystems in which one of the components, especially perhaps the consumers, is missing, but such as may exist are rare and exceptional. The essence of the concept is that of a *system*, in which there is a balance between consumption, growth and decomposition, between, if we like to draw an analogy, anabolism and katabolism, such that the system endures. For the most part the balance requires all four components to harmonize. If one or other gets out of balance it may destroy the system. Grazing animals may be necessary to maintain a grassland system, but they may also, in excess, destroy the vegetation more or less completely. In the absence of an appropriate consumer one plant species may become so rampant that it smothers the rest of the plant community, as the climber, *Calamus* may smother an area of rain forest.

Another term which is relevant here is the adjective *holocoenotic*, which is used to describe the totality of environmental action in the ecosystem. The environment is holocoenotic in the sense that no factor is wholly independent, that all factors interact so that the environment is an integrated whole.

Ecotypes

Turesson has produced evidence to show that in large species there are often strains with special habitat preferences. He called these *ecotypes* and he showed that they were also genotypic segregates which were presumably genetically predisposed to certain types of habitat. This naturally tends to isolate them from one another and in some cases may be the basis for the development of endemic species.

The existence of ecotypes as subdivisions of species is a phenomenon of extensive importance in Ecology. Ecotypes are specialists in their relationship to particular environments, with which they are genetically compatible and in which natural selection has segregated them. They may be to some extent morphologically distinct from other races of their species or they may not. Their differences may be wholly physiological and not reflected in their outward appearance. They may be compared with the 'physiological races' of pathogenic fungi which are each separately compatible only with certain hosts.

Every species population possesses a degree of genetic diversity. It thus retains flexibility and a balance with the forces of natural selection. The widest distribution implies the greatest genetic diversity and vice versa. Various races, local or regional, are subject to environmental selection and thus become ecotypes. All have come from the same ancestral strain. They owe their segregation to the chances of dispersal which have brought them into environments optimum to their requirements, in which they preferentially survive. They are not products of direct adaptation, but are constitutionally prepared for their environment, if or when they find it.

Ecotypes may be either climatic or edaphic. The former naturally have wider ranges. If any species has a considerable range in latitude it will usually be found that individuals from the northern areas differ ecologically from the southern types and cannot be interchanged with them. For example, *Pinus sylvestris* has a northern or arctic race or ecotype which differs in its habit of growth from the southern race and has a distinct area of distribution. Similarly, there are alpine ecotypes of lowland species, which have found their congenial home at high levels. They can usually be cultivated in the lowlands, but they will not spread naturally below a certain level where they would meet the competition of the lowland race. An ecotype is thus a special kind of race, for different races can and do co-exist in a population without having been segregated as ecotypes. Whether ecotypes should be regarded as 'beginners' in specific evolution is not a question to which a general answer can be given. In some cases it probably has been so.

Polyploidy has been considered as one of the genetic features associated with climatic ecotypes and there are valid correlations of polyploidy with geographical and climatic distributions (see also p. 3865), but the environmental differences between polyploid and diploid races are not apparently fixed or predictable.

Edaphic ecotypes are generally related to soil preferences or tolerances, often with respect to the chemical constitution of the soil. Some striking examples of such ecotypes are related to the presence of toxic heavy metals in the soil and these we have described later in relation to special soils (see p. 3667). Some of these ecotypes are so distinct that they have been given specific names and are treated as vicarious with the related species on ordinary soils.

Climatic ecotypes may include edaphic ecotypes. Kruckeberg grew *Achillea borealis* at a series of stations along a transect across the various climatic zones of California from the Sierra Nevada to the sea. He found climatic ecotypes among his populations, but, in addition, these ecotypes showed varying soil preferences in different climates. A climatic ecotype might therefore be compounded of several edaphic ecotypes.

Compensation

This term expresses the compensation of one environmental factor by another. This implies some degree of replaceability, generally partial, probably never complete. It shows itself in some unusual distribution of a species, which is more obvious near the natural boundaries of its area, either in a general extension of that area or in local disjunct occurrences in special habitats beyond the boundaries. The effect is made more striking by the fact that species usually show a restricted range of tolerance in areas near their geographical boundaries. In such cases they may be under some ecological strain, which makes them more sensitive to any change of environment and more responsive to any favourable factor.

Compensation for climatic factors is often given by topography. A wet slope facing north will allow the southward extension of northern species and conversely, a dry slope facing south, and consequently well-warmed, allows southern species to extend their natural boundaries northwards. Southern species can also extend northwards in mild, moist oceanic areas, without extremes of temperature. Frequent mists may compensate for a scanty rainfall or, perhaps better, supplement it, as in the coastal redwood forests in California. Conversely a porous, sandy, soil may serve to compensate for a rainfall greater than the optimum for a species. Water courses may compensate for a dry climate as is well shown by the 'gallery forests' following water courses in semi-desert areas, for example in north Africa and in South Brazil. The former are extensions of the mid-European deciduous forests, the latter are extensions of subtropical rain-forest.

Factorial compensation may occur in all sorts of ways and edaphic factors sometimes compensate for climatic factors, or else the reverse. One of the edaphic factors which is very influential in this respect is a calcareous soil. Limestone or chalky soils very often offer an extension of habitat to plants which are not otherwise restricted to calcareous soils. Thus *Fagus sylvatica* in central Europe shows no marked soil preferences, but its westward extension in Britain is confined to calcareous soils. This may also apply to altitude as well as latitude. Billings cites pure stands of *Pinus aristata* on limestone at altitudes of 3048–3350 m, where adjacent quartzite soils carry only a subalpine community. In Britain the outcropping of the highly calcareous Lawers–Canlochan schist controls the distribution of a group of rare montane species. A dry cliff with a southerly exposure may thus be rich in these species if it consists of the calcareous schist, while a neighbouring cliff, facing north and dripping with water, but formed of an acidic rock, is destitute of them.

A deep, moist soil may compensate for the lack of an oceanic climate and allow the continental dispersal of oceanic species, while as we have seen above, a shallow, dry limestone soil may compensate continental species for the heavy humidity of an oceanic climate.

No factor operates in isolation and the effect of one cannot be predicted without consideration of others. Thus the effect of rainfall depends upon the prevailing temperatures. In the low average temperatures of Britain a minimum of 50 mm of rain annually permits the growth of forest, whereas in the more extreme climate of eastern North America a minimum of 1118 mm is required. An annual rainfall of 559 mm, which provides wet conditions in the west of Ireland, only permits a semi-desert in California.

From this principle of compensation we may draw the conclusion that different combinations of factors may in effect add up to closely similar habitats and that when a community appears to be developed on two distinct habitats closer analysis may very well show that compensation is providing *effectively* similar conditions, at least within the plant's tolerance ranges.

The restrictive effect of competition may negative compensation, and it is well known that where competition is eliminated a plant will tolerate a wide range of conditions which it could not support in the wild state.*

Compensation to be effective must apply to the factor or factors which are otherwise limiting. Unless the shortage of a limiting factor is compensated, no amount of compensation in other directions will have any effect, but compensation may be accomplished in more than one way. For example, if shortage of water in the soil is the limiting factor, its effect can be compensated either climatically, by decreased transpiration, or edaphically, by decreased drainage in the soil, which will better conserve the supply; but if water is the limiting factor then only compensation for that factor will have any effect in modifying the distribution of species.

Some part in compensation also depends upon the plants selectivity. Turesson has produced evidence to show that in large species there are often strains with special habitat preferences. He called these *ecotypes* and he showed that they were also genotypic segregates which were presumably genetically predisposed to certain types of habitat. This naturally tends to isolate them from one another and in some cases may be the basis for the development of endemic species.

Clausen, Keck and Hiesey in California have shown that two local species of *Achillea* are made up of a number of ecotypes, varying in their reaction to climate, and it seems likely that such biological races may be widespread among flowering plants. One or other of these ecotypes may find a place where only it can survive and thus obtain a growth area disjunct from that of the rest of the species. Instances of such disjunct 'islands' of distribution have been recorded. Mutation and gene exchange, where more than one ecotype is involved, may cause a new race or sub-species to evolve in such island distribution and the origin of endemics may sometimes have lain in this process.

Endemism

This is the condition of a species which is restricted in its distribution to a single area. There are many degrees in this, depending on the size of the area. An endemic may be restricted to one country or, in the extreme case, to a single locality in one country and the frequency of endemic species in any country is a function of the size of the flora. The huge flora of Brazil contains a high proportion of endemic species, Britain on the other hand, can only claim one or two. It is also related to the geological history of the flora, since long

---

* Rübel has humorously pointed out how similar reactions appear in human society, where the limiting factor of lack of brains may be compensated by wealth. Likewise a man can tolerate an idle and primitive life on a tropic island which he could not maintain in the press of civilization.

undisturbed possession of the soil gives more chances of the indigenous evolution of species, many of which will remain limited, for one reason or another, to their native locality.

Geographical isolation has always been a powerful factor in promoting endemism. This may operate over an entire continent, like Australia, or on a small island.

The high proportion of endemic species on oceanic islands or isolated mountains has long been known and wondered at. They are mostly species belonging to genera with a distribution on adjacent continents, species which have originated on the islands presumably from ancestral migrants from the continents. There are relatively few endemic genera, diversification has usually not gone beyond the species.

A most exceptional family of endemics is the Podostemaceae with the closely related Tristichaceae. In them diversification has proceeded to an extreme, so that vegetatively they are practically unrecognizable as Angiosperms. They grow attached to rocks in tropical waterfalls and, as Willis showed in Ceylon, practically every waterfall has its own endemic species. If they originated, as Willis maintained, by the spread of formerly riparian land plants into these new habitats, they display a plasticity and a range of tolerance quite unparalleled. Moreover, it appears to involve direct adaptation, which it is very difficult to accept. They still have some characteristics which are not those of typical aquatic plants, such as the production of large numbers of small seeds, indicating a land-living ancestry in comparatively recent times, but their evolution remains mysterious.

On oceanic islands it was early remarked that endemic species belonging to genera that were also found on neighbouring continental areas often showed characteristics that were considered primitive for the genera concerned. This gave rise to the idea that such species were isolated relics of types which had died out on the mainlands and created a tendency to interpret all endemics as relics of species formerly widespread.

This was vigorously combated by Willis with his theory of 'Age and Area', in which he maintained that statistically the area occupied by a species was directly proportional to its age as a species, with the saving clause 'other things being equal', which unfortunately they seldom, if ever, are. As a logical proposition this is doubtless true, but it ignores all differences in such matters as rates of dispersal, fecundity, longevity or ecological restriction, which invalidate its application in many individual cases. In other words it is only statistically true. It led Willis, however, to deny the relic view of endemics which he regarded as beginners whose spread was still to come.

The truth probably lies, as so often, between the two views. There are some endemics, especially some survivors of the Glacial Period, which are unquestionably relics. Some of these species owe their survival to finding themselves in *refugia*, local areas which were for topographical reasons protected from the full severity of the climate; they did not, however, escape unscathed. Of the biotypes present in the original population only perhaps one survived and thus genetically impoverished they have lost the adaptability to spread into new environments and remain anchored, as it were, to their refuge areas.

Competition

This term covers the effect of species upon one another, which in almost every case except that of parasites takes the form of a modification of the environment. Many of the distributional patterns of species which we shall have to speak of later (p. 3835) can be traced ultimately to the modification of the environment by other species. Competition between species must be distinguished from competition between individuals, possibly of the same species, which is often the sharpest and most direct. Competition also differs at different stages of the life cycle and may affect any stage from germination to seed-dispersal. In fact, plant societies differ from animal societies in that competition is omnipresent and universal and in that there are none of the devices for modifying its effects that many animal societies have adopted.

Competition is often viewed simply as a division between individuals of the supplies of water, nutrients and light, so that the amounts available to each individual are reduced, and the incorrect corollary is drawn that competition only exists where the supplies are insufficient for all. On the contrary, an excess of available supplies, by stimulating growth, may well intensify competition. When did increased wealth eliminate competition in human societies?

Competitive behaviour may be primarily divided into shoot competition and root competition. Shoot competition takes many forms apart from the obvious demand for light. There may be a simple mechanical struggle for growing space where the community is dense. Rosette plants, such as the grassland weeds, *Plantago lanceolata* and *Bellis perennis* ensure their living space by simply smothering their neighbours, and *Juncus squarrosus*, by the decumbent spread of its shoots, literally elbows away the surrounding plants of *Festuca* and *Agrostis*. Mechanical competition is also seen where strong climbers like *Hedera*, or the strangling *Ficus* species, take hold of and eventually smother some pre-existing tree. Again, broad-leaved herbs may intercept rainfall or may, on the other hand, conserve a high degree of humidity, either of which may be injurious to competitors. Vegetative propagation may be an important agent in competition for space. In this there are many degrees of what gardeners call 'aggressiveness', but the completeness with which vegetative spreading, fast or slow, can occupy an area of soil is impressive. The most effective counter to it is the opposite type of life-form, the mobile monocarpic form with rapid maturation and efficient seed-dispersal. Underground shoots, such as close-growing rhizomes, often exert mechanical pressure on their neighbours slowly to acquire a living space. The competition of fast-growing rhizomes, such as those of *Agropyron repens*, is more akin to that of roots, but they build up new centres of growth and of spreading most effectively.

Buttery and Lambert have observed an interesting example of phasic shoot competition between the two aquatic grasses *Glyceria maxima* and *Phragmites communis*. In spring *Glyceria* overgrows *Phragmites* due to more rapid growth and consequent smothering of the young *Phragmites* shoots. In late summer *Phragmites* has the advantage due to the ability of its shoots to remain erect after the *Glyceria* lodges. The result is to produce mutually exclusive areas.

Shoot competition may also affect reproduction phases, through competition to attract the visits of pollinating insects or through interference with seed-dispersal.

Individual competition is severe in the seedling stages, for establishment of the seedling is perhaps the most vulnerable stage of a plant's life. If a plant produces collective fruits such as those of *Beta* or *Fragaria*, or else produces a multitude of seeds, as in *Papaver*, a close growth of seedlings may result, between which there is a struggle for survival, operative through both roots and shoots. Experiments with *Papaver* have shown clearly that even under experimental conditions, with no interspecific competition, increased density of sowing leads to a diminution in the number of mature plants through individual competition. Where interspecific competition has been tested the effects on germination may be even more marked, as Lazenby showed. Seeds of *Juncus*, sown alone and together with four combinations of *Molinia coerulea*, *Trifolium repens*, *Lolium perenne* and *Agrostis tenuis* showed that even in the most favourable mixture (*Trifolium* + *Molinia*), the number of *Juncus* plants was only 70 per cent of the number when *Juncus* was grown alone, while in the most unfavourable case (*Trifolium* + *Agrostis*), the number was only 2·3 per cent. Such marked suppression may not be due only to seedling competition, there may be actual inhibition of germination by competitors, either by physical or chemical means. In the above tests the reduction of *Juncus* was not explicable by simple starvation, since it occurred at all levels of soil nutrition. It was a true plant-to-plant competition.

The importance of life-form in competition is most clearly evident in the comparison of trees and herbs. Seedlings of trees have to compete with the ground vegetation on equal terms and may not infrequently be suppressed, e.g. by *Calluna* or *Vaccinium myrtillus*, but once past this stage their arboreal form asserts its predominance and they are no longer vulnerable to the competition of the ground flora. Indeed, the position is rapidly reversed and the ground flora becomes subordinate to the trees.

Differences of height and particularly the rate at which height is reached may be important agents of competition even between plants of comparable life-form. The aggressive character of *Lolium italicum* in grassland has been attributed by Stapledon and Davis largely to the rapidity with which its foliage develops. The effect of height is also shown negatively by the susceptibility of cut lawns to the invasion of weeds which find no place in a hayfield. If superior height is combined with persistence, as with the fronds of *Pteridium*, the efficiency of their suppressive power is almost equal to that of an evergreen tree. Perhaps this is why *Pteridium aquilinum* is one of the very few truly cosmopolitan species.

In comparing trees with one another consideration must also be given to the density of the leaf canopy as well as height. As between two species of approximately equal density, it is the taller which will tend to predominate.

Evergreen trees such as *Taxus* and *Ilex aquifolium* obviously have a greater density and persistence of shade than deciduous trees. Their suppressive effect on smaller plants is often complete, yet they themselves remain nearly always subordinate to *Fagus* or *Quercus* respectively, because of their slow growth and lack of height. A tree with a very light canopy, such as *Fraxinus excelsior*, may have the advantage of height, but it allows the development of a vigorous undergrowth of shrubby species, with which the seedlings of the tree may find it difficult to compete.

We see here the interplay of four important factors in shoot competition: rapidity of development, height, density and persistence, which are combined in various degrees in different species, fortunately seldom all together in one species. It has been well said that every species has some strength or it could not survive and that it also has some weakness or nothing else could survive.

Some very interesting attempts to define more clearly the elements of 'competitiveness' in plants have been recorded by Harper (1964). In these experiments comparisons were made of growth and yield between species or agricultural varieties grown singly and in mixture.

From pure stands of one variety the first point to emerge is the importance of *plasticity* as shown by adaptation to different densities of population. High plasticity confers a degree of constancy in yield per unit area independent of density, and the range of densities over which constancy of yield is maintained is a measure of the plasticity of the variety. Plasticity is not uniform throughout a plant's structure. Vegetative parts in general are highly plastic and the ability of a plant to vary such organs both in size (which is usual in plants of determinate growth), or in number (as in plants of indeterminate growth) is a vital part of their competitive equipment. Seed-production is plastic in some but not in others. Mean seed size and germination time are under genetic control and are strikingly non-plastic. Cultivated varieties within one species, e.g. soya bean, may show marked differences in plasticity, an important matter to the cultivator, in balancing density of sowing against yield.

Mixed cultures were made by Harper in three different ways. (1) In constant proportions, but at different total densities. (2) At constant densities but in different proportions. (3) In different patterns.

Under (1), with a fixed fifty/fifty proportion it was found that the balance of the components was often changed by increasing density, due to differences of plasticity. Thus, for example, *Rumex obtusifolius* was ahead in seed-production at low densities, but at high densities *R. crispus* was the major seed-producer.

Under (2), at constant density, there are two alternative results. Either the total yield remains predictable from the yields of pure stands, though the proportional contribution of the two species may change, or the total yield is significantly above or below the expected amount. In the first case, the species may be said to be complementary. In the second, there seems to be either a synergistic influence, or, if the yield is less, that some complementary influence has been lost in the mixture of species.

Experiments with patterning have shown that the differential between

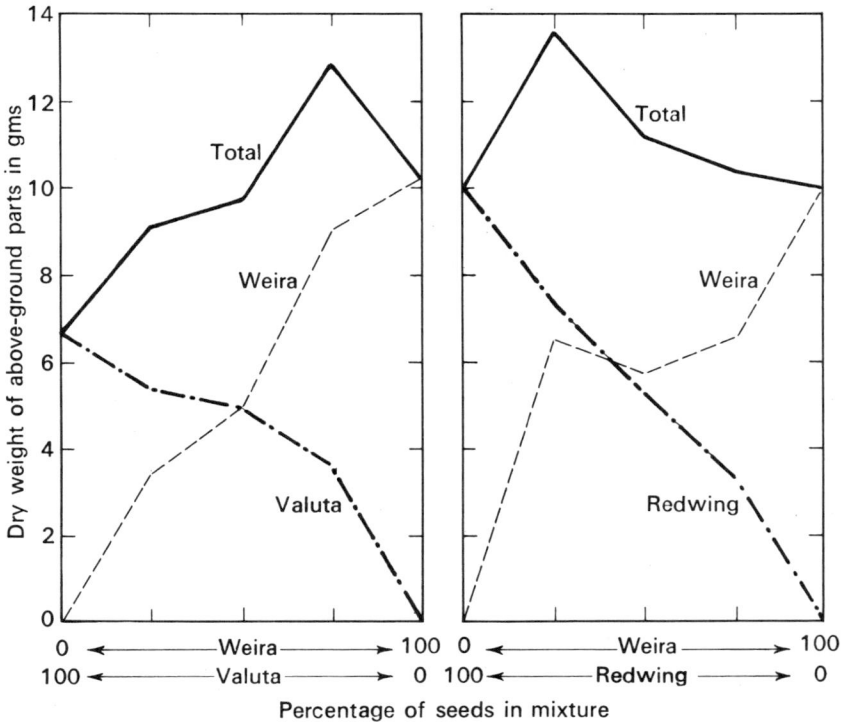

Fig. 4.3. Interaction between species after Harper. Three varieties of *Linum usitatissimum* were sown in graded mixtures at a constant overall seed rate. The graphs show the dry weight of aerial parts produced by pure stands and by different proportionate mixtures. The total production in the mixtures is in excess of expectation. A similar excess was found for weight of seed produced.

competing species is much reduced if they are over-dispersed, i.e. clumped, and that their mutual influence is a function of the extent of their contact with each other, i.e. is maximal with random dispersal.

Priority in development seems to be a highly important element of competitiveness. The species which establishes first, which puts its canopy highest, and is the first to expand its root system, will tend to gain the ascendancy. Seed weight, which implies embryonic capital, is also a positive factor, which is operative even in individual competition within the same taxon. For example, *Trifolium subterraneum* produces seeds of variable weight and a population from large seeds in competition with a population from small seeds, completely overcame the latter and after eighty days was capturing 97 per cent of the incident light.

Root competition, though hidden from the eye, is no less active than shoot competition. We know relatively little about its methods and recognize it chiefly in its effects, but it would seem to be a direct attack upon the soil's resources of water, nutrients and, perhaps equal important though often overlooked, soil air.

Weaver has revealed how remarkably extensive the root systems of even small plants can be. The development of a root system has two aspects; first, spreading into new areas of soil by means of fast-growing and relatively unbranched roots and, second, the intensive exploitation of the occupied volume by small branch roots. The spread may be both lateral and downwards. The first brings opposed root systems into contact, the second takes the roots into zones where there may be more available water, but in which the physical and chemical nature of the soil may be different from that of the upper zones. (See Vol. I, Fig. 785, p. 790.)

The lateral spread may be so extensive that it has been shown to have a determining effect on the spacing of forest trees. It is also very great in semi-desert communities in which the plants may appear to be widely spaced but in fact the ground between them is fully occupied by their root systems. Deep-rooting may sometimes be a positive advantage and will give predominance over shallow-rooted competitors in sandy soils or under drought conditions.

The effect of competition in restricting the development of root systems was shown in experiments by Dittmer and by Pavlechenko with *Secale* and *Triticum*. Competition between plants of the same species, sown in drills, reduced the total length of the root systems to 1 per cent of the total produced by an isolated plant, while severe weed competition in addition, could bring the figure down to one-tenth of this latter amount. (See Vol. 1, p. 789.) No plant can maintain a monopoly of a soil volume and competition is close and intimate, not just a matter of marginal contacts.

As a result of experimental cultures with *Hordeum*, Brenchley concluded that when nutrients are in limited supply it is the amount of available Nitrogen which is the dominant factor in competition. She also showed that crowding diminished the power of the plants to make use of the food available. Equal numbers of plants well spread out, collectively absorbed more Nitrogen than those close together. When there is no competition the growth of the plants approximates to a constant standard, but with overcrowding this approximation entirely disappears

Some of the competitive effects of roots can be attributed to induced shortages of water or of Nitrogen and can sometimes be reduced or eliminated by increasing the supply of either of these. Not all effects are thus explicable. Shallow-rooted herbs in a *Pinus* wood benefited by being surrounded by deep trenches which cut through the root system of the trees, though supplying extra water or Nitrogen had little effect.

This and other experiments have long suggested the diffusion in the soil of substances having a toxic effect, produced or excreted from the roots of some species, which would be potent in competition. The classic tests were made by Spencer Pickering, following-up observations made by himself and the Duke of Bedford on the bad effects of a grass cover on young apple trees in plantations. The original field observations were not conclusive because the effect could be almost eliminated by supplying the trees with extra water and Nitrogen. Pickering, however, tried extensive pot cultures with a variety of species in various combinations as subjects and agents. The subjects, growing in large containers, were surrounded by ring-shaped troughs, resting on the soil in the container and themselves containing the same soil, in which the agent plants were grown. All water was given through the culture in the ring-trough and in every case there was a positive result in the diminution of growth of the subjects. The average reduction of growth was from one-half to two-thirds of that of the control plants, watered in the same way through a trough of soil without growing plants. The effect was not only interspecific. The most striking case was the effect of mustard plants as agents on subjects of the same species, where growth was reduced to 1 per cent of the controls.

The fact that these effects may be largely counteracted by extra supplies of Nitrogen does not in itself negative the existence of a toxic effect on the root systems, since a reduced or enfeebled root system might be unable to absorb sufficient Nitrogen from the supply normally present in the soil but could do so from an enriched supply.

Since Pickering's experiments in 1917, however, the literature of apparent toxic action in the soil has multiplied to a considerable extent, but without producing any certainty. Clear evidence for the excretion of organic poisons from intact roots is wanting, though small quantities may be released by the decay of detached root-hairs, calyptra cells, etc. Decomposing plant residues in the soil are another likely source, for some non-toxic substances, such as the bark constituents amygdalin and phloridzin, can be broken down into toxic products by microbial action. Cold water extracts of decomposing straw yielded several toxic phenolics which inhibited the growth of roots in *Secale* and *Triticum*.

Roots are not the only possible sources of toxins. Rain wash from leaves may also contain plant poisons, e.g. absinthin from *Artemisia absinthium* and juglose from *Juglans regia*. Three species among those tested in water culture, *Franseria dumosa*, *Thamnosma montana* and *Encelia farinosa* (all xerophytes) produced substances in the water which were toxic to the roots of *Triticum*, but in the case of *Encelia* the substance excreted, 3-acetyl-6-methoxybenzaldehyde, has been shown to be ineffective in soil. It should be noted that any such effects which have been observed are more marked in sand or water cultures than in soil, for the toxic substances appear to be very short-lived in soils and are probably quickly broken down by micro-organisms. Alkaloids and antibiotics have also been regarded as possible agents in direct competition between plants, but without definite proof.

A positive demonstration of the possibility of transfer of an organic substance from one plant to another through the soil has been given by Preston. He inoculated bean plants in pots with alpha-methoxyphenyl acetic acid (MOPA) and found it present in other bean plants growing in the same pots within nine hours.

Chemical effects of one plant upon another have been called *allelopathy* by Molisch. The phenomena have been extensively studied by A. G. Winter, but he had to confess that we were in no position to explain allelopathic effects in chemical terms. He did, however, produce evidence for three important aspects of the problem. (1) The occurrence in soils of metabolic products which were physiologically active in significant concentrations. Some of these substances are phenolic, e.g. p-hydroxycinnamic acid and p-hydroxybenzoic acid, which are leached out in considerable amounts from leaf litter. (2) Although most metabolic products are broken down by microbial action in the soil, their concentrations are kept up by fresh production. Moreover, the substances made available to the micro-organisms are equally available to plant roots. Adsorption in the soil can also prevent losses by leaching and can lead to the concentration of active substances in certain layers, especially near the surface. (3) Antibiotics, such as penicillin and streptomycin, can be absorbed by plant roots and reach the aerial parts of the plants. Streptomycin has a molecular weight of 581·6 and is therefore a fairly large organic molecule. Further, alkaloids with molecular weights between 162 and 413 can be absorbed by roots and stored in the aerial parts. These substances remain stable after absorption, but phenolic substances are generally turned into glucosides. Hydroquinone, for example, becomes arbutin in the shoots and leaves and remains stable in this form. Winter was also able to show that a variety of organic substances could be absorbed by the roots of *Triticum* from soils which had been mixed with the material of plants known to produce them. The evidence for the absorption by plants of organic metabolites released from other plants is thus pretty complete. This does not by itself prove that such allelopathy is a mode of competition, but it proves its possibility.

Several workers do seem to have demonstrated an allelopathic competition in *Agropyron repens*. The growth of many cultivated plants can be severely reduced by the presence, even in small amounts, of this weed; the effect is, however, selective. Some species are very sensitive, among them some *Brassicas*; other species may show some stimulation. Oats are severely affected only if the weed is previously present in the soil. Water extracts of the rhizomes show similar effects. An aromatic substance, agropyrene, is peculiar to *Agropyron* and may be the source of the toxic effects.

While the role of organic toxins as agents of competition in the soil still remains doubtful, there are other possibilities which may be important. Simpler substances such as ammonia, ethylene or hydrocyanic acid may be released into the soil with poisonous effect. Nor is there any need to postulate excretions of any kind. It has been plausibly suggested that the excess of carbohydrate derived from decaying plant remains may raise the microbial population to the point where they compete seriously with the higher plants for Nitrogen supplies. In such circumstances plants with a high N demand would suffer most. Another possibility is that plants with a high kation exchange rate might take up so much Ca and Mg from the soil that they increase its 'acidity' and deplete it of nutrients to a degree adverse to some of their competitors.

There need not be adult competition between two species at all, if the presence of one species suppresses the germination of the seed of the other, but, although there may be various degrees of suppression in this way, comparatively little is known about it. It has been found that *Festuca rubra* can inhibit the germination of the seeds of various crop plants and weeds, which may account for its ability to form pure stands, while a perhaps extreme case is that the presence of seeds of *Lolium perenne* inhibits the germination of seeds of *Anthemis arvensis* and *Matricaria inodora*.

Fungal relationships in the soil, whether mycorrhizal or parasitic, can hardly be described as 'competitive', though it is conceivable that the mycorrhizal fungus of one host species might be a parasite of another host species, in which case it could be an agent of competition. This is, however, quite speculative, though such an ambivalent relationship on the part of a fungus is known to occur.

Angiospermic parasitism, on the other hand, may fairly be called 'competitive', since the action of the parasite is definitely to weaken or destroy the host species to the benefit of the parasitic species. It is a very one-sided competition, however, for the elimination of the host automatically eliminates the parasite also, at least in that spot, though it may have allowed the parasite to seed and so perpetuate itself elsewhere. Angiospermic parasitism is, however, rarely total; it does not destroy the host, but only weakens it and hence depresses its aggressive power. The host has no defence against the parasite, its only resource is to escape by rapid seeding.

A common action in competition is the reduction of the available water supply to the roots, and here height again presents advantages, since the taller plant can shield the soil from rain. In deciduous woodland at maturity only about 50 per cent of the incident rainfall reaches the soil, while in tropical forests, which are many-layered, very much less does so. This may sometimes be a positive advantage in preventing soil erosion, and, even where it is a deprivation from the point of view of the ground flora, it is set off by the return of nutrients to the soil by the forest litter. This is a picture of a balanced ecosystem, but the balance is not always even. Thus the heavy fall in *Fagus* woodland covers the soil with a comparatively impermeable layer, which is responsible, more than the shade of the trees, for suppressing the ground flora.

In communities with no such great disparity of stature as is seen in woodlands, the nature of the litter produced by one species may be important. Plants such as *Vaccinium myrtillus*, which cover the soil with a peaty, spongy layer of organic debris, may cause most of the rainfall to be retained near the surface, where only shallow-rooted species can benefit. The deeper roots of young trees may be so far deprived of water that they cannot survive and tree growth is precluded.

Another similar effect is the suppression of *Picea sitchensis*, Sitka spruce, by *Calluna*: the growth of the young plants of the spruce is either very slow or stopped, but when the spruce is planted in a mixture with either *Pinus* or *Larix* there is little or no check. The other trees are technically called the 'nurse crop'. Investigation shows that the nurse trees are more effective in their early stages in suppressing *Calluna*, and the roots of the spruce develop in the layer of organic litter beneath the *Larix* where they are free from the effects of the *Calluna*. As the spruce root system is shallower than the *Larix* root system there is little competition between them.

Differing levels of root development can often reduce root competition in closed communities and allow more efficient exploitation of soil resources. Woodhead first called attention to this in 1906. The *Quercus petraea* woodlands on sandy soils in the north of England often have a rich ground flora, in which *Pteridium aquilinum*, *Holcus mollis* and *Scilla non-scripta* are prominent. Among these three the *Holcus* is shallow-rooted, the *Pteridium* rhizomes are intermediate and the *Scilla*, thanks to the development of 'droppers' from the bulbs, is the deepest, so that the root systems do not compete, though the plants are growing in close association. Woodhead called such a state of affairs a 'complementary association' and there is evidence that it is not uncommon, though not always as clearly marked as in the above instance. Weaver showed that in prairie soils many grasses tend to keep their root systems compactly in the first half-metre or so of soil, while perennial herbs and some woody plants often have a double system, of widely spreading, superficial roots combined with tap roots which penetrate as deeply as the soil will allow.

Other things being equal, spatial competition by roots between species with similar rooting systems may often be decisive in keeping species apart, that is to say, in producing a negative correlation in their distribution in a community, because the more vigorous of two species will tend to weaken or suppress the growth of the second species in its vicinity. Positive correlations in distribution are less common, though the *Picea-Larix* affinity, which was previously mentioned, might be cited as an example.

An important aspect in competition which is often overlooked is what may be called 'selective repression'. A particular factor may be inimical to both of two competing species, but bears more heavily on one than on the other, with the paradoxical result that the less heavily affected species is actually promoted by the factor which threatens it. Farrow gave a good example of this in the East Anglian heaths. Grazing by rabbits was at that time very heavy and practically ubiquitous, to such an extent that in some areas vegetation was practically destroyed. Rabbits ate both *Calluna* and *Carex arenaria*, but preferred the *Calluna*, with the result that the rabbit burrows were often surrounded by a pure stand of the *Carex*, which owed its success to the suppression of *Calluna* where the 'rabbit pressure' was strongest.

The same effect may be produced by toxic substances in the soil, which injure some species more than others, though all may be affected.

The reproductive capacity of a species is obviously an important factor in competition, whether it is vegetative or by seeds. Between the two methods there is a certain proportion. Spreading by seeds tends to be more important in open communities which are generally associated with early stages in the development of vegetation. Short-lived species which have no means of vegetative spreading must rely on seeds, which they produce rapidly and often in quantity.

It is not only quantity that counts, but also germinability of seeds, which is often very low, especially with small seeds. There is an obvious correlation between seed output and germinability. Some species with a small output, like *Mercurialis perennis*, have also low germinability, whereas *Digitalis*

TABLE 2. WEIGHTS OF SEEDS AND INDEHISCENT FRUITS

Class	Open Ground	Short Turf	Meadow	Scrub	Herbs	Woodland Shrubs	Trees
24							1
23							1
22							2
21						1	
20						1	
19							2
18						1	3
17						2	3
16					4	5	1
15	1	1		3	3	2	3
14	3	3	3	4	7	5	1
13	2	1		9	10	4	1
12	7	7		9	4	1	
11	10	10	4	8	2	2	1
10	15	10			4		
9	17	4	1	6			
8	18	9		2	1		1
7	8	1		9			
6	4	1					
5	6	2					
4	6	1					
3	1	1					

Horizontal columns in ecological order. Vertical columns in order of weight. The weight classes are in ascending order. Each class has an upper limit which is twice that of the class below. In class 3 this is 8 µg, in class 24 it is 16g. (From SALISBURY, *The Reproductive Capacity of Plants*, Bell, 1942.)

purpurea with an annual output of about three-quarters of a million seeds has a high rate of germination. (But cf. *Verbascum*, below.) There is, however, some degree of negative correlation with the power of vegetative propagation, those species which are active in this respect tending to have a less reliable seed product. Seed production is most important in the colonization of new and unoccupied ground and so is most important to species which form the early stages in vegetation development. In the later closed states of a plant community the chances of the establishment of a new plant from seed are small and vegetative propagation is relied upon extensively. Seeds produced in the later stages of the community's development are chiefly of importance for long-range dispersal and the colonization of new ground.

In spite of high germinability, the mortality of seedlings can be immense. Populations of *Digitalis*, for example, remain fairly constant from year to year so long as the conditions are unchanged, though they may show an upsurge when a local clearance gives greater opportunity for germination. Salisbury described the case of a plant of *Verbascum thapsus* growing on waste ground without close competition. It produced about 700,000 seeds, which were 88 per cent germinable. Nevertheless, six months after seeding there were only 108 offspring left out of a potential 600,000. *Verbascum* and *Digitalis* are both biennials, and it may well be that this habit is related to the tremendous drain on the plant of such large seed crops. Most perennials have much lower outputs, but the output is repeated annually, with an aggregate of production depending on the lifetime of the plant.

Seeds may, however, remain dormant in the soil for a considerable number of years, only germinating when some disturbance of the soil occurs, normally associated with clearances which may give them a better chance of establishment. Many wild species show delayed, discontinuous or continuous germination the effects of which are to spread germination over a period of time, which may be several years. This avoids the risk of the whole year's crop of seedlings being destroyed by some natural disaster and it increases the possibility of wide dispersal. Cultivated plants have been subject to selection for ready germination and they differ very markedly in this respect from the products of wild plants.

One disadvantage of very large production of seeds is their necessarily small size and meagre supply of reserves for the embryo plant. At the critical period of establishment this tells heavily against their survival. Plants with large seeds show a marked advantage in rapid development of height and of leaf spread which are of great competitive importance.

There is thus a balance of advantage and disadvantage with regard to seed production. On the one hand we have a large output of small seeds, which are more readily dispersed but which suffer from a high mortality though they still may make possible rapid short-term multiplication of the species. On the other hand we have a smaller production of large and relatively heavy seeds, usually with good germination and lower mortality, which yield more robust seedlings. Which method produces the best results will depend on the conditions of life, the niche, with which the species is associated. The first method is generally best-suited for short-lived plants in the early stages of vegetation with open communities. The second method is more generally attributable to long lived perennials in closed communities.

We can, in fact, trace a reproduction sequence in the development of communities, the early stages being populated largely by short-lived annual and biennial plants depending chiefly on rapid seed dispersal and, as development proceeds towards stability, with an increasing percentage of species which are perennials of increasing stature, depending more upon short-range vegetative propagation than upon seeds.

The native flora of a region has achieved an equilibrium in competition, each species having found the niche wherein it is possible for it to live. This equilibrium can be upset by the introduction of foreign species which sometimes achieve a spectacular or even catastrophic success, usually because they possess powers of competition that the native species are not equipped to resist. In other cases their success is due to their being particularly well adapted to the climate and soil, which they would have colonized in the course of nature but for geographical reasons. Plants in this latter category may settle down and become indistinguishable from native plants, like *Opuntia* in Mediterranean lands, a process that must have occurred by natural invasion many times in the history of the British flora. Plants in the first category have a varied history, depending on their biological equipment: they may continue unchecked until man is forced to intervene with a campaign against them, not always quite successfully, or they may lose some of their original vigour, as happened to *Elodea* in Britain and settle down into the native equilibrium.

The concept of competition has taken many forms in the history of biological theory. In Darwin's time it was often confused with the 'struggle for existence' as if the two were synonymous, which they are not, since an organism in complete isolation may have a struggle for existence against an adverse environment.

Etymologically the word *compete* comes from the Latin *competere*, which may be translated as 'to seek or to demand in common' and the emphasis in most definitions of competition has been on common demand, with the implication that it can only occur between organisms which make the same or closely similar demands on the environment. The root idea here is rivalry, as in the German word, *Bewerb*, which leads to the conclusion that competition involves 'success' on the one hand and 'failure' on the other, with the weakening or even extinction of one of the rivals. There is almost always expressed in such definitions a limiting condition, namely, that the resources competed for are in 'short supply', that is that the total demand of the competitors is in excess of the available resources of whatever is being competed for.

There is too much anthropomorphism in such views, a perhaps unconscious leaning on the analogy of human society, in which competition is presented to our immediate consciousness. We should try to look for a broader, more fundamental concept with no human overtones, if such a thing is possible; a concept which is at least applicable to the plant world, while admitting that for animals a different concept may be needed.

Let us first eliminate the requirement that the subject of competition must be in limited supply. We have only to think of light to realize the fallacy of this restriction. Light is almost everywhere in superabundant supply, yet every green plant must obtain an adequate ration, which often involves competition. The point here is that competition in such a case is not *necessarily* rivalry, though it *may* sometimes be so. Plants vary considerably in their adjustment to light and competition for light in a community may be simply a degree of interaction which enables each member of the community to get what it needs. The crux of the question is what the plant needs, and these needs may differ in plants living together. Let us suppose two plants A and B have different Nitrogen requirements, that of A being the greater. If these plants cohabit in a certain area, their root systems will compete for Nitrogen, whatever its level in the soil may be. Both may ultimately satisfy their individual requirements, or alternatively A may go short and suffer in consequence. Nevertheless, in both cases, there has been competition between them, without any question of one taking a disproportionate supply.

To put the matter in the most general terms, may we not say that whenever two plants which have similar requirements interact, there exists a state of competition between them. Competition may either be balanced, resulting in an equilibrium or it may be unbalanced, resulting in injury to one of the competitors, but in both cases there is competition, though the second alternative is probably the commoner.

Interaction need not involve physical contact. We are too apt to concentrate attention upon the higher plants on land, but in aquatic habitats and among algae, protophyta and bacteria there exist most complex interactions, favourable and unfavourable, mostly due to excreted metabolites (ectocrines), which may in some cases produce exclusion of species or may, on the contrary, create conditions of mutual dependence.

We have mentioned favourable interactions above. That these exist implies that interaction between plants is not invariably hostile rivalry. Conscious co-operation we cannot assume, but it is possible for one plant to benefit another notwithstanding. There is the obvious case of one plant providing shade and moisture which benefits others, but there are many degrees of such beneficial action, *commensalism* as it is called, leading up to complete symbiosis which is its extreme expression. Since it has been shown conclusively that plant roots absorb quite complex organic molecules there are many cases known of one plant absorbing excreted metabolic products from another. One interesting case is the liberation of amino-acids from the nodules of leguminous roots, which can raise the Nitrogen level in poor soils quite appreciably, to the benefit of non-leguminous species.

Enough has been said, we hope, to show the difficulty of forming a clear concept of competition or of formulating one in a definition. Interaction of organisms, even among plants, can take so many forms that it is often a matter of opinion whether a particular case is legitimately to be considered as competition. Among animals the predator-prey relationship makes it still more difficult.

It may be well to be specific about one point in our own attempt at clarification. We regard competition as being between individuals. The competition of one species with another species is an abstraction from the reality of a myriad of individual competitions, with varying results, but in which an advantage may lie statistically on the side of one species.

We have tried to stress the view that interaction is the root of competition, but perhaps it has not been made sufficiently clear that competitive interaction need not be reciprocal. A may be in competition with B, yet not B with A. If this sounds paradoxical, let us take a simple case: *Salix repens* overgrows and suppresses a colony of moss. From the point of view of the moss, *Salix* is a dangerous competitor, but the *Salix* is completely indifferent to the moss, which has no effect upon its growth. Yet both are making a similar demand on the environment, i.e. for space, and this situation, which is by no means uncommon, can hardly be excluded from our ideas of competition. It is certainly not a predator relationship, for the *Salix* gains nothing from the suppression of the moss; it would occupy the space whether the moss were there or not.

Perhaps competition is too complex for any complete definition, yet it is an omnipresent and important reality. Nor is there any acceptable substitute for the word itself. Its use is inescapable. It may be burdensome, when we refer to competition, to have to state precisely what we have in mind, but it would greatly assist clarity if we did.

Migration

This is not quite the same thing as dispersal. The latter is universal among plants and implies simply the release from the parent plants of the seeds, spores, vegetative offshoots or whatever means of propagation they employ. Collective names such as 'propagules' or 'disseminules' are sometimes used to cover all natural units of dispersal. Migration means more than that, it means the movement of propagules from one habitat to another. It is only of value if it is followed by the establishment of the species in the new habitat, but it is an essential element in the succession of communities in any habitat, which is due to the arrival and establishment of new species which replace those already there by virtue of greater competitive power and greater permanence. Such successions are a fundamental feature of natural vegetation and are practically ubiquitous. They arise from purely biological causes, without necessarily involving any change in the characters of the habitat, apart from those changes which the new plants themselves may bring about, and they all depend on migration.

The extent of migration naturally depends on the reproductive capacity of the species and on the means of dispersal. It may take the form of a slow but steady extension of the area in which the species occurs, moving its frontiers forward, as in the westward extension in northern Europe of *Picea excelsa*, quoted by Warming, or it may create a rapid and dramatic spread into new areas, as with such species as *Matricaria matricariodes* or *Chamaenerion angustifolium* in Britain. Relatively slow migration is generally characteristic of long-lived species, especially of trees, while rapid spread is usually found among herbs or annuals which inhabit disturbed soils and are in consequence among the beginners in the colonization of denuded areas, where they have only a precarious tenure.

There are very few species, perhaps only three or four, which are genuinely cosmopolitan; for others one may safely say that their absence from any area is probably due to the fact that they have not yet arrived there in the course of migration. This is shown by the astonishing spread of some species which have been introduced, usually by man, into new regions. Three-quarters of the flora of Entre Rios in Argentina is said now to consist of introduced European species. Australian species are rapidly displacing native species over large areas of the Cape district in South Africa. *Pinus radiata* is far more vigorous in the Southern Hemisphere than in its native California. The vicious spreading of the American *Opuntia* in Australia is celebrated and may be contrasted with its wide distribution round the Mediterranean, where, however, it is held in check by the native flora. European blackberries (*Rubus* spp.) are major pests of cultivated lands all over the Southern Hemisphere. Among water plants the massive invasion of British waters by *Elodea canadensis* in the middle of the nineteenth century and the present pan-tropical spread of *Eichornia crassipes*, the Water Hyacinth, show how almost unchecked success can attend the arrival of a plant in new areas if it does not meet immediate competition or if its niche is unoccupied. In fact the introduction of a plant to a country where it does not occur naturally can be dangerous and destructive and is checked by legislation in many countries.

Where the vegetation has reached stability, especially in geologically old countries, migration is reduced to a minimum, for it is forbiddingly difficult for a species to find any new area where it can establish itself. It is only possible for the most aggressive species and then only to a limited degree. Human interference, even in tropical countries, often provides cleared sites which are open to migrants, and on such denuded sites the succession of colonizing herbs and grasses, shrubs and trees will, in a comparatively short time rebuild a stable forest which it may be difficult to recognize as a secondary growth. In the extreme case of tropical rain forest, density and stability may be so high that even normal regeneration by dispersal may be restricted and may have to await openings caused by the death of trees.

While the absolute range of migration depends on the mode of dispersal, with wind-borne spores in the highest class of mobility, yet, relatively, the incidence of dispersal is remarkably constant. If a graph is constructed with the logarithms of the distance units, whether centimetres or kilometres as abscissae, and the number of propagules of the species found as the ordinates, the plot will show a descending straight line. In other words the effectiveness of dispersal is inversely proportional to the logarithm of the distance.

Migration may be important where the position is not that of entering a new area for the species, but of entering another population of the same species. In this case gene exchange between the old and the new populations may have a marked effect upon the future of the mixed population.

We have said above that the value of migration depends upon subsequent establishment in the new locality. The 'double coconut', *Lodoicea seychellarum*, has been carried over the Indian Ocean by currents for centuries, but it has not yet succeeded in establishing itself outside the Seychelles Islands. The reason is that it is an inland plant, growing on river banks, and only the improbable event of its finding its way up a river in a new country would give it even a proximate chance of establishment.

If the new locality is a denuded area, the settlement there of migrant species is called *colonization*, if there is already a settled plant community present then the establishment of a migrant is called *invasion*. An invader does not necessarily menace the existing order; it may become integrated as a balanced member of the community, usually as a subsidiary member. On the other hand its establishment may so change the existing balance as to provide opportunities for the establishment of other invaders and thus pave the way for the succession of a different type of community.

Establishment in a settled, i.e. 'closed' community, inevitably involves competition and an important effect of the new competition is that it restricts the range of tolerances of all the species concerned. While this affects the indigenous species, often to their detriment, it also affects the incoming species and militates against its chances of establishment, unless the conditions are unusually favourable or unless it exploits a niche which gives it a competitive advantage, especially if there is no other species present with a niche which markedly overlaps that of the newcomer.

The process of establishment may therefore call for adjustments in growth and physiology, which begin with germination and are continued through life to successful reproduction and so to further propagation in the new area. This whole process of adjustment to novel conditions has been called by Clements *ecesis*, and plants which show recognizable modifications in relation to abnormal environments he has called *ecads*.

Ecesis is by no means always successful. The conditions in the new locality may even preclude or greatly delay germination, so that the process does not even get a start. If the migration has been over a considerable distance the plant may find itself outside the climatic region where life is possible for it. After germination the most critical phase is that of the young seedling, when the plant is most vulnerable and may be destroyed by any number of accidents. Once the seedling has got its roots well down into the soil and has begun to unfold leaves, its expectation of life is greatly increased. For these reasons a successful invasion increases in probability proportionately to the number of propagules of the species arriving. In the case of a wind-borne propagule, successful migration is much more probable in the direction of the prevailing wind than in any other direction, though occasional propagules may find their way otherwhere.

Reproduction is the final test in ecesis. If a species is highly specialized in reproduction, if, for example, it depends on a single species of insect as the pollen vector then the plant cannot establish itself permanently outside the area of distribution of the insect, however well it grows. Only if the plant can maintain a high and regular reproductive capacity will it have a chance of maintaining its hold in a new community.

Although long distance invasion is possible, it can rarely be successful because the number of migrants is necessarily small and their arrival infrequent. By far the greater amount of invasion is between contiguous communities and is to some extent mutual. This often produces an intermediate zone in which the two communities are mingled and in which competition between them is active. One of the two communities may thus take over ground from the other. Surveys or photographs of such boundary zones in successive years will often show surprisingly rapid changes in the balance of the two communities.

The effect of an isolated invader on a community depends largely on its stature and its endurance. If it is of the same growth form as the prevailing vegetation, it may become integrated into the community but only rarely will it displace the existing species. An example of this occurring is the spread of *Bromus tectorum*, accidentally introduced from Europe, over very large areas of semi-arid sagebrush vegetation in western North America. Not only did it replace the existing grass cover, but, aided by its quick recovery from fires, it has also displaced the sagebrush (*Artemisia tridentata*) until it forms practically a pure stand of *Bromus* in many areas.

True succession, however, depends on the introduction of higher life-forms, with increased coverage and longevity. Invasion by shrubs or trees, though it may be slow, does introduce increased stability in the vegetation until something like permanence is reached.

Invasion will naturally have its maximum effect on a community which has already been rendered unstable by some physiographic change in the area or by the attacks of predators or disease. Relatively stable communities, which are well balanced against the environmental factors, can absorb a considerable amount of invasion without marked change of character, provided that the invaders are of the prevailing life-form or forms. There is a tendency to think of invaders as if they were always species alien to the existing community, but they may also be reinforcements of species already established. If an established species has been weakened by attacks, such immigrations from a new source may preserve it from extinction. On the other hand, by increasing the density of that species in the community they may seriously upset the inter-species equilibria and render the community unstable without changing its species composition. There are many examples of this occurrence among animal populations as well as among plants.

Succession

This is one of the most important concepts in Ecology. That natural vegetation showed natural changes was noticed by the earliest observers and must, indeed, have been part of the nature lore of primitive man, whose observation was sharpened by his close dependence on his natural environment. The observation that there was order and method in these changes came only recently. Hult, in 1885, was one of the first to show that, in the district of Blekinge in southern Sweden, all the various forms of vegetation tended to develop into one or other of six types of woodland, whose distribution was determined by the soil conditions. These woodlands showed no tendency to develop further and were apparently permanent: this was true succession, as now understood.

Succession was adopted by Cowles in 1899 as the leading principle in his exposition of the sand-dune vegetation of the Lake Michigan area, and it was subsequently elevated by Clements into the position of the primary law of natural vegetation. This claim was largely, though not entirely, accepted by Moss and Tansley in Britain and became a characteristic trait of British Ecology, more or less in line with the American school.

In his voluminous writings, especially in *Plant Succession* (1916),* Clements introduced a considerable number of new terms for various ideas connected with succession, some at least of which have passed into general use and are of great convenience.

Clements's views were certainly guided by his experience of the American West, where there are, or were half a century ago, vast expanses of natural landscape scarcely touched by man and providing in mountainous areas many

* Carnegie Institution of Washington. Publication 242.

occasions of physiographic change which enabled all stages of succession to be seen. In Europe, where vegetation is much more stabilized and very little has escaped human activities, succession has never assumed the primary importance that Clements attributed to it. Its reality is conceded, but it is regarded more as a matter of history than as a subject of practical importance. In Britain it is perhaps the phenomena of our unstable coast line which have impressed on ecologists the immediate relevance of succession, for in such areas it can be seen in action, observed, analysed and measured.

The marrow of Clements's doctrine is this. Over the whole of any area within which a substantially uniform climate exists, the progress of development will inevitably tend towards a uniform, stable type of vegetation which is the highest that the prevailing climate will support. In this connection 'highest' means having the most complex structure and physiognomy that is compatible with stability. Every type of climate is envisaged as associated with its own particular type of permanent vegetation.

The stable type of climatically conditioned vegetation is called the *climax* (no pun on 'climate'). In time, the given area would develop into the climax all over. The process of development Clements called the *sere*, implying the whole chain of events from start to finish. In his view all vegetation is a part of some sere; either as a stage or as a climax all vegetation is involved in this universal stream of life. A sere may be very prolonged, with many stages, or it may be short and simple, observable in a human lifetime. In every climate which permits it the climax will be some kind of forest, since this is the highest and most enduring type, but it may be any of the principal types of vegetation other than forest: grassland, savannah, bushland or desert.

A sere which begins on a bare soil is called a *prisere* (primary sere) and among such there are various subdivisions, one of the chief being the *xerosere* developing on bare rock soils, such as the debris of eruptions or landslides, which slowly become the matrix of true soils as the vegetation develops towards the climax, which will only be possible after a long period of soil building.

A second division is the *hydrosere*, which begins on flooded or submerged soils on which only aquatic plants can grow. By the steady accumulation of plant remains below the surface of the water, aided often by deposits of silt from floods, the level of the soil is eventually raised to the extent that marsh plants can colonize it. As the process of accumulation continues, shrubs like *Salix* and *Alnus*, which can tolerate waterlogged soils, may appear and their more massive contribution to the litter will accelerate the rise of the soil level until eventually it is above the water table and mesophytic trees can root in it.

Both seres therefore, starting from opposite poles, converge in developing a mesophytic soil with a mesophytic vegetation.

As reference has been made already to our coastal vegetation it may be pointed out that the succession on a salt marsh is a *halosere*, while that on sand dunes is a *psammosere*.

While the climax is always some kind of forest wherever the requirements of soil, water and heat are adequate, yet it is obvious that in any climatic region there will be many areas where this is not the case. The coastal zones mentioned above afford an example. Permanently waterlogged soils with an insufficiency of soluble minerals available will develop only acid peat, not a mesophytic soil. Escarpments of dry, hard rock, areas of shallow, barren gravel, high wind-swept ridges, on none of these would forest be possible. One might also add mountain heights above tree level, were it not that it could be argued that these constitute a different climatic region. Confining ourselves to lower levels, however, the kind of areas we have mentioned will not allow development of vegetation to proceed all the way to its natural climax and the plant cover will remain arrested at some lower level of development, more or less permanently. Such arrested stages are called *sub-climaxes* or sometimes *disclimaxes*. True sub-climaxes are, however, relatively rare, that is to say, either that they seldom correspond exactly to stages in the true prisere or that they are not often true *pre-climaxes*. Most commonly they represent specialized modifications, corresponding in physiognomy and to some extent in floristic composition to prisere stages, but with characteristics which show that they are not part of the general succession. This is spoken of as a *deflected succession* or a *plagiosere* and if it becomes apparently permanent it constitutes a *plagioclimax*. The work of man has been prolific in producing plagioclimaxes. The whole of the pasture land in Britain is a deflected succession or plagioclimax, held artificially in this condition. If totally abandoned it would return to the prisere and in less than a century would probably bear woodland, as it did in primitive times.

The distinction between sub-climax, pre-climax and plagioclimax is somewhat difficult, and Clements himself is not wholly consistent in their respective uses. It seems best to apply the term sub-climax to all cases of arrested succession. If the stage of arrest is an actual stage of the prisere it is then also a pre-climax, though this may be difficult or impossible to decide unless the stage is one very close to the true climax. If, however, the arrested stage becomes modified into a community with its own special characteristics, which is almost certain to happen with lapse of time, then it is a plagioclimax, but it is cardinal that if the arresting or deflecting conditions are removed, the prisere will re-establish itself.

An essential feature of Clements's general theory of succession is that it is irreversible, that there can be no autonomous or spontaneous retrogression from a higher to a lower type of vegetation. Where this appears to occur it can be traced either to a climatic change or a physiographic change, such as flooding, or to animal interference, including man. The principal case of such apparent retrogression is that of the disappearance of forest, especially its replacement by heath. This took place on a large scale in the northern part of Britain in the disappearance of the sub-Boreal coniferous forest at the onset of the much wetter sub-Atlantic period in late post-glacial time. Peat formation increased under the oceanic climate and the tree limit in altitude was lowered to about 450 m leaving bare large mountainous areas previously forested, where the tree stumps are still plentiful, embedded in peat. This was definitely due to the start of a new climatic cycle, initiating a new sere and cannot be called retrogression.

Other causes of the destruction of forest may be flooding, fire or felling, and grazing by man and his animals. The last-mentioned has been a very important factor, caused either by bark-grazing or more permanently by the destruction of seedling trees and the stoppage of natural regeneration. The disappearance of oak wood over large areas in northern England and its replacement by grassland or heath was due to large-scale grazing by sheep under the hands of the big Cistercian monasteries, who were the great wool producers in the Middle Ages. The destructive effect of forest fires needs no emphasis. It has enormous effects in drier climates where its occurrence is an ever-present menace. It would seem that Clements was right in arguing that the destruction of a forest climax and its replacement by a more lowly-organized vegetation is always due to extraneous interference.

FIG. 4.4. Wistmans Wood on Dartmoor. A small relict of *Quercus petraea* woodland at about 1400 feet, the only remnant of extensive woodlands which owes its preservation to the rocky ground or 'clatter' on which it grows. (Photograph by L. J. Watson.)

The disappearance of a climax community opens the way for a new sere and a secondary succession of this kind is called a *subsere*. If the cleared area is relatively small the subsere may repeat the development of the prisere and in time re-establish the original climax. In the tropics the process may be rapid and a great deal of rain forest is actually the product of a subsere rather than of the original prisere. In regions of heavy rainfall, however, the removal of the forest cover lays open the soil to changes which may deflect the subsere into new courses.

In hot countries direct leaching of the soil and its impoverishment in mineral nutrients is inevitable if any considerable area is left exposed. A secondary effect is the destruction of soil fungi by the desiccation of the soil in dry seasons, thus depriving tree species of the mycorrhizal fungi on which so many depend. In cold countries leaching leads to podsolization (p. 3599), which not only impoverishes the topsoil but leads to the formation of a layer of humus-cemented sand at a depth of a foot or more. This layer is known as 'hard-pan', or by the German name of 'Ortstein'. It is practically impenetrable to roots, and it limits plant development to shallow-rooted plants like grasses and Ericaceae or to shallow-rooted trees such as *Pinus*. For such reasons the subsere may follow a different course and result in a different climax from the prisere, frequently replacing the forest climax by a grassland or heath climax or plagioclimax.

The question whether retrogression does or does not occur has been strenuously debated, for on the respective views held depends the whole concept of the nature of succession. That retrogression *does* occur is certain, but the argument is concerned with whether such retrogression can be included in the normal process of succession or whether it is always due to some extrinsic influence. We may be in a better position to consider this when we have given some attention to the mechanics of succession.

The concept of vegetational succession is in some ways like that of organic evolution. In both, a large time-scale is involved so that direct observation is rarely possible. Both are concepts which we cannot dispense with, in face of all the evidence which supports them, yet both are mental inferences from observed facts in the grand landscape of life. In both we infer development along a great variety of channels, conditioned by environment, towards conditions of greater stability in equilibrium with the environment, yet in neither case can we directly observe this development except upon a minor, sometimes a minuscule, scale. From these observations we extrapolate our ideas towards infinity, taking as guide the comparatively stable points which we can see along the line our curve must fit.

It is time now to look at the supposed processes of development more closely. Tansley did a great service to Ecology when he discriminated clearly between two methods of development which have been greatly confused with each other. One he called *autogenic*, due to intrinsic causes, the other he called *allogenic*, due to extrinsic influences. Autogenic succession will occur, even in the hypothetical case of a completely stable external environment, arising from causes which are inherent in the nature of the plant community, chiefly the reactions of the plants upon each other (competition in the wide sense) and on their immediate environment. It also involves the arrival and ecesis of new elements in the community as a motive force of change. These are vectors which are universal and inescapable.

The environment acts upon the plant, controlling its growth and in return the plant reacts upon the environment, modifying it in a variety of ways, usually complex. Some of these reactions upon the physical environment are readily appreciable. The reactions of the plants upon each other we have already spoken of under the heading of 'Competition', but they are, in general, less understood than the reactions upon physical factors.

Although reaction is a function of the individual plant, the effects are sometimes only appreciable in the community as a whole, sometimes in fact they are only exercised by the community as a whole. A single tree casts a temporary shade on the ground, but a canopy of trees casts a continuous shade, which has quite a different effect upon ground vegetation. A single plant of *Salicornia* has little or no effect in gathering silt from tidal waters, but a stand of *Salicornia* plants is an effective strainer and collects silt rapidly. A single plant is quite ineffective in stabilizing a mobile substratum, but a community of mat-forming plants may be a complete stabilizer. The community not only emphasizes the individual effects, but may change them in kind by total action.

Reaction begins with individual pioneers in a bare area and is unfavourable to the pioneers themselves since it opens the way to the ecesis of more permanent plants, which will displace them. Plants of the earliest stages cannot ascend the sere, they can only seed and escape.

First come reactions on the substratum, which include weathering, i.e. the breakdown of the rooting medium by the mechanical action of roots and in some cases by chemical weathering due to CO_2 excretion. There is a steady accumulation of plant remains which results in the incorporation of humus in the substratum. Wind-borne dust is trapped and added to what is now becoming a soil. This is by no means an inconsiderable factor, as Jenny has shown in the Swiss Alps. These reactions have marked effects upon the structure of the soil. Over-rapid drainage is checked and the water-holding power of the soil is increased; the soil is stabilized and compacted and at the same time a crumb structure is built up in the soil which facilitates aeration. These processes are cumulative as succession progresses and tend towards the formation of a mesophytic soil.

Chemically, the most obvious effect should be the decrease of soluble nutrients by absorption from the soil, but except in very poor soils the amounts taken up by plants are only a small percentage of the soil reserves and this loss is counter-balanced by the return of materials to the soil in the form of leaf litter and plant remains, so that in a natural community there is an approximate balance of withdrawal and return and the soil is not impoverished by plant reactions alone. Leaching by rainfall is, of course, a different matter; this may remove nutrients to a disastrous extent and its effects are only partially hindered by the plant cover. Nitrate especially is readily lost in drainage water, as it is not adsorbed by the soil to any extent, hence the importance of the nitrification process in the soil. Leaching may even be encouraged by plants, since the formation of acids by roots may render mineral nutrients soluble and thus lead to the complete impoverishment of the upper zones of the soil profile, especially on sandy soils where drainage is rapid. This is part of the process called podsolization which we shall speak about later (p. 3599).

The contribution made by dustfall to the maintenance of a nutrient balance in natural soils should not be overlooked. Even in the pure air of the Alps Jenny showed that it amounted to several tons per acre annually, of which a considerable proportion consisted of Calcium carbonate. Rain itself is by no means pure water, but contains significant amounts of salts dissolved from the fine dust of the atmosphere, notable among these being sulphates.

Acid production in the soil can have very notable and widespread effects. In regions of high rainfall, areas where drainage is deficient become water-logged, aeration is diminished and in consequence the humification of plant remains is imperfect and abnormal. This leads to the accumulation of acids in the substratum which further interferes with humification by suppressing bacterial action. The process is cumulative, the more imperfect humus accumulates the more acid the soil becomes, sometimes reaching pH 3·0. The result is the formation of a growing bed of acid peat which limits colonization to a comparatively small group of plants which are tolerant of acid soils and are able to cope with reduced root absorption. The physiology of these 'physiological xerophytes' is still only partially known. Some of the plants of peat bogs appear to be true xerophytes, but many are not. The supposed effects of an acid medium and deficient aeration in reducing root absorption are still an open question.

Peat accumulation once started may continue indefinitely, unless there are climatic or physiographic changes which alter the normal succession, and it would be straining language to describe such a state of affairs as a 'stage' in succession or anything other than a 'climax' vegetation.

The climatic reactions of plants are rather less direct than their reactions upon the soil, but are nevertheless important. One reaction which at once attracts attention is that of cutting off a proportion of the daylight, or in other words, shading. This is conspicuous in forests and has a governing effect on the development of shrubs and herbs, but in any community where woody plants are present or even herbs with large leaves, some elements of the community will suffer shading. In close canopy the shade-forming species may even prevent the growth of any lower layers, except for some mosses and ferns which have great powers of shade tolerance.

Relative degrees of shading may also determine succession among the major plants themselves. For example, *Betula* can be superseded by *Quercus petraea*, which is slower growing but with greater canopy shade than the former. Similarly, *Fagus* can supersede *Fraxinus* and, in turn, *Taxus* may supersede *Fagus*. Light-demanding trees like *Fraxinus* and *Betula* form a light canopy, because their leaves cannot adapt themselves to shade conditions, and their regeneration becomes impossible in the deeper shade of the competitive species.

There are also reactions upon humidity, wind and temperature. The first is affected by increased evaporation of water in transpiration and the effect is increased by shelter from wind, as the moisture is not effectively dispersed.

Shelter from wind is increased, like shading, by the growth of trees and 'shelter belts' of wind-resistant trees are commonly planted in exposed districts. Increased humidity also tends to prevent frost by raising the dew point.

The effect of vegetation in increasing rainfall is fairly well established, though the extent of the effect may be disputable. The amount of water transpired by a close plant cover is very considerable. Wholesale disforestation has been held to account for the semi-desert conditions existing over large parts of the Middle East and there are numerous records of increased rainfall following on extensive forest plantations. Increases of up to 305 cm annually have been recorded in various parts of India. In tropical countries it is not uncommon to see a condensation cloud stationary over forest, formed by rising air carrying the transpired moisture with it. The amount of water absorbed by an acre of mature oak forest has been estimated by von Höhnel as 10,000–11,800 litres per day, which is enough to produce a rainfall of 76–100 mm per month.

A forerunner in the field of physiological Ecology was the novelist Robert Louis Stevenson who contributed a paper 'On the Thermal Influence of Forests' to the *Proceedings of the Royal Society of Edinburgh* in 1873. It is a closely reasoned analysis of the thermal behaviour of a forest in its relation to the external atmosphere with a critical examination of published observations. He concluded from the evidence that the thermal effects of forests on climate were slight and that, in principle, they behaved like any other radiating surface such as grass, though with quantitative differences which were as yet unknown.

Climax

It is now time to examine the nature of the terminal state towards which all succession may be supposed to be directed. Under constant conditions, that is to say under a degree of constancy (since there is no absolute constancy outside of laboratories) which does not exceed the range of tolerance of the plant species concerned, every community must, sooner or later, reach a condition of equilibrium with its environment, which will have, by human standards, some semblance of permanence. Such communities we recognize as units in the vegetational mosaic, or, if we carry through the idea of universal succession, as top notes in the vegetational scale.

Again, permanence is only a relative term, like constancy. On the geological time-scale nothing is permanent. In regions where there has been little tectonic movement, such as some continental areas in the tropics and parts of Russia, which have not been upheaved since Permian times, vegetation has had a long run of many millions of years without revolutionary changes. Yet even in these ancient communities there have been alterations: long-term ones of climate may have occurred, due to the shifting of ocean currents or the elevation of distant mountain ranges, or to continental drift, or to variation in solar radiation, or to any one of many possible causes. The continuous processes of erosion must have caused big changes in topography, in drainage and the nature of the soil. Failing any kind of geological effects in such ancient communities the process of biological evolution has had time to alter the nature of the plants themselves. Indeed, the geological record shows us, however imperfectly, the succession of one type of vegetation by another more advanced type, in the evolutionary sense, all through the earth's history.

In the fringe areas of continents, such as Britain, where elevation, erosion and subsidence of the land are continually active, no such permanence of vegetation is to be expected. Yet even here we have some evidence of a surprising degree of persistence, if not exactly permanence, of the same type of vegetation. In California some beds of fossil plants of Pliocene Age show not only species still living in the same area, but the same associations of species which can be recognized in the living vegetation. The interglacial deposits in Britain show repetitive patterns or vegetation, covering periods measurable in tens of thousands of years, changing only by the dropping out of some species and the incoming of others, as we have described under the 'Quaternary Era' in Vol. 3 of the present work.

The secular succession of one type of climax by another has been called by Clements a *clisere*, the longest and most imposing succession of all, indeed coeval with the world itself.

The concept of the climax in vegetation is one which has been widely and critically discussed. At the one extreme we have Clements's theory of the monoclimax and at the other extreme the view of vegetation as a continuum which discounts the validity of the plant association as an objective reality.

Clements's theory is fundamentally simple. Within an area of uniform climate *all* vegetation must eventually develop by successional stages into a single uniform, stable community, the climax, the characters of which will depend on the type of climate. Even in regions where a high degree of uniformity exists, as in the *taïga*, the northern coniferous forest, it is evident that there are aberrant localities, such as bogs or areas of bare rock, which the forest does not cover but which have characteristic communities of their own. These have been called 'edaphic climaxes' to distinguish them from the general 'climatic climax', but Clements will not allow this. To him, such aberrant areas are merely delayed stages of the general succession and as such are *sub-climaxes*, destined some day to follow the inevitable line of succession to the sole monoclimax.

This is admittedly a simplified presentation of Clements's views, omitting the subsidiary theories and the highly complicated terminology with which it is adorned, but it expresses the essence. One significant addition should be mentioned. Clements allows that in certain localities, where special factors favour it, there may be an advance to a type of vegetation higher than the climatic climax. Where the latter is already of the most advanced type, as in tropical rain forest, this is scarcely possible, but where the climatic climax is of a lowly type, such as grassland or semi-desert scrub, such a *post-climax* (which sounds paradoxical) may appear. The classic example is the growth of woodland fringing water courses, a riparian association, in an area which is climatically grassland or prairie.

The logical simplicity of Clements's ideas gained him many adherents who were glad to find some guiding light through the maze of ecological phenomena. Not so many, however, accepted the theory just as the leader proposed it. Although his views have greatly influenced the development of Ecology in North America and in Britain, there have had to be reservations and amendments as knowledge has extended. Clements's views are, in fact, too nakedly diagrammatic to fit the facts of nature without considerable distortion. Though they can be applied in broad outline in his own country of the western United States, when we try to apply them in other regions doubts arise. The principle of succession seems to be soundly based and experience bears it out as a general phenomenon. It is the Clementian 'climax' that has raised most controversy.

Clements, however, went further. He maintained that the plant association was actually a super-organism, the constituent species being, in his view, as closely integrated as the organs and cells of an organic individual, and the association was conceived as going through a life cycle which was in important respects parallel to that of a biological individual. This imaginative concept has found little support, and it is fair to say that it was generally received with silent astonishment rather than with serious criticism.

Tansley, while in agreement with most of Clements's views, stopped short of this one. He agreed that there were parallels between the plant community and individual organisms, but maintained that they did not prove more than an analogy and he therefore qualified the concept as that of a *quasi-organism*, a position which has been tacitly accepted by many British ecologists.

Philip, who has accepted Clements's theory of the super-organism, relates his belief to the philosophical doctrine of 'holism', as expounded by Smuts. This is a doctrine which maintains that all compound entities possess characters as 'wholes' which are not found in any of their constituent elements separately and are additional to the sum of the constituent characters. These new properties, which only appear when the constituents come together, are regarded as the distinctive mark of 'wholes', a doctrine which is closely similar to that of 'emergent evolution' put forward long before by Lloyd Morgan.

One difficulty of the monoclimax theory is that of defining what is meant by an area of uniform climate. Strictly speaking there is no such thing. Climate, in the sense of the aggregate of sub-aerial factors affecting the plant, is highly variable and investigation has clearly established the reality of what are called *microclimates*, which are intensely local and almost particular to individual plants. Although it is possible to classify climate types, which show some statistical constancy over considerable areas, yet, within these areas, topography can cause significant departures from the climatic mean, quite sufficient to alter the character of the vegetation. Mountain and valley or the north and the south slopes of a hill, for example, may have climatic differences which allow the development of forest in one area and only grassland or bushland in the other. Clements's avoidance of the difficulty might be called begging the question. If there is only one climatic climax then an area of uniform climate is one which has a uniform climax, a species of argument for which some early plant geographers were criticized.

If we can accept this, then we can agree that broadly Clements is right, in that vegetation maps show the prevalence of certain forests and other vegetational types over large areas, which are much too large to be controlled by physiographic factors and can only be attributed to climate.

Where departure from Clements's strict doctrine of the monoclimax has been general is in the view taken of the aberrant types of vegetation which occur within such uniform areas of climate. To Clements these were all sub-climaxes, communities which were in process of developing into the climatic climax but were delayed by local difficulties, or else dis-climaxes, where the natural succession had been distorted by animal or human interference.

However this may be in theory, in practice many of these sub-climaxes appear to be just as stable and as permanent as the climatic climax itself and in some the nature of the controlling factors is such as to preclude, within a geological erosion cycle, the development of forest or any other higher type of vegetation. For these reasons the *polyclimax* theory has found more favour than the pure milk of the Clementian doctrine. Tansley called all these aberrant types sub-climaxes, but he used the term differently from Clements. He considered them as subordinate in rank to the climatic climax, but, nevertheless, in their own right climaxes, stable under their own conditions and classifiable according to the governing conditions, i.e. as edaphic, physiographic, biotic, anthropogenic, etc., climaxes. This is a much more flexible theory to work with and has been widely accepted. It is easier to reconcile with natural facts and it avoids the necessity for the forced interpretations which are sometimes required to link up all stable communities with the climatic climax. It does not deny the existence of the latter as the highest type of prevailing vegetation within a climatic area, but regards it as limited to localities where there are no countervailing circumstances and not as potentially ubiquitous.

Another difficulty with regard to the climax arises from its supposed uniformity. In northern Europe, where woodlands with a single dominant species are widespread, their inherent diversity of structure is at a minimum, but in North America, with a richer tree flora, the lack of uniformity becomes manifest. In the broad-leaved forests of the United States the discrimination of 'forest types' is standard forestry procedure. Curtis and McIntosh analysed the distribution of the leading dominant species in the mixed conifer-hardwood forests of northern Wisconsin. They used the criterion of 'importance value' based upon a summation of relative frequency, density and dominance (basal area) and classified the various woodland stands into groups. If only one leading dominant was considered there were 15 groups of stands, and if four dominants were considered there were 105 groups. The forest climax is in fact a many-stranded pattern, with populations differentially distributed along natural environmental gradients, forming a continuum (see p. 3409).

If the temperate forests are thus complex patterns of distribution, how much more so are tropical forests. The rain forest covers a multitude of entities, among which little possibility of population analysis yet exists, but it is generally agreed that primary, lowland forest in tropical regions constantly humid is the highest type of living vegetation and should be considered a climax, if there is any such thing.

One of the most detailed analyses of an area of tropical forest is that by the Oxford expedition to Moraballi Creek in Guyana. Reporting on this in 1933–1934 Davis and Richards showed that there were five distinct forest types or associations recognizable in the area of primary forest examined. They are all under the same climatic conditions, all appear equally stable and each depends on a different, but equally permanent, combination of soil conditions and topography. All are of equal status and Richards maintains that no change of soil conditions is reasonably imaginable which would change one of the five types into another. This is clear evidence of the heterogeneous structure of a climatic climax vegetation, which is destructive of the monoclimax theory. It reinforces the evidence from temperate forests, which we have already mentioned, that the climax vegetation of a single climatic area is not a uniform entity.

If the climatic monoclimax is an illusion the question then arises: does the same sort of destructive criticism apply to the other types of climax, the subclimaxes which the polyclimax view accepts as true climaxes in their own right, controlled by other than climatic factors?

Stability has usually been employed as a criterion of climax vegetation, but stability is a purely relative concept depending on our human experience of time. Absolute stability is impossible. Can we find other means of recognizing or defining the outcome of sucession?

Succession is generally rapid in its earlier stages and gradually slows down. In later stages more than one line of succession may *converge* towards a common type before stability is reached. Stable vegetation may therefore be the outcome of what Cooper has called 'a braided stream' and compound in its very nature. The retardation in the rate of succession means that equilibrium is approached slowly and Cowles maintained that it is never actually reached. Long-term climatic fluctuations can induce successional changes in even apparently stable vegetation and Cowles, again, spoke of succession as 'a variable approaching a variable, rather than a constant'.

We have previously spoken of succession as chiefly motivated by biotic causes, the reactions and interactions of the plants themselves, and this is generally true except in areas of progressive physiographic change, as in the silting up of a lake. In this connection we have valuable evidence from the work of Greig-Smith and of Kershaw. (See also p. 3409.)

The former investigated areas of secondary and of primary rain forest in Trinidad and also sand-dune areas in Britain, while Kershaw examined populations of *Calamagrostis neglecta* and of *Rhacomitrium* in Iceland. In each case the object was to test the degree of correlation in the distribution of species, that is, the amount of interaction of species in the communities at different stages of development, as shown by the pattern of specific distribution. Greig-Smith pointed out that Clements's theory of the climax commun-

ity as a super-organism implied a high degree of interaction of species. All
the above investigations, conducted on widely different communities in
widely different climates, showed that the opposite was the truth. Interaction
of species was maximal in younger communities and decreased to a minimum
as stability was approached, the terminus being marked by almost complete
randomization in the distribution of species. In other words the outcome of
succession was to produce a community in which change was minimal and the
randomization of the constituent species was maximal. Such a community
will be relatively stable and form a 'climax', but these criteria are much more
positive than stability as diagnostic characters (see also p. 3404).

There is another property of a terminal community which is diagnostic. If
it is destroyed it regenerates a community of the same kind, whereas if a seral
stage is destroyed the sere may be cast back to its beginnings or may be
deflected into another channel of development. A good example of this
property of a stable community could be seen some years ago at the foot of
Mount Rainier in the State of Washington. A large area of mixed coniferous
forest had been killed by a mud-flow several feet deep. After two or three
years hardly any herbaceous plants had appeared on the surface of the mud
but there were numerous seedlings of the conifer trees, which would, in time
regenerate the original forest.

It would appear to be wrong, as Lippmaa argued, to place climatic and
edaphic communities in opposition, since both sets of factors are operative
throughout natural vegetation. Edaphic factors are operative, as we have seen,
in creating differentiation within the 'climatic' climax, and if we abandon the
idea of a uniform climatic climax, then there is no need for the opposition.
The factors operative in bringing about the terminalization of succession work
in all communities alike. The over-emphasis of the rigid doctrine of the stable
climax has fostered an artificial outlook in Ecology and it is time it was
dropped.

Cooper in 1926 made a strong appeal for fluidity in ecological concepts. In
particular he rejected Clements's concept of the super-organism as well as
other concepts and definitions which tend, as he said, to ossify and give
rigidity to our views. He reminded us that all existing vegetation is only a
three-dimensional section of a four-dimensional entity, whose fourth dimension, time, goes back to the beginning of plant life. In this immense pro-
cession vegetation has followed a great variety of courses and successional
change has been universal and perpetual, motivated basically by evolution
itself. He holds that the concept of development in succession is irrelevant; in
his view all change is succession, whether progressive or retrogressive.

He does show, however, from examples, that the course of a unit succession, a prisere, follows a sigmoid curve, starting slowly, then accelerating as
larger and longer-lived plants take part in it, and eventually slowing down towards stability, as the number of extraneous species available to enter the
community drops towards zero. Such a succession is, in fact, developmental,
tending towards a condition of minimal change, though never reaching

finality. He would retain the term 'climax' for this latter phase, though clearing it of all suggestion of being a terminus.

Cooper also objects to the separation of autogenic and allogenic causes in successions, though in this case rather less logically, since he admits that both sets of causes, biotic and physiographic, are constantly active and himself quotes cases in which either one set or the other predominates in a succession.

His pragmatic judgement of the validity of concepts by their usefulness and his plea for greater elasticity in our definitions are most praiseworthy, but have found too many ears clogged with tradition.

We shall have a further discussion of these topics in connection with the nature of the plant association and its analysis (see pp. 3401 and 3434).

Clements also tried to stabilize the fluctuations of meaning attached to the term 'Formation' by identifying it with his climatic climax. Unfortunately this too must be sacrificed and we shall see later what, if anything, can be put in its place (p. 3428).

There are one or two terms relating to the structure of plant communities which should be discussed.

Synusia (pl. -ae)*

This is a word introduced by Gams in 1908 to designate assemblages of plants within a community which have similar life-forms and similar relationships to the environment.

Gams laid down three grades of synusia:

Grade 1. Communities of plants or animals whose independent components belong to the same life-form and, within one district, to the same species.

Grade 2. Communities whose independent components belong to various species of the same life-form class and substantially to the same class of physiognomy.

Grade 3. Communities whose independent components belong to different life-form and physiognomical classes, but which through close correlations are bound together into an ecological unit in a unitary habitat.

Grade 1 is therefore mono-specific; grade 2 is pluri-specific and grade 3 is substantially the same as a plant association as envisaged by Braun-Blanquet.

The concept implies that a synusia forms an ecological unit which cannot be replaced by another within the same district, unless in special circumstances. Such a unit need not necessarily occur by itself, more often it is united with others into a topographical complex unit or phytocoenose.

There has been a tendency to equate synusia quite simply with the layers or strata of a complex community, but this is not strictly correct. Sometimes it is so, but not always. For example corticolous epiphytes on trees form a synusia, but not a stratum (Rübel).

* This is the direct translation of the German *Synusie*, but some authors prefer synusium (-a) which is probably more correct etymologically.

Warming described the following synusiae in a reed swamp under the term 'ecological unions', which some may prefer. The reed-swamp association is made up of the following unions: in the soil a union of mud-living Schizophyceae; on the water the free-floating *Hydrocharis* union; rooted in the mud the *Limosella* union; rooted in the mud but rising in the air, the marsh-plant union.

Braun-Blanquet also quotes as examples: a covering of crustose lichens, a carpet of moss or dwarf shrubs, the herb layer in a beech wood, the tree layer in a wood of *Abies* or *Pinus*, the agaricalean fungi in a wood, the lianas or the epiphytes of a tropical forest. Sometimes, as these examples show, the synusia is a definite layer in a community, sometimes it is the whole community. On the other hand it may be simply a biological rather than a sociological unit, as, for example, the lianas or the agarics, in which case it corresponds to a *guild* as defined by Schimper.

Synusiae have been the objects of some attention on account of the question as to whether they should be treated as sociological units. They exist, but their status is disputable. The question is bound up with two other concepts which we should look at first.

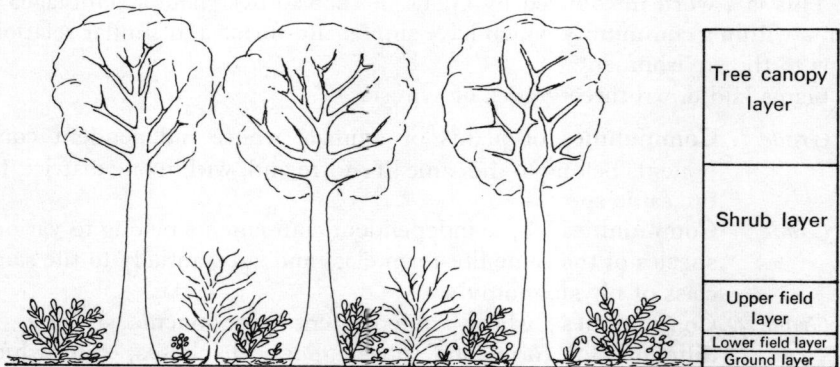

FIG. 4.5a. Typical layering of a temperate woodland. (*After Lundegardh.*)

Stratum

Many types of vegetation consist of distinct layers of plants of differing stature and life-form. This is conspicuous in temperate woodlands, where we commonly have a *ground layer* consisting mostly of Bryophytes; a *field layer* of herbaceous plants; a *shrub layer* of such forms as *Corylus* and *Ilex*, and finally a *tree layer*. Lowlier forms of vegetation may also show stratification. *Calluna* on a heath may shelter a subordinate stratum of Bryophyta and lichens with some small flowering plants like *Galium hercynicum*. There are also *unistratose* communities, of which some grasslands are conspicuous examples, The plants of such a community also form a synusia.

In tropical forests the layering is multiplex and often confused, for here we find not only shrubs and trees, but trees of many different sizes and heights

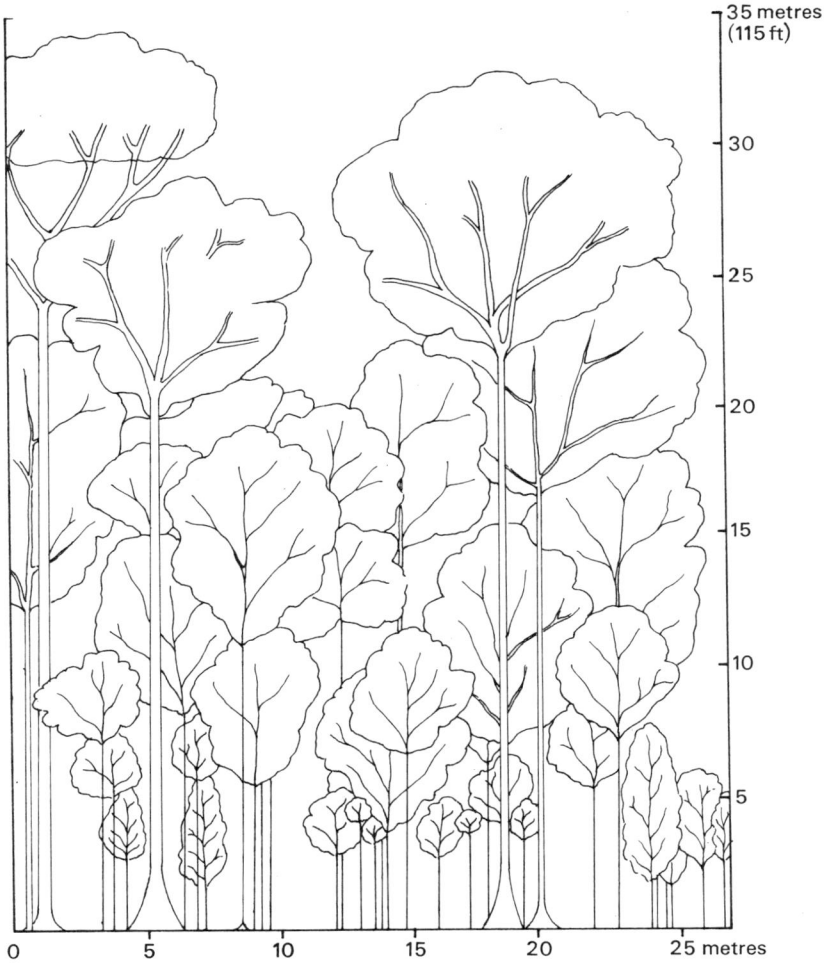

FIG. 4.5b. Diagrammatic profile of layering in a primary tropical rain forest, at Moraballi Creek, Guyana. Trees less than 5 m high are omitted. (*After Davis and Richards.*)

which themselves may constitute several layers, not easily separable. Both temperate and tropical woodlands also contain *transgressive species*, which pass, in the course of growth, from one layer to another. Thus, tree seedlings in their earliest stages belong to the field layer, later they form part of the shrub layer and finally join the tree layer.

Some lower strata are ecologically dependent on the shelter provided by the higher strata so that they occur in constant association. In other cases the field stratum of a woodland is comparatively independent and can flourish either with or without the tree cover. The *Rosmarinus-Lithospermum* association in southern France occurs widely as an independent community and also as a shrub layer under *Pinus halepensis*. In Britain the *Allium ursinum* field layer of *Fagus* woodland is dependent and does not long survive the removal of the trees, but the *Anemone nemorosa* field layer of *Quercus* woodland increases in density and spreads vigorously without the tree cover. The *Rhododendron-Vaccinium* heath of the Alps shows a high degree of independence, occurring extensively as a separate association, but also forming a shrub layer under various trees, *P. cembra*, *P. mugo* and *Larix europaea*.

Similarly a single tree species may have different field layers below it in different localities or different circumstances. Thus in the Harz mountains in Germany the field layer of *Fagus* woodland varies with the direction of slope. On northerly to easterly slopes the field layer is dominated by *Dryopteris perennis*; on north-westerly slopes the field layer is chiefly *Festuca sylvatica* and *Mercurialis*; while from north-west to south it is dominated by *Luzula nemorosa* with *Deschampsia flexuosa*. Lastly, south-easterly exposures carry a field layer of *Luzula* with *Calamagrostis arundinacea*.

The degree of independence of each other frequently shown by the strata in a complex community, together with the existence of unistratose communities, led Lippmaa to propose in 1934 that each stratum should be treated as a separate association, the sociological unit. According to this view all plant associations are essentially unistratose and the complex many-layered community is a group of associations. This is an idea which could only arise in northern Europe under fairly rigorous climatic conditions and with a relatively small number of species. In such circumstances there is only a limited number of dominant species and they tend to be constant. The proposal to break down the association into separate stratum-associations has not been found applicable to the richer temperate forests with an increased number of dependent layers, nor *a fortiori* does it apply to tropical forests.

Kerner in 1863 was the first to describe stratification and Hult in 1881 used it extensively in his descriptions of vegetation. He also observed the interchangeability of strata and described sixteen *twin formations*, as he called them, in which a common or 'combining' stratum was associated with two or more 'alternating' strata. Thus two communities linked by a combining stratum of *P. sylvestris* had alternating field layers of *Vaccinium myrtillus* on the one hand, combined with a ground layer of *Hylocomium*, or a field layer of *Calluna* with a ground layer of *Cladina* on the other. This, again, is a distinctively northern phenomenon and it is noteworthy that it has been exploited chiefly by Russian ecologists who have found it useful. Katz, for example, found not only twin associations but twin series of associations. He described two such series in which *P. sylvestris* was the dominant stratum common to all. In one series there were three different field layers, dominated by *Ledum palustre*, *Cassandra calyculata* and *Eriophorum vaginatum* respectively, but with a common ground layer of *Sphagnum recurvatum*.

In the other series, again under *P. sylvestris*, there were the same three field-layer plants, but with a common ground layer of *Sphagnum medium*, thus making up three sets of twins which were distinguished from each other only by the moss in the ground layer.

When, however, Cain applied this system of analysis to the richer forests of Virginia, which have five strata, he found he could distinguish no fewer than 256 twin associations under one combining stratum of *Picea rubens* alone. In these circumstances it would seem that Braun-Blanquet's warning of the inutility of too much subdivision of vegetational units is applicable.

Stratification results from a long process of natural selection. The subordinate layers are subject to varying amounts of light deprivation, but they are also sheltered from extreme variations of temperature, wind velocity and humidity. They must also harmonize with the reactions of the dominant layers, especially with their root activities and leaf fall. This implies a balance of conditions which sensitizes, so to speak, some of the species to relatively small changes, for which they serve as indicators, and foresters have found by experience that some species in the ground and field layers are valuable indicators of soil conditions. In the Swedish conifer forests, for example, the occurrence of *Anemone hepatica* is an indicator of the best planting soil.

Aspect

This is used to denote the seasonal changes in the appearance of vegetation in a single area, caused by the natural differences in the growth and flowering periods of different species. Such changes of aspect occur even in unistratose communities, such as marshes and meadows, as spring-flowering species are overtopped and hidden by summer species and these, in turn, by autumn-flowering species. The most marked changes, however, are to be seen in woodlands where they are related to the changes of light intensity due to the successive vernation and fall of the tree foliage and even more important to the growth and leafage of the shrub layer.

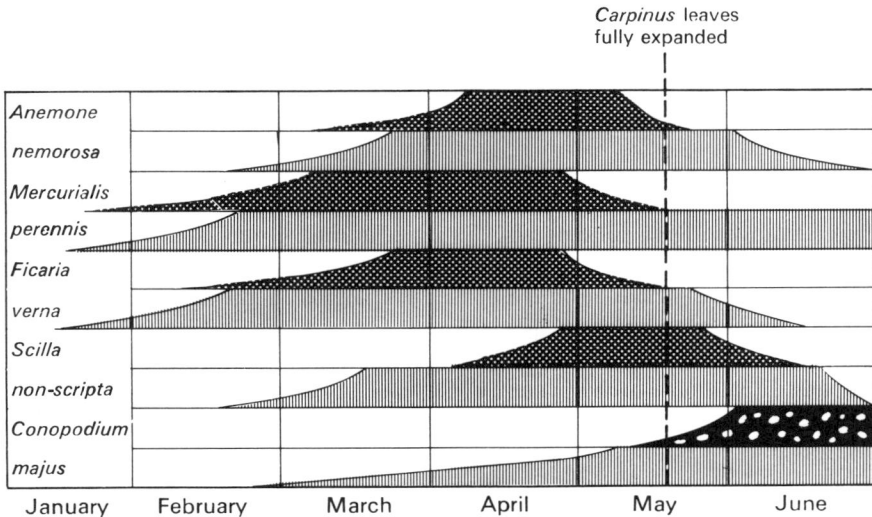

FIG. 4.6. Flowering and leafing periods of some typical members of the pre-vernal flora in English woodlands. Flowering period in black, foliage development striped. *Conopodium majus* flowers only in coppiced woods.

In 1916 Salisbury distinguished the pre-vernal and the vernal aspects of deciduous oakwoods. In the *Quercus robur-Carpinus betulus* woods of Hertfordshire, with the oak as standard trees and the *Carpinus* coppiced, he showed that the 'light phase' was sharply separable from the 'shade phase', the change coming usually at the beginning of May. The pre-vernal aspect, during the light phase, was conspicuous for the flowering of *Anemone nemorosa* and *Ficaria verna*, the foliage of which disappeared in June. *Scilla non-scripta* also began flowering in the pre-vernal phase, but continued rather later than the other species and carried its foliage longer. *Mercurialis perennis* which flowers very early in the pre-vernal aspect, carries its foliage throughout the summer. The onset of the shade phase marks the end of flowering for most field-layer plants, though *Oxalis acetosella* and *Viola* spp. continue to form cleistogamic flowers. It is in the pre-vernal stage that most photosynthesis is possible so that we find that the species of the vernal aspect, which follow slightly later, *Primula acaulis*, *Galeobdolon luteum* and *Conopodium majus*, produce their foliage very early. The extent of the decrease in light with the onset of the shade phase is illustrated by the following figures given by Salisbury for various *Quercus-Carpinus* woods. These show how the shade effect increases with the period since the last coppicing* and the building up of a new canopy by the coppiced *Carpinus*. The light values are given as percentages of the diffuse lighting outside the wood.

YEARS SINCE COPPICING	LIGHT PHASE (per cent)	SHADE PHASE (per cent)
4	61	1·3
6	41–55·5	0·78
7	33–40	0·36
10	33–54	0·77
13	40–46	0·44–0·74
17	38·4	0·83–1·1
20	24	0·83–1·1

A further cause of aspect changes can be found in major variations of rainfall, as in semi-desert communities where the rainy season brings out an immense flowering of ephemerals. This is also marked in its effects on mixed grasslands, where drought years and flood years promote the growth of different species, which produce effective changes in the appearance of the vegetation.

Aspect is also sometimes used in a more conventional sense to signify the compass direction of the exposure in the case of communities on slopes. It is often expressed in degrees from o (north) to 360, but the usual points of the compass may be sufficient.

---

* Coppicing is a distinctive forestry practice which originated in times when oak for ship-building was in great demand. The standard oaks are limited to about twelve to the acre, which allows them to develop a full branch canopy. The subsidiary trees and shrubs (*Corylus, Acer, Fraxinus, Carpinus, Populus*, etc.) are cut down to stumps every twelve to fifteen years. The long straight stool-shoots which they develop are useful for making fences and gates and for firewood.

Ecotone

This is the term for the boundary zone between two well-marked communities. It may be a comparatively sharp line of demarcation where an abrupt change of habitat conditions occurs, such as the line of contrast between two geological formations of contrasting character. Often, however, it is a zone rather than a boundary line, a zone of transition in which there is some intermingling of the two plant communities, either as a whole or, more often, by the interpenetration of certain species from both communities, which have a wider range of tolerance than their associates.

Ecocline

This is an extension of the idea of the ecotone, applicable to cases where habitat conditions change only gradually over a considerable distance. At each end of the transition there may be two distinct communities, but between the extremes there is a gradual transition from one type to the other, without any marked ecotone.

Topocline

This is an axis of transition between two states of some topographical or physiographic factor. Sloping ground is the most obvious case, where the factor chiefly affecting the plant population is that the water table in the soil comes nearer to the surface as the slope descends, and vice versa. A topocline may be associated with an ecocline or it may not. If the change of factor or factors along the topocline axis, though measurable, is not outside the range of tolerance of the community species there need be no change of community along the axis, though there may be changes in the relative frequency of some of the constituent species. Some authors, however, would include such internal rearrangements in one community as ecoclines, if they are associated with a change of habitat conditions.

The *catena* is a special case of the topocline. The term is applied to a series of related changes in a soil along an axis, accompanied by corresponding changes of vegetation (ecocline). The classic case is that of a soil lying on a slope between hill top and valley where the hill topsoil loses soluble materials by rainfall leaching (eluvial), the soil on the slope receives these materials in passing (colluvial), and the valley soil accumulates them (illuvial).

It has sometimes been put forward as a paradox that a gradual change of conditions along a topographical axis can be associated with sharp lines of demarcation between communities. Such, for example, are the clearly limited zones of vegetation around a freshwater pond, where the topocline is the gradual increase in depth of water. This result is not surprising if it is interpreted as due to a balance between range of tolerance and biotic competition or between physical tolerance and biotic tolerance. The former is always restricted by competition, as is well illustrated by the disappearance of all but a few garden plants in an abandoned garden. The plants of each zone around the pond are restricted to their optimum range of water depth by the competition of those whose optima in this respect lie on each side. The occasional stray individuals are insufficient to extend a zone or to confuse its relatively sharp boundary. Changes in such a case can only arise from a change of physiographic conditions, either the drying up of the pond or the accumulation of silt. It is the latter factor which has enabled the zone of *Phragmites communis* in the Norfolk Broads to extend in the most formidable way over the last century, since the cessation of commercial barge traffic allowed the channels to silt up and the reeds are no longer cut in large quantities for thatching.

CHAPTER XLIII

THE PLANT COMMUNITY IN ITS SOCIOLOGICAL CONTEXT

WE have hitherto referred, as far as possible, to examples of natural vegetation under the neutral or undefined term 'the plant community' which with its German counterpart of '*Gesellschaft*' or the French '*groupement*' has been generally retained as a convenient label for all types of vegetation of whatever status.

We must now, however look more closely at natural plant communities, we must examine the principles on which they may be distinguished from each other and raise the question of their classification. In this we shall unfortunately be obliged to use certain vernacular words, like 'association' and 'formation', in a restricted and technical sense, a procedure always liable to cause confusion, particularly in international discussions where the vernacular words may have quite different connotations in different languages, which makes the ascertainment of equivalences very difficult. It is a pity that it has not been possible to use Latin, generally the most precise of languages.

In surveying the organic world it has been agreed to adopt and adapt the ancient concept of the species as the unit of classification. Observation confirms that the idea has its validity for nearly all sorts of organisms, yet it has not been possible to frame a complete and watertight definition of a species. Very much the same situation confronts us when we survey the communal world of plants. There is general agreement that the unit of natural vegetation is the *association*, but how an association should be defined has given rise to prolonged controversy which has not reached finality. Indeed, Tansley argues that it is best not to attempt a logical definition, but to leave it, like the species, in the position of a natural datum which, in spite of variation, hybridization, and so forth, remains a recognizable 'appearance'. This, however, brings us to the verge of the ancient philosophical problem of the relation of the 'real' to the 'apparent', which may profitably exercise gymnastic minds, but not, we think, here.

Definitions of natural entities are rather apt to smell of the study rather than the field, and Tansley compared a definition of the association to a Bed of Procrustes to which natural appearances had to be forced to fit.

How then are we to recognize a natural plant association? What are the marks by which we may know it? In discussion of the problem there has been one major line of cleavage between those who maintain that the floristic composition is the essential quality of an association and those who maintain that, in addition, the nature of the habitat is an essential part of the concept. The latter view was stated in the most famous definition which won international support at the Brussels International Congress in 1910.

This was proposed by Flahault and Schröter and ran thus: 'An association is a plant community of definite floristic composition, presenting a uniform physiognomy, and growing in uniform habitat conditions. The association is the fundamental unit of synecology.' This is referred to by its critics as the 'double definition', placing uniformity of habitat alongside of uniformity of composition as the criterion of a genuine plant association. Schröter's term 'synecology' has since been largely displaced by the term 'plant sociology' which has a rather more restricted and definite meaning. It might be remarked perhaps that if an association always has a definite floristic composition it must of necessity have a uniform physiognomy, but we must notice that the floristic composition is described as 'definite' not as 'uniform', the word used elsewhere in the definition. In nature the community of species involved may well be the former without being the latter, since species with different life-forms may predominate in different portions of the community and the physiognomy will not therefore be altogether uniform.

Against the view embodied in the definition, du Rietz and others have insisted that constancy of floristic composition is the true criterion of a plant association, and that consideration of the habitat is no more essential to the concept of the association than it is to the concept of a species. In other words they take a morphological view rather than a physiological view. This does not mean that they ignore the study of habitat as a vitally important part of Ecology, and they would agree that in two different habitats you will find two different associations, but they consider that an association regarded as an assemblage of species can be studied sociologically in the relations of its components to each other even if, as may happen, we know little or nothing about the habitat conditions. This is 'pure' plant sociology.

The Swedish school, of which du Rietz is an exponent, go so far as to claim that an association can occur in more than one type of habitat, to which Tansley retorts that that only shows the habitat has been conceived too narrowly. Looking at the matter detachedly, it might be argued that neither school is very clear about what they mean by habitat. In a former section (p. 3342) we pointed out that what is called the 'communal habitat' is really a plexus of individual habitats and only exists as an abstraction. Every plant in the association has its own relationship to habitat factors and these relationships are all subtly different. How can we therefore speak of a 'uniform habitat'? Tansley gets over this difficulty by his doctrine of 'master factors'. Accepting the diversity of individual habitats he maintains that 'uniformity' consists in the uniform operation of some single overriding factor over the whole area, a factor abnormally developed locally, such as a particular component of the soil or the water content, or climatic exposure, which imposes an element of similarity which all the individual habitats have in common.

We have only quoted above one definition of an association, but there have been many others. The Brussels definition deserves its prominence as the first attempt at concerted agreement, but Carpenter's *Ecological Glossary* quotes three and a half pages of individual definitions by both plant and animal ecologists, some of them flatly contradicting others.

The plant sociologists seem to have carried their point in Europe, for the International Congress at Amsterdam in 1935 adopted a recommendation for the use of the term 'association' which makes no mention of habitat. It ran as follows: 'To use the term association for vegetation units characterized mainly by characteristic and differential species in the sense of the Zürich–Montpellier plant sociologists or at least for units of the same sociological value.' We shall come back later to the meaning of this recommendation.

No matter what view we hold about the nature of the association there is no doubt whatever about the necessity for a thorough knowledge of the flora involved, down to the smallest taxa. Indeed, a knowledge of the genetics of the plants is a consummation to be hoped for. Differences in ecological behaviour cannot always be resolved at the species level. Two genotypes which react differently to habitat may be morphologically almost indistinguishable.

So far, then, all we seem to have gained in pursuit of the association, is that it is a recurrent, not an accidental or temporary assemblage of species, with a limited and defined constitution. We may add a rider to the effect that it is defined by the interrelationships of the component plants, i.e. by their sociology. It is an idea born from experience of north temperate vegetation and it remains to be seen how far it is applicable to other parts of the world. By no means all plant populations qualify to be given the name. One essential requirement is replication, that is that we find the same assemblage, with similar interrelationships, repeated in different areas. This cuts out casual assemblages in localities, such as waste ground, subject to human interference (called in German *ersatz Gesellschaften*). Another requirement is that there must be some degree of interaction between the plants, some sociological integration. How close this integration must be to fall within the limits of our concept is a matter of judgement.

To take an extreme case; in the South China Sea there is a recognizable community of peredinians in the plankton, dominated by *Ceratium dens* and *C. schmidtii*, with *C. saltans* and *C. breve* as accessories, but here surely there is no sociological integration between the members.

Analogous taxonomically-distinct plankton populations have been found to characterize certain areas of the North Atlantic. They apparently owe their existence to direct environmental reactions. They can hardly be called associations unless interspecific biotic relationships can be proved.

More difficult cases for judgement appear when we look at vegetation in some other parts of the world. Desert vegetation often consists of characteristic species forming as well-defined group of plants constituting at least an elementary community. In some places these plants may be close enough to attain root contacts with each other and therefore there is some biotic competition, but in others it is not so. Throughout the south-western United States and northern Mexico there are many separate desert areas which have a common flora. The same group of species, including *Prosopis glandulosa*, *Covillea tridentata*, *Franseria dumosa*, *Olneya tesota* and species of *Parkinsonia*, constantly recur and therefore fulfil the association requirement of replication in similar habitats. These plants are widely separated, amounting to twenty to sixty plants per acre, and their occurrence is entirely due to their individual powers of remaining quiescent during long periods of drought and of rapid growth and reproduction during the very short and sparse rainy periods. There is no competition, no sociological structure, they are simply a collection of individuals selected by the habitat from a large number of possibilities, whose only association is their geographical propinquity. To call such a grouping an ecological *association* would surely be a perversion of language. For these and many similar cases the term *settlement* (Ger. *Einsiedlung*) would seem appropriate.

Quite different is the condition of the ephemeral population of therophyte herbs which spring up thickly and disappear promptly in many desert areas, coincident with the rains. Although they form but one stratum of the community they may fairly be considered as independent of the perennials, but whether they are themselves a true association is an open question.

What we have said about settlements of independent plants may raise the question whether *all* plant associations are not just what we have called 'a collection of individuals selected by the habitat'. The answer is in the negative, for wherever sociological interaction exists between the plants there is biotic selection as well as habitat selection. The nature of the habitat may give precedence to certain species, but only those species able to compete with them will find a permanent foothold in their company. It is just this element of biotic selection which marks off the true association from a casual assemblage and ensures its definite floristic character and relative constancy.

Certain areas, such as the ballast heaps of Cardiff Docks or the banks of the Tweed below Galashiels, have yielded rich harvests of adventive species from many different parts of the world, but these constitute no true association, for the assemblage has neither constancy nor permanence.

We seem here to have uncovered two more characters of an association to add to its definite floristic composition. The composition must be *constant* and there must be some degree of *stability*, if perhaps permanence is too strong a word. The Brussels definition says nothing about this and the wording as it stands might apply to a temporary covering of common weeds on a rubbish tip.

When we speak of constancy we do not only mean constancy in one particular locality. Wherever the conditions are propitious we should expect the same association of species to recur, at least in essentials. This is a further association character, *local replication*. If the term association is to be used collectively for the community type, wherever it is found, then we need a word for the single, local manifestation of it. Many terms have been used in different languages, for example, *Bestand*, *Flāch*, association-unit, *Siedlung*, site, *locale* and stand. None is wholly satisfactory, but where we have occasion to use it we will refer to the single *stand* of an association.

The question of stability or permanence must be judged relatively to our own time-scale, as we have previously said, but it is a question of considerable importance. In Clements's monoclimax theory only the climax community had any claim to permanence. All other communities, with few exceptions, were destined with lapse of time to gravitate towards the climax type. He therefore logically restricted the term 'association' to climax associations only; this binds it to the concept of the universal prisere in a climatic region. For all other association-like communities he used the term *associes*.

Those who prefer the polyclimax view naturally reject this limitation of use. They maintain that many of the communities which Clements called subclimaxes are just as permanent and distinct as the climatic climax itself and they apply the term association to all such edaphically or topographically determined communities which have the appearance of being end-phases in the local development of the vegetation. Clements's term *associes* can, however, be profitably used for all communities which are manifestly temporary stages in a sere.

Pavillard has pointed out that it is not a complete conformity of ecological requirements which is the binding force between the species of an association, but their diversities. Close similarity of requirements implies sharp competition and is more likely to keep species apart than to draw them together. Although they must have some fundamental similarities, particularly in their reaction to the master factors of the habitat, it is the diversity of specific requirements (or niches) which enables the species, like the multiform pieces of a jig-saw puzzle, to fit together and achieve sociological cohesion.

Most ecologists agree that the character which most clearly marks a natural association is the degree of organization which it displays among its constituent plants. This is as much as to say that their relationships, as expressed in their relative constancy, dominance, frequency, abundance, dispersion, etc., terms which we shall endeavour to clarify later (p. 3439), have been so far regulated during the long process of development that they are found to be more or less uniform in different stands of the association when these are compared and it is this replication of the organizational character, taking it as a whole, which establishes the claim for recognition as a natural association.

The expression 'more or less uniform' obviously leaves a good deal to the judgement of the ecologist. Complete uniformity is not be expected in nature and, indeed, as we have pointed out, such words as 'uniform', 'identical' and 'permanent' do not accord with fact and their use in descriptions or definitions conveys a sense of exactitude which is quite spurious.

Many attempts have been made to reduce the element of subjective judgement in plant sociology by methods of closer and especially of quantitative analysis of community organization, some of which we shall deal with in the next section (p. 3442), but judgement inescapably plays some part in all of them, and when Tansley says that associations 'are the most unmistakable vegetational units we find in nature', he claims too much. Poore is nearer the truth when he says that we can only arrive at an objective understanding of associations by a series of successive approximations, by which the element of judgement is reduced to a minimum.

The high degree of organization shown by natural plant communities has impressed ecologists from the start. Clements carried the idea to its extreme in claiming that the plant community was a complex super-organism (see p. 3389.) Tansley modified this claim to something more in accordance with British ideas, by allowing that there were a number of parallels between the natures of organism and community and calling the association a quasi-organism. If this means anything, it must imply that its characters arise out of itself, not out of its habitat, and that consequently the latter should not appear in the description of the quasi-organism. This, however, Tansley did not allow and in happy defiance of logic, but in reliance on their own incommunicable instincts, most British ecologists have followed his example, to the despair of those on the continent who are apt to hold that British vegetation has never been properly described. Viewed strictly from an analytical standpoint this is true, but whether the analytical systems of the Uppsala and the Zürich–Montpellier schools or the free and easy naturalism of the British approach gives a truer picture, who can say? Each misses something of the whole truth, each needs to know more of the other side. Two superb presentations exist. On the British side there is Tansley's *The British Islands and their Vegetation* (Cambridge U. P., 1949) and on the continental side Braun-Blanquet's *Pflanzensoziologie* (Springer-Verlag, 2nd edn, 1951; 3rd edn, 1964) and both should be required reading for all plant sociologists. The British ecological rather than sociological approach to vegetation is undoubtedly nearer to Warming's view. Writing sixty years ago he said of plant communities: 'The greatest aggregate of existence arises where the greatest diversity prevails.'

An almost diametrically opposite view of the community was put forward by Gleason in 1926, in a very cogent criticism of the idea of plant associations as units in a classification of vegetation. He called his view the 'individualistic concept' because, in brief, he assigned the leading part in the formation of plant groupings to the individual species. He developed a severe analysis of the realities of plant associations as contrasted with the ideal definitions which imply that associations are fixed entities. He pointed out that in some geographical regions associations are so vaguely delimited in space that their assigned boundaries are purely arbitrary; that their composition varies internally from year to year; that no two areas of the same association type are exactly the same; that their time relationships from origin to disappearance are so variable that it is impossible to draw a line between 'valid' associations and transitional stages; that fragmentary associations and mixtures of associations often occur; and, finally, that similar associations may occupy different environments or different associations similar environments.

His conclusion is that while plant associations undeniably exist yet they show none of the properties of abiding identity which those who look upon them as classificatory units assume. The idea of fixity in associations comes from relying on isolated snapshots rather than a long film.

The origin of the association, he claims, lies in the selection by the environment of certain species, from among those which arrive on the spot, which find there conditions near enough to their optima to enable them to support the environment, and it is the reactions of these species which constitute the association.

Such a view lays all the emphasis on the governing effect of the habitat. The weakness of his argument is that he almost ignores biotic competition or the mutual relationships of the species as important factors in constituting the association. He regards each plant as an isolate, directly and individually related to the habitat, which is a negation of the views of those who hold that the sociological relationships of the plants are the essence of the idea of association, as the very name implies. To Gleason the association is simply an appearance, part of the data of nature, a legitimate subject for ecological investigation but unsuitable for either definition or classification.

We have treated these somewhat nihilistic ideas at some length, partly because the predominance of sociological literature might disguise the fact that opposition exists, and partly because they illustrate the difficulties which workers in other parts of the world have often felt in trying to apply European ideas of vegetation to their own regions.

It cannot be denied that European ideas of the plant association are based to a large extent on northern montane vegetation, either unistratose or having a relatively simple stratification. If one looks through the illustrations in Braun-Blanquet's *Pflanzensoziologie* one cannot but be struck by the high proportion which deal with montane vegetation and how few with vegetation of complex character.

In dealing with mixed forests difficulties begin to appear. The mixed forest community is foreign to Britain, where almost all our woodlands are dominated by one or sometimes two species of trees. At the present day, complex, mixed forest is represented by tropical rain forest (in the wide sense), by mixed deciduous forest in the eastern United States and by some areas of mixed deciduous forest in the Caucasus and Asia Minor, which last two are probably survivals of the mixed Tertiary forests of Europe, swept away by successive glaciations (see Vol. 3, p. 2570). Britain saw a revival of mixed forest on a reduced scale of complexity in the mixed oak forest (*Eichenmischwald*) which covered large areas of western Europe at the period of the climatic optimum in Post-Glacial times.

In its highest expression the mixed forest appears to be a unitary association in which no species achieves more than a very local dominance and, in tropical forest at least, the floristic composition is so rich that no species is represented by more than a few individuals in any sample area. All mixed forests show some areas of physiographic segregation within which some favoured species becomes dominant and even forms pure *consociations*, that is associations with a single dominant, comparable with the temperate woodlands. Are these associations, to be reckoned as distinct from the general body of the forest, into which they often merge? Furthermore, apart from these ecological variants, the general body of the forest association shows segregation, variety of expression, forest types or whatever we call them, (see p. 3430) due to segregation and regrouping of the dominants, often very dissimilar, and simply expressing the dynamic equilibrium between dominants, which shifts from place to place and from time to time. These segregates really form a continuum, as Curtis and McIntosh have shown (p. 3429), and it would not be possible to classify them as distinct associations within the mixed forest.

Another kind of difficulty may arise when dealing with stratified woodlands, and that is the apparent independence of the field vegetation and the tree cover (see p. 3396). That shading, seasonal aspection, root competition from the trees and other factors can markedly affect the growth and abundance of the lower layers is well known. Watt and Frazer in 1953 showed that the isolation of the ground flora from root competition by *Pinus sylvestris* allowed weight increases of up to one hundred per cent in the crop of *Deschampsia flexuosa* and *Oxalis acetosella*. This, however, is a quantitative, not a qualitative effect and sheds no light on the presence of these species as part of the field layer. It has long been observed that the composition of the field layer may vary quite independently of the tree layer. A widely ranging tree such as *P. nigra*, as von Beck records, may have under it a field layer of either Baltic, Central European or Pontic affinities in different parts of its range. On a smaller scale Poore describes the vegetation of the Keltney Burn gorge, which comes down to the north side of Loch Tay in Perthshire. Here there is a mixed, variable woodland, growing on soils of variable stability, depth and moisture content and consisting of *Quercus robur*, var. *intermedia*, *Betula pubescens*, *Fraxinus excelsior*, *Ulmus glabra*, *Acer pseudoplatanus* and *Sorbus aucuparia*. This is a relict type of mixed oak-wood which appears to be peculiar to the Scottish Highlands. The ground flora varies independently of any variation in the composition of the tree layer and the variation seems to be imposed on the ground flora by an underlying mosaic of micro-habitats. Poore considered that the only possible way of treating such a complex sociologically was to abstract the tree layer and consider it simply as a biotic factor in the habitat of the ground flora.

These examples illustrate the contention of Lippmaa (p. 3396) that each stratum in complex layered vegetation should be considered a separate association for sociological analysis. This is probably a valid approach in some cases like the above, but it breaks down in the most complex vegetation where layering, if it exists at all, is so interwoven and interdependent that it becomes impossible to consider any layer separately from the complex as a whole.

The structure of any community rests primarily on the relative distribution of the species which compose it. Truly random dispersal implies random dispersal of seeds over the whole area, which cannot often occur, together with a degree of vitality which allows seedlings to develop in any situation within the habitat. It is not surprising therefore that the distribution of many species is significantly non-random. Random distribution is expressed by the Poisson Series (see p. 3465) which enables one to calculate the expected number of samples (quadrats) in which 1, 2, 3, 4, etc., individuals of the species will occur, for comparison with the observed number. In 1952 Greig-Smith compared the distributions of trees in rain-forest stands of different ages in Trinidad. He was thus able to work out the degree of association of the important species with each other and to show that the number of samples in which certain species occurred together was significantly greater than expectation, which we might call a coefficient of association between them. He took this as an index of the extent of integration in the community and found the very interesting fact that the degree of integration was greatest in young secondary forest on areas recently disturbed and was *least* in undisturbed primary forest. Similar results were obtained by Greig-Smith in the *Ammophila arenaria* association of sand dunes, where the time series from young dunes to stabilized dune heath is particularly clear. This was confirmed by Kershaw for an association of *Calamagrostis neglecta* and of *Carex bigelowii* in the *Rhaconitrium* heath of a mountain plateau in Iceland. They both found that the maximum of association patterning came at about halfway in the series and again was least in the mature stabilized vegetation.

These results are obviously antagonistic to the concept of the association as a complex organism, in which the degree of internal organization should be maximal in the climax community. From this we may conclude that patterning in the dispersal of species throughout a community is characteristic of the dynamic, developmental stages of a community and that maturity and stabilization are associated with a minimum of interaction between component species. This is not the same thing as saying that individual competition is minimal in the climax phase, which is obviously not true, but it does mean that competition has been stabilized into an equilibrium in the long process of development, during which all aggressive or disturbing elements have been either suppressed or absorbed and integrated. One might also conclude that any evidences of patterning in a mature community are due to habitat factors only, which may be true but goes rather beyond the available evidence.

The two authors mentioned above both express a preference for Gleason's individualistic view of the plant association as in better accord with facts as they found them, rather than for the organismal view of Clements. Their results agree with Gleason's idea of the mature association as being in a high degree individualized in its relationship to the habitat. If this view is accepted, it follows that as the environment is infinitely variable, associations must be so also. The grouping of stands into abstract associations can only be approximate and the abstract association is undefinable and *ipso facto* unclassifiable.

Such a view of the world's vegetation as an infinitely variable continuum, though revolutionary, has its attractions and many may be inclined to accede to it as a philosophical principle while remaining convinced that for practical purposes some kind of classification, however arbitrary, is a necessity. A consequence of this attitude would be that the principles of our classification may be, first, simply convenience and, second, not violating natural appearances more than can be helped. Once again, we can see a parallel with the systematics of species at the historical stage at which a confessedly artificial system prevailed over a natural system, which was regarded as a chimaera. At present current practice in the classification of communities depends inescapably on subjective, i.e. personal judgements, which may be well or ill founded. Whether it will ever be possible to reach a truly objective classification depends on continually deepening our understanding of plant life and on looking at the vegetation itself with eyes unclouded by theory.

Before trying to deal with some of the proposed classifications of vegetation communities, there is one concept which is integral with the idea of the association as a definite sociological unit and that is the concept of the *minimal area* of an association, or, the ideally smallest unit area in which the character of an association is fully expressed. It might be argued that the stand is in itself such an area, but individual stands vary enormously in size. If they are very large their limits may be difficult to define and to treat them as minimal units is impracticable. On the other hand if the stand is very small we need some standard of judgement as to whether it really comprehends an association or is only an *association-fragment*.

The idea of a minimal area was first expressed by Braun-Blanquet in 1913. He intended it to mean the smallest area in which all the species, called by him the 'characteristic species of the community', may be expected to occur. The idea has therefore a statistical basis and it is on statistical grounds that its validity must be judged.

There have been two usual methods of assessing it. Braun-Blanquet's school (Zürich-Montpellier) have used the species-area method. Quadrats of increasing size are used and the number of species in each size of quadrat is plotted against the quadrat area. As the areas are enlarged there is at first a rapid rise in the number of species included, but later the increase in the number of species diminishes and the curve flattens out either to the horizontal (no increase) or to a very low slope (slow increase). Ideally there is a point of inflexion of the curve which can be regarded as the minimal area for the inclusion of all except rare or accidental species.

The other method is that of the du Rietz (Uppsala) school who estimate on a constancy basis. Here again quadrats of two or three different sizes are used and a considerable number of samples are taken with each size of quadrat. The species are then listed according to the percentage number of quadrats in which they appear and those with percentages between 90 and 100 are reckoned as constants. If a quadrat size is then found beyond which increased area produces no further constants, this is taken as the minimal area.

Both these methods have been weightily criticized and it seems that they will work with some kinds of association and not with others. In the species-area curves the point of inflexion is often vague, being merely indicated by a change of slope, and in many cases there is no stop to the increasing number of species included, for each increase of area brings in a few more. Much the same objection applies to the constancy-area estimation, that it is imprecise and that eventually it depends on individual judgement where the minimal line shall be drawn. Cain has tried to increase the precision by stipulating that the minimal area be that area beyond which an increase of 10 per cent in area brings in less than 10 per cent of new species.

If vegetation were truly homogeneous, that is if all the component species were randomly dispersed, then the minimal area would be a valid entity and could be estimated with some precision, but this is only true in a small minority of cases which probably explains why both methods apparently work satisfactorily in some associations only.

Hopkins (1957), in a critique of the methods used for distinguishing minimal areas, gave a number of representative estimates taken from a variety of plant communities. It is noteworthy that the species-area curves which agreed best with the theoretical bent curve were from low-growing vegetation with little or no stratification, while two which departed most strikingly from the norm, in that the curve continued to rise, up to sample areas of 400 m², were from woodlands. This corroborates du Rietz' contention that, in a woodland, the minimal area, so far as the field layer or the ground layer are concerned, is very much smaller than that from the tree layer, and that this tends to be true of all stratified vegetation. He claimed that a minimal area (constancy-area) estimation, made from the upper layers only, was not changed by the inclusion of the lower layers and that, in general, the minimal area for the ground layer was smaller than that for the field layer and this again smaller than that for the tree layer, making the minimal area a function of the size of the plants. It is not true, as has been said, that the Uppsala method only gave a good result from one association, as du Rietz used his method on a wide selection of types of vegetation and claimed good agreement both from his own results and those of other workers.

Hopkins concluded from his investigations that the minimal area was a useful idea, but that it could not be determined objectively, at least in most cases, by either the species-area or the constancy-area methods. He proposed another type of area which might be considered as a unit feature of an association, namely, the *basic unit*. The fieldwork involved in this is relatively simple, but the calculations subsequently are rather complex. These units are regarded as the elements in a mosaic or pattern in the vegetation, however caused. The question whether these elements should be regarded as distinct communities in themselves depends upon whether they are always part of the same pattern, or whether they occur separately elsewhere or associated with other elements in a different pattern or mosaic.

The first step is to sample the community with numerous, small, random quadrats, listing the species present in each. Species occurring in less than one per cent of the quadrats are ignored. All possible pairs of species in the combined lists are then compared for the number of times they are recorded together, and this figure is compared with the calculated number of expected occurrences together if there were no association between them. The significance statistically of the deviation between 'observed' and 'expected' numbers is then calculated from a 2×2 contingency table and tested by the χ^2 test as follows:

	Species A present	Species A absent	Total
Species B present	a	b	w
Species B absent	c	d	x
Total	y	z	n

$$\chi^2 = n(ad - bc)^2/wxyz$$

Only associations for which $p < 0.001$ are considered significant.

The basic units are then made up thus:
1. All species with negative associations are listed in decreasing number of their negative associations.
2. To these individual species are added any species positively associated with them.
3. When two or more of these groups contain a species in common they are added together.
4. Extra groups of any other positively associated species are added to the list.
5. These are then grouped with the others as in 3.
6. Each remaining group of species is regarded as a basic unit.

As a further step individual quadrats were allocated to separate units according to the number of basic unit species they contain. Here arbitrary judgement is bound to enter to some extent, even if an empirical rule be adopted. The further attempt to ascertain the mean size of a basic unit is equally arbitrary.

Some of the examples of basic units which he gives from observation appear to be ecologically comprehensible, but in most cases the best correlations appear to be between plants of similar life-form. When the groups are heterogeneous in this respect they look much less probable. What ecological significance can attach, for example, to the inclusion of *Lonicera periclymenum* in a unit with nine assorted Bryophytes and Lichens except that *Lonicera* is a climber? The idea of basic units is an attractive one, but the method of their discrimination needs further investigation.

The species comprising a basic unit can be grouped in the form of a diagram as shown in Fig. 4.7, where the lines indicate significant positive correlations between species.

Hopkins compares his basic unit with Watts' pattern patches (p. 3459) which are quite definite units of an ecological nature. He further suggests that the ideal 'minimal area' is one that includes 'at least one of all the important basic units'. As some basic units are widespread and form a matrix within which the others are distributed, this does not seem possible.

One difficulty which has caused some confusion in these discussions is the double use of the term 'constancy' (see also p. 3439). As used by Braun-Blanquet it implies the constancy of occurrence of a species in different representative stands of an association. As used by du Rietz and the Uppsala school it implies constancy of occurrence in sample quadrats of the same stand, a very different thing and more closely related to 'frequency'.

A somewhat different approach to the discrimination of vegetation types is provided by the concept of the *catena* (p. 3399). This was primarily defined by Milne in 1935 as a descriptive term for a sequence of soil types repeated in similar topographical situations within a region. The most frequent example of a catena is provided by a slope connecting a hill with a valley. The drainage water from the hill carries dissolved and suspended material from the soil on

Spergularia
salina

Salicornia
stricta

Armeria
maritima

Aster
tripolium

Suaeda
maritima

Triglochin
maritimum

Limonium
vulgare

Festuca
rubra

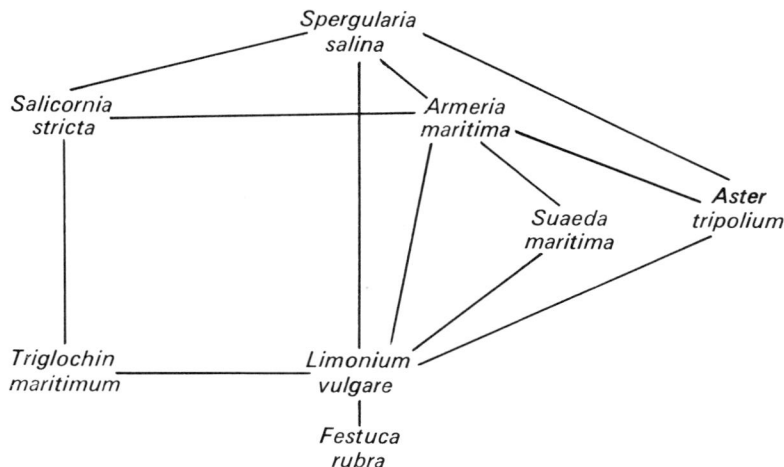

FIG. 4.7. The *Limonium* 'basic unit' (8 species) of the salt-marsh community at Blakeney, Norfolk, after Hopkins. The lines show significant
positive associations between species. This unit comprises 63 per cent.
of the area of the community, i.e. it is the matrix unit.

the hill, enriching in graduated measure the soil of the slope and finally
coming to rest in the valley soil, from which its further removal depends on the
drainage slope of the valley itself, and may be very prolonged. Three types of
catena have been distinguished: first, the simple or drainage catena, which we
have described above. Second, the geological catena, where soil types are
related to a series of rock outcrops. Third, the mixed catena, where both of
the above factors are operative.

Although originally applied to soils, the term has spread to cover the
sequence of vegetation which is associated with the soil sequence, it is, in fact,
a special kind of ecocline, soil factors and vegetation being closely and
causally linked in direct connection. To try to analyse such vegetation on
purely sociological lines would be to miss the whole meaning of the situation.

Ideally the vegetation on a drainage catena should form a series of belts
or zones running transversely to the line of drainage, but diagrammatic
regularity is seldom approached. The main catena may be broken up topographically into a number of sub-catenas or the vegetation may be segregated
into patches of any size or shape, forming a *mosaic*, dependent on differences
of surface relief or on the incidence of soil phases (the latter being understood
to mean the most narrowly defined soil-type unit which is distinct and forms
part of a soil complex).

The catena interpretation of vegetation originated in Africa, in regions
where a medium rainfall (812–1524 mm) and large-scale topography combine
to make the catena-mosaic a prominent feature. The soil phase-units are
complex in these regions because the medium rainfall in a tropical climate
gives importance to every variation of surface relief, even where it appears

vanishingly small. As a result each catenary zone tends to be broken up into a mosaic in which the species groups of particular patches are endlessly replicated in similar patches, controlled by soil-phase and micro-relief, throughout the region.

Although on a smaller scale than in Africa, catenas are frequent in temperate regions. Descriptions of several such drainage sequences have been published with regard to sites in Britain. The soils at the top are usually shallow or truncated by erosion and are base-impoverished by leaching, while the lower soils are deepened by downwash and base-enriched by drainage downwards. Where a calcareous soil is involved there will thus be a tendency for the frequency of calcicolous species to increase towards the bottom of the catena. As Bird has pointed out, the occurrence of patches of vegetation which do not accord with the general catena sequence may be a valuable indication of local conditions which deserve ecological investigation.

The extensive survey of the south-western Sudan in 1939, by Morison, Hoyle and Hope-Simpson, showed that the apparent medley of vegetation types encountered could be adequately understood and explained by the application of the drainage catena concept, which illustrated in the clearest way the causal dependence of the vegetation types, not only physiognomically but floristically as well, on the habitat conditions. A purely sociological analysis of communities without reference to the habitat, in this region at least, would have produced no enlightenment whatever, whereas the catena concept provided the key to the whole complex.

A sociological concept which enters into every consideration of vegetation, from whatever point of view, is that of *dominance*, as applied to certain species in the community. It is an idea easily grasped in essentials but somewhat difficult to define, and, moreover, the word is not always used consistently by different authors or even by the same author at all times. It is often used as if it were synonymous with abundance though this is fallacious. A truly dominant plant must, it is true, have a certain degree of abundance, but it may well be outdone in this respect by lowly cryptogams which have no claim to dominance. Tansley speaks of a dominant as being nearly always the tallest species (in a given community) which, by its stature controls many of the conditions affecting the other members. This is true in some cases but not in others, as he himself indicates by referring to dominant species in mixed grassland. Braun-Blanquet makes dominance equivalent to *Gewicht* by which he implies the social weight or importance of a species, to be estimated by a combination of frequency with area covered. This would be open to the same objection as simple abundance, if he did not add the proviso that dominance should be estimated separately in each layer of a stratified community. Several other definitions may be summed up as, 'Dominants are organisms which control the habitat', with the rider that they so alter the habitat as to affect the habits of their associates. Several other definitions simply stress that they are the chief constituents of the vegetation of their own stratum, which is too vague as it stands. If they control the habitat then they are by the same token the chief constituents of the community in a sociological sense, but we must have, if possible, some standard of judgement about those words 'control' and 'chief'.

Curtis (1947) introduced the use of an index figure, which he called the 'DFD index', as a numerical value for the social importance of a species. This index is the sum of the three factors, density, frequency and dominance, which are given the following values. *Density* is the number of individuals per unit area expressed as a percentage of the total number of individuals of all species in the same area. *Frequency* is the number of quadrats per 100 in which the species appears and *Dominance* is estimated by the basal area of the species as a percentage of the sample area. There has been some variation in usage. Density has been expressed simply by the number of individuals in the unit area and frequency by the number of quadrats within the same area, in which the species occurred. Such differences matter little when a single worker is comparing the composition of two communities on the same basis, but it is a different matter where general use is concerned.

This index showed its usefulness in sorting out the sociological differences between communities which were floristically the same but ecologically distinct. In 1951, however, Curtis and McIntosh put forward a revised form of index, the IVI index, on a more strictly defined basis of relative density, frequency and dominance. (See p. 3430.)

Braun-Blanquet bases his standard of judgement on a *combined estimation* of abundance and covering power, which he designates as the *Artmächtigkeit* or specific potency. This seems indistinguishable from *Gewicht* or dominance. For this he uses a subjective scale:

r Extremely scarce, with very slight covering power.
+ Scarce, with very slight covering power.
1. Plentiful, but with slight covering power; *or* somewhat scarce, but with greater covering power.
2. Very plentiful or at least covering 5 per cent of the area.
3. Numbers variable, but covering 25–50 per cent of the area.
4. Numbers variable, but covering 50–75 per cent of the area.
5. Numbers variable, but covering more than 75 per cent of the area.

Only the two highest classes would be likely to satisfy the concept of dominance. While this deals with the question of 'chief species' it does not settle the matter of control. Does 'control' imply that the presence of the species is essential to the existence of the integrated community, that without the species in question the associated species would have to vacate the site? If this is so, there can be no question as to the status of the species as dominant, but it is an extreme case and not very frequent. There are other grades of dominance below this. The case of woodlands at once comes to mind. Is the tree cover beneficial or inimical to the lower strata? Can it be said to control the environment of the lower strata? This is not always an easy question to answer. When the tree cover is removed we commonly find that some species of the field layer disappear, but that others multiply and become more abundant, species which were subordinate before now assuming dominance. It might be said that the former field layer association has given place to a new association and that therefore the trees had controlled the habitat, but such a decision must depend on circumstances and be a matter for judgement. It could also be argued that the new combination of field-layer plants is not a true natural association, but only a temporary and accidental community, but this again is not always so. In montane woods of *Quercus petraea* in Britain the field layer is often simply an extension downwards of the alpine heath or grassland from above the tree limit, differing little except perhaps in density from the open grassland: if the trees are removed, it does not change its character.

An interesting reversal of the usual relationship has been sometimes observed, where the density or character of the field-layer prevents the growth of tree seedlings. Here the field-layer is controlling the tree layer.

Braun-Blanquet cites the very interesting case of the widespread community characterized by *Rosmarinus* and *Erica* in the western Mediterranean region. In woodlands of *Pinus halepensis*, which provides only a light canopy, this community flourishes even better than in the open. The covering protects it from the worst effects of flood, wind and frost and it becomes denser and more closed. But this is a quantitative change; it is still the same community as outside and it is not controlled by the trees. This is not the whole story, however. The shade of the *Pinus* allows the invasion of shrubs from the *Quercus ilex* woodlands and eventually the entry of the *Quercus* itself, so that in the long-run the former community is gradually suppressed.

It is not the *tallest* species which always dominate, but species of the most highly developed life-form. In tropical jungle the tallest species are the emergent trees which contribute little to the canopy.

There are, however, the communities, such as grassland, where there is only one life-form and it is questionable whether in such cases we can separate dominance from abundance or whether we should speak of dominance at all. In a grassland there may be some species of outstanding vigour, stature and covering power which deserves to be called a dominant, but even so it can seldom be said to control the habitat nor to be essential to its associates, if those are characteristics of dominance. There is a danger here of confusing dominant with predominant. Common usage has certainly not distinguished them, but it is open to question whether it is necessary to maintain the distinction in practice or whether by stressing the idea of control we are not taking an untenable position. Control of the habitat must always be difficult and sometimes impossible to assess. We have seen that even in the case of the tree layer it is not always clear. It might be wise and more practical to say that *we designate as dominant any species which by virtue of its superior life-form, abundance, frequency and covering power, or by its reactions, defines the essential character of the community of to which it is a member*. All these factors are necessary ingredients of the concept. The trees in a savannah define the character of their community, but they are not dominants if they lack abundance, frequency and covering power. The latter factors may well be variable from place to place so that we must accept the idea of *local dominance* as well as general dominance.

In stratified vegetation where the strata are clearly distinct there will be dominants in each distinctive layer. Where the strata are intimately associated, as in a *Calluna* heath, there may be one overall dominant; in the open communities on acid soils there may be no qualifying dominant.

We have been speaking as if one species only came into question as dominant, but there is frequently more than one and dominance is then shared between two or more *co-dominants*. We must also recognize *subdominants*, of somewhat lesser weight than the dominants, but still standing out in social importance from the majority of the constituent species.

Pavillard has put forward an interesting classification of the social role of different species. He classifies them socially into constructive species, stabilizing and preserving species, neutral and destructive species. Constructive species are naturally most active in the early stages of the establishment of a community, those which prepare the way for the entry of the stabilizing and preserving species and themselves tend to disappear as the latter multiply. Destructive species are those whose entry so alters conditions that they prepare the way for a new combination of species which may constitute a further stage in succession.

When we look at the various schemes which have been proposed for the classification of plant communities we find somewhat the same divergence of views which exists in plant taxonomy, namely, between 'lumpers' and 'splitters'. The former prefer large, widely embracing units under the name of 'associations'. Such is Tansley's acceptance of the whole of the deciduous *Quercus-Fagus* woodlands of Europe as one association with local segregates, or Smith's Arctic-alpine grassland for the whole of the montane grassland of Britain above the tree limits. This we might describe as the native British attitude, uninfluenced by techniques of quantitative analysis of vegetation. To the splitters such wide concepts of associations belong rather to the order of *association complexes* or even *formations*. Their opinions are based upon close floristic analysis of communities, which is apt to lead to the extreme of separating as associations all communities which do not exactly agree in analysis. Vegetation is thus parcelled out into relatively small and strictly limited associations and it then becomes necessary to group these into a hierarchy of larger units based partly on groups of dominant species and partly on predominant life-forms, i.e. on physiognomy.

This procedure is more logical and more in accord with accepted scientific method, but it is not free from some taint of artificiality. It is adapted to precise statement and the convenience of this has led to its wide application in Europe, which ensures for it full consideration. Compared with this analytical system the British usage seems vague, but this may be rather a merit than a defect since it may be closer to nature. It leaves open the question of a continuum in natural vegetation which we shall return to later (p. 3429). There may be also some doubts as to whether vegetation can in fact be parcelled out into neat units without doing violence to facts. One is reminded of the sexual system of Linnaeus which was also logical, neat, convenient and popular, in comparison with the somewhat nebulous concept of the natural system.

We would here mention the valuable and significant practice by Poore, who calls each well-defined community a *nodum*, a knot in a meshwork, as it were, having multilateral links with other similar communities, an idea which accords very well with natural fact (see also p. 3432).

The earliest efforts in the classification of vegetation were based on physiognomy, as we have described earlier (p. 3319) and the possession of a uniform physiognomy still enters into the concept of the association. This phase was followed by classification on the basis of habitat, which Warming adopted. This was, and still is, only possible in a very wide sense, for the knowledge of habitat conditions is too scanty to allow the discrimination of small units occupying closely related habitats without a good deal of *ad hoc* research. Moreover, it tends to bring together communities which are floristically and physiognomically different. As the plants themselves are by far the best judges of habitat conditions, it seems obviously best to base our units in the first place on their floristic composition, and to keep together those units in which the composition is most similar.

We begin therefore by listing all the species present, identifying each in full detail, since many genera are rich in ecotypes with distinctive habitat preferences. If now we were furnished with such lists from several localities we would be faced with the question to which Rübel says no answer has been given: how many species in two lists must be different before we decide that they represent different associations? Fortunately not all species in a community are of equal importance sociologically and systems for grading or weighting species in order of importance have been devised, especially by the Montpellier ecologists, which greatly assist in making the decision, even if they do not entirely remove Rübel's difficulty (p. 3455). That such a system should originate in the Mediterranean region is not surprising if we remember that the region is very rich in species and has a very varied topography which presents communities well-differentiated on a floristic basis. This is very different from the immense stretches of comparatively uniform landscape north of the Alps, relatively poor in species and with the effects of underlying rock largely disguised either by beds of loess or by glacial drift.

We shall deal with the details of the Montpellier system in Chapter XLIV on community analysis (p. 3437).

Warming proposed a system based on habitat characteristics which we have already outlined on p. 3323. This is a large-scale classification, and is really a classification of formations rather than associations, but Warming divided these formations on the basis of dominant species which was the criterion he used for distinguishing different associations of a formation. For example, under Helophytes or Marsh Plants he distinguishes two formations: reed swamp and bush swamp, then under reed swamp he designates associations dominated respectively by *Phragmites*, *Scirpus* and *Typha*. Under the formation of deciduous dicotyledonous forest he groups as separate associations: beech forest, oak forest, birch forest, ash forest and several minor sorts, all distinguished by their dominant trees. This is a simple and useful method where it can be applied, and the habitat terms Hydrophyte, Helophyte, Mesophyte, etc., partly adopted from earlier writers, have largely passed into general use. Unfortunately the system breaks down in face of more complex communities such as the mixed temperate forests of North America and south-eastern Europe or coniferous forests, tropical rain forest, or grass steppe where there are either no consistent dominants or a mosaic of 'types' of differing floristic composition. These Warming leaves unclassified and it is precisely in this kind of case that subsequent workers have found most difficulty in distinguishing associations.

Perhaps next in importance if not in point of time is the physiognomical classification proposed by Brockmann-Jerosch and Rübel in 1912 and elaborated later by Rübel in his work *Die Pflanzengesellschaften der Erde* in 1930. He reviews previous proposals by Schimper, Drude and Diels, which were all based on a mixture of ecological and physiognomical characters. Schimper's general division of climatic formations from edaphic formations, to which Clements offered such opposition, has already been referred to on p. 3325. Rübel professed himself dissatisfied with all these efforts, and put forward a scheme, intended to be worldwide, which is of great interest geographically. It is based on three main divisions, Lignosa, Herbosa and Deserta, with Phytoplankton thrown in as a rather unnecessary appendage.

LIGNOSA

Pluviilignosa	Rain forests and bushlands.
Laurilignosa	Forests of *Laurus*-like trees.
Durilignosa	Sclerophyllous forests.
Ericilignosa	Ericaceous heaths.
Aestilignosa	Summer green, deciduous forests.
Hiemilignosa	Forests green in the wet season. Monsoon forests.
Aciculilignosa	Needle leaved (coniferous) forests.

HERBOSA

Terriherbosa	Steppe and prairie.
Aquiherbosa	Herbaceous marsh and water vegetation.

DESERTA

Siccideserta	Dry deserts.
Frigorideserta	Cold deserts.
Litorideserta	Shore-line deserts.
Mobilideserta	Deserts of wind blown sand.

Sub-division: *Petrideserta*

Rupideserta	Dry rock faces.
Saxideserta	Talus, gravel and shingle.

To these classes were attached rather unusable classical names, such as Chomopetreremia for Rupideserta.

A uniform descriptive treatment of all types of world vegetation is naturally not possible for the lack of full knowledge of many types. Rübel therefore adopts a mixed system, naming associations or association alliances by their dominants where possible and otherwise describing less-known communities on a geographical basis, in the manner of Schimper.

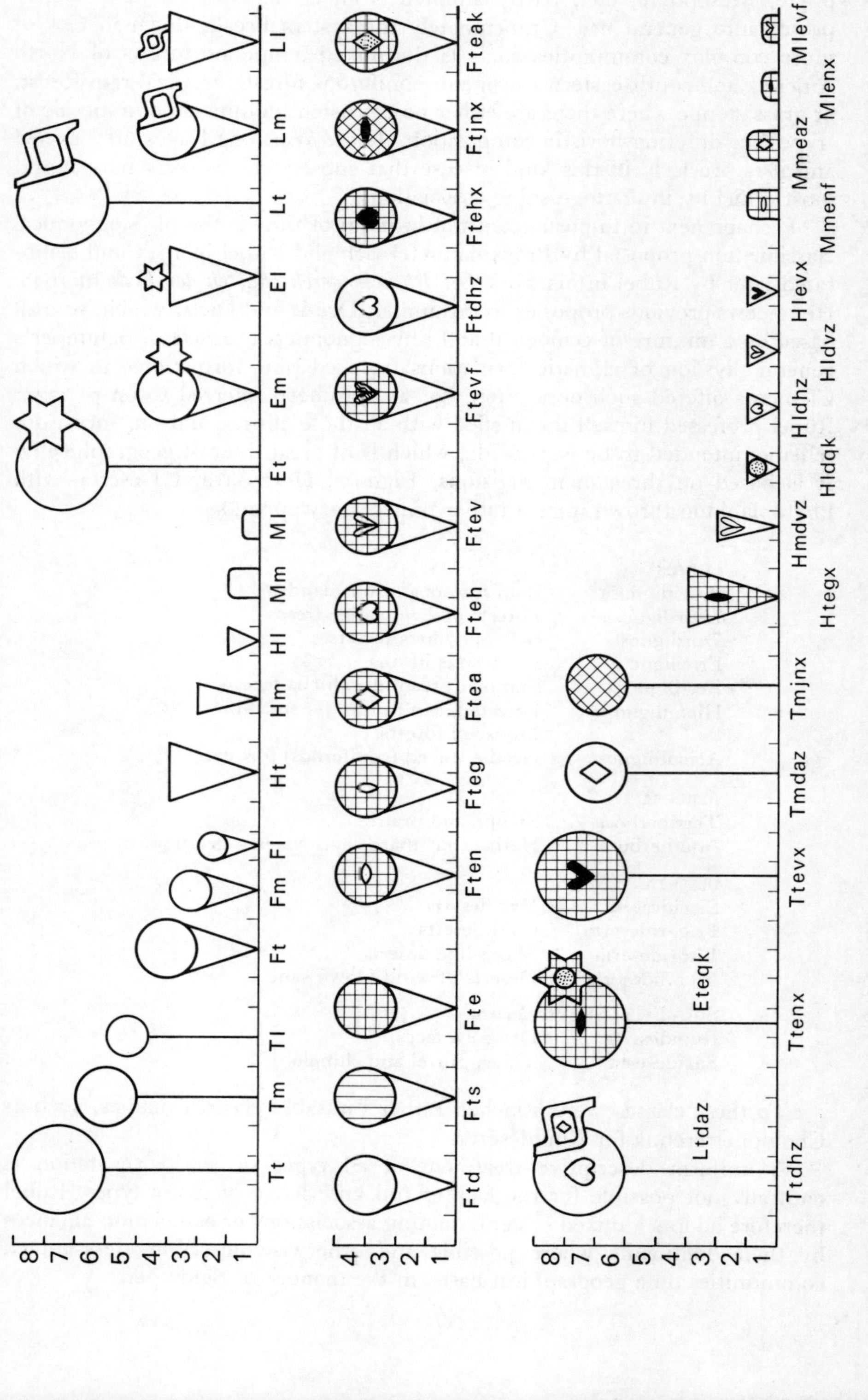

8 7 6 5 4 3 2 1
Tt Tm Tl Ft Fm Fl Ht Hm Hl Mm Ml Et Em El Lt Lm Ll
4 3 2 1
Ftd Fts Fte Ften Fteg Ftea Fteh Ftev Ftevf Ftdhz Ftehx Ftjnx Fteak
8 7 6 5 4 3 2 1
Ltdaz
Ttdhz
Eteqk
Ttenx Ttevx Tmdaz Tmjnx Htegx Hmdvz Hldqk Hldhz Hldvz Hlevx Mmenf Mmeaz Mlenx Mlevf

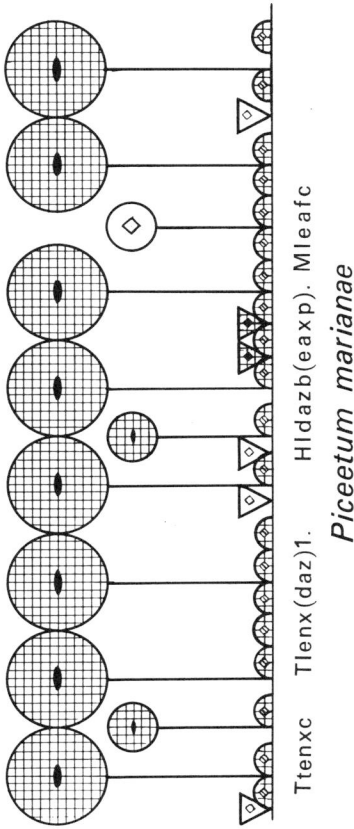

Ttenxc   Tlenx(daz)1.   Hldazb(eaxp). Mleafc

Piceetum marianae

FIG. 4.8. The Dansereau scheme for the symbolic representation of life-forms in a community. Key on p. 3422. Application of the symbols to several types of plants, plotted on squared paper with a maximum of eight in height and three in breadth. The upper group shows all possible combinations of series 1 (life-form) and 2 (size). The second line shows various combinations of series 3 (function) for Ft (shrubs, tall); of series 4 (leaf shape and size) for Fte (shrubs, tall); of series 5 (leaf texture) for Ft (shrubs, tall). The lower line shows various applications of the five series. Each one of the formulae presented here will correspond to a large number of plant species. Suitable examples are as follows: Ftevf (*Cyathea gairdneri*), Ftdhz (*Corylus cornuta*), Ftehx (*Rhododendron maximum*), Ftjnx (*Cephalocereus arrabidae*), Fteak (*Crassula arborescens*); Ttdhz (*Fagus grandifolia*), Ltdaz (*Celastrus scandens*), Ttenx (*Tsuga canadensis*), Eteqk (*Usnea longissima*), Ttevx (*Cabralea eicheriana*), Tmdaz (*Betula populifolia*), Tmjnx (*Opuntia brasiliensis*), Htegx (*Cortaderia selloana*), Hmdvz (*Desmodium grandiflorum*), Hldqk (*Amanita phalloides*), Hldhz (*Clintonia borealis*), Hldvz (*Oxalis montana*), Hlevx (*Coptis groenlandica*), Mmenf (*Sphagnum fuscum*), Mmeaz (*Raoulia lutescens*), Mlenx (*Polytrichum juniperimum*), Mlevf (*Hypnum crista-castrensis*).

In 1921 du Rietz put forward a classification along similar lines to that of Rübel with particular reference to the vegetation of Scandinavia. He assigned to each community a symbol made up of the initial letters of the chief and sub-classes to which the community in question belongs, with serial numbers added where required. This he claimed gave a great saving of time and space in writing descriptions. Whether it is easily intelligible to the reader, unless he has the table of classification before him, is questionable (see Table 3).

TABLE 3. THE SIX CATEGORIES OF CRITERIA TO BE APPLIED TO A STRUCTURAL DES-CRIPTION OF VEGETATION TYPES, ACCORDING TO A NEW SYSTEM.

TABLE 3

1. LIFE-FORM

T	trees
F	shrubs
H	herbs
M	bryoids
E	epiphytes
L	lianas

2. SIZE

t	tall	(T: minimum 25 m)
		(F: 2–8 m)
		(H: minimum 2 m)
m	medium	(T: 10–25 m)
		(F, H: 0·5–2 m)
		(M: minimum 10 cm)
l	low	(T: 8–10 m)
		(F, H: maximum 50 cm)
		(M: maximum 10 cm)

3. FUNCTION

d	deciduous
s	semideciduous
e	evergreen
j	evergreen-succulent; or evergreen-leafless

4. LEAF SHAPE and SIZE

n	needle or spine
g	graminoid
a	medium or small
h	broad
v	compound
q	thalloid

5. LEAF TEXTURE

f	filmy
z	membranous
x	sclerophyll
k	succulent; or fungoid

6. COVERAGE

b	barren or very sparse
i	discontinuous
p	in tufts or groups
c	continuous

Dansereau has more recently devised a symbolic system of describing the biological structure of communities by using six categories of specific criteria in each of which letters are given to denote characteristics, combined with conventional diagrams. This yields a qualitative formula which is at once a shorthand description and record of the species while the diagram shows the structure of the community (Fig. 4.8).

Braun-Blanquet arranged associations into: first, alliances (*Verbände*), second, orders (*Ordnungen*), and, third, classes, to which he assigned a system of nomenclature which has been widely accepted. The association is designated by the name of the most characteristic plant or plants, the generic name with the termination *-etum* and the specific name in the genitive case, as first used by Schouw in 1823. The alliance is similarly named but with the termination *-ion* to the generic name. The order has the generic termination *-etalia* and the class the generic termination *-etea*. As an example we have the class *Salicetea herbaceae* including the orders *Salicetalia herbaceae* and *Arabidetalia coeruleae*, each with one alliance: *Salicion herbaceae* and *Arabidion coeruleae* and each of these can be divided into several associations of which the chief are the *Salicetum herbaceae* and *Arabidetum coeruleae*. When the names of two dominant plants are used together the first name is given the termination *-o* or, for euphony sometimes *-eto*, thus: class *Vaccinio-Piceetea* or alliance *Pineto-Ericion*.

This closely organized hierarchy of groups arises from Braun-Blanquet's strictly limited floristic concept of the association, which leads him to divide what most British ecologists would view as mere local variants of one association. This is not a criticism of Braun-Blanquet's views, but simply to point out a difference of outlook which markedly persists. His alliances consist of groups of associations agreeing in predominant life-forms and ecological character but geographically separated within one floristic region and each having its own special combination of species.

Braun-Blanquet also provided for the grouping of larger units, orders, etc., along 'natural' lines, that is in groups based upon the general biological characteristics of the communities rather than on floristic similarities. The following is an outline of his proposals.

A. Non-regional, non-layered communities of ecologically simple lowly organisms in loose association.
   1. Floating communities without constant components: water and air plankton.
   2. Communities with more or less constant local components. Snow and soil communities of bacteria, algae and fungi.
B. Regionally limited, layered communities mostly of more highly organized plants with stronger mutual relationships.
   1. One-layered communities without root competition.
     (a) Free floating. *Lemna*, *Azolla*, etc.
     (b) Attached.
       (i) Dependent corticolous communities without developmental possibilities.
       (ii) Soil-attached communities with developmental possibilities. Soil and rock-living algae and lichens.
   2. Two or more layered, mostly rooted communities.
     (a) Open communities of weakly-connected components. Competition for germination and nutrition.

(i) Climax communities without developmental possibilities. Desert and semi-desert vegetation

(ii) Edaphically controlled, mostly pioneer communities. Vegetation of sand dunes, rocks and crevices.

(b) Closed communities, competition for germination space and nutrition.

(i) Communities weakly stable, consisting mostly of Therophytes and often under human control. Cultivated ground, pond margins, etc. *Chenopodietalia*.

(ii) More stable communities with closer integration.

First, communities of few species, biologically uniform. Water and marsh communities. Floating, *Potametalia*: Emergent, *Phragmitetalia*.

Second, richer and more complex communities of ordinary soils.

α) Few-layered communities, without woody plants (*Herbosa*, Rübel).

* Sub-aerial layering, weakly developed. Layers with little mutual influence. *Caricetalia, Molinetalia, Brometalia*, etc.

** Layering permanent with well-developed mutual influence between layers.

† Ground layer open or rarely absent, under the influence of the field layer. Dwarf shrub communities. *Rosmarinetalia*.

†† Ground layer close and well-developed, controlling the field layer. *Sphagno-Ericetalia*.

β) Communities with several layers, the lower strata more or less controlled by the upper layers. (*Lignosa*, Rübel).

* Mostly three-layered, often edaphically or biologically controlled. Epiphytes absent or rare. Bushlands.

** Mostly more than three-layered, often as climax in equilibrium with the environment. Epiphytes often abundant. Woodlands.

Clements, on the other hand, was more interested in associations, their varieties and subdivisions, than in units of higher order, which was due to his preoccupation with succession. The only unit higher than the association that he recognized was the formation, which he regarded as being 'the climax community of a natural area in which the essential climatic relations are similar or identical' (1916). He quotes only a few examples from North America, such as the deciduous forest climax and the prairie-plains climax, but he is only concerned with describing their successional development, not with characterizing them as communities. As he only recognized one climax association in a climatic area (to which he limited the term 'association'), the definition given above would make association and formation equivalent. Moreover, under desert climaxes he quotes MacDougals' description of a number of communities which are clearly edaphically determined. It is thus dubious what Clements meant exactly by 'formations' or why he used the term at all.

In describing associations Clements uses the generic name of a dominant plant or the names of co-dominants together. To this he appends a descriptive tag, usually referring to habitat, framed from Greek roots. Of these there are a great number. Three are used to distinguish grassland, forest and desert vegetation respectively, *poium, hylium* and *eremium*, and these may be combined with generic names to designate particular communities in the group, e.g.

Picea-Pseudotsuga hylium, Deschampsia-Carex poium. Many others designate special habitat relationships, e.g. *chalicium*, a gravel slide community; *lochmodium*, a dry thicket community; *oxodium*, a humus marsh community; *coryphium*, an alpine meadow community. Other terms apply to successional seres, not to particular communities, e.g. *pyrium*, a succession following fire; *xerasium*, a succession on drying-out soils. Those interested will find the system explained in *Research Methods in Ecology* (1905).*

It is worth remarking, as illustrative of the historical difficulties of Ecology, how two experienced men, working in isolation from each other in different parts of the world, could come to such radically different views about vegetation as did Clements and Braun-Blanquet.

Apart from his rather ambiguous 'formations', Clements only proposed one other unit of higher rank than the association, the *panclimax*, a grouping of related climaxes, of similar climatic relationships, life-forms and dominant species, regarded as descended from an earlier ancestral climax, a concept more theoretical than practically useful.

When, however, we turn to Clements's subdivisions of the association, his impact has been much greater and his nomenclature is widely used in America and in Britain, but there is a difference of opinion on the use of the term association itself. In accordance with his monoclimax view, Clements confined the term to the climax itself, this being, in his opinion, the only community which was in permanent equilibrium with the prevailing climate. All other communities were to his mind either quasitemporary, or else diverted or frustrated developments (dis-climax or sub-climax units) for which he proposed separate names (see below). The unitary association he called the *consociation*, i.e. an association with a single dominant species. In so far as the association is a climax community this would include the majority, since mixed dominance in climax associations is, he claimed, uncommon, a conclusion open to question.

Below the rank of association or consociation comes the *society*, 'a localized or recurrent dominance within a dominance' (Clements) or, 'subordinate communities dominated by species which are not general dominants of the main community' (Tansley). Such localized dominance may occur in any of the strata of a layered community. A society of *Taxus baccata* may occur in a wood of *Fagus sylvatica* and it may sometimes qualify as a consociation where it occurs by itself; or a society dominated by *Mercurialis perennis* may occur in the field layer of an oak wood, or a society dominated by mosses such as *Porotrichum alopecuroides* or *Leucobryum glaucum* in the ground layer of calcareous woods. Quite often society dominants are only seasonal in their growth and die down during a part of the year, yielding place to other species. These form *aspect societies*. Such are the pre-vernal society of *Anemone nemorosa* or the vernal society of *Scilla non-scripta*, which is often followed by an aestival society of *Pteridium aquilinum*.

Still more local is the *clan*. It is a limited cluster or constellation of plants

* The full account is given in CLEMENTS (1902) 'A System of Nomenclature for Phytogeography'. *Engler's Botanische Jahrbuch*, Vol. 31, 1902.

of a secondary species with little or no effective dominance. It may be part of a society or it may occur independently. Actually it is difficult to lay down a clear distinction between society and clan, though the latter is usually relatively small. The term *colony* is also sometimes used for a temporary group of species invading a bare space or another community. A *family* is only a special case of the clan and is so-called when the occurrence of the species concerned is quite isolated and the individuals may be regarded as the offspring of one mother plant.

All the above terms apply only (in Clements's original sense) to units of the climax association. For similar units in earlier stages of a sere Clements proposed terms ending in -ies. Thus a definite stage in the sere is an *associes*, with *consocies* where it has a single dominant, and society becomes a *socies*. As Clements regarded all communities other than the climax association as actually or potentially stages of a sere, he applied these seral terms pretty widely, but if we accept the polyclimax view or if we doubt the concept of the climax as a universal idea, we will naturally apply the term association to all communities which are in equilibrium with an environment and the term associes only to those which are discernibly developmental stages in a recognizable sere. This has been largely the practice in British ecology.

May we once more re-iterate the standing plea for the use of Latin names in ecological writing. A resident in New England may recognize a 'blue cohosh', but is it familiar in California? Westerners may know the differences between yellow pine, loblolly pine, digger pine, etc., but does a worker in Florida? Certainly as they stand they mean nothing to Europeans. Germans, too, seem pretty indiscriminate in their usage of *Kiefer*, *Föhre*, *Fichte*, *Tanne*, etc., which need to be carefully checked. How would others like it if British writers started drawing upon their rich store of colloquial names and referred to Cleavers, Gromwell or Dyer's Greenweed?

This is only one example of the diversity of opinion and practice regarding the use of the term 'association'. The attempts since 1931 to obtain international agreement on the definition and classification of natural vegetation units have foundered chiefly on two obstacles. First, the desire in some quarters to make the association the sole fundamental unit of classification, defined according to certain principles. Second, the belief, widely held in Europe, that criteria which are suitable for that continent must necessarily be applicable everywhere, and that it is possible to establish a universal classification on that basis. The situation is also exacerbated by the passion for *netteté et exactitude* in a realm which is markedly wanting in both. The attempt to impose them simply produces a false situation.

Watt has pointed out the reasons why British practice must differ from the purely floristic criteria used so much in Europe. Because of its comparative isolation and for historical reasons Britain is floristically poor and communities have to be made up of the species available, most of which have wide ranges. Floristic lists are not enough to distinguish communities which are ecologically different. Characteristic species are uncommon or often wanting, so that communities cannot be sharply defined by this means. Moreover, Britain is a focus of different climatic types, permitting the co-existence of species from widely separated vegetational zones in Europe.

The method adopted by northern ecologists who have to face similar conditions has been to use the term *sociation* instead of association for the majority of their distinct communities. Sociation, like association, is an abstract concept. The difference between the two lies in this, that the concept of the association (in the mid-European sense) is founded on fidelity relationships, that is on characteristic and differential species, recognized by their occurrence in many distinct stands, whereas the sociation is based upon the relative abundance and dispersion of the constituents, which can be numerically expressed. Stands of a sociation must agree together in the physiognomical and floristic characters in all strata; as to how far they must be homogeneous is left to the judgment of the ecologist.

The advantage of the sociation as a unit is its greater flexibility and applicability to units of diverse sizes. In some cases the sociation is in fact equivalent to a floristically determined association, in other cases it is simply a distinguishable part of an association. Although a single sociation may not provide characteristic species in the Braun-Blanquet sense, yet it is possible to combine floristically related sociations into alliances in which characteristic species may appear and which are thus comparable to units founded on the main European system.

That the sociation is marked out by the dominance of species rather than by their fidelity relationships is the idea embodied in the resolution adopted in 1935 by the sixth International Botanical Congress, which runs as follows:

1. To use the term *sociation* for vegetation units characterized mainly by dominance in the different layers, in the sense of the Scandinavian plant sociologists.
2. To use the term *association* for vegetation units characterized mainly by characteristic and differential species in the sense of the Zürich–Montpellier plant sociologists, or at least for units of the same sociological value.

This concession to the Scandinavians from the pure gospel of the Zürich–Montpellier association is very well as far as it goes, but as it leaves out of consideration all the rest of the world it does not go very far towards a truly international agreement, which, with du Rietz, we consider to be hopeless under the prevailing dogmas.

Nevertheless, the idea of the sociation is a valuable one which deserves wider application. Poore, for example, found it applicable in analysing the Scottish mountain vegetation.

An attempt has been made by Rietz to make comparisons between the application of the chief descriptive terms used in plant sociology. The 'society' as used by American and British ecologists he equates with the 'synusia' of Gams and Rübel. As a society is only a local feature of a stratum this cuts across the loose identification of a synusia with a stratum, which has become common. The two things are sometimes identical but not always. The *Cetraria islandica* layer in the *Vaccinium uliginosum* alpine heath of Central Europe is both stratum and synusia, but the *Mercurialis perennis* society in an oak wood in Britain is only part of the field layer. The 'sociation' he compares to the taxonomic species and the 'association' with the genus, which is somewhat misleading as the association is not necessarily built up from sociations as the genus is from species. The sociation would apply to the unitary *Vaccinium uliginosum—Cetraria islandica* alpine heath. The alpine *Loiseleuria—Empetrum—V. uliginosum* heaths of mid-Europe would represent an 'association alliance' (Zürich–Montpellier), while the mid-European ericaceous heath in general would be an 'association order' from the same viewpoint, but only an association to the wider concepts of Clements or Tansley.

The grouping of associations into alliances and orders has the advantage of dispensing with the disputable concept of the *formation*. This term was first introduced by Grisebach and was adopted and used by a succession of writers, including Warming, in very much the same sense. It described vegetation which was uniform in physiognomy and prevailing life-forms and was adapted to the definite external characters of a habitat. It was expressly not concerned with floristic differences and might embrace comparable types, based on the same life-forms, in different parts of the world with entirely different constituent species. The basis of the formation was in fact the *form* of the vegetation.

A part of the blame for the miscellany of meanings which have been attached to the term lies with the use of vernacular names such as heath, moor, meadow, etc., which have no exact meaning in popular usage and of which the supposed equivalents in different languages have often quite different applications. This was pointed out as long ago as 1863 by Kerner, but nevertheless the mistake has persisted. Kerner's own remedy was to propose a set of new terms in German (*Gehölz, Gekräut*, etc.) which did little to commend his views internationally. Formations have been constructed out of habitat types, successional seres, association alliances and on various other foundations. Rübel used the term for a grouping above the association alliance, that is as an equivalent to Braun-Blanquet's association order. Clements and some other American writers have used it to cover a group of allied climax associations within a major climatic area, which were related in physiognomy and life-forms.

Tansley used 'formation' in a sense similar to that of Warming, but narrower. It implies 'a mature eco-system (i.e. biocoenose) dominated by distinctive life-forms'. Such formations, occurring in different geographical areas and composed of different floras, are grouped as 'formation types'. The association then becomes a major community dominated by distinctive life-forms *and* by distinctive species. The terms are applied in senses similar to those of Clements but more widely, in accordance with the polyclimax concept. To Tansley, as to most British ecologists, a climax represents any community of relative stability which may persist indefinitely in equilibrium with relatively stable environmental conditions.

The emphasis on distinctive life-forms distinguishes this use of formation from Braun-Blanquet's association alliance, which is its nearest equivalent, and formation type from his association order, both of which have a floristic basis. The difference, however, is more formal than real, since floristic composition usually defines the life-forms which prevail, but Braun-Blanquet gives preference to floristic composition as being more precise than a life-form classification and this must be conceded from the point of view of scientific principle. Life-forms are, nevertheless, much more easily recognized than floristic composition and reliance on them may be defended as of practical value in comparisons of world vegetation or as steps in what Poore calls our 'successive approximations' to an understanding of plant communities.

Tansley's concepts as expressed above and laid down in his last great work (1949) are surprisingly in contradiction with his earlier attitude, which expressly rejected life-forms as a criterion of the formation. He then (1920) accepted the developmental concept of the formation proposed by Moss who applied the term to the whole sequence of stages developing upon an essentially uniform habitat and culminating in a stable association, in other words to what Clements later called the 'prisere'. This was widely criticized as classing together communities which were radically distinct, such as chalk grassland, chalk scrub and chalk beech forest. A further criticism is that the habitat does not remain essentially uniform while the vegetation develops and that, except in some special cases, such as salt marsh, it is not possible to define the limits of a habitat.

On the whole, in spite of Tansley's attempt to rehabilitate the formation as a sociological unit, it can be dispensed with. It is perfectly possible to give a conspectus of world vegetation units on a life-form basis without using the idea of the formation, and the scheme outlined earlier (p. 3419) by Brockmann-Jerosch and Rübel has been very successful in this respect. Since the floristic characterization of the association also defines the prevailing life-form it still seems to offer the best means of distinguishing associations (see Chapter XLIV, p. 3434).

A different approach to community study from any that we have hitherto mentioned, is that embodied in the concept of the *continuum*. This is the view taken by the American school associated with the names of J. T. Curtis and G. Cottam, which shares Gleason's views that associations are not discrete entities and looks upon vegetation as a variable continuum. This work is comparatively new and has chiefly been applied to forest stands. It has, however, provided an entirely new means of analysing vegetation and reveals relationships which would not be perceptible by inspection. (See also p. 3431.)

The method depends on the use of a series of index values assigned to each species, which enable the contributions of all the species to the construction of the community to be quantitatively assessed. It is somewhat laborious, but the results are striking and present an entirely new scene for our consideration.

The idea of the continuum does not postulate that all vegetation everywhere is part of a universal continuum, though this is a development which we may yet see.

Many sharp changes of environments exist which seem to create equally sharp boundaries between communities, but for the most part habitat conditions occur as gradients and the boundaries between communities related to the habitat are set by the average tolerances of their dominant species as modified by biotic competition. This means that the boundaries of communities can often only be determined statistically and it is on this sort of situation that the idea of the continuum is founded, which calls in question the accepted concept of distinct plant associations.

The classic example, in which the ideas of Curtis and his colleagues were fully worked out, was that of upland, broad-leaved forests in south-western Wisconsin. Curtis and McIntosh examined ninety-five separate stands of forest, covering all the south-western area of the State. Each stand was chosen for its natural, undisturbed condition and adequate size, i.e. not less than fifteen acres (six hectares). Every species, in tree, shrub and field layers, was assigned an *Importance Value Index* (IVI), which is a modification of the DFD Index we have previously referred to (p. 3415). This Index is made up of three factors, like the DFD Index, but each factor, density, frequency and dominance, is expressed as a percentage, relative to all the species present. Thus, for each species in a stand:

$$\text{Relative density} = \frac{\text{Total number of individuals of the species}}{\text{Total number of individuals of all species}} \times 100$$

$$\text{Relative frequency} = \frac{\text{Frequency of the species}}{\text{Sum of the frequencies of all species}} \times 100$$

$$\text{Relative dominance} = \frac{\text{Total basal area of the species}}{\text{Area of the stand}} \times 100$$

Taking the tree species first, they can be graded on three bases. First, their maximum IVI, i.e. the highest figure they reach in any stand, out of the theoretical maximum of 300. Second, their constancy, i.e. the number of stands out of the total of 95 in which the species is present. Third, the average IVI, over the number of stands in which they occur.

Taking each stand separately the IVI of each species picks out the dominants for that stand. The authors found that if the stands were classified on the basis of one leading dominant they fell into 9 groups or types. If two co-dominants were used they formed 30 groups, three co-dominants required 75 groups and if four co-dominants, in the same relative order, were used then all 95 stands were different, each stand would be a type by itself.

The stands were then grouped according to their leading dominant. Four species were dominant in 80 of the 95 stands. These were *Quercus velutina*, *Q. alba*, *Q. rubra* and *Acer saccharum*. Importance values were then recalculated for each of the four with reference to the stands in which it or one of the other three were dominant. When these figures were tabulated they showed an ordination, with *Q. velutina* at one end of the scale and *A. saccharum* at the other, the two dominants being mutually exclusive, with *Q. alba* and *Q. rubra* as intermediates, in that order. It was found that this order corresponded to the amounts of available Calcium and Potassium in the A_1 layer of the soils and to the water retaining capacity of the soils, which indicates a sequence of increasing mesophytism. The stands could now be arranged in a sequence beginning with the stand in which *Q. velutina* had its highest IVI to that in which *A. saccharum* had its highest IVI and the curves of the respective IVI of each of the four species over this sequence showed wide overlapping. This sequence was given numerical value by means of a *continuum index*. Each species of tree was assigned a *climax adaptation number* (CAN) on an arbitrary scale from 1 to 10. *A. saccharum* received the number 10 and *Q. macrocarpa* 1.* *Q. velutina* received the number 2. These figures are adjudged on the character on the plant as dominant either in early successional stages or only in terminal stages of a succession. The continuum index is the product of the IVI multiplied by the adaptation number, thus lying between a maximum of 3000 (IVI $300 \times$ CAN 10) to a minimum of 300 (IVI $300 \times$ CAN 1). Each species can then be assigned its place in the continuum by its continuum index. The continuum index of each stand, i.e. its place in the continuum, is the total of the continuum values for the species of that stand, which gives it its continuum rating. As all species are taken into account, this places the rating of the stand on the basis of maximum data. If curves are drawn of the importance values of each dominant species in the different stands in which it occurs, lying at different points on the continuum scale, a series of curves is obtained of Gaussian form, some complete within the continuum range, others only partially contained with the continuum, but all widely overlapping. There is no separation into discrete associations. The same type of curves and the same continuous over-lappings are also shown by graphs of the non-dominant species.

This continuum analysis has not been widely extended as yet to other forms of vegetation, but Cain, Castro and others showed in 1956 that it was possible, even in the absence of species dominants, to assign an IVI to the leading species in a tropical rain forest and thus grade them hierarchically in a manner which could not otherwise have been arrived at. A further comparison of low-lying rain forest with montane forest, based on the DFD index, showed clearly marked differences in the relative importance of certain species in one forest type or the other.

A serious drawback to the purely floristic designation of an association is

* This is based on the observation that the climax vegetation in south-western Wisconsin is the association *Acer saccharum-Tilia americana*.

that it depends for its validity on the subjective assessment of a community stand as homogeneous. An attempt to define homogeneity, so as to minimize the purely personal nature of such an assessment, expresses it as the condition in which the probability of a species, or of the group of species considered to be characteristic of the community, occurring in quadrats of equal size is the same at all points in the area. The methods of pattern analysis, which we shall attempt to describe in the next section, show that such homogeneity rarely, if ever occurs. In fact homogeneity is not assessable by inspection except as a matter of degree and the subjective rejection of areas as being obviously not homogeneous involves ignoring very considerable areas of vegetation which demand investigation of a different sort. Associations segregated on such a basis may be real entities, but they only represent a fraction of the totality of vegetation and are to that extent arbitrary abstractions from nature.

No resolution has been achieved of the controversy between the concept of associations as absolute, classifiable units and the concept of vegetation as a continuum, the phases of which are subject to ordination but not classification. Each may be true in different circumstances, for the concept of a universal continuum is certainly unproven. In their extreme forms both concepts are open to serious objections and neither is likely to be the whole truth.

The laborious methods of analysis of correlations between species-occurrences in various communities has tended to direct attention to internally correlated groups of species rather than to larger groupings and has done a good deal to weaken the traditional idea of the association. (See p. 3474). Whether such groups should inherit the name is an open question. They are certainly, in many of the ascertained examples, related to habitat conditions in a significant way. The term *sociation* might be used for them as it has an international tradition of use for minor units delimited by local species groups. Their importance depends on their constancy and this has been little investigated.

More fruitful, however, is Poore's concept of the *nodum*. Accepting that vegetation is variable and that variation may occur in many directions, so that many variation series are traceable, Poore imagines the following model. In a three-dimensional matrix the communities are placed in relation to each other so that each line of variation forms part of a three-dimensional lattice connecting communities together. If variation is continuous the community points will be randomly dispersed in the matrix and they can only be grouped arbitrarily, but if not, the points will tend to be gathered in what Poore calls 'nebulae', or, in other words, they will be clustered, and such clusters between which the variation-linkages are close, form a nodum. This model recalls closely the statistical procedure in analysing species-correlations in one community (p. 3474) which also bring out nodal groups among species, as we have mentioned above. It is the expansion of this correlation method to a different scale, the correlations of communities. At present such a model remains imaginative only, but it is possible in practice to pick out by experience of a region and by 'successive approximations' the nodal types of vegetation, which have multilateral variation-linkages with other types. Not infrequently they will be found to correspond with 'types of vegetation' which have been recognized on a descriptive basis and are familiarly utilized in descriptive accounts. The difference lies in the way in which they are regarded, not as discrete entities, but as focal points in a network of variations. How far such networks extend is another question. It should be emphasized that they are ideal constructions, not topographical features, and the existence of a network of relationships between communities is not incompatible with the existence of sharp discontinuities on the ground. It is not postulated that there is a universal continuum of vegetation, each region probably has its own network, dependent on its own floristic character as well as its own environmental peculiarities.

The floristic basis for discriminating communities from one another is essential, but it has its limits. First, communities so separated can only be treated as absolute units by the arbitrary rejection of intermediates, which would appear to be unscientific. Second, as Braun-Blanquet recognizes, floristic communities are only valid within one floristic region, which he calls a *Gesellschaftskreis*. When we look further afield and try to take a world view, we find that there is a strong tendency for structural patterns to recur in different parts of the world, that, in fact, the number of such structural patterns is limited, and that to separate structurally and ecologically similar communities on purely floristic differences would be artificial. On the world scale physiognomic and ecological criteria override floristic considerations, as has been recognized by Schimper, Rübel and others who have attempted worldwide classifications.

This highlights the confrontation of the two ideas of 'community' or phytocoenose and of 'ecosystem' which runs through the whole of Ecology. Despite international differences of opinion, the two ideas are not in opposition but complementary. As the ecosystem is the more comprehensive idea and has the deeper significance, its understanding is bound to be the ultimate aim.

Ecosystems are not, however, immediately recognizable, because of their complexity, and for their recognition the diagnosis of communities and some system of community classification is essential. For this floristic analysis is required, but classification of communities should follow as the next step. Its value is in giving a wide conspectus of the vegetation of a region, which allows the deduction of generalizations, whose accuracy will depend on the accuracy of the classification. Such a survey, even though it is based on qualitative characters, is most valuable in depicting vegetation and showing what is significant and what is accidental, thus giving pointers to where intensive investigation and quantitative analysis will be most fruitful.

Qualitative and descriptive characters are not to be ignored. They are part of the truth about a community. Accuracy may be attained at the expense of truth, if it is too limited, too partial. Reliance wholly on statistical data may result, quite literally, in 'not seeing the wood for the trees'.

CHAPTER XLIV

THE ANALYSIS OF THE PLANT COMMUNITY

WE now relegate to a secondary place, for the time being, questions of habitat relationships, for what we are concerned with is the nature of the plant population, a word which raises in mathematical minds visions of many delightful problems. Accordingly, a great deal of mathematical ingenuity has been devoted to the development of population statistics, which has tended on the one hand to illuminate and on the other hand to obscure ecological situations.

Classificatory procedures are deficient in that their information content is at a low level. They leave untouched the internal conditions of the plant communities and they give no answers to a host of questions which are the main interests of the ecologist. It may be justly argued that classification was never intended to do so, but only to lay a foundation for intensive investigations which must be experimental.

The methods of mathematical analysis, on the other hand, do concern themselves with the internal affairs of the community. They are not, as is often represented, an opposition to classification, which seeks to define the existence of distinct communities, they go a step beyond it in that they are concerned, above all, with the patterns of species in a population and with giving these patterns a mathematical expression by means of which they may be handled and compared. Where this succeeds in revealing patterns hitherto unobserved it does deepen our scientific understanding.

The doubts felt by critics of mathematical analysis arise, however, from their belief that the data employed contain so large an element of probability and of transience* that they are intractable and that the objective precision which is sought may be deceptive, that while mathematically sophisticated they may be biologically naïve. Whether the information they retrieve is of a sufficiently high order to justify the labour involved remains an open question.

The opposition of classification versus ordination is a false problem. Neither has an exclusive right. There are areas where physical causes segregate vegetational communities so markedly that that classification is mandatory but there are other areas, for example, where a particular floristic assemblage is widespread, in which the attempt to classify local differences becomes arbitrary. Such assemblages are best regarded as a continuum and it is in a continuum that ordination and mathematical analysis prevail. Generally the local differentiation in a continuum is of a biotic rather than a physical nature and boundaries are ecoclines rather than ecotones. It should be made clear,

* This is familiar to every biologist who has had intimate knowledge of a stand of natural vegetation over a considerable period and not just while making a set of observations. To present a single still from a ciné film as if it were the whole picture will not do.

3434

however, that the idea of a vegetational continuum does not necessarily involve physical continuity. For example, the *Fagus sylvatica* woodlands of Europe may be regarded as dispersed stands of a widespread continuum.

Anderson has suggested that where natural selection has operated in a 'gross and continual' manner the result is to produce clearly differentiated vegetation units which are susceptible of classification. Where, on the contrary, natural selection has been weak or intermittent at a low intensity, the result will be a continuum which can best be analysed by ordination. This thesis might be modified by saying that where natural selection has operated through physical factors the result is differentiated vegetation, but that where natural selection has been chiefly through biotic competition the result is a continuum.

We are not to be understood as decrying the applications of mathematics to ecological problems. Quantification is an essential element in all science. In ecology as in other studies it has yielded important information which would otherwise have remained unknown, but we feel obliged to warn that a purely mathematical approach, unless in firmly critical hands, can lead to false simplifications and distorted views. The universe is not wholly describable in quantitative terms. It is replete with qualitative characteristics which are of fundamental importance but for which there is no numerical expression, and these characteristics must be kept clearly in the foreground of any comprehensive mental picture of natural vegetation. We should always bear in mind that vegetation, like a kaleidoscope, offers a constantly changing series of patterns, even if the changes are sometimes slow. The universal cycle of birth, death and reproduction makes this inevitable. Homogeneity and stability are myths, not to be found when sought for. Pattern and change of pattern are inseparable and practically universal. One thing alone picks out a pattern from among abstractions and gives it ecological validity and that is the possibility of assigning to it an environmental cause, using environment in the widest sense. That this cannot often be done does not detract from its importance as an objective. It is the necessary link between analysis and synthesis.*

Before we can subject a population to analysis we must first know what the population consists of. This means that the complete floristic list (*relevé*) is the foundation of everything else, it is the signature of the community. It has also been used, as we have already seen, as the basis of systems of com-

---

* Teilhard de Chardin has drawn attention to the specific energies inherent in arrangement. The obvious illustration is that of mechanical devices in which the energy developed depends on the arrangement of the parts; but the biological application is more valuable. In living organisms the production of energy depends on arrangement at all levels from the arrangement of nucleotides in the RNA molecule upwards in scale to the arrangement of the organs upon the body of animal or plant.

It does not stop there, however. The energy developed by a community depends on its social arrangements. Only if the units are most harmoniously arranged (Driesch's 'harmonious constellation') will the collective energy of the community be maximal.

If this be true of animal communities, why not also of plant communities? Is not this perhaps the significance of 'pattern' in the plant community, that, as between one pattern and another, between one system of species-linkages and another, there are differences in collective energy production, slight perhaps but possibly of decisive importance? A line of inquiry suggests itself.

munity classification widely adopted in Europe. For classification, quantitative analysis is of little use, for it deals with intrinsic characteristics of a community which are never exactly replicated in other communities; but a floristic list may be replicated, which gives rise to a *prima facie* assumption that where this occurs the two communities concerned may be grouped together as parts of a wider whole.

This reminds us of the query raised by Rübel (p. 3418) as to how different two lists must be before they should be separated as representing different communities (see p. 3455). In that form the query may be called naïve, because, of course, mere lists as such are not enough on which to base a judgement. As Jaccard early pointed out, lists merely record the presence of certain species and the same species may have an entirely different importance or weight in each community where it is present. To form judgements we must go much further and apply a qualitative or semi-quantitative analysis to the various elements in our floristic lists, to ascertain what kind of pattern they make together and whether any two patterns are comparable.

This is the basis of the Zürich–Montpellier system, to which we have made previous reference, associated with Professor Braun-Blanquet of Montpellier who has been its chief exponent. As we have already pointed out it is far from being strictly objective and rests on the subjective choice of areas and of criteria, which depends ultimately on the experience and judgement of the worker for its success. As Ashby has forcibly said, when the ecologist stops his car and decides that here is a suitable place for study, he has already formed a subjective estimate of the status and comparative homogeneity of the community before his eyes. In other words he has already adopted *a priori* certain important assumptions, only from the guidance of his experience, which is individual and incommunicable. The Braun-Blanquet system stands therefore in contrast to the methods of statistical analysis, which seek to remove these subjective elements and to base our appreciation of patterns on strictly objective criteria from which we may, if possible, draw unbiased conclusions. That this is the more scientific method is unquestionable. The difficulty of applying it lies in those two words 'if possible', for the ecological meaning of much statistical data is unfortunately very obscure and we may be misled, as we have said above, into making false simplifications.

Although the Braun-Blanquet system depends importantly on experience, it has several advantages. It is fairly rapidly applied and demands little in the way of apparatus. It is therefore very suitable for travellers in unfrequented places. Granting the need for experience in making initial choices and in drawing final conclusions, the intermediate steps are definite and can be carried out with comparatively little training. Also it provides the kind of data needed for comparing and classifying communities. In the hands of a few masters therefore it has enabled the systematization of our knowledge of the vegetation of Europe and some other extensive areas in a way that could not have been done otherwise.

While the associations which have been recognized and described by the Montpellier system are undoubtedly real and important entities, the subjective element involved in the choice or rejection of sites leads to an over-simplified picture, since sites rejected as non-homogeneous are no small proportion of the whole. The accepted associations may be regarded as 'noda' in Poore's sense (see p. 3432) connected multilaterally with other noda by chains of intermediate variants. This concept gives a truer representation of the facts, which are much less precise and sharply defined than the Montpellier system suggests.

The foundation of operations in the Montpellier system is the *Association Table*, which is an assemblage of floristic lists tabulated for comparison. In order that these lists may be comparable certain subjective decisions of a preliminary kind must be taken. First, the ecologist must decide that the vegetation to be surveyed is probably a unitary plant association. This is by way of a first approximation and it is left for the survey to show whether it is justified or not. Second, he must choose a number of other areas, or stands, which he judges to belong to the same association. These stands are chosen as apparently uniform, on the basis of uniformity in physiognomy, the dispersion of the dominant or most abundant species, uniformity of slope, topographical aspects, soil and habitat, so far as this can be judged by observation. The progress of the survey may reveal whether some of these stands do not fulfil the necessary conditions, when they may be rejected or assigned to a different association.

As complete a list as possible of the species in the population is then written vertically down the left-hand side of the table, placing Angiosperms, Gymnosperms, Pteridophytes, Bryophytes and Lichens in that order. Parallel with this list of species are then arranged the analytical *relevés* ('reviews' is perhaps the best translation) from each of the selected stands. In these lists the entries against each species are not simply a mark of presence or absence, which gives no information as to the relative importance of species, but figures are used to denominate a rating according to a predetermined scale. Braun-Blanquet uses two such figures, one representing a combination of abundance and coverage and the other being a mark of the mode of growth or 'sociability' of the species.

> $+$ = sparsely present, insignificant coverage.
> 1 = plentiful but with small coverage.
> 2 = very numerous, coverage at least $\frac{1}{20}$ of the area.
> 3 = coverage between $\frac{1}{4}$ and $\frac{1}{2}$ of the area.
> 4 = coverage between $\frac{1}{2}$ and $\frac{3}{4}$ of the area.
> 5 = coverage more than $\frac{3}{4}$ of the area.

In stratified vegetation each stratum should be reviewed separately and the scale value given should refer to the stratum occupied by the species in question.

Although a visual estimate of coverage is liable to considerable personal error, the above groups are sufficiently wide to enable a visual classification to be made with reasonable accuracy.

The sociability scale is as follows:

Soc. 1. Isolated individuals, growing singly.
Soc. 2. Small groups or tufts.
Soc. 3. Small patches or cushions.
Soc. 4. Extensive patches or carpets.
Soc. 5. Pure populations of one species.

An indication of the density of growth is added by writing a line under the sociability figure (dense growth) or a dotted line (open growth).

In making these estimates an effort is made to reduce the limits of personal bias by using a series of contiguous quadrats so arranged that each quadrat in turn doubles the combined area of the previous quadrats (see p. 3462). This nest of quadrats is used for the compilation of the species list and the abundance and sociability indices. It is also used by Braun-Blanquet for estimating the 'minimal area' of the association, a concept to which Braun-Blanquet attaches considerable importance. We have already discussed this to some extent (p. 3410) and will have more to say about it below.

A modification of the Braun-Blanquet scale of sociability is that proposed by Domin and often preferred, as it provides more classes.

+ = Isolated, coverage very small.
1 = Scarce, coverage small.
2 = Very scattered, coverage small.
3 = Scattered, coverage small.
4 = Plentiful, coverage about 5 per cent.
5 = Abundant, coverage about 20 per cent.
6 = Coverage 25–33 per cent.
7 = Coverage 33–50 per cent.
8 = Coverage 50–75 per cent.
9 = Coverage more than 75 per cent but less than 100 per cent.
10 = Coverage 100 per cent.

On the whole, considering the variability of subjective estimates, the wider classes of the Braun-Blanquet scale are probably more suitable, though it is possible for an experienced observer to maintain a reasonably consistent level of estimation by either scale.

The lists comprising an association table should not be arranged in a haphazard fashion. They may be arranged in order of decreasing number of species, or according to their geographical position, or the nature of the underlying rock, or according to the degree to which some ecological factor, such as wetness, is developed. Any logical arrangement will be helpful. Some particulars of altitude, aspect and degree of slope should be given for each stand listed. If trees are included their average height and general coverage should also be noted.

With the compiled table as a basis, which may comprise a score or more of reviews from different stands, the answers to certain questions must be formulated. Are all the lists consistent with belonging to one association, or should some of them be separated off as representing a different association? What are the distinguishing characters by which the association or associations may be recognized and defined?

To do this the status of each species must be considered and they must then be arranged in a hierarchy of groups according to their importance for our purposes.

Dominance, in the usual sense of physiognomical importance, is not by itself the most significant character of species in the analysis of a stand, for it is a fluctuating character which may vary between stands of the same association or even change with the seasonal aspect. Moreover, dominant species may well be widely distributed and may occur, with varying importance, in several associations.

Braun-Blanquet himself lays the chief emphasis on two categories of species: first, the characteristic species; second, the differential species.

The characteristic species, which are the most important, are assessed on two distinct qualities, both of which are important. The first is *constancy* (*Stätigkeit*) and the second is *fidelity* (*Treue*), which are denominated by classes in two scales. *Constancy*, as we have previously remarked has two meanings. Among Scandinavian writers it is generally used in the sense of frequency and is assessed in the same way, by the percentage of sample quadrats in which it occurs in any one stand.

This may be called the local as distinguished from the general constancy of the species. The method adopted by du Rietz is to use quadrats of increasing size, usually from 1 m² to 16 m². The percentage frequencies tend to rise, as would be expected with increasing size of quadrat, but as du Rietz insists, and supports with many examples, the number of species in the 90–100 per cent class does not rise continuously according to mathematical expectation, but tends to level out beyond a certain area of quadrat. When this point is reached he designates those few species in the highest frequency group as constants of the association. The number of constants found depends on the richness of the flora. In a rich association it may be under 10 per cent of the total, in a poor flora it may rise to 20 per cent. This may be associated with the fact that a poor flora generally implies a more open community and with the probability that a poor flora implies relatively unfavourable conditions and that its members are therefore plants with considerable vitality and a wide range of tolerances.

From the special frequency or constancy we may proceed to the general constancy by extending the frequency observations to as many different stands of the association as possible, to find by comparison which species are the general constants of the theoretical association, i.e. those which are present in at least 80 per cent of the stands examined. Braun-Blanquet regards these as the most important means of characterizing the association. Apart from them, each other species is assigned its appropriate constancy percentage, which Braun-Blanquet, like Raunkiaer (p. 3445), groups into five classes covering the range of 1–100.

Some confusion has arisen between constancy, as thus defined, and fidelity.

Fidelity

This represents the quality of exclusiveness and is in fact a measure of the tolerance range of a species. It is assessed in five grades, the highest of which applies to species with a narrow range of tolerances which are limited in their occurrence to one association only. This would of course provide an infallible index of the association if such species were invariably present, but apart from being exclusive they are often rare and may sometimes be absent, which diminishes their value as characteristic species.

The fidelity grades used by Braun-Blanquet are as follows:

A. Species which have value as characteristic species.
  (v) *Treu.*   Confined exclusively or almost so to one association.
  (iv) *Fest.*   Species with a clear preference for one association. If they occur elsewhere, it is only exceptionally and sparsely.
  (iii) *Hold.*   Occurring in several associations, but with an optimum of abundance in one.
B. Allied species.
  (ii) *Vag.*   Species without a clear preference for one association.
C. Accidental species (outsiders).
  (i) *Fremd.*   Rare and occasional wanderers from other associations, or sometimes relics of a former successional phase.

Obviously the most valuable characteristic species are those which combine a high degree of constancy with high fidelity, but of the two qualities constancy is the more reliable and generally useful. Poore has also pointed out that to assess fidelity accurately requires extensive knowledge of the whole vegetation of the region, whereas constancy can be assessed from several stands of a single association.

The procedure for assessing constancy by the comparison of quadrat analyses of different stands presupposes fairly large and more or less level stands, which should be maturely developed, as the number of constants tends to increase with the age of the vegetation. There are a number of plant communities where the quadrat method is not applicable, e.g. aquatics, lichen epiphytes, or cliff plants, in which cases constancy can only be estimated subjectively with the aid of comparative lists of small, well-separated areas, estimated by eye, which will afford a useful guide.

In listing species, it is not enough only to use Linnean species. Sub-species, varieties and often races must be taken into account, as such small taxa may have genetic differences which are expressed in habitat and community preferences. Some polymorphic genera, such as *Hieracium, Euphrasia*, etc., contain numerous taxa which have ecological limitations.

Fidelity of the highest grade is rare, as it implies a narrow and disadvantageous range of tolerances. Moreover, it varies with the region. Towards the boundaries of its area of geographical distribution a species will often become more exacting in its requirements and show a higher degree of fidelity, though possibly with a lower constancy than it does near the centre of its area.

Alongside the important characteristic species, Braun-Blanquet also recognizes *differential species*, which are not sociologically restricted in distribution but which have special ecological affinities for particular substrata or microclimates. They may occur in any of the associations of a particular type, but within an association they may often serve to distinguish two floristically similar sub-associations which differ ecologically or geographically.

It is perhaps not unfair to point out that the logic of this procedure is not very clear. The original choice of stands is made on the subjective criteria which are used elsewhere, in Britain particularly, namely, habitat, dominance, prevailing life-forms and floristic composition. These stands are then analysed and the association is redefined on the basis of its characteristic species. This, despite its greater precision, does no more than confirm the original conception of the association, as indeed it is bound to do. It is no substitute for the subjective choice. Logically, of course, the analyses should be applied ubiquitously and the associations be extracted by pure induction from the accumulated data but this is obviously impracticable. The results are just as 'real' as the preliminary choice is 'real'. The analysis may result in a subdivision of the original concept and to this extent it is advantageous, but it should not be idealized as a purely inductive process, which it is not.

We have already spoken of the various characteristics of the dispersion of species in the community which are used in assessing their status in the community, such as abundance, density, frequency, sociability and vitality (see pp. 3437 to 3439), and we shall have more to say about frequency below.

There is, however, another descriptive aspect of the community as a whole which seems to have been overlooked and that is what we shall call *closeness*. Communities are often described as either 'open' or 'closed', but there are grades of closeness which deserve recognition. This is a community character and is quite different from density or coverage, which are specific characters.

We propose the following scale of grades:

0. Plants remote, without any contacts. Open ground over 80 per cent of area.
1. Plants remote but with sub-surface contacts. Over 50 per cent open ground.
2. Plants with occasional branch or foliage contacts. Open ground less than 50 per cent.
3. Plants with general branch or foliage contacts, but rooting centres separated.
4. Plants in close array, contacts universal, very little open ground.
5. Plants in closest possible array, with interpenetrating growth. Ground completely occupied.

In stratified vegetation the estimated degree of closeness should be applied to each stratum separately as it may be very different for a tree layer and for ground vegetation, which is more likely to reach a high grade of closeness than are the trees. If a tree stratum is considered separately the expression 'open ground' may mean 'ground not occupied by plants of that stratum', whereas with low growing or unistratose communities 'open ground' has its plain significance.

Order of closeness is also to some extent an order of maturity. If we regard climax vegetation as that which makes the maximum use of all the resources of the environment, such vegetation will, in mesophytic conditions, tend towards maximum closeness, but in specialized or unfavourable environments even climax vegetation may be unable to reach further than a low grade of closeness.

Associated with closeness and also with estimations of frequency is the concept of the *mean area* of individuals of particular species. This is a theoretical quantity derived from density, i.e. the number of individuals of a species per unit area. The mean area is the reciprocal of the density, or area divided by the number of individuals. Except as related to density it has no ecological significance, but it is an important concept in the determination of frequency.

Density and mean area are not measures of closeness in the sense in which we have used the term, as closeness depends on the physiognomy, growth-form and size of the individuals, which the above characteristics do not tell us.

Frequency

Ever since it was first expounded by Raunkiaer, frequency has been a prominent feature and one very widely used in vegetation analysis, partly because it is relatively easy to estimate and partly because it is basic for the estimation of certain other features.

Frequency is an expression both of density and of pattern in the dispersion of species, two features which it integrates in a handy manner. Pattern is a fundamental feature in vegetation analysis and techniques for its detection are numerous (see p. 3457), but they are all complex and pattern is rarely definable in mathematical terms, so that frequency is still, and will remain, a valuable characteristic. It is, however, because it is estimated by random sampling over a whole stand, not sensitive to small-scale patterning of species, which may often be significant of local ecological conditions. When such small-scale patterning is suspected, it may be useful to estimate *local frequency*, also on a small scale, by using one fairly large quadrat, say 1 m², divided into square decimetres, or in some cases 0·25 m² divided into areas 5 cm by 5 cm, recording the presence or absence of the species in each unit area. The results of several such quadrats taken at random should be averaged as the expression of local frequency.

Another question in frequency estimations is the recognition of the individual plant and the related question of what constitutes 'presence' in a quadrat sample. In most cases the nature of the individual plant will be obvious, but in the cases of mat-forming species or species with vigorous vegetative propagation it is not so simple. Raunkiaer got over the difficulty by using the rule that the presence of any part of a plant carrying a perennating bud within the quadrat counted as presence for that species, even if it was not rooted within the sample area. Whenever possible, frequency should be based upon rooted individuals or rooted parts of individuals. If this cannot be done, resort may be had to 'shoot frequency', either according to the Raunkiaer plan or by counting any part of a plant which falls within the quadrat. This should be separately expressed, as the results will usually be higher than the root frequency.

There are three important questions regarding the practice of frequency estimations. These are (*a*) The appropriate size of quadrat samples, having regard to (i) its effect on the results and (ii) the size of the plants concerned. (*b*) The number of samples required and (iii) the distribution of the samples over the area.

The size of quadrat employed obviously affects the percentage occurrence of the species sampled. The larger the quadrat the greater the number of species that will appear in the high frequency classes. If we imagine the quadrat expanded to include the entire area, then every species would have 100 per cent frequency. Assuming random distribution the relation of the frequency at one quadrat size to that at any other quadrat size is expressed by the equation:

$$F^2 = 1 - (1 - F_1)^{Q2/Q1}$$

where F_1 is the frequency with quadrat size Q_1 and F_2 the frequency at size Q_2.

The effect is shown in Table 4, which is condensed from Curtis and McIntosh.

TABLE 4

Frequency at Quadrat Size x	Calculated Frequency at Other Quadrat Sizes (per cent)					
	$2x$	$5x$	$10x$	$20x$	$50x$	$100x$
1 per cent	1·9	4·9	9·6	18·2	39·5	63·4
3	5·6	14·2	26·3	55·5	78·2	95·3
5	10	23	43·5	64·2	92·4	99·4
10	19	41	65	86·8	99·5	
20	36	67·3	89·3	98·8		
40	64	93	99·4			
60	84	99·9				
80	96					

When tested with an artificial population these figures gave a very close approximation to observed numbers except for the lowest (rarest) frequency class. They show that with an expansion of the quadrat to 100 times its area practically everything in Raunkiaer's lowest (1–20 per cent) frequency class has moved up into the highest (81–100 per cent) class.

Curtis and McIntosh relate these figures to the mean area, i.e. to the density of species, in the generalization that all species will be in the 1–20 per cent frequency class if the quadrat is approximately 0·20 times the mean area of the most numerous (highest density) species and will be in the 81–100 per cent class when the quadrat is 2·0 times the mean area of the least numerous (lowest density) species. The effect of quadrat size on frequency is greater when densities are great than when densities are low.

Density is expressed as the number of individuals of any species occurring per unit area, without regard to their dispersion in that area, which may be very variable for any given density.

The mean area is the reciprocal of this value, namely, the area divided by the number of individuals of a given species that it contains and if their dispersion were regular it would also be a measure of the mean distance between individuals. It is also, plainly, an index of the relative size of individuals within the same frequency class, but it cannot be used as a standard of comparison between different frequency classes, for species of low frequency will all tend to have large mean areas and vice versa (Fig. 4.9).

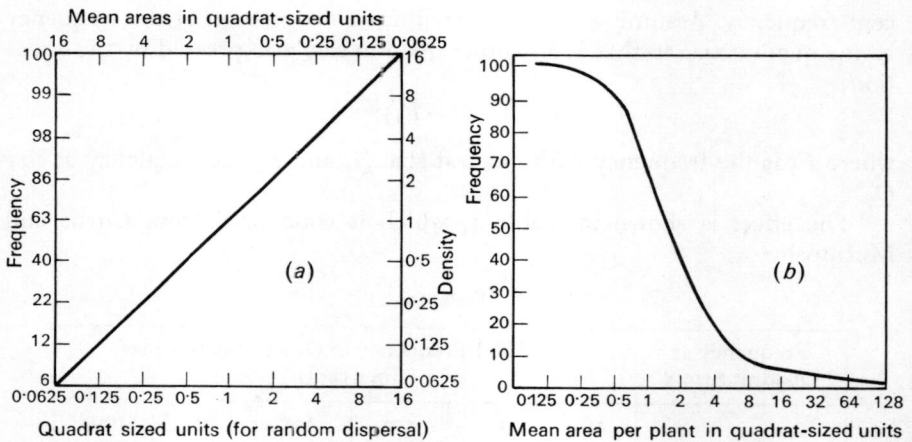

FIG. 4.9. (a) Theoretical interrelations of frequency, density, mean area and quadrat size. (b) The relation between frequency and mean area per plant in quadrat sized units expressed on a log basis. (*After Curtis and McIntosh.*)

As frequency is related to quadrat size on the one hand and to density on the other and as mean area is the reciprocal of density, it is possible theoretically to construct tables or a graph embodying the relationships of all four quantities, but since these relationships only hold good for truly random dispersion of species, which rarely occurs, they are of little practical use. It has been said that frequency data represent not so much the density of a species as its ubiquity, that is the regularity of its dispersion over the area. Non-regular or non-random dispersion can alter the frequency data to a very great degree.

To take an extreme case, Grieg-Smith postulates an area of 5 m^2 divided into 25 quadrats of one square metre. If a given species is evenly distributed, i.e. ubiquitous in the area, it will appear in every quadrat and its frequency will be 100 per cent. If it is aggregated in one spot it may only occur in 1 out of 25 of the quadrats and its frequency will only be 4 per cent, although density and sampling are the same in both cases.

Frequency data being then subject to such uncertainty we may ask, are they any use? Yet they are constantly used by field ecologists. We should therefore inquire whether they have any compensating advantage and what empirical conditions should be applied to minimize the inherent disadvantages.

Raunkiaer proposed a law of frequencies based on a comparison of over 8000 frequencies in northern Europe. He divided the scale of 0–100 per cent into five classes labelled A to E. These classes and the average frequencies in each were as follows:

A. 0–20 per cent   53 per cent of all frequencies
B. 21–40 ,,   ,,   14 ,,   ,,   ,,   ,,   ,,
C. 41–60 ,,   ,,   9 ,,   ,,   ,,   ,,   ,,
D. 61–80 ,,   ,,   8 ,,   ,,   ,,   ,,   ,,
E. 81–100 ,,   ,,   16 ,,   ,,   ,,   ,,   ,,

On this result he based his 'law': $A > B > C \geqq D < E$, which is in accordance with the general field experience that there are relatively few common species compared with uncommon ones. A similar extensive comparison by Kenoyer in North America confirmed this pattern of distribution between classes, but gave a larger class A (69 per cent) and a small class E (9 per cent). This difference may be attributed to the greater number of species in the American flora and to the use of 'rooted frequency' instead of 'shoot frequency' as used by Raunkiaer.

When the figures of class frequencies are plotted in the form of a histogram they yield a diagram in the shape of a reversed J. Raunkiaer considered that this was characteristic of natural plant associations. He further claimed that it could be used as a test of a natural association and that a marked departure from the 'normal' diagram showed that the vegetation was ecologically heterogeneous. As can be understood from what we have already said, this is too sweeping a claim to stand. Indeed, if quadrats of increasing size are used in a sample of vegetation which does yield a Raunkiaer diagram with quadrats of small size, the form of the diagram can change progressively towards uniformity between classes, as we have pointed out above. This is not simply the effect of sampling a larger area, for increasing the number of small quadrats gives a higher proportion of the 'rarer' species in class A, while the use of a smaller number of large quadrats, with the same total area, increases the number of species in class E.

Apart from its size, the shape of the sampling unit has a considerable effect. The standard unit has always been a square (quadrat), but, as we have pointed out, square units are comparatively insensitive to non-random dispersion of species, which can only be improved by using a larger number, not less than 100, of samples. Several authors have examined the use of rectangular samples instead. The comparison of the methods is based upon the calculation of *variance*, which is the average value of the square of the deviations (disregarding the sign of the deviation) of individual observations from the mean of the series.

This is expressed thus:

$$\frac{S(x-\bar{x})}{n-1}$$

where x is a single observation and $\bar{x}$ the mean of the series. S signifies summation and n is the total number of observations. The term $n-1$ is preferable to n, to compensate for the unavoidable fact that we are only using a relatively small number of samples out of an indefinitely large 'population'. It will be seen that the variance is the square of the standard deviation.

Clapham (1932) investigated the variance in the counts of five species over an area of grazed turf on old sand-dune, taking counts both by squares and by strips of the same equivalent areas. The squares measured 4×4 units and the strips 16×1 units. The average variance for strips in both directions was 219·77, while the variance for squares was 400·80, nearly twice as great. Shorter strips, 4×1 units, gave a higher variance than long strips, 348·65, but even this fell considerably below the variance for squares.

Bormann (1953) made a similar investigation with larger units adapted for forestry work. He used squares of 1×1 units and 2×2 units and rectangles of 1×2, 1×4 and 2×4 units corresponding to Clapham's shorter strips. The greatest variance was in the 1×1 unit squares and the least in the 1×4 and 2×4 rectangles reduced to 1 unit values. These were counts of individual plants, i.e. measures of density, not of frequency. For density estimations, apparently, long strips are preferable, but there are two drawbacks. One is the difficulty in randomizing the distribution of long strips; indeed, Bormann advises against this and claims that the long axis of strips should be laid across the (ecological) contours of the area. Another disadvantage is the increased 'edge effect' in rectangular strips, that is the difficulty of deciding in individual cases whether a plant is or is not included in the sample.

The proportion of edge to area enclosed increases as the ratio of the short to the long axes increases and becomes maximal when the short axis is zero, in other words when the sample strip becomes a *line transect*.

The line transect can be very useful in several ways. It is, in effect, a cross-section of a community and is generally adopted as a means of revealing ecoclines, the gradual transitions in vegetation which occur in relation to a factor gradient, or to distinguish units in a mosaic, or to mark the boundaries between communities. The line is laid out along the presumed gradient or transversely to boundaries and the species of plants contacted by the line are recorded in linear succession. If the plants are very close or small it may be desirable to limit the number of recorded contacts by using short intervals, say about 10 cm, and recording only the species met with at each interval point. This procedure is closely related to the pinpoint analysis, especially of grasslands, which we refer to on p. 3469.

Results may be shown as a linear chart of the transect or pictorially by stylized or symbolic representations of the species drawn upon a base line in their order. Such a drawing gives vertical information about the plant growth which is not given by any other analysis and illustrates the relative prevalence of life-forms, relative heights and the degree of spreading, contact and shading effects along the sectional line.

In some circumstances it may be preferable to use a narrow belt instead of a line, that is a narrowly elongated quadrat, if this will bring out the desired particulars more clearly.

The use of line transects for frequency estimations has been advocated on the ground that long lines yield results with less variance than do square plots when species are non-randomly dispersed, as is most generally the case. The method is to divide the line into a number of equal unit lengths, along which the plants intercepted by the line are recorded and the units treated as a series of contiguous samples.

The frequency values express the proportion of samples in which the individual species occur, or, putting it another way, the probability of any given species occurring in a particular sample. This means that the frequency estimates in a number of series of samples taken in the same area will be binomially distributed, provided that the samples are truly random, and this will hold good whatever the dispersion of species in the vegetation. This provides a means of estimating variance and standard error directly from the binomial series $(p+q)^n$, where n is the number of samples taken, p is the probability of the species being present in any given sample (the frequency figure) and $q = 1 - p$. The variance of the series is then npq. Supposing 200 samples have been taken and a species gives a frequency figure of 25 per cent, the variance is $200 \times 0.25 \times 0.75 = 37.5$. The standard error is $\sqrt{37.5} = 6.12$ and the standard error of the 25 per cent figure is 3.06.

There are narrow limits to the use of this procedure. The binomial distribution is not symmetric, and it is not possible to determine the confidence limits of the figures unless n is large (100 or more) and p is near 0.5 (50 per cent), so that it is only applicable to the middle range of frequencies.

In 1948 Preston analysed the theoretical distribution of frequencies in populations as compared with the theoretical 'universe' of which the populations are samples. He employed both the distribution of species in collections of moths caught by a light trap and in quadrat analyses of plant populations. He concluded that the universes from which the samples were drawn had the form of the normal Gaussian probability curve drawn upon a logarithmic base. The samples follow the same form, but the curves are truncated. The logarithmic scale is practical because in some samples the range of numbers is very large. He arranged the frequencies into sequences which he called *octaves*, an octave being an interval of 2 to 1. Group A consists of species only represented once. Group B, 2; group C, 4; group D, 8; group E, 16; group F, 32; and so on. The octaves increase as the series progresses, but species in any one octave were regarded as having the same degree of commonness. A complication arises with regard to species whose numbers fall on an octave boundary; these are credited one-half to the group below. In the case of the first two groups, half of those in group B (individuals) are assigned to group A and one half of the single individual species (group A) to groups below A, which are supposed to represent species too rare to be represented regularly in a sample. If one of these does get included it is therefore over-represented.

Preston showed that if a Gaussian curve is constructed on a base of octaves with the mode taken as representing 100 species, then the sum of all species in the octaves of the graph is 674 2. If these are distributed into Raunkiaer's frequency classes they give the following figures:

Class A	368·0 species
Class B	94·8
Class C	59·3
Class D	47·3
Class E	104·9
Total	674·2

If these figures are reduced to percentages of the total we get the result shown in Table 5.

TABLE 5

Class	Theoretical Values (%)	Raunkiaer's Values (%)
A	54·6	53
B	14·1	14
C	3·8	9
D	7·0	8
E	15·6	16

The agreement is sufficiently close to show that the universe of frequencies represented by Raunkiaer's 8087 samples had a Gaussian probability distribution, and that the changes of frequency and frequency classes noticed as resulting from a change in the size of the quadrat, will also follow the probability distribution.

This work of Preston's has its practical importance in that it gives a theoretical basis for Raunkiaer's law of frequencies, which is shown to be not merely an empirical or accidental result and that we are justified in treating it as a standard in frequency analysis of plant population.

In carrying out an analysis of frequencies there are several practical conditions which must be decided. First, the number of sample units must be as large as possible. Raunkiaer used only 25 units, but statistical analysis of the confidence limits show that with such a small number of sample units the probability spread of the frequency estimates is much too great for reliability. Not till 100 sample units are reached do the tables show confidence limits that in general fall within Raunkiaer classes. This being so then obviously the size of quadrat unit has practical as well as theoretical importance, since the larger the unit the more time will be taken by the analysis in the field.

There has never been a consensus of opinion about the size of units. Individual ecologists have usually had to decide for themselves what is most suitable for the vegetation they are engaged with and its character will direct them to a choice. Hitherto we have spoken as if the vegetation consisted of a single unstratified layer, but this is exceptional. In stratified communities, it is usually necessary to make estimates of the strata separately. For this, units of several sizes may be required, for a unit of a size suitable for the trees of a forest canopy would be impossibly large in dealing with ground level Bryophyta. As the strata are usually distinct synusiae this is quite legitimate.

Cain has suggested the following as appropriate sizes:

Ground layer, mostly cryptogamic: 0·01 or 0·1 m².
Field layer of small herbs: 1·0 or 2·0 m².
Field layer of large herbs and small shrubs: 4·0 m².
Shrub layer and small trees: 16·0 m².
Large trees: 100·0 m².

These may be generally applicable, but no hard and fast rules can be laid down. A universal scale would not be possible.

The shape of the units must also be left to choice, with a preference, as we have shown above, for rectangular units rather than squares. The large size of some of the above units may suggest difficulty, but it must be remembered that in the larger sizes there will be far fewer of the large-size individuals to be recorded.

To make a choice of a suitable size of sampling unit there are two courses open. One is empirical and has been widely used. It is to accept the Raunkiaer law of frequency as a standard and select by trial a size which gives results in accordance with the law. This, as we have seen above, has a theoretical sanction. The sizes suggested by Cain may often be found to fulfil this aim. It has also been urged that a size which ensures that if all the most important species are included in class E the rest of the species will also be appropriately classified. The question is not simply one of the total area sampled by quadrats, for if we sample the same area by means of a larger number of small quadrats and by a smaller number of large quadrats the results are different. A larger number of small quadrats increases the probability of meeting with the uncommon species and Raunkiaer class A is increased, whereas a smaller number of large quadrats of the same total area as before shows an increase in class E and a proportionate drop in class A. Estimates taken on the same area are shown in Table 6.

TABLE 6

Relative Size of Quadrats	Class A	Class E
1	34·5	28·6
2	23·9	40·0
1	43·1	24·9
5	20·2	48·9

The second course which may be taken is to estimate the species/area curve for the particular community. This has already been described in connection with the determination of 'minimal area' (see p. 3410). The object in this case is to relate the total area covered by the sampling series with the minimal area of the community and the ratio of sample unit size to the number of sample units to be taken can then be adjusted so that the total will at least cover the minimal area, in which case few if any of the species present will be left unsampled. If this area proves to be unworkably large an inspection of the series of quadrats used will show which size within practicable limits, gave the most effective coverage of species recorded.

It is highly recommendable that sample units should be distributed over the area in a truly random manner. Random is not the same thing as haphazard. The latter only gives an uncertain degree of randomization. The best method is to mark off two sides of the area at right-angles, into a number of equal lengths which are then numbered. These provide the coordinates of the position of each sample unit, the numbers being taken successively from a table of random numbers to indicate the position of the unit. Alternatively numbers may be drawn from a box containing numbered slips of paper, with due precaution that all numbers are equally represented.

If units are not randomized, they are generally arranged by a pattern, either in contiguous lines or set at equal intervals in a rectangular pattern. Such arrangements are easier to lay out than random samples, but they are not susceptible of statistical treatment, for example, estimates of variance or standard error, which apply to truly random observations.

It must be admitted that average frequencies estimated by random and by regular samples in comparison are remarkably close. The difference between them is sometimes less than the difference between two sets of random samples. The low frequency species however, are, better represented in the random series.

We have already pointed out that a proportional relationship between frequency and density only holds good in the theoretical situation of perfectly random dispersion of a species. Density may then be calculated, as number of individuals per 100 quadrants:

$$D = 100 \, \log_e \left(\frac{100}{100 - F} \right)$$

where F is the percentage frequency.

This being so, the theoretical density so calculated may be used as a measure of non-randomness, i.e. over-dispersion (aggregation), in any species. In a case where the true density of a species can be known by direct observations over a given area and its percentage frequency is measured, the ratio of the calculated to the observed density is a measure of the extent of departure from random dispersion. There are, however, other tests of non-random dispersion which we shall refer to later (p. 3464).

Whatever exception may be taken on theoretical grounds to estimations of frequency, it remains a useful and widely employed datum of the composition of a community and is in any case preferable to subjective estimates of frequency by inspection, which were commonly used in earlier days but which Hope-Simpson has shown to be liable to flagrant errors.

Closely linked to frequency, and affecting the results of its estimation, is the concept of *diversity* in the population. Diversity has been described as the chance that any two individuals picked at random from a population will belong to the same species. Obviously the chance is high when there are few species with many individuals and vice versa. It is thus a measure of the ratio between the number of individuals present and the number of species into which they are divided or, in other words, of the specific richness of the population. If there are few species but many individuals, the frequency will be heavily biased towards the top class E; if there are many species, each with a few individuals, the bias will be towards class A. Field experience shows, however, that species with an intermediate number of individuals are much the most numerous.

The concept of an *index of diversity* is based upon the conclusion by Fisher, Corbet and Williams (1943) that frequency distributions in a population form a logarithmic series, which they show to be in accordance with observation. In the present case the series expresses the distribution of the numbers of species represented in the sample by different numbers of individuals, i.e. 1, 2, 3, etc., individuals. The series can be written thus:

$$n_1, \quad n_1 x/2, \quad n_1 x/3 \quad n_1 x/4, \text{ etc.}$$

The successive terms are the numbers of species with 1, 2, 3, etc., individuals, and n_1 is the number of species with one individual. There is no zero as there is no means of knowing the number of species which have no individuals in the sample. x is a constant, always less than unity (since otherwise negative numbers might result and there are no logarithms of negative numbers). It is a property of the sample and is dependent on its size. In random samples of any size from one population the ratio n_1/x is a constant and Fisher calls this ratio α or the *Index of Diversity*, which is a property of the population as a whole.

The log series can then be rewritten

$$\alpha x_1, \quad \alpha x^2/2, \quad \alpha x^3/3, \quad \alpha x^4/4, \text{ etc.}$$

If we know α and x, we can write the series. The series is convergent and the sum of the series, the total number of species in the population, is given by:

$$S = \alpha \log e(1 - x)$$

If N is the number of individuals in the whole sample, then:

$$\alpha = N(1 - x)/x \quad \text{and} \quad x = N(N + \alpha)$$

According to the value of N/S, that is the average number of individuals per species in the population, x ranges from zero to 0·999999, which is only reached when the average number of individuals per species is over 70,000, a situation only likely to be reached in micro-plankton.

We can, however, evaluate α without reference to x. Following the log series the number of species present in a random sample of N individuals is:

$$S_1 = \alpha \log_e \left(1 + \frac{N}{\alpha} \right)$$

If, however, the sample is sufficiently large the 1 may be neglected and the formula becomes

$$S_1 = \alpha \log_e \left(\frac{N}{\alpha} \right)$$

By taking successive samples, each p times the size of the preceding one then:

$$Sp - S_1 = \alpha \log_e p$$

by plotting these figures with areas or number of quadrats on a log scale as baseline we obtain a straight line the slope of which depends on α, increasing as α increases. Thus, for example, if successive samples (quadrats) are each double the area of the preceding sample the increased number of species included will be $\alpha \log_e 2 = 0.69\alpha$ and if the size of the sample is increased e times (i.e. 2.718) the increased number of species will be equal to α, the index of diversity.*

The straight-line graph obtained thus cannot be extrapolated backwards, for it will be found that it then reaches the base line of zero species at a point short of zero area or in other words there would be an area containing no species. On the other hand, if numbers of individuals are plotted on a log base line against number of species as ordinates, the straight-line graph prolonged downwards will cut the base line at a population of individuals equal to α, the index of diversity. This is because at smaller areas the 1 in the formula for S cannot be neglected but must be introduced as a correction, so that the straight line becomes curved at its lower end.

The same relationship can also be used for calculating the theoretical curve of the increase in the number of species included with increasing size of quadrat samples, i.e. the species-area relationship.

There are, however, other ways of estimating α. Williams has suggested substituting areas for numbers of individuals, the enumeration of which is open to some uncertainty with plants. In a reasonably uniform vegetation it is at least approximately true that equal areas contain an equal number of individuals. Thus, if we take a number of quadrat samples, doubling the area each time and counting the number of species in each sample we get a table relating size of quadrat to number of species, from which we can obtain the average increase of the number of species in doubling the area. The number of species plotted against the logarithm of sample area should give a straight line. If we

---

* This may be carried out by comparing quadrats measuring 1 m² and 165 cm square (1.65 = $\sqrt{2.72}$) respectively. The average difference in the number of species included gives α directly.

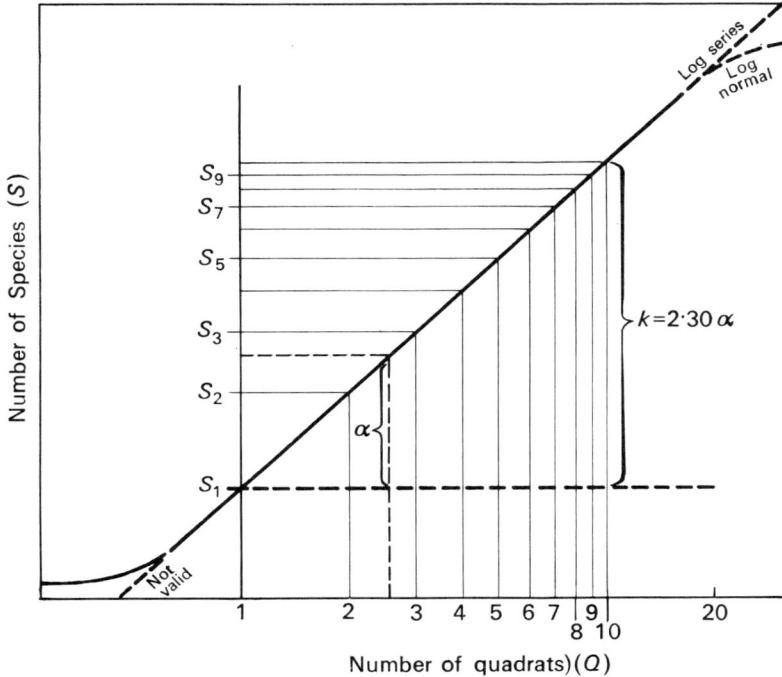

FIG. 4.10. Relationship between the number of species included and the number of quadrats in the range where $S = k \log Q$. The relation of the index of diversity α to the log series is also shown. If $S_1 =$ the number of species in a single quadrat then $S_Q - S_1 = k \log Q$. (*From Williams*, 'Pattern in the Balance of Nature', *Acad. Press*, 1964.)

assume the number of individuals involved to be proportional to the sample areas then:

$$\alpha = \frac{\text{Average increase of species on doubling area}}{\log_e \alpha(0 \cdot 69)}.$$

Williams has suggested another way in which α may be obtained which may also be used as a *measure of similarity* between two populations, which is probably more reliable for large than for small areas, where non-random dispersion of individuals is a vitiating factor. (Compare 'coefficient of community', p. 3455.)

Let A and B be two areas measured in the same units. They include a species in A and b species in B respectively. Let T be the total number of species involved. Then:

$$T - a = \alpha \log_e \left(\frac{A+B}{B} \right) \quad \text{and} \quad T - b = \alpha \log_e \left(\frac{A+B}{B} \right)$$

from which T and α can be calculated. The expected number of species in common between the two populations will then be $a + b - T$. The ratio be-

tween the expected and the observed number of common species can be used
as an index of similarity.

The use of the logarithmic series as a basis of computation is open to certain objections. It assumes random dispersion of individuals and any marked
degree of non-randomness will cause serious discrepancies between calculated and observed figures. Another point is that the log series carries the
implication that n_1 is the largest figure, or in other words that uncommon
species are the most numerous. Calculated figures for n_1 are therefore sometimes considerably in excess of the observed figures.

The alternative view of a log-normal distribution has been argued by
Preston, as we have already remarked (p. 3447). This implies distribution over
a Gaussian normal curve on a log basis, which Preston achieves by grouping
species into octaves, according to the number of individuals of each present in
the sample, each octave doubling the number of individuals, < 1, $1-2$, $2-4$,

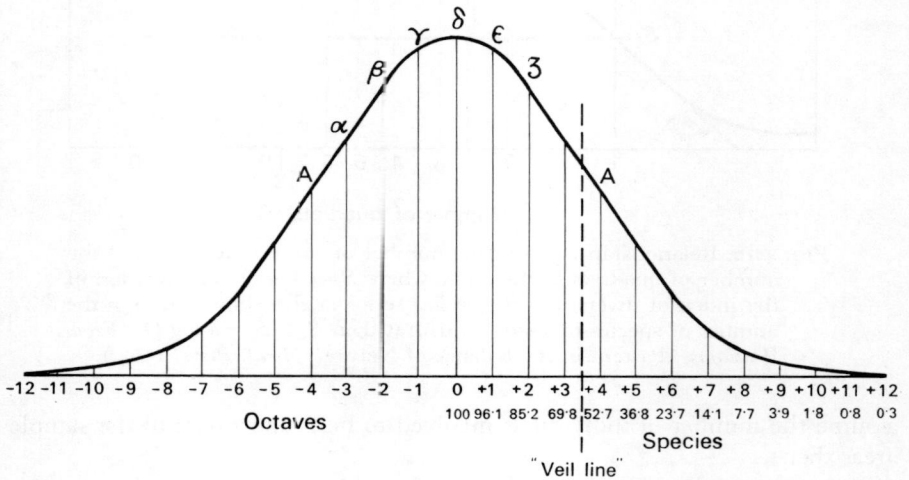

FIG. 4.11. Preston's Gaussian curve of distribution of frequencies on a log-normal
basis. Veil line p. 3455.

4–8, etc. Species with numbers falling on a dividing line between octaves are
divided between two octaves. Such a distribution implies that species with
intermediate numbers of individuals are more numerous than either the uncommon or the very common species. There are cases where the numerical
distributions observed seem to fit this arrangement better than the log series,
but in most cases the expectations derived from both series do not seem to
differ significantly so that a decision between them cannot be made. If the
sample is small relative to the population there will be a maximum of species
in the one individual class, as in the log series. The Gaussian curve is therefore

* This lowest term represents species so uncommon that they may not appear at all in a
given sample, but may appear in larger or in repeated samples. This means that their expected
occurrence in any sample is only fractional, their fractional probability diminishing with their
rarity.

truncated by what Preston calls the 'veil line' and is only completed by the hypothetical octaves below 1 (see footnote, p. 3454) with a diminishing series of fractional probabilities (Fig. 4.11).

Allied to the index of diversity is the *coefficient of community*, the name given by Jaccard (1908) to an index value of similarity between two populations. Jaccard was looking for a means of objective judgement in comparing any two stands of vegetation with one another.* His proposed coefficient was expressed thus:

$$\frac{\text{Number of species common to the two samples}}{\text{Total number of species in the two samples}} \times 100$$

If the two populations are identical the coefficient reaches 100 per cent.

Greig-Smith calls this idea 'naïve', and it is true that the situation is not quite so simple as it sounds. The number of species included in a sample depends on the area of the sample, so that in comparing two stands the same area must be sampled in each. Furthermore this area must be as large as possible, otherwise rare species may be included in one sample but missed in the other. The sample area must be large enough to include all the rising limbs of the species-area curve (p. 3453), that is, it must be at least as large as the minimal area of the community. Even so it will be affected by the pattern of dispersion of species in the two populations and by their index of diversity, since one population may be a more intimate mixture of species than the other. In fact, if the two samples were to be taken from one and the same population, the coefficient would generally fall below 100 per cent. It might be possible to enumerate all the species in each population, regardless of area, and then reduce them to the same area, which would be permissible only if the two numbers lie on the straight-line portion of a log series distribution.

Alternatively, we may abandon the use of the total of species and use only those of most sociological importance, as for example, the constants, in the du Rietz sense (p. 3410) or those with a high DFD index (p. 3415). It will be appreciated that the calculation of a reliable coefficient may well involve much more than at first appears. (See p. 3452 for Williams's method based on α.)

Besides Jaccard's, several other proposals for a coefficient of community or similarity index have been put forward, notably by Gleason (1920) and Sörensen (1948). The latter is very similar to Jaccard's scheme and is open to similar objections. Gleason criticized the omission of any consideration of frequency or abundance, pointing out that two populations might consist of the same species, but in such different relative abundance and relationships that the two populations were really distinct communities. He proposed therefore that the species should be weighted with an abundance figure. This amounts to a selection of species such as we have suggested above. Gleason's scheme was used by Bray and Curtis in 1957 to obtain similarity values between different stands of hardwood trees, as the basis for their 'ordination'

---

* This would be the answer to Rübel's question, see p. 3418.

(see p. 3480), that is the arrangement of vegetation units in a dimensional order rather than into classes. Gleason's coefficient is given by the expression $C = 2w/(a+b)$.

This was interpreted quantitatively in a somewhat arbitrary manner. Twenty-six species were selected, twelve of the most important trees and fourteen shrubs and herbs. For the trees the data were: (a) Absolute density per acre. (b) Absolute basal area per acre (i.e. in square inches at breast height). (c) For shrubs and herbs, simple frequencies. These figures are transformed into a relative (percentage) basis for comparison. Thus:

$$\frac{\text{Relative}}{\text{density}} = \frac{\text{Number of individuals of a species}}{\text{Number of individuals of all species}} \times 100.$$

$$\frac{\text{Relative}}{\text{basal area}} = \frac{\text{Basal area of the species as}}{\text{percentage of total area.}}$$

$$\frac{\text{Relative}}{\text{frequency}} = \frac{\text{Frequency of the species}}{\text{Sum of the frequencies of all species}} \times 100.$$

Then in applying the Gleason formula:

a = sum of the quantitative characters for all the species in one stand;
b = similar sum for the second stand;
w = sum of the lesser of the two values found for species common to both stands.

For two identical stands $C = 1$, for two stands which are completely dissimilar $C = 0$. In this way an attempt has been made to evaluate the importance of species in making a comparison of two populations instead of using undifferentiated lists of species, and it is certainly an advance on previous efforts to evaluate similarity. The utilization of the Gleason coefficient thus expanded, in the technique of the ordination of stands will be dealt with later on (see p. 3481).

We have already (p. 3354) described the Raunkiaer system of classification of life-forms. The 'spectra' based on species lists alone are obviously an incomplete picture of the constitution of a community, since rare and common species all count alike. The species-based classification is valuable on a large scale, for example, in the correlation of whole floras with climate, but if we wish to apply the system to a single community or in the comparison of two communities, the species must be weighted in some way to give them their relative importance. Raunkiaer himself used frequency to weight the species, totalling the frequencies of the species in each life-form class and then expressing the results as percentages of the total frequencies for all species. The result, in the examples given by Raunkiaer, show considerable fluctuation. In some cases the percentages of a class is hardly changed and the order of classes in the spectrum is unchanged, implying that members of a life-form class tend to have the same frequencies. In other spectra there was a sharp up-grading of one class with a corresponding down-grading of others.

Frequency is not the only, or perhaps the best criterion to use for quantitative weighting, as it is very sensitive to the size of unit sample. Coverage might be better as it is independent of sample areas and can be directly measured. Coverage estimated from quadrats or from the intercepts of species on a line transect give closely comparable results. Life-form spectra calculated on a percentage-cover basis can give a very different picture from those on a species basis; the percentage of some life-form classes being more than doubled, emphasizing their weight in the constitution of the community. More work is needed on the use of weighted life-form spectra of various types in the internal analysis of communities, since existing examples are few.

Pattern

The word 'pattern' is applied to any dispersion of species which is non-random. It may be on any scale of dimension, from the pattern of vegetation over the surface of the globe as a whole, down to the distribution of individual plants within a single quadrat. From the point of view of sociological analysis the most significant patterns are those of the dispersion of species in a stand of some particular community, or, occasionally, in the community as a whole. The simplest non-random dispersion would be a regular arrangement, with equidistant individuals all over the area, but such an arrangement is quite unnatural. It has been said that to find four trees in a straight line is strongly indicative of human action and if on measurement they prove to be equidistant the conclusion is certain.

Ashby has claimed that the scale of the social pattern is related to the stability of the community, since it usually means that fewer sensitive species are present and hence a greater change of environment is required to change the pattern. A prolonged period of reaction by the plants tends to equalize conditions over an area and hence to increase the scale of pattern of the established species. The distribution pattern of invaders, which are likely to be nearer to their tolerance limits, will tend to be on a small scale, though this may enlarge with time and selection.

Patterned, i.e. non-random, dispersion may be due to a multiplicity of causes. Some are due to obvious physical factors, such as the zonation by depth of water round the edges of a pond, but even where we, perhaps rashly, conclude that the causative factor is obvious, we may have no knowledge at all of its intimate effects on different species, that is to say of how the factor works. In other cases, where, for example, physical or chemical gradients are operative, the causation may be quite hidden and require extensive research before we get even an inkling of its effects. Intensive studies of the ecological physiology of a species are all too few and a wide extension of knowledge in this direction is badly needed. Such studies should not be regarded as over-specialized, since every addition to our scanty knowledge helps to make the general pattern clearer and so build up that 'theory of vegetation' which the practical needs of mankind require.

Biotic factors, too, are often decisive. Attacks of predators or of parasites often follow a spatial pattern, the origin of which lies in the ecology of the predator or the parasite respectively and may, or may not, be unconnected with anything in the ecology of the plant victim. There are also the complex and obscure interactions of the plants themselves, which we describe as competition and which often lead to multiplex patterns affecting not one species only, but whole groups of correlated species, which may constitute the 'basic units' that Hopkins has postulated (p. 3411). The analysis of such groupings is one of the latest refinements of sociological analysis, which we shall comment upon later (p. 3472). Groupings of this kind may arise, of course, from a similarity of response in a number of species to the same factor in the environment as well as from competition.

Factors seldom operate independently and peculiarities of dispersion may arise from the complex interaction of two or more factors varying in such a way that only at certain places do they achieve a balance which is favourable to one or more species. Such a situation may be very difficult to unravel. Conversely, a factor complex may act negatively by excluding a species from all except certain areas, which may not be, in themselves, especially favourable but are the only ones left available. Again, the action of a causative factor may not be compelling but only permissive; that is so say, that it does not obligate a certain pattern of dispersion, but simply allows of more development in some places than in others. In this way we may often find a random dispersion over parts of an area with notable concentrations in others, or the other way round. In fact the circumstances or combinations affecting the pattern of dispersion in a community are so numerous that cataloguing them may become tedious.

We must not, however, overlook factors of a different kind, those which may be called 'intrinsic' to the species. This covers all those genotypic features which give the plant its specific character. These include not only the physiological peculiarities which govern the plant's response to environmental influences, but also, and very importantly, its reproductive capacities, for example, its capacity for producing and dispersing germinable seeds, or the balance of its reliance on seed reproduction and vegetative reproduction. The latter can be very important in some cases, where the chance arrival of seeds starts centrifugal areas of vegetative spread; the patches may be dispersed randomly (see p. 3463).

Seeding is a factor too often overlooked in questions of dispersion, for obviously before a plant can grow in a particular place a seed must have arrived, germinated and established itself there, and the determining factors in these early, critical stages of existence may have more weight than anything affecting the plant's adult life. Owing to such influences seedlings and young plants may show a degree of aggregation which disappears later. Mature trees, for example, are more nearly random in their dispersion than are young tree seedlings. They represent random survival from an initially much larger aggregated population.

While this is generally true of single, fairly uniform stands, when we come to compare different stands we may find that the survival of trees in certain stands has not been random but is positively correlated with some physical factor such as the depth or water-holding power of the soil.

There is also the question whether non-random dispersion is related to the age or state of maturity of the community or to its position in a successional sere. Little work has been done on dispersion in early successional stages, as they are usually temporary, but Raunkiaer compared life-form/frequency tables of various stages of the dunes at Skagens Odde in Denmark and Pichi-Sermolli (1948) has done the same for stages of succession in the upper Tiber valley. These studies appear to show a general rise in percentage frequencies with increasing maturity and a diminution in the number of sporadic species, which might be interpreted as implying a general drift towards uniformity of dispersion with time. The latter author proposed an *index of maturity*, namely, the total of the percentage frequencies (Raunkiaer's 'frequency points') divided by the number of species involved, the index ranging from 1–100, increasing with maturity. According to this view the most mature community would be one with a small number of uniformly dispersed species all with frequencies in the highest class.

Randomness is generally associated with early occupation of the ground. Weeds on open ground are generally random while their density is low, arising from random arrival of seeds. As density increases groupings arise around seed parents, which may be accentuated by ecological diversities in the habitat, which were inconspicuous when the plants were thinly dispersed. An inequality which was hidden when only one plant was concerned becomes the basis of a pattern when a score are involved. We thus have randomness giving place historically to non-randomness, but if the observations of Raunkiaer and Pichi-Sermolli quoted above are accepted, there is a subsequent decay of non-randomness and an approach to uniformity in mature, stable communities. This is the broadest sort of generalization and it assumes a major degree of uniformity in the habitat. Any marked ecological diversity in the habitat will maintain patterning indefinitely.

This is where the idea of the homogeneous association runs into real difficulties. If the diversity of habitat conditions in a stand of vegetation is such as to maintain a permanent pattern in the dispersion of species, how can we decide whether it is a single association or a mosaic of association fragments? There is no objective test of homogeneity and subjective judgements are quite unreliable (see Poore's noda, p. 3432).

Fortunately such puzzling conditions are not always present, though it should be remembered that they do occur. Patterns of dispersion in a community are of very diverse kinds and may exist on any scale from the largest to the smallest and in the majority of cases they are no threat to that idea of an integrated community which is generally called an association.

Watt (1947) has described a type of pattern of this latter sort which occurs in cyclic phases and is due to the interaction of species with differing modes of growth. He emphasizes the impermanence of individuals and of groups of individuals, and suggests that this is often due to a cyclic change of phases in the growth and development of a chief species, whereby an ever-changing pattern may be created over a large area.

Watt describes several cyclic developments which create pattern, among them a very good example from the grasslands in the East Anglian Breckland. The micro-topography shows a widespread system of hummocks and hollows which are due to the cyclic development and decay of *Festuca ovina*. The hollows are floored with a pavement of flints and chalk stones. Among these the seedlings of *Festuca* establish themselves. As the plants grow they collect mineral soil carried by ants and by the wind. This gradually builds a small hummock around a robust, branching *Festuca* plant. By the time the hummock reaches a height of about four centimetres the main stock of the grass plant has decayed and instead of a single plant we find a large number of small prostrate and less vigorous plants which are the detached and rooted branches of the parent. Among these the fruticose lichens *Cladonia alcicornis* and *C. rangiformis* establish themselves and form a mat beneath which only dead stems and roots of the *Festuca* persist. The lichens are unable to hold the soil together and erosion begins. The lichen mat breaks up and is replaced by crustaceous lichens, *Psora decipiens* and *Biatorina coeruleo-nigricans*, on the eroding surface. Finally, these also break up and erosion exposes the original stony surface, on which *Festuca* again invades.

This cyclic pattern resembles the 'regeneration complex' described by Godwin and Conway in 1939 from the wet peat bog at Tregaron in Cardiganshire. Here there is also a cycle of hummock-building and erosion which forms a mosaic pattern of hummocks and pools over an area of the bog. The cycle starts with the growth of *Sphagnum cuspidatum* floating in open water, which it eventually covers with a carpet. This is joined, especially at the margins, by other Sphagna, in particular *S. pulchrum* and *S. papillosum*. The latter builds up peat hummocks which may rise to 40 cm above the pool bottoms. Along with it occurs *S. medium*, which is the chief peat-builder in many other bogs, though not at Tregaron. As they are built up these hummocks become dry enough at the top to allow the growth of *Calluna*, with some associates, such as *Erica tetralix*, *Trichophorum caespitosum* and *Eriophorum vaginatum*. These tussocks are invaded by *Cladonia sylvatica*. This starts a degeneration phase which results in the erosion of the hummock to water level and comes back to the beginning of the cycle.

The above description of cyclic patterning suggests competition between the flowering plants and lichens, especially *Cladonia*, as initiating the downgrade phase. Competition, however, may only be secondary in some cases, possibly in many. Changes in the vigour or 'performance' of individuals with age can be itself an effective cause and if the species is one with a well-developed system of vegetative spreading it may effect the balance of a whole population. Such changes can only be readily detected in species where the morphology allows the age of the individual or of different parts of the individual to be ascertained.

The phasic nature of such changes of vigour can often be seen in plants which are extending by rhizome growth, such as *Pteridium aquilinum*, which show what is called the 'marginal effect'. At the extending margin itself there is a building zone, in which the vigour of the shoots increases with the distance from the periphery. This passes over into the mature zone where the shoots reach their maximum vigour and this is succeeded by a degenerative zone with comparatively dwarfed and scattered shoots. These changes are not due to the exhaustion of the soil, since applications of fertilizers have little or no effect. They represent a process of senescence in the plant itself. The decay and break up of the rhizomes in the degenerative zone releases side shoots which are physiologically younger than the parent plant and these may then pursue an independent course, building up into new centres of spread and repeating the phasic history.

This is patterning on a large and obvious scale, but the same process can be detected where it is not obvious, if some morphological criterion of performance, such as the length/breadth ratio of leaves, is carefully measured. Several species have been investigated in this way by Kershaw and in each case a curve of performances was found showing a rising phase, a mature phase and, with age, a declining phase.

Competitive ability will be affected by the incidence of these phases, which are continually going on in what appears to be a stable population, so that the declining phase of one species may be associated with a local increase of another species which is in a mature phase. Such changes may be too slight to be readily seen, but they are brought out as realities by careful density measurements. It is only exceptionally that the decline of a species is so drastic as to leave the ground bare. The performance-cycle results only in changes in the balance between species in a permanently closed community and thus in changes of pattern.

Mathon has given the name 'autodynamism' to spontaneous internal changes in a community, associated with statistical stability.

A possible suggestion of the cause of the phasic growth in a perennial plant is the decreasing ratio of photosynthetic to non-photosynthetic (but respiring) tissues which may come to a point where, for the plant as a whole, the rate of photosynthesis is little, if at all, above the compensation point.

A not infrequent cause of patterning is differing environmental requirements by different species. This may often be a differing requirement for water supply. Species with a high water requirement will tend to colonize the lower-lying areas of the surface and vice versa. A very small difference of level may be enough to cause differentiation of dispersion and hence patterning and the underlying cause may not be at all evident to the eye but is only discerned by careful levelling.

Small-scale patterns may often be superimposed on a large-scale pattern. Thus a widespread pattern perhaps due to an ecological factor may have a small-scale pattern or patterns superimposed upon it by the morphology of growth of individual plants, especially such as spread vegetatively, and embodied in the growth pattern of rhizomes, runners, etc. These features may show themselves in the morphology of the individual clump or cluster, in the distribution of clumps due to one season's expansion, or in the pattern of the whole connected spread of the plant as revealed by exposure of the rhizome system.

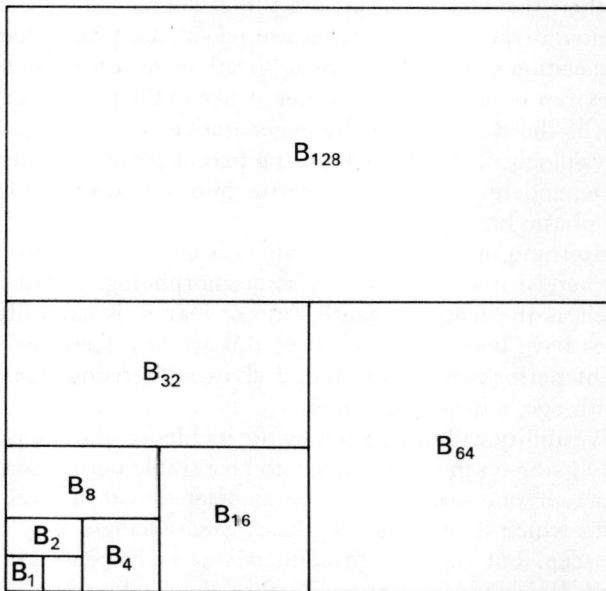

FIG. 4.12. Arrangement of quadrat blocks on a grid (256 units) for analysis of variance. (*After H. R. Thompson.*)

The scale of pattern can be tested by the block method proposed by Greig-Smith. A grid of contiguous quadrats is laid down, each side of the grid measuring a number of quadrat units which is a power of 2. The size of the grid must be chosen to enclose what is presumed to be the largest-scale pattern. Thus, grids may be 2×2 units, 16×16 units or larger. The number of individuals in each quadrat is counted. The variance of the differences between blocks of different sizes or numbers of quadrat units, i.e. 2×2, 4×4, 8×8 units, is then calculated. If dispersion is truly random, the ratio of variance to mean number of individuals will be unity for blocks of any size. If the dispersion is non-random, there are two alternatives. In the case of regular dispersion the variance will fall with increasing size of blocks, but with a non-random pattern or 'patches' of dispersion, the variance will rise to a maximum at that size of block which corresponds to the area of the patches. When there is more than one scale of heterogeneity present there may be more than one peak of variance, corresponding to the different scales of pattern.

We have already mentioned that pattern may be due to intrinsic factors in the growth of a plant or it may be due to variation of environmental factors. It may also be due to the interaction of species and this may be effective in either of the two preceding ways. The spreading of one plant may have a direct effect of exclusion on other species or it may have an indirect effect by the way in which it modifies the immediate environment. The latter is more common. Direct exclusion must not be hastily assumed, since it has been shown in some cases that two species which appeared to be mutually exclusive (i.e. were negatively correlated in their dispersion) were in fact separated by a difference in their environmental preferences.

If random dispersion be considered as statistically 'normal', then dispersion of greater regularity has been called 'under-dispersion' and dispersions which show local aggregation 'over-dispersion'. The last term has often been misunderstood and has largely been replaced by the term 'contagious dispersion', introduced by Polya, which implies some kind of dependence or influence of the occurrence of one plant on others of the same species, tending to produce local aggregations. This is the strict application, but it is generally used in a looser sense to cover any aggregation, even if it is due to some local peculiarity of the environment.

One obvious case of contagious dispersion is that of a clone or family of plants produced by limited seed-dispersal from a parent plant, but certainly it cannot always be explained by reproductive behaviour. If this were so we would expect the same species to show the same kind of aggregation wherever it occurred, but this is not the case. A species may be aggregated in one area and randomly dispersed in another. Similarly, if aggregations arose through seed or spore dispersal, the density of individuals should be related to the distance from the centre of the aggregation, following possibly the law of inverse squares, but there is no evidence that this is so, and if it does occur it is exceptional.

Although individuals may show non-random, contagious dispersion, the groups which they form may be random. It has been suggested that in some cases the random dispersion of groups or clusters may be due to the random arrival of seeds from elsewhere, and that the subsequent development of the groups followed some principle of contagion.

'Over-dispersion' in the sense of 'aggregation' is a purely statistical concept. From the point of view of plant life it obviously means exactly the opposite, namely, 'under-dispersion', since aggregated species are failing to realize their full potential towards random distribution.

We may therefore distinguish four hypothetical types of aggregation:

	GROUPS	INDIVIDUALS IN GROUPS
1.	Random	Random
2.	Random	Non-random
3.	Non-random	Random
4.	Non-random	Non-random

All four types are possible, but we have no information on which to judge the relative frequency of their occurrence.

Non-random aggregation is too often considered as if it applied only to occurrences in clusters, but aggregates may also take the form of drifts of individuals or of zones, which are only contagious in a wide sense, since they are almost always the result of the distribution of one or more environmental factors.

Apparent clumping can sometimes result from the overlap of points randomly distributed on a plane, but the expectation of this occurring is mathematically abstruse and has only been solved for the simplest models. The possibility may be borne in mind, but it has little practical weight.

Random dispersion of plants does not necessarily mean a random dispersion of environmental factors, which is improbable in any case and in most cases is clearly not true. Some factors, such as rainfall, may be constants for a given area, other factors vary from point to point or from time to time. If however the range of their variation is well within the limits of tolerance of a species, then, other things being equal (e.g. interspecific competition), the occurrences of the species may be due to pure chance, that is random. This will hold good even though there may be some predominant factor, such as a lime-free soil, which is determining in regard to the existence of the species in the area. In general we are justified in believing that in most cases non-random dispersion is due to the determining action of some environmental factor either climatic or edaphic or biotic or of several factors together. While the effect of the variations of such factors may fall short of the complete exclusion of a species, it may so affect its vigour as to render it incapable of successful competition at some points but not at others. Its survival at certain points may be due therefore either to a favourable value of the determining factor or to the absence of a particular competitor at that point.

Species do not exist in isolation, however, and more than one species in the population may be similarly affected. They will therefore show a positive correlation in their occurrences, whether the individuals are randomly dispersed or not and a study of interspecific correlations may have great value in indicating predominant factors.

Several statistical tests for contagious dispersion of species in a community have been proposed. The statistical validity of some of them is open to question and in general their value may be gauged by the amount of labour involved compared with the ecological importance of the information gained. The object of these procedures is not mathematical entertainment, but simply to uncover a situation the ecological meaning of which can then be explored.

To enter upon a full discussion of the statistics of dispersion is impossible within the limitations of a general textbook and those who are concerned should consult the excellent books named below,* in which the mathematical possibilities are explored at length. Statistical analysis is a most valuable means to an end, but the end must be to deepen our ecological understanding not just to establish a table of figures.

The general principle involved is to test the assumption that the figures

---

* GREIG-SMITH, P. (1957) *Quantitative Plant Ecology*. Butterworths.
KERSHAW, K. A. (1964) *Quantitative and Dynamic Ecology*. Arnold.
WILLIAMS, C. B. (1964) *Patterns in the Balance of Nature*. Academic Press.
  More ecologically descriptive work:
CAIN, S. A. and CASTRO, G. M. DE O. (1959) *Manual of Vegetation Analysis*. Harper.

observed for frequency or density correspond to some mathematical 'model' that is, in this case, that they have a distribution corresponding to some understood numerical series. The number of such possible series is probably infinite, but only a few are based on principles which could apply to the conditions of living populations, where all numbers of groups (species or samples) and of units (individuals) are positive integers and neither fractions nor negative integers can occur.

The basic series is the *binomial* $(p+q)^k$, where p and q are the chances of two alternatives occurring (e.g. the presence or absence of a species in a particular sample, of which the frequency is an expression) and k is the number of cases involved (e.g. the number of samples).

If $p=q$, i.e. the chances are equal, we get the *normal* or *Gaussian series* expressed by the continuous 'normal curve of variation', but which can also be approximately represented by a histogram with integral intervals. A variant of this is the *log-normal series*, in which the basal ordinates are at geometric intervals, not arithmetic.

Another variation is the case where p is greater than q, that is one alternative is more probable than the other. This constitutes the *Poisson series*, which takes the form:

$$e^{-m}, \quad me^{-m}, \quad m^2/2\,!e^{-m}, \quad m^3/3\,!e^{-m}, \quad m^4/4\,!e^{-m}, \text{ etc.}$$

where $m=$ the mean density of individuals.

This can be given another form. An area A is sampled by quadrats of area a and contains n individuals. The mean density $x=n/A$. Then:

$$e^{-ax}, \quad axe^{-ax}, \quad \frac{(ax)^2}{2\,!}e^{-ax}, \quad \frac{(ax)^3}{3\,!}e^{-ax}, \quad \frac{(ax)^n}{n\,!}e^{-ax},$$

of which the successive terms give the number of quadrats expected to contain 0, 1, 2, 3, etc., individuals in a random dispersion.

The proportion of quadrats expected to contain at least one individual is the frequency and it is given by $1-e^{-ax}$, or in the percentage form $F=100(1-e^{-ax})$. Density per quadrat will then be:

$$m= -\log_e\left(1 - \frac{E}{100}\right)$$

Density is thus related to frequency logarithmically, not linearly.

A further possible distribution is the *logarithmic series*, which may be written:

$$n_1, \quad \frac{n_1 x}{2}, \quad \frac{n_1 x}{3}, \quad \frac{n_1 x}{4}, \text{ etc.}$$

which is also a series of positive integers, the successive terms of which represent the numbers of samples with 1, 2, 3, etc. individuals. This series is defined by n_1, the number of samples containing one individual and the constant, x. The terms form a geometric series with x as constant multiple.

x depends on the size of the total sample of the population and is defined by the ratio N/S, i.e. the total number of individuals recorded divided by the number of unit samples, or otherwise, the average number of individuals per unit sample, which will naturally depend on the size of the sample. (See Index of Diversity, p. 3451.)

If α, the index of diversity, can be found by observation of the increased number of species included by increasing the size of the sample unit by 2·718, (p. 3452) then x can be calculated from, $x = N/(N + \alpha)$. A very exact estimation of x is not necessary in view of the many sources of error in sampling procedure with living populations.

Bray and Curtis in their ordination procedure in Wisconsin forests (see p. 3481) estimated their overall sampling error to be of the order of 10 per cent. In many procedures there is room for some degree of individual judgement, particularly on the inclusion or exclusion of individuals on sample boundary lines. Estimates of coverage by using the method of contacts with pins along a transect showed differences of as much as 22 per cent between different observers. Everyone has some degree of bias and the single observer, ignorant of what his particular bias may be, can only hope to minimize it by replicating his observations as much as possible, though some effect of bias will always remain. Many provisos, many pitfalls, many uncertainties are recognized by the statisticians, but without much guidance about avoiding them, apart from generalities such as 'reasonable' or 'practicable'; for the very good reason that it is impossible to lay down rules which will cover the almost infinite range of possible situations which the observer may face. With only a few general principles of sampling to guide him, he must rely on his 'reasonable' judgement on the treatment of his observations. Unfortunately it is very possible that without expert advice the observer may apply a mathematical treatment which is, in the circumstances, illegitimate and thus come to wrong conclusions.

We cannot help expressing the opinion that without expert and experienced guidance, of which there is little enough available, the more intricate methods of analysis are best avoided and analysis limited to the simpler of the elements we have endeavoured to describe.*

As regards sampling we have only the following generalities to guide us: that samples should usually be distributed randomly; that their total area should at least cover the rising portion of the species-area curve (p. 3410); that their form may be either square or rectangular (p. 3445) with preference to the latter in some cases; that their size should be proportioned to the scale of the vegetation (p. 3449); that the most appropriate size of individual samples

---

* Those who possess mathematical ability are especially liable to become obsessed with the ambition of analysing all natural phenomena mathematically (cf. 'God is the great mathematician') which appears to promise a revolution in the natural sciences. Undoubtedly the mathematical approach (by which we do not mean simply a quantitative approach) has sometimes led to a new knowledge, but it is not to be accepted as the be-all and end-all of research. Just as the domination of logic in the minds of mediaeval schoolmen often led them to mistaken conclusions, so, a passion for devising mathematical models may lead, not to the heart of nature, but into a metaphysical realm of subjective constructions.

may be found to be that equal to the mean area (p. 3444) of the most numerous species.

The consideration of what is involved in the selection of a mathematical model has led us into a considerable digression. We must now return to the application of such models in determining non-random dispersion as an element of pattern.

Several tests have been used. One which has a fairly wide range of applicability depends on the fact that in a Poisson distribution the ratio of the variance to the mean is unity. By observation there will be available a series of figures representing the number of individuals from zero upwards found in a series of quadrats. If the ratio of variance (p. 3447) to the mean in this series is greater than unity, a contagious distribution is indicated, if it is less than unity the distribution shows an approach to regularity.

When the difference from unity is small its significance must be tested by a *t* test.

Thus, if N equals the number of samples and Sx = the total of individuals counted, then Sx/N = the mean and the variance is

$$\frac{S(x)^2 - (Sx)^2/N}{N-1}$$

sometimes also called the mean square. The standard error of the ratio, variance/mean, depends only on the number of samples taken and is expressed by

$$t = \frac{\text{Observed ratio} - \text{expected ratio (i.e. } 1\cdot0)}{\text{Standard error}}.$$

From tables of *t* values* the probability, *p*, of the result being due to chance may be obtained. The smaller the value of *p* the less likelihood is there of the result being accidental. Thus if $p = 0\cdot05$ the chances against an accidental result are 20:1 and this is generally accepted as a good level of probability. If *p* rises as high as $0\cdot5$ the difference in the ratio may be neglected as being possibly due to chance, which would mean in such a case that the distribution of observed figures probably accorded with a Poisson series, i.e. was random. The divisor $N-1$ represents what are called the 'degrees of freedom' of the sample, and it is used instead of N because the mean is calculated from a sample or samples, not from the entire population. The point of entry into the tables of *t* is governed by degrees of freedom.

The test of the ratio of variance to mean has certain limitations. Its results become erratic if the mean is very small and it is not applicable to species with very high relative densities, for in these cases the Poisson distribution is not reliable. Nevertheless, if the dispersion of a species is indicated as contagious under these conditions there is a probability that it is truly so and it may be tried by means of another test.

---

* See FISHER, R. A. and YATES, F. (1951) *Statistical Tables for Biological, Agricultural and Medical Research*. Oliver and Boyd.

If the mean of the observed series is unusually high *Moore's ϕ test* can be applied. This takes into account the first three classes only of frequencies, that is the numbers of quadrats containing 0, 1 and 2 individuals respectively, which are denoted by n_0, n_1 and n_2. Then

$$\phi = \frac{2n_0 n_2}{n_1^2}$$

which for a Poisson distribution is equal to 1. If the value of ϕ is significantly greater than unity a non-random distribution is indicated.

David and Moore have proposed an *index of clumping*, which estimates the degree of departure from random dispersion.

$$\text{Index of clumping} = \frac{\text{Variance of the observed distribution}}{\text{Mean of the observed distribution}} - 1.$$

For a random dispersion the index is zero.

This index also offers the advantage of a ready means of comparing the degree of non-randomness in two sets of samples, e.g. of the same species in two different stands, provided that the same number of quadrats is used in each case. If m_1 and m_2 are the means of the two sets respectively and v_1 and v_2 are their variances, then their ratio

$$r = \frac{1}{2} \log_e \frac{v_1 m_2}{v_2 m_1}$$

The significance of the ratio is tested by the factor $\pm 2 \cdot 5 \sqrt{N-1}$ where N is the number of quadrats in each set. If r lies outside this $+$ or $-$ range, a difference in the index of clumping in the two sets is significant.

One of the simplest tests we have already referred to on p. 3450. This is the *ratio of observed density to the expected density* for a random distribution. The density expected is calculated from:

$$D = 100 \log_e \left(\frac{100}{100 - F} \right)$$

where F is the percentage frequency and D is the number of individuals per 100 quadrats, with a limiting value of 700, which corresponds to a 100 per cent frequency. The ratio of observed density to expected density will give a measure of the degree of non-random dispersion and will be greater than unity for a contagious dispersion, and less than unity for a regular dispersion.

Lastly there is the χ^2 *test of the goodness of fit* between the observed distribution and a Poisson distribution. This is best used where the number of quadrats is large, of the order of 100. The numbers of individuals per quadrat are tabulated against the numbers of quadrats in which they occur, from which the mean number per quadrat can be calculated. This is m. Using this mean the corresponding Poisson series is calculated for the number of quadrats containing 0, 1, 2, 3 and > 3 respectively. It is usual to group together all the numbers above 3, that is, all the groups with a low expectation. This is to keep the expected number above 5 in each class, which is the arbitrary minimum used. Then, e^{-m} is the first term of the series, and

$$me^{-m}, \quad \frac{m^2e^{-m}}{2!}, \quad \frac{m^3e^{-m}}{3!}, \quad \frac{m^4e^{-m}}{4!}$$

give the terms above this. If each term is multiplied by the number of quadrats, say 100, this will give the expected percentage (or number) of quadrats containing 0, 1, 2, 3 and 4 (or > 3) individuals.

$$\chi^2 \text{ is totalled as } \frac{(d_0)^2}{e_0} + \frac{(d_1)^2}{e_1} + \frac{(d_2)^2}{e_2}, \text{ etc.}$$

where d_0 is the difference between the expected and observed numbers of quadrats containing no individuals and e_0 is the expected number, with the other terms in sequence. The χ^2 tables give the probability of the result and they are entered with degrees of freedom which are 2 less than the number of terms in the calculation, i.e. 3 in the above example. The smaller the value of p in the tables, the greater the probability that the differences represent a real departure from random dispersion. If $p > 0.05$ it may be taken that the dispersion does not differ significantly from random.

All these methods of analysis are based upon the assumption that individuals have a single-point location and are countable, but for many species this is not true. Species which spread vegetatively may have distinct rosettes or tillers which are countable as units, but many procumbent herbs, especially those which root adventitiously at many of their nodes (such as *Ranunculus repens*), are not separable into units and if nothing but the entire mat is counted as a unit the mean number per quadrat will be much too low for reliable estimates of non-randomness. Each plant is in fact a non-random unit of pattern and unless the size of quadrat can be practically increased so that it is considerably larger than the scale of the pattern the observations will tend to group themselves into two classes only, namely, quadrats with 0 and quadrats with 1 individual, which would indicate an illusory regularity.

When we consider the coverage of such plants they may represent a very important element of the population and if frequency or density estimates cannot be used then their dispersion must be measured on a coverage basis.

This may be achieved by using a frame of pins with which to enumerate the number of 'hits' upon some portion of the plant concerned. Here, as with quadrats, the dimensions of the sample affect the result. If pins are too close together, the tendency is for two neighbouring pins either both to touch the same plant or both to miss it, which gives a reading biased towards contagion. If they are too far apart, the bias will be towards random dispersion. In short grassland a distance apart of 51 mm has often been used, but this will depend on the scale of the vegetation and a greater distance may be desirable. When prostrate mat-plants are investigated, a square frame with pins at each corner may be used, instead of pins set in a line, the sides of the frame being approximately the average diameter of a single mat.

When the coverage of a single species is being examined the records will take the form of the number of pins in a set which touch the plant in question.

These are then tabulated to show how many sets did not touch the plants at all, i.e. gave zero occupancy, and how many sets recorded touches with 1 pin, with 2 pins, etc. The relative frequency is given by the number of contacts per 100 points, as with rectangular quadrats.

The expectancies form a binomial rather than a Poisson series. The mean coverage is given by the total number of hits recorded divided by the number of sets with one or more hits. It is advisable to pool the numbers for sets with more than 3 hits. In the binomial $(p+q)^k$, $p+q=1$, the mean cover $=q$. Suppose the mean cover to be 12·53 per cent, then $(p+q)=(0·8747+0·1253)$ and $k=$ the number of classes with 1 hit, 2, 3, and >3 respectively (in this case 4). From these figures χ^2 can be calculated, as before, and its probability found from the published tables, which will show the likelihood of the differences between observed and expected figures being significant of a departure from random distribution.

Point analysis with pins has been widely applied in the examination of grassland, either natural grassland or artificial pasture. In the latter case the object may be to follow the fate of the different species included in the seed mixture and gauge their value as herbage. Quadrat analysis of close-growing plants in turf is very time-consuming and liable to error; pinpoint records are much more practicable, and, indeed, the use of a single pin has been recommended. How the samples are arranged does not appear to be subject to a general rule. Often they are arranged in lines, as transects, or they may be randomized, or arranged in groups of, say, 10 samples within quadrat areas. More important is the number of samples taken, which should be as large as possible for the sake of statistical confidence. To reduce the standard error of the mean of the observations to less than 10 per cent for abundant species and less than 5 per cent for those less abundant, a number of points of the order of one thousand is required. This should not be impracticable, as each observation can be made very quickly.

Pasture analysis is generally concerned with the species composition of the population and the relative importance of species, rather than with the dispersion of single species. In these circumstances it is necessary to record all of the species touched by a pin on being lowered to the ground, whether they be touched on leaf, or stem or on the crown of the plant. The coverage of a species is directly proportional to the number of hits recorded.

Vertical probes afford no means of calculating the foliage area per unit volume of space or per unit area of the ground, the leaf/area index. To meet this it has been proposed to add horizontal point quadrats to the vertical point quadrats.

Frequency of vertical contacts $= F_{90}$.

Frequency of horizontal contacts $= F_0$.

Then the mean foliage angle α is given by:

$$\tan \alpha = \frac{\pi}{2}\left(\frac{F_0}{F_{90}}\right)$$

and the foliage density by: $F = F_{90} \sec \alpha$.

Successive layers may be analysed by a series of horizontal probes and the total area of foliage per unit area of ground is the sum of F for all layers.

Pinpoint analyses of coverage agree reasonably well with those made by quadrat, but the estimates of frequency are liable to be biased by the morphology of the plants, broader leaved species being overestimated and plants with erect leaves underestimated. The method is well adapted for discriminating between species on such important particulars as persistence, seasonal changes or the effects of grazing, the last of which may have profound effects on the composition of grassland, both natural and artificial.

Non-random dispersion is, of course, only one element of the wider problems of pattern, but it is a very important element and it challenges ecological investigation. It occurs in very various degrees from conditions where it is obvious to the eye, to those in which it consists of variations of density which it requires painstaking analysis to reveal. Whatever its nature, the recognition of non-random dispersion is only the first step, a pointer towards the ecological problems which require solution. Contagious dispersion has too often been dismissed as due to reproductive peculiarities, as if each aggregation of individuals was necessarily a hereditary clone. While such considerations undoubtedly enter into some forms of aggregation, especially where the centres of density are clearly defined and well separated, with a dispersion among themselves which approaches the random, they clearly do not apply to the more obscure differences which analysis alone reveals. There we must look for the effects of environmental factors, including, not least in importance, the biotic factor of competition.

It is widely regarded as axiomatic that competition is strongest between species which make the same or closely similar demands on their habitat and in this form the proposition has not been disproved. The corollary, however, which has been frequently drawn, that it applies between species belonging to the same genus, is demonstrably wrong. Williams, in examining this assumption statistically, showed clearly that in comparing the flora or fauna of a large area with small areas within the large one, there was no evidence for the exclusion of congeneric species in the small areas. Such an apparatus of analysis is hardly necessary in this case, since every cultivator knows that species within the same genus may differ widely in their requirements for growth and that indeed in some large genera the various species may span the whole gamut of possible conditions from submerged aquatics to desert plants. The limitations of the representation of a genus in one community to a single species or a small number of species is not therefore necessarily the result of intra-generic competition, but is more likely to be due to selection by the habitat conditions from among all the species of the genus which might be available.

In the opposite case, in which a genus as a whole is associated with some special habitat condition, there may be positive selection in favour of congeneric species, on the logical ground that where one species flourishes its related congeners may also flourish. Williams found indications that such a positive selection might occur fairly widely, as would be expected if the species of genera (usually small genera) have related demands on the environment.

One aspect of analysis, which arises out of such considerations, we have not yet touched upon, namely, *correlations* between the dispersions of species in a population, which is also called the degree of association between species. Here we meet a problem fundamental to the whole idea of plant association: Is the supposed association of species in a community measurable? for if it is measurable it must exist and we would have an objective proof that there really is a community of plants.

Basically the investigation of specific associations consists of a survey of an area or areas by means of quadrats in which the presence of all species is recorded. As this is simply a presence or absence record it does not differ from a frequency estimation and it is similarly sensitive to the size of quadrat used (see p. 3443). In some cases the whole apparent trend of association, from negative to positive will depend on the size of quadrat adopted. If the scale of pattern is appreciable to the eye, the quadrat size may be adapted to it. Alternatively, where a certain type of community is of widespread occurrence, whole stands may be used as samples, which will not, of course, show anything about correlations within a stand, which is often ecologically the most important thing to know.

As the labour of calculation is considerable the lists of species present should be pruned of all those which have only low frequencies, say below 5 per cent. The remaining species are compared two by two in all possible pairs. As a species list of only 20 involved 188 pairs, the number should be restricted as far as possible unless a computer is available.

Each pair is entered in a 2×2 contingency table as shown in Table 7.

TABLE 7

Species A

		+	−	
Species B	+	a	b	$a+b$
	−	c	d	$c+d$
		$a+c$	$b+d$	n

a = number of quadrats containing both A and B,
b = with B but not A,
c = with A but not B,
d = both A and B absent,
n = number of quadrats.

First, the expectation of the number of quadrats with A + B is calculated thus:

$$a = \frac{(a+b)(a+c)}{n}$$

the other expectations can be obtained by subtraction from the marginal totals, thus:

$$b(\exp) = (a+b) - \frac{(a+b)(a+c)}{n}$$

The expected and observed numbers are now compared by the χ^2 test:

$$\chi^2 = \frac{(ad-bc)^2 . n}{(a+b)(c+d)(a+c)(b+d)}$$

Using χ^2 tables with one degree of freedom we can find the probability of the difference between the expected and the observed figures being significant. If a (observed) is greater than a (expected), then the sign of association is positive, and vice versa.

This test is based simply on the criterion of presence or absence and is thus, as we have pointed out above, sensitive to differences of quadrat size. For this reason it may sometimes be advisable to replace the criterion of presence in a quadrat by a quantitative measure such as density or coverage and calculate correlation coefficients between the species in pairs. This involves a more laborious calculation, but gives a result which is free from the dubiety about quadrat size. With two sets of values to compare, say, of the respective densities (or the respective coverage or, indeed, the density of one species with coverage of the other if this seems appropriate) of the two species in a pair (represented by x_1 and x_2), the correlation coefficient is calculated from the formula:

$$r = \frac{S[(x_1 - \bar{x}_1)(x_2 - \bar{x}_2)]}{\sqrt{[S(x_1 - \bar{x}_1)^2 . S(x_2 - \bar{x}_2)^2]}}$$

The expression $S(x_1 - \bar{x}_1)$ implies the sum of the differences between each value of x_1 and the mean of the series $\bar{x}_1$.

This may be simplified by substituting for the numerator expression in the above formula the algebraic equivalent:

$$S[(x_1 - \bar{x}_1)(x_2 - \bar{x}_2)] = Sx_1x_2 - \frac{Sx_1x_2}{n}$$

and

$$S(x_1 - \bar{x}_1)^2 = Sx_1^2 - \frac{Sx_1^2}{n},$$

where n is the number of observations in the series. The range of the coefficient is from $+1$, complete positive correlation, to -1, complete negative correlation.

The variance of r is $(1 - r^2)/(n - 2)$ and the standard error is the square root of the variance.

The degree of probability may be obtained from a t test, i.e. the ratio of the correlation coefficient to its standard error. The number of the degrees of freedom for entry into the t tables will be $n - 2$, one degree having been used for the r estimation. Fisher and Yates (see p. 3467) give probability tables for values of r over a range of values of n, which may be referred to directly.

Cole (1949) has emphasized that quantitative assessments of interspecific association must be interpreted in the light of a study of the biotic relationships of the species and of the methods of collecting or recording the species. He points out that great disparity of size, range or abundance between the species concerned may produce a false picture of association since any gathering large enough to include A will necessarily include B also. Interspecific association is only to be regarded as supplementary to other methods of studying biotic relationships and is only applicable where the species are biotically comparable.

Forbes in 1915 said that 'a biological association is comprised of species which are more frequently associated with each other than they are with other species'. From the ambit of such a definition one must obviously exclude cases in which the association is biologically obligate, as in symbiosis or parasitism. What we are looking for in such studies is an indication of the existence of some common requirement or some interaction between species which may demand further direct inquiry. The uncovering of such associations is not an end in itself.

In a review of the various methods of calculating coefficients of association which had been proposed before 1949, Cole recommends the use of the 2×2 contingency table with a formula which will yield a straight-line relationship between $+1$, complete association, through zero, which represents purely chance association, to -1 or complete negative association, i.e. over the same range as the correlation coefficient r. The size of the samples also requires careful consideration, for the method does not apply if either of the species is present or absent in all the samples.

Setting out the observations in tabular form we have, as on p. 3472:

TABLE 8

		Species B		
		Present	Absent	
Species A	Present	a	b	$a+b$
	Absent	c	d	$c+d$
		$a+c$	$b+d$	n

n = total number of samples.

Cole proposes three formulae applicable in different circumstances, thus:
When $ad \geqq bc$

$$C = \frac{ad - bc}{(a+b)(b+c)} \quad \text{with standard error} \quad \sigma \pm \sqrt{\left[\frac{(a+c)(c+d)}{n(a+b)(b+d)} \right]} . \quad (4.1)$$

When $bc > ad$ and $d \geqq a$

$$C = \frac{ad - bc}{(a+b)(a+c)} \quad \text{with standard error} \quad \sigma \pm \sqrt{\left[\frac{(b+d)(c+d)}{n(a+b)(a+c)}\right]} . \; (4.2)$$

When $bc > ad$ and $a > d$

$$C = \frac{ad - bc}{(b+d)(c+d)} \quad \text{with standard error} \quad \sigma \pm \sqrt{\left[\frac{(a+b)(a+c)}{n(b+d)(c+d)}\right]} . \; (4.3)$$

Significance may be tested with the χ^2 test with one degree of freedom, as above, or by utilizing the relationship $\chi^2 = t^2$, where $t = C/\sigma$.

Using the data of χ^2 or of correlation coefficients thus assembled, the species may be arranged diagrammatically according to their associations. As central species those which have the greatest number of positive correlations may be chosen, and their correlated species grouped around them connected by lines of lengths inverse to the degree of correlation, i.e. the greater the degree of correlation the shorter the line. In this way the nuclei of several constellations of species begin to appear, which will also be linked together by correlations between some of their constituent species, but the correlations between species in one constellation are more numerous and closer than those with species in other constellations.*

These assemblages of associated species are of considerable interest. We may find it difficult to draw a line between them and the larger assemblages of species elicited by means of an association table (p. 3437) as forming an association in the classical sense. Their grouping may be due to a direct effect of one species on another, or to the effect of one species on the environment, which influences another,† or due to a collective similarity of response to a controlling factor in the environment. An example of direct effect may be seen when species differ much in size. If plants of species A are big enough to exclude species B and C from the area of a quadrat, then A will show a negative association with B and C and the last two will, so to speak, be squeezed into an apparently positive association with each other. This also illustrates the need for quadrats of a size greater than the mean area of the largest (or most covering) species concerned. Smaller quadrats will produce false negative associations simply by reason of spacial exclusion. Apparent negative associations may also arise from rarity, species with too few occurrences to make their correlation legitimate. All species with 5 per cent or less of occurrence can therefore be omitted.

An element of doubt remains when a large number of comparisons are involved, namely, that a small percentage of apparent associations may be the result of pure chance. Too much importance should not therefore be attached to single estimates if their probability approaches the 0·05 limit (5 per cent). Species with only one correlation at this level may be omitted.

---

* The ordination technique described on pp. 3455 to 3456 might be used here, but does not seem to have been tried.

† Martin Jones (1933) attributed the positive association of *Juncus effusus* with high densities of grasses or *Trifolium*, to the better germination of *Juncus* seeds in conditions of shade and high humidity.

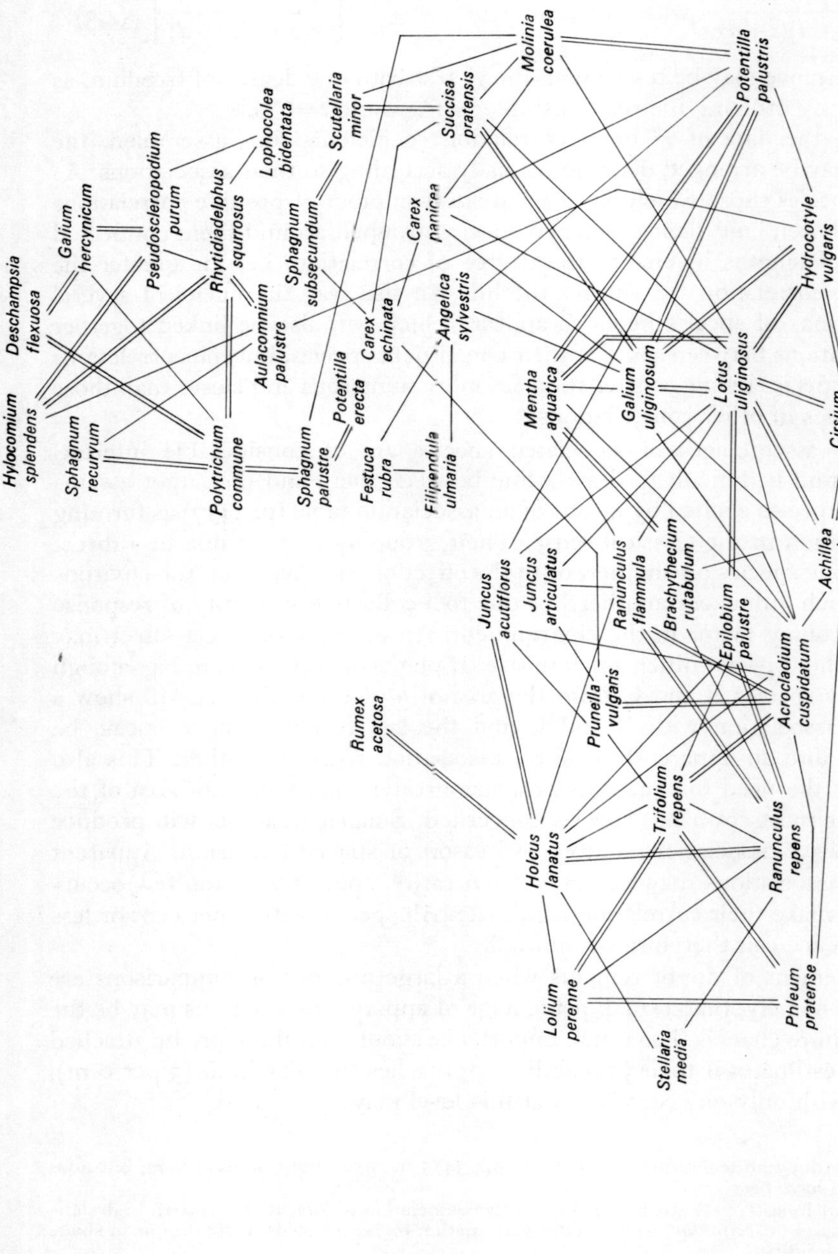

FIG. 4.13. Diagram of species relationships, showing the positive correlations between species found in ninety-nine samples of communities in which *Juncus effusus* occurred. Single lines represent *p*-values of 1–5 per cent, double lines show *p*-values of 1 per cent or less. (*After Agnew.*)

When an association between two prominent species is alone in question the relationship can be illustrated by a 'scatter' diagram. This consists in plotting the density or coverage or abundance of one species against the similar value for the other species in all the quadrats examined. If correlation were complete the points would fall along a straight line, but generally they form a scattered group. If the tendency of the group is upwards it indicates positive association and if downwards a negative association. An irregular group of points with no apparent trend usually indicates absence of association. Many species appear to be able to co-exist thus independently even in small areas.

The use of the χ^2 test to prove the significance of the relationship between expected and observed occurrences is a valuable tool of analysis, and it can be applied to extract information from a series of data, either of association between species or between species and the values of an environmental factor. It has also great value as a means of giving quantitative precision to indications given by qualitative inspection. Frequently, the large-scale groupings in the vegetation are obvious and do not need any analytical procedure to bring them out, but an association analysis brings to the surface groupings of associated species which are not at all obvious, but which may be ecologically significant. It may be permissible to see in these groupings the counterparts of the noda which Poore has postulated as elements of vegetation (p. 3432).

Grasslands, whether natural or artificial, present special difficulties with any method of quadrat sampling because of their high densities and the growth form of the dominant grasses. Analysis of grasslands can, however, be of considerable practical importance in assessing the effects of climate and soil or of grazing or manurial treatments on their composition. They have usually been analysed by pinpoint sampling, as we have mentioned before (p. 3469), abundance and coverage being measured by the number of 'hits' made on plants of each species by sets of long pins, or better, with single pins, which need fewer samples, lowered to the ground through the grass cover. Since total coverage varies widely with the season, Coupland has used percentage comparisons between species, that is the proportion of the total existing cover attributable to each species. Sampling is simplified by this method, as, instead of a very large number of points, only enough need to be taken to assess the relative coverage of each species as a percentage of the whole, which is enough to ascertain the state of composition of the grassland.

Herbaceous dicotyledons and other casuals in the grassland cannot be thus estimated because of their relative scarcity and the small number of hits recorded for them. They must be separately examined by quadrats, if necessary.

Another valuable method with grasslands is that of measuring yields. Small, equal, random areas are marked out and clipped clear. The clippings are then sorted into species and the crop weighed. This may be done at various dates to assess the relative rates of growth, using the last of the series to give the total for the growing season, if there is one.

If an estimate of coverage has been made with pins, the weights per unit area can be compared by expressing them in terms of percentage coverage, that is, so much weight for each 1 per cent of coverage. A series of random samples should be taken on each occasion of sampling and variance and mean calculated and tested, on the assumption that all samples are being drawn from the same normal population.

This assumption can be tested when two series of samples taken from different localities are available. Their respective means are calculated and the variance of each, V.

$$V = \frac{1}{n} \cdot \frac{S(x - \bar{x})^2}{n - 1},$$

the sum of the differences of each observation in the series from the mean of the series, divided by the number of samples in the series, n, less 1, and the variance of the difference between the two means is:

$$\frac{1}{n} \cdot \frac{S(x_1 - \bar{x}_1)^2 + S(x_2 - \bar{x}_2)^2}{n - 1},$$

of which the standard error is the square root. The ratio of the difference of the means to its standard error gives t, with $n - 1$ degrees of freedom, if n is the same for both series.

Another special case is that of the trees in a woodland. Their distribution can be analysed by large quadrats, but this can be very laborious in a large wood. A quicker method is by measurements, from random points. Line transects are laid out across the wood and random sampling points established along each line. At each point a line at right-angles across the line of the transect is laid down (a long bamboo could mark this) and the distance to each of the nearest trees in each quadrant of the cross is measured. The species and basal area of each tree measured is also recorded. Basal area is rather an arbitrary quantity, but it is generally found from the diameter of the trunk at a fixed height above the ground. This is often quoted as *dbh*, or diameter at breast height. The mean of all the distances recorded for trees of each species is equal to the square root of the mean area for the species, from which the density is given by the reciprocal. The densities and basal areas thus found are relative to the total density and basal area of all species, in other words they measure the percentage contribution of each species to the tree population. In this form they are not affected by departures from random dispersion.

Absolute values may be obtained by multiplying by the estimated total density or total basal area, but if this is done the results are affected by non-random dispersion and may be erroneous if the departure from random dispersion is considerable. The relative figures are, in any case, the more informative. The variance and standard error of the series of measures, either for each species or for total trees may be calculated in the usual way (p. 3447). If the four measures at each point are treated as unit samples the association between species may also be tested by 2×2 contingency tables and a χ^2 test of significance (p. 3472).

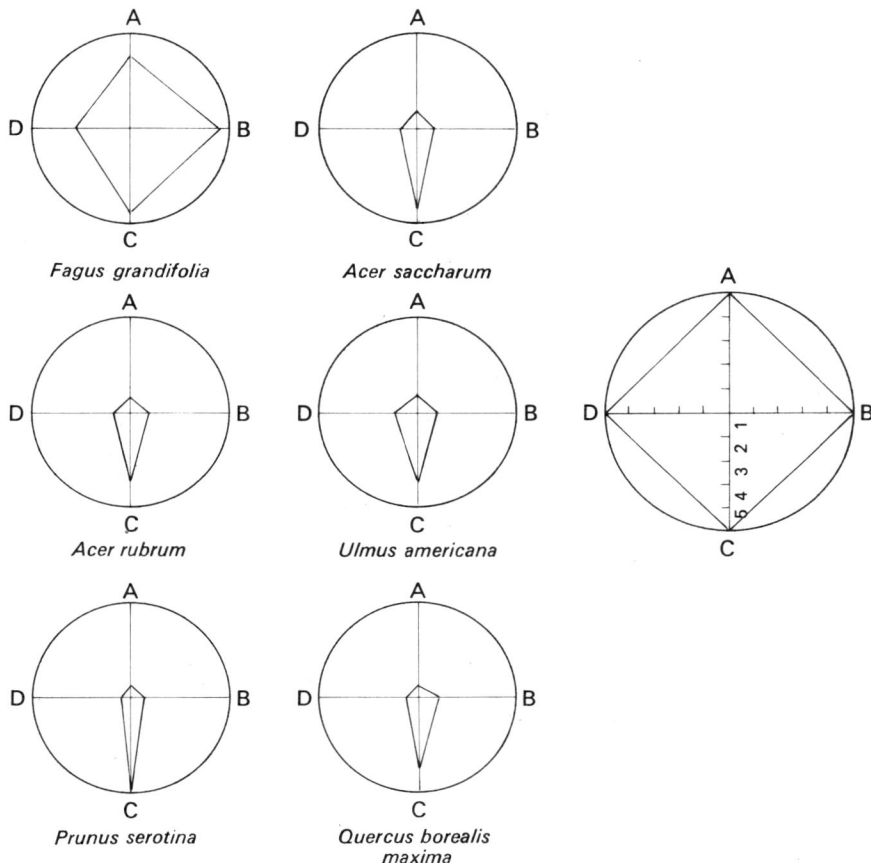

Fig. 4.14. Lutz phytographs of trees in virgin hardwood forest in Michigan. The four axes represent: A. Percentage of total density of trees attributed to the species. All trees of 25 cm or more diameter at breast height. B. Percentage frequency of the species. C. Size classes: 1. Regenerative units up to 30 cm. 2. Ditto 30 cm to 4 m. 3. Saplings 2·5 to 7·5 cm. 4. Poles 10 to 23 cm. 5. Trees above 25 cm dbh. D. Dominance as measured by percentage of total basal area attributed to the species. (*After Cain, from Cain and Castro*, 'Manual of Vegetation Analysis'.)

Valuable as statistical analysis is with regard to the composition of communities, it does not help us with the larger question of the classification of communities which still rests on very much the same kind of subjective judgements as those employed in the classification of individuals, with successional development as a guiding principle. The attempts which have been made, with varying success, to apply quantitative procedures in classification (p. 3437 *et seq.*) with the view of minimizing the subjective element, produce, it is true, well-defined classes, but at the expense of rejecting intermediates, in other words, the classes are more well-defined than is the natural vegetation. Although classification is such a well-established procedure in Ecology, the criticism has been more and more heard in recent years that it is based upon arbitrary and subjective limitation of classes, that it is in fact a man-made framework imposed upon nature While there is no doubt some substance in this objection, it is doubtful whether any advocate of classification would claim that any and every form cf vegetation can be comprehended within limited classes. His reply to the above criticism would be that vegetation corresponding to the definitions cf his classes does exist in nature and that a classification of such natural entities is legitimate, even admitting that inter-mediate and aberrant types exist. His position is very much that of the taxonomist who works on the foundation that certain discontinuous units called 'species' exist, the essential characters of which must be described, despite the facts of variation. Ecological classification is likewise founded on the idea that it is possible, by comparing many stands of vegetation, to recognize groups possessing a constellation of characters in common, in which they resemble each other more than any other group. These are his discontinuous 'species' units or associations. Such a concept does not imply identity of stands in the same group any more than a species implies identity of individuals.

The opposite view, which has spread widely since it was introduced by Gleason in 1920 (p. 3406), involves the denial that discontinuities exist, except where there is some marked change of physical conditions. This view is intended to remedy the omission of intermediates from classification. Instead it lays most stress upon them and regards large areas of vegetation as being essentially a *continuum* (p. 3409). To replace classification it is proposed to arrange stands along one or more axes of variation connecting extremes, the relative positions of stands on the axis being determined by statistical analysis. This is the process called *ordination*. It is laborious and has not yet been widely applied, though there have been a few thorough-going essays with the method.

Orloci (1966) has described ordination in the following general terms:

> Ordinations imply a summarization of the information content of a matrix whose elements, distances or angles, define the spatial relationships between ecological entities. These entities may represent either species, stands, environmental factors or habitats. The entities to be ordinated, called individuals, are visualized as points in space with their attribute scores as co-ordinates. Summarization is achieved by projection of the points into a space which has lesser dimensions than the original. The simplified geometric model so obtained is interpreted in terms of the major ecological gradients and the clusters of the individual points.

If stands are taken as individuals then the species are their attributes and vice versa.

This approach is radically different from that of classification. Like the latter it recognizes that natural entities occur in vegetation, but maintains that they are not discontinuous and therefore are not susceptible of being divided into classes. Poore (p. 3432) amended this extreme view by calling such entities 'noda' and regarding them as foci in the several axes of variation.

As usual, when two antithetic views are urged, there are those who hold that neither view can be universally applied and that circumstances may dictate where one or the other is preferable. They believe that classification may be a valid approach in some cases, but that in others there is undoubtedly a continuum.

Anderson (1965) is one of the latest to give expression to these views. He relates the difference in circumstances to the action of selection during the successional development of the vegetation. If, during development, the action of natural selection has been intense, development may result in distinct and classifiable communities. If, on the other hand, the influence of natural selection has been at a low level, then 'probability' becomes the ruling factor in development and results in a continuum.

The fundamental technique is that used by Bray and Curtis (1957) in the ordination of upland forest communities in southern Wisconsin. Fifty-nine stands of hardwood forest were compared, each with each, on the basis of their *index of similarity*, the index being derived from Gleason's formula $C = 2w/(a+b)$ (see p. 3456) expressed in the terms used by Motyka, where each value is expressed relatively with a sum = 100.

a = the sum of the various quantitative measures for the plants of one stand;

b = the similar sum for the second stand;

w = the sum of the lesser of the two sets of values for species common to both stands.

The index ranges from complete identity = 1·00 to complete dissimilarity = 0·0. This gives a good linear measure of identity.

The quantitative measures used in each comparison (1711 in all) of the stands, were applied to a selection of species, the 12 'most important' trees and 14 herbs and shrubs, as follows:

1. Absolute density per acre.
2. Absolute basal area per acre, in square inches at breast height.
3. Simple frequency of shrubs and herbs.

The trees were assessed by the measurement of random pairs, 40 random points and 80 trees per acre. The herbs were assessed by quadrats of 1 m² taken at alternate points. The inverse of the Gleason index (index subtracted from 100) was used as a measure of the separation of individual stands along a linear axis.

Score sheets were made up of the 38 tests on each stand, that is, 14 frequencies of herbs and shrubs, 12 tree densities and 12 basal areas of trees. As these measures are in different units they were reduced to percentages of the maximum value (called 100), attained by that test in any stand. The individual results for each species on all the score sheets were summed, and the contributions of individual species were expressed as percentages of that sum. These were the 'adjusted' scores used for the comparison of stands on a strictly uniform basis.

The first or x-axis is divided into 100 parts. Replicated tests on the same stand showed a reproducibility of results within 20 index units on the axis. For this reason it was considered safer to assume a maximum index of 80 (complete similarity) instead of 100 units and the indices of similarity were inverted by subtraction from 80 instead of 100. A table or 'matrix' of the adjusted indices for all 59 stands was drawn up and a pair of stands chosen with the maximum separation. One of these was placed at 1 on the axis and the other marked in at its appropriate distance in units. Actually three pairs were found to have the maximum separation of 80 units (i.e. a similarity index of zero). The correlation coefficient between the three members of each set (three low and three

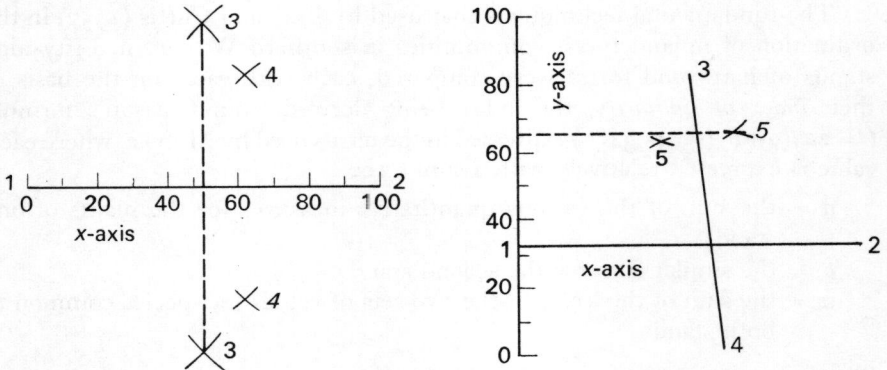

Fig. 4.15. Construction of the ordination diagram showing the positioning of stands on the x- and y-axes respectively. Stands 1 and 2 are the selected distant reference stands on the x-axis. Stand 3 is located by the intersection of arcs with radii of 70 units from centres at 1 and 2. This gives two points which are above and below the axis respectively. The line joining these points gives the position of the projection of stand 3 on the x-axis.

Similarly with the y-axis. Stand 5 is located at one of the two intersections of arcs struck at 20·4 units from 3 and 64·8 units from 4. The distance of 50 units from 2 shows that the righthand intersection is the correct position, from which the stand is projected on to the y-axis. The z-axis is constructed at right-angles to y. (*After Bray and Curtis.*)

high) was high, so they were all used as reference points on the axis. Although their respective indices of similarity were zero, the members of these three pairs had positive indices of similarity with every other stand, so they were not considered to be entirely unrelated to each other.

Subsequent positions on the axis were inserted with reference to the first two stands, by striking arcs, with the appropriate radii, from each reference point and dropping a perpendicular on to the x-axis from their point of intersection. Where, as above, there were multiple reference points, the final x-axis position of a stand was the median of the positions in relation to each of the reference stands separately.

Ordination on one axis is, however, incomplete. Two stands which coincide or lie close together on the x-axis, as fixed from the reference stands, may have a large separation between themselves. They can only be correctly separated by spacing them on a second, y-axis at right angles to the first. Selected as reference points for this axis were the pair of stands which had the highest ratio between their x-axis positions and their separation relative to each other. Let us suppose that stands 3 and 4 are chosen. Arcs are struck for 3 and 4 from the reference points. The intersection of the arcs for 3, above the axis, is then joined to the intersection of the 4-arcs below the axis. This gives the reference line for the y-axis. The position of other similar pairs of stands on the y-axis are fixed from the reference points on this line. Similar discrepancies on the y-axis are referred in the same way to a third or z-axis. The aim is to place every stand at its correct interstand position from every other, but the arrangement is really three-dimensional and can only be approximated in a two-dimensional diagram. A final ordination arrangement is made by calculating the ordination separation of all pairs of stands. If two stands have axis positions of x_1, y_1, z_1, and x_2, y_2, z_2 respectively, then their ordination distance is:

$$\sqrt{[(x_1 - x_2)^2 + (y_1 - y_2)^2 + (z_1 - z_2)^2]}.$$

The correlation of ordination distance with inverted similarity index gives a coefficient of $+0.73$. Since the first was based upon the second, it is surprising that the coefficient was not 1.0 or identity. The difference looks like a measure of the loss of precision involved in the complex indexing and approximations of the ordination process. Some loss of exactitude need not, however, be a fatal objection. Confirmation of the validity of the ordination may be obtained by relating the axes to environmental factors, viewed as a field of interrelated units and events. This involves plotting the changes of physical factors and biotic events over the area of the system. Comparison of the spatial patterns thus obtained may indicate the degree to which factors participate in a mutually determined complex of ordination and environment, from which the causal interactions of any part of the ordination may be deduced.

For example, Bray and Curtis found that the x-axis appeared to represent a history of recovery from major disturbance. Progress along the axis was associated with an increasingly mesic environment and development of canopy. This corresponded with increasing depth of the A_1 layer in the soil, with increasing organic matter in the soil, pH, Calcium, Phosphorus and the ratio $(0.1 \text{ Ca})/\text{K}$. The y-axis was correlated with drainage and water-holding capacity and Ammonium in the soil. The z-axis was correlated with recent disturbance, the filling of gaps and with organic matter, Potassium and $(0.1 \text{ Ca})/\text{K}$ in the soil.

By comparison with the continuum analysis of Curtis and McIntosh in 1951 (see p. 3430), it will be seen how far the idea of ordination had progressed by 1957.

The Bray and Curtis method of ordination within a vegetational continuum is now generally referred to as the 'Wisconsin' method and may be regarded as foundational, but it is open to several objections from a statistical point of view. Nevertheless, the original method has shown its validity in that the ecological interpretation of the stand relationships is in accord with data of environmental gradients.

If the individual stands in the array covering a community are represented as points in space whose co-ordinates are the individual species-scores, then a maximum of $n-1$ axes may be required to account for all the distances between n stands. This requires three-dimensional space and cannot be achieved by axes in one plane. Simplification involves limitation of the number of axes, so positioned in relation to the first axis that each accounts for a maximal portion of the residual variation, after the establishment of the principal axis of maximal variation between stands (see p. 3482).

The principal axis *must* coincide with the direction of maximal variation between stands. Only if this is so is the inevitable distortion in the relative positions of stands due to projection, reduced to a minimum. The Wisconsin method of choosing two extreme stands as the reference points on the principal axis does not necessarily ensure this. Furthermore, the Wisconsin-type axes lie at oblique angles to each other, and if plotted as perpendiculars further distortion is introduced. Efficiency in these respects is increased by the use of a weighted coefficient of similarity between stands instead of the Czekanowski–Gleason coefficient used by Bray and Curtis (see p. 3481). The weighted coefficient proposed by Austin and Orloci (1966) is:

$$\text{Wsc} = \sum_{i=1}^{n} (x_{ij} - \bar{x}_i)(x_{ih} - \bar{x}_i)$$

where j and h represent any two stands containing the species i; x_i the species scores and $\bar{x}_i$ is the mean of the scores for species i in all stands.

The authors claim that the use of this coefficient provides axes which coincide with the axis of maximal variation between stands. A further difficulty arises from the use of $C_{max} - c$ as a measure of geometric distance between stands on the axes, C_{max} being the highest value found in interstand comparisons. A definition of the linear distance D between any two stands j and h is expressed as follows:

$$D_{jh} = \sqrt{\left[\sum_{i=1}^{n} (x_{ij} - x_{ih})^2 \right]}$$

which allows for orthogonal projection of the stands as points in a three-dimensional space with perpendicular axes.

We have presented some of the refinements of the Wisconsin technique as an indication of the development of the subject since 1957. It must, however, be added that the full procedures are laborious and usually require the use of specially programmed computers. They are still experimental.

Gittins has usefully pointed out that the full rigour of the ordination technique is not always necessary, and that where variation is limited and the number of species involved is small, satisfactory ordinations can be obtained with only two axes or even with one.

Analytical techniques, though not as strictly objective as is sometimes claimed, can provide critical standards against which subjective judgements may be assessed.

Reference must be made here to the form of community analysis which has become known as 'hierarchical' grouping. A large number of quadrats are recorded and the species associations estimated by 2×2 contingency tables, as before described. A matrix table is then drawn up showing the χ^2 value of association between all pairs of species. The species with the largest aggregate of χ^2 values is then chosen to divide two quadrat groups in which this species is (a) present and (b) absent. Each of these groups is then successively divided on the basis of the species having the next highest $S(\chi^2)$ and the subdivisions are thus continued until groups are obtained showing no significant associations. Williams and Lambert have shown that these groups, when applied to mapping, show close agreement with environmental data, but whether the process reveals any fresh information of ecological value and whether the analysis of the community could not equally well have been accomplished by direct ecological examination in the field, is open to question.

Williams and Lambert have lately (1960) expanded their methods to utilize a digital computer, which will make practical the analysis of larger communities of species. There is no doubt intellectual satisfaction to be obtained by the demonstration that observed data, when subjected to a remote process of analysis, produce results which can be shown to accord with observable facts in nature, but some might consider it more scientific to stay close to nature throughout.

Watt, also in 1960, showed by observational methods on permanent plots, the remarkable changes of pattern and relative abundance of species in the composition of grass-heath over a period of twenty years. He insists on the indispensable nature of such continued observation if the dynamics of the vegetation are to be understood and he shows clearly the small value of static data from a single set of observations. (Autodynamism, see p. 3461.)

We have already mentioned (p. 3458) the importance to be attached to seed distribution as a basis of patterns. Although patterns cannot all be referred directly to data of seed dispersal, yet situations in which pattern is related to variations of environmental factors inevitably mean that seeds of the species concerned had been provided in sufficient amount to form the background for selective survival and development, since it cannot be supposed that seeds were only provided in the situations where the plants were able to grow. The reservoir of seeds required to allow the development of differential patterns must be, in fact, immense.

We have some partial data about this reservoir from the reports of the Weed Research Organization at Begbroke near Oxford. In one experiment the seeds produced by a dense stand of weeds approached 5500 million per acre, of which a large proportion are either eaten or decay immediately; only a

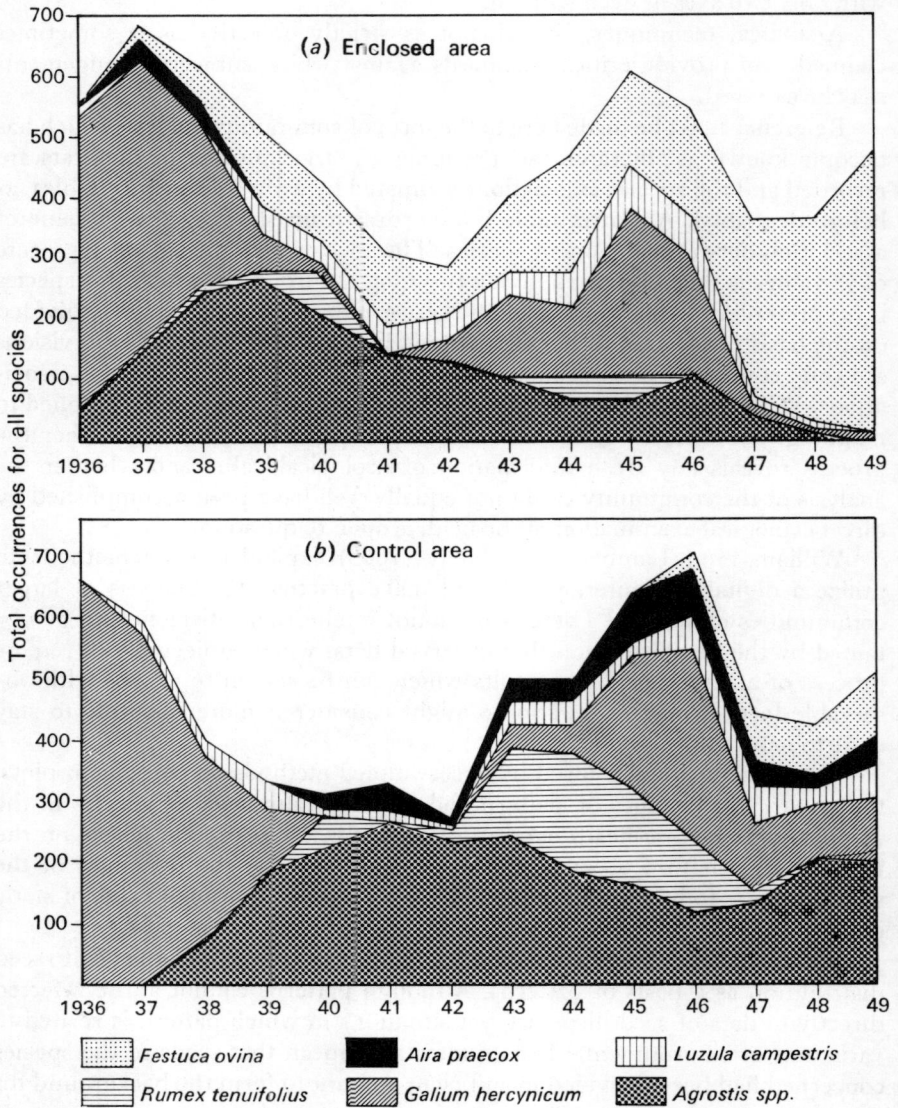

Fɪɢ. 4.16. Changes in the composition of a grass heath on a truncated podsol in East Anglia, over a period of thirteen years. The enclosed area measured 6 × 6 metres and was free from interference by rabbits. The control was unenclosed. *Festuca ovina* became dominant in the enclosure. All records were made between 28 June and 28 July. (*After Watt.*)

small proportion of the remainder germinate in any one year. The germination of *Matricaria chamomilla* in one year was only about 4 per cent of the reservoir of viable seeds in the soil, about 125 million to the acre. Among those buried viable seeds the rate of natural decay is about 50 per cent per annum. Even after five years, therefore, there would still be nearly 4 million *Matricaria* seeds available per acre. While not all plants have the reproductive capacities of common weeds, it is clear that the assumption of a fairly comprehensive background of seeds in the soil is not improbable.

Harper, Williams and Sagar have recently investigated the germination behaviour of seeds in soil. They show that the relations of seed-size and of special germination requirements to the micro-topography of the soil surface are very heterogeneous. Populations resulting from germination were a function of the frequency of safe micro-sites on the soil. When a mixture of species was sown, the relative densities of seedlings was independently determined for each species because their requirements were separately satisfied at different sites on the soil surface. These differentials will determine the constitution of the population resulting from germination in a mixture of species and will markedly influence the capacity of the species concerned for cohabitation or, in other words, the specific associations.

Not only in germination, but also in the early seedling stages, the plants are sensitive to relatively minor influences to which the mature plant may be indifferent. The origin of a patterned dispersion may thus be very difficult to trace, since it may not be due to any existing accord between plant and environment, but depend on the conditions of survival of the seedling at a much earlier stage. A wide field of investigation remains open here.

Much of what we have said with regard to analysis of populations and its importance in the understanding of vegetation, may have given the impression that the traditional descriptive methods of classification, based on the 'association' as a unit, are obsolete. This is not the case. Descriptive methods have rendered, and still render, great services to Ecology. They have covered vast areas to which analytical methods have never been applied and perhaps cannot be applied for generations to come. They have blocked out, as it were, the pattern of the world's vegetation, as an atlas gives a survey of world geography. The point is that descriptive methods are not the last word, but only the introduction to vegetational study. They are necessary in the same way that descriptive taxonomy is necessary. Studies in the cytogenetics of species populations do not supersede descriptive taxonomy, they supplement and enlarge it, just as descriptions of vegetation, if they are accurate within their own limits, are supplemented by experimental and analytical studies. Although the 'species' has been shown not to be the simple monolithic entity it was once held to be, it remains a useful concept and so does the plant 'association'.

Qualitative information is an essential ingredient of natural knowledge. Some empiricists ignore it and try to reduce all information to quantitative expressions which can be firmly (not to say rigidly) grasped. Such machine-mindedness in dealing with nature only helps to cover over reality with artificial constructions of spurious simplicity. Only a resolution to stick closely to facts, all kinds of facts, will serve Ecology in the future. If we do not know 'What' we cannot answer 'How?'. Nevertheless, theoretical constructions of How will not teach us What.*

* We would not have it thought that because we have expressed a critical attitude towards mathematical analysis of plant populations we are hostile to such procedures. This is not at all the case, but we are concerned that the lure of mathematical deduction may lead some minds to regard it as intellectually superior to the practice of observation and experiment. Without the latter, the former only 'bombinates in a vacuum' from which it cannot escape.

This is a very old controversy, which affected Physics long before Biology. Reporting to the British Association in 1835, William Whewell, himself no mean mathematician, referred to mathematics as 'that deep and charmed labyrinth' and added that 'the direction which the speculations of our mathematicians concerning heat have thus taken has not been in all respects favourable to the progress of the subject as a branch of experimental and inductive science.' Indeed, as Sharlin has pointed out, the subsequent experiments of Joule 'uncovered facts about the nature of heat that no mathematical analysis could have found'.

CHAPTER XLV

THE NATURE OF THE ENVIRONMENT

WHAT is implied by the term 'environment'? Is it the same as 'habitat'? In the strictest sense the plant's environment is localized at its external surface and is the physical and chemical situation actually in contact with the plant. This is true so far as it goes, but it is not the whole truth and it would not be correct to regard the plant as unaffected by conditions beyond the area of contact. Obviously many events which are more or less remote from the plant may occur to change the situation at its surface of contact. The upgrowth or the disappearance of neighbouring plants, changes in soil drainage or periods of drought, these are only a few of such possible events which may alter the strict environmental situation. We may, if we like, retain the strict application of 'environment' and call the whole volume of space enclosing the plant its 'ambience', but there seems little need to use a different term. The direct environment is superficial, the indirect environment or ambience is spatial, but in one way or another the living plant has to contend with the influences of both and we would be justified logically in saying that the environment is the sum total of the influences bearing upon the life of the plant.

An environment in this fuller sense is not therefore simply a condition, it is dimensional, and it implies a spatial volume extending indefinitely in all directions from the plant as centre. This is a concept which is not equivalent to 'habitat'. The latter is essentially a topographical term, implying a place of growth, like the word 'home', and although both these words carry with them overtones which are evocative of certain conditions there obtaining, they are not descriptive of those conditions but remain essentially terms of position.

Looked at in this way, we have two variables in our equation of nature, the plant and its environment, clearly separated by a morphological surface. The internal constitution of the plant should not thus be considered as part of its own ambience. This is, however, too sharp a distinction. Our variables are mutually dependent. The environment affects the plant's internal constitution and, in turn, the plant affects the environment, on which it reacts. Moreover, some influences of the environment, such as sunlight or temperature, may permeate the plant's interior and become part of its internal condition, and this is the environment of each of its cells.

We must not try to carry this analysis too far or get involved in metaphysical difficulties about the limits of the individual, but it is important to get the situation clear. We have spoken before, in anthropomorphic fashion, of the plant 'contending' with its environment and in a limited sense this is true, for unless an equilibrium is established the plant cannot continue to exist; but it is necessary to grasp that a plant cannot be considered apart from some sort of environment, that it is largely a product of its environment and that the individual (and this holds for animals too) and its environment form an indivisible functional system in which, as in a system of machinery, the living individual plays the part of a prime mover.

Hitherto we have been speaking of the environment of the individual as if it were a special and peculiar thing and it is a fact that every individual has a unique environment, but is it not permissible to enlarge the concept to include a community? Yes, within wide limits. Over the area of any community the variations of environment must lie within the boundaries of tolerance of the constituent species or they would not be there, and the lowest common multiple of these ranges of tolerance defines the communal environment. Such communal environments are as real as the communities are real and the relationship of communities to them is as intimate and as inexorable as is the relationship of the individual to its peculiar share of the common environment. Indeed, in this relationship, the community *is* a unity.

Between the individual and its environment the equilibrium is dynamic, for the environment changes with every hour of every day. The plant is also changing as it develops. The young seedling establishing itself amid dense herbage is in a very different environment and in a very different equilibrium from the mature plant towering up in comparative isolation. The plant's own reactions to influence also change and its range of tolerances changes as it develops. The remarkable differences in the temperature optima for the various stages in the maturation of bulbs (see Vol. III, p. 3179) show how changeable the plant's reactions can be. At certain stages it may be more difficult to arrive at an equilibrium than it is at other times, and these stages are called critical. Germination, establishment, flower initiation and pollination are such critical stages, and are specially sensitive to environmental influences and may be decisive regarding the continued existence of the plant. At the earlier critical stages some influence, such as shading, may be fatal, although the mature plant is comparatively indifferent to it. Thus, the key to aberrations of dispersion in a community may have to be looked for in antecedent rather than in existing conditions.

The existence of an equilibrium between plant and environment implies the idea of adaptation, a battleflag of the early evolutionists. In the widest sense there must be some harmony between the structure and functioning of the plant and its ecological niche so that an equilibrium may be reached, but this does not mean that this equilibrium is the only one of which the plant is capable or that a unique lock and key relationship is involved. There are a number of well-known cases in which plants which have a very limited natural habitat have shown themselves widely adaptable in cultivation. *Polygonum baldschuanicum*, *Cupressus macrocarpa* and *Metasequoia glyptostrobcides* are examples. These, and many others, possess what has been called 'latent adaptation', which might be interpreted as a wide range of tolerances except for competition, which is presumably the factor that prevents their natural spread. This sort of situation is by no means uncommon and the localized distribution of a species or its absence from certain areas may be simply due to the accident that its seeds have never reached far enough.

The older concept of adaptation was an active one. The plant was supposed to react directly in response to its environment, by such alterations of structure or function as would better 'fit' it for its environment, a teleological idea. That the environment can indeed act directly by encouraging or suppressing the expression of genic characters, is well known, but this involves no change of the genotype and is only accidentally related to adaptation. Plants whose growth or structure are thus altered by direct environmental influence are called *ecophenes*. The changes are strictly phenotypic and non-hereditary. The changed forms of water plants growing on drying mud are well-known examples.

The basis of all adaptation is genotypic. Every Linnean species is a congeries of biotypes, groups of genetically concordant individuals, the number of which can be multiplied by free intercrossing. Biotypes may have different potentialities. Some of these differences have a survival value and the spatial distribution of biotypes can be determined by their suitability or otherwise to different environments. Single biotypes or sometimes a group of biotypes may thus be associated with a particular ecological environment. This was first recognized by Göte Turesson in 1922. He collected specimens of certain wide-ranging species from different habitats and found that even under cultivation they retained certain characteristic differences. These ecological races he called *ecotypes*. In some of the more extreme types of habitat an ecotype may be genetically uniform, that is it consists of a single biotype. Even where this is not so, the biotypes involved must be uniform with regard to the ecologically critical characters, though they may be heterozygous for other alleles which do not affect their survival capacity as they are not critical with respect to the particular environment.

Species with wide geographical ranges may also contain climatic ecotypes which differ in their climatic tolerance even though they are morphologically indistinguishable. A northern race or one from high altitudes may be more winter-hardy than a southern race and neglect of this difference of provenance has often led to failure to acclimatize a species in a new area. Stapledon, who worked with races of *Dactylis glomerata* collected from many countries, found numerous ecotypes with markedly different habits of growth, and he found similar ranges of ecotypes among populations collected as far apart as New Zealand and the U.S.A., though the proportion of the ecotypes was not the same in each population. He pointed out that in some less severe habitats the conditions might not limit a species to one well-marked ecotype, but that the species in such habitats showed reduced variability (that is contained fewer biotypes) in spite of the possibility of continual new introductions.

Ecotypes are liable to phenotypic variation and if one is moved to the habitat of another ecotype, the *ecophene* it there produces may closely resemble the second ecotype, though retaining at least some of its own distinctive characters. This has often been observed in the transplantation of lowland races into alpine habitats. As the alpine ecotypes of different species all tend to show certain characters in common, it may even happen that an ecophene of a transplanted species may resemple the alpine ecotype of a related species more than it resembles the alpine ecotype of its own species. This has given rise to claims for the adaptive transmutation of species, which do not, however, stand up to close examination.

Differentiation of species into ecotypes originates in the selective action of particular environments on the biotypes of all the species present in it. The ecotype may therefore consist of only one biotype or of several. This is the *modus operandi* of adaptation, arising from inherent differences. As ecotypes are freely inter-fertile, hybrids may arise, but as the survival of the ecotype depends on its adaptation to one particular environment, the chances are against hybrids finding suitable habitats, unless it be at the boundary zone between two ecotype habitats.

We do not know how widespread among species is differentiation of ecotypes, though it appears to be not uncommon among wide-ranging species. Ecotypes may exist without any external mark of differentiation, the distinguishing features being purely physiological, which can only be revealed by extensive experiments in transplantation. Important physiological differences may thus exist without external indices or, on the other hand, they may be linked in the genotype with apparently trivial morphological differences, which, though they appear to have no selective value, are yet strongly correlated with a particular environment.

A severe or peculiar environment may select very few or even a single biotype from among the many a species contains. If the population is small, continual inbreeding may in time reduce the population to a homozygous state. This is a condition of affairs characteristic of 'relict' species, small populations of endemics confined to a limited area. The intense severity of the glacial periods left a considerable number of such relicts, in small areas where the topography afforded a refuge. The genetic impoverishment of these populations by the extinction of most of their biotypes so limited their genetical resources of variability that only in a few instances have they shown any ability to spread from their remaining localities. The same is true of species occurring as outlying populations well away from their general area of distribution. Small colonies of a species may thus remain isolated for very long periods without any tendency to spread, even where other apparently suitable sites are available.

The harmony between many plant species and their environment may be associated with aberrant forms which naturally appeared at first to be the result of direct adaptation. Such, for example, are the stem succulents of desert areas. These are predominantly members of the Cactaceae, Euphorbiaceae and Asclepiadaceae, with a minority of Compositae, which show a remarkable convergence of form, in harmony with their peculiar environment. They did not descend upon the desert full-blown in character, nor were they created by the desert. They originated as new biotypes in areas round about the desert; they had a genotypic tendency to succulence, and while their slow growth handicapped them in a mesophytic vegetation they found a possible home in the empty spaces of the desert, where the greater the tendency to succulence, the better would they survive selective pressures.

The population of deserts is often regarded as very open and scanty, yet it has been found that the root systems of desert perennials are often extensive enough to occupy the available space very effectively. Walter has, indeed, pointed out that vegetation naturally tends towards saturation, towards the condition in which the vegetation is utilizing all the resources of the environment with maximum efficiency, a condition identified with resistance to change, i.e. relative stability.

In general there is a direct relationship between the extent of range of a species and its genetic variability, but there are some exceptions, where, for example, a species inhabits a special type of habitat, such as maritime sand-dunes, which is itself wide-ranging, and in which species may each be represented by a single ecotype but one which is as wide-ranging as the habitat. In the same way the ecotype may be genetically homogeneous, i.e. contain only one biotype, or it may contain several biotypes. In the former case its ecological range will generally be limited, in the latter case it will show a certain adaptability to differing environments, though it may display ecophenic modifications in adapting itself to them.

Ecotypes of limited range may be excellent indicators of habitat conditions and observation has designated a considerable number for this purpose. At the other end of the scale are the species which contain many genetically heterogeneous ecotypes. In this category are the chief crop plants, the heterogeneous nature of which has enabled man, in their long history of cultivation, to segregate their ecotypes and the many biotypes they contain, and by selection and cross-breeding to provide himself with races suitable to many different climates and conditions.

Where morphology cannot provide characteristics for recognition, as is often the case, it follows that such ecotypes and biotypes cannot be separated except by genetic analysis and this introduction of genetical structure into ecological problems has been called 'genecology', a study of increasing importance. Such considerations emphasize the importance in ecological experiment of using plants of either a pure line, i.e. the offspring of homozygous, self-fertilized parents, or else a clone, which is the vegetatively propagated posterity of a single individual.

Many efforts have been made to formulate an ecological classification of environments, often with the object of providing an environmental framework for a classification of plant communities. In the past, these efforts have mostly depended on the relative interaction of important factors or on the predominance of a single 'master-factor' with a supposedly overriding effect, either in the climate or in the ground. Such systems are workable only with the more extreme types of environment and they fail to distinguish the wide variety of conditions to be found in environments lumped together as 'mesophytic', or moderate. Two things militate against the success of such classifications. One is the recognition of the close interaction of all the factors in the environment and the extent to which compensation occurs between them.

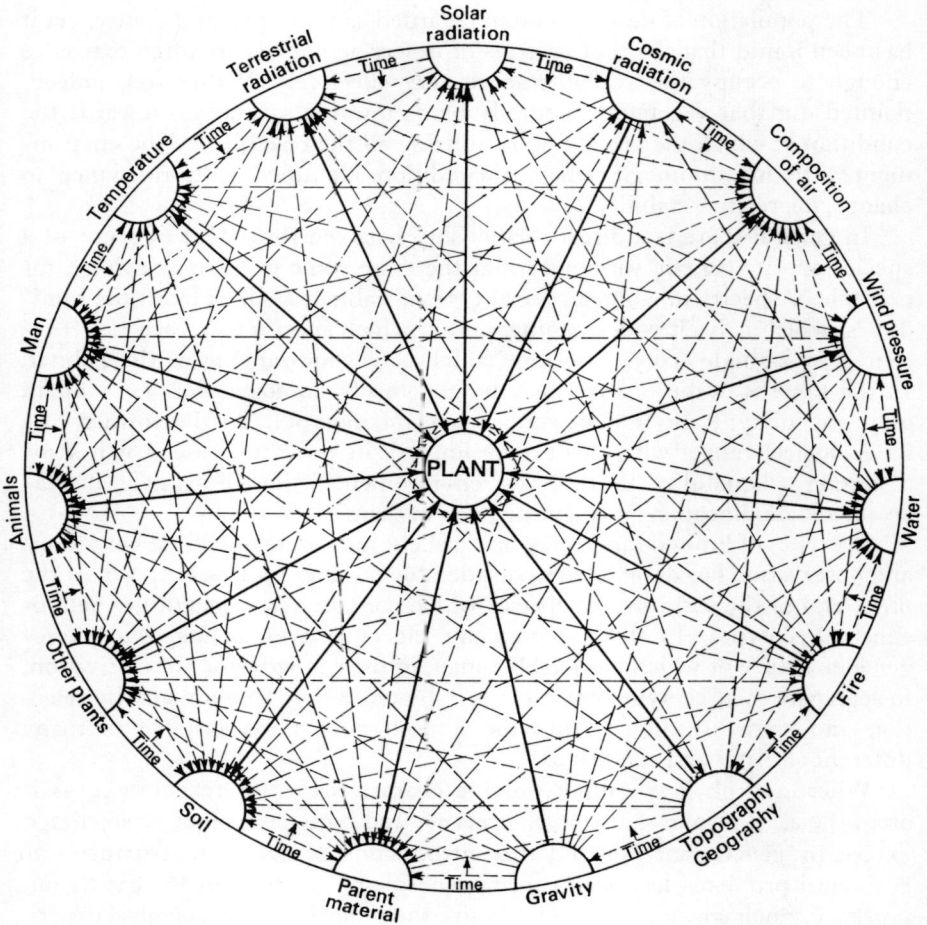

FIG. 4.17. Diagram of the interrelationships of a plant and its environmental factors in an ideal holocoenosis. Solid lines show plant-factor relationships, broken lines the influences of factors on each other. (*After Billings.*)

The second is the recognition of the importance of the microclimate, the immediate ambience of the plant, which may vary remarkably, even in small areas. On the broadest lines a classification such as Schimper's, which runs the gamut from 'Areas perpetually hot and wet' to 'Areas perpetually cold and dry', may be valid in plant geography, but it is on much too large a scale to be helpful in local Ecology.

Although experimental limitations often oblige us to study single factors of the environment as single entities or perhaps the interrelationships of a pair of factors, this cannot be regarded as a fully satisfactory approach. No factor acts wholly in isolation, though one may be more powerfully influential than others.* It is now widely accepted that the environment forms a *holocoenosis*, that is to say a complex and indivisible whole, the two complex systems, plant and environment, interacting as functional units. This concept is distinct from that of the *ecosystem* as envisaged by Tansley, for the latter is wider, including the idea of development, with time as a significant dimension. The holocoenosis might be said to be, at any given moment, one element of the ecosystem. A theoretical diagram of the interrelationships in an ideal holocoenosis has been given by Billings and is reproduced in Fig. 4.17.

It must be confessed that to comprehend an entire holocoenosis in one analysis is beyond our present range. Fortunately much can be learned which is of ecological value by limited, partial analysis, for all factors of the environment are not of equal weight and some, as Billings points out, influence the plant only indirectly, through other factors.

In some specialized environments the main ecological determinant is a single factor or a small group of factors which is so overwhelmingly developed that its influence counterbalances almost all others. These are what Tansley called 'master factors'. Even in the majority of environments, when no master factor is present, there are usually one or two factors of greatest influence. For example, the climatic factors of temperature and humidity are often of greatest ecological weight and the chief vegetational features can be related to them (Fig. 4.18).

The idea of factors having different weights was first expressed in Liebig's Law of the Minimum (1840) resulting from his studies of plant nutrition. Originally it was limited in its application to nutrient substances, and it laid down the principle that the growth of the plant is controlled by the nutrient which is present in minimum quantity. This was a limited statement of a situation which was later expanded by Blackman into the general principle of *limiting factors* according to which the rate of physiological processes in any given situation is controlled by that factor (the limiting factor) which is furthest removed from the amount required for the maximum rate. In this generalized form the principle can be applied either to a deficit or to an excess of the factor in question. Either condition may be limiting, though the former is more frequent. The effect is that growth, or whatever is in question, will only respond to an appropriate change in the limiting factor and that without such a change, alterations in the other factors are without effect. Thus there are two values of most factors, a minimum value and a maximum value beyond which the factor becomes limiting. The range between is the range of tolerance for that factor, with an optimum value somewhere between the limits. The optimum is not, however, a fixed point, but will depend on the situation created by the balance of other factors. Obviously any effective change in a limiting factor must be in the direction of the optimum, not the reverse and not

* The term 'factor' of the environment was presumably first used in the arithmetical sense, but etymologically the word implies action. Most factors are indeed active, but some are neutral or only effective through the intermediacy of other factors. Many geographic and topographic factors are in this category. The word 'factor' has now become well-established, but if one were pedantic it would be more correct to speak of the 'properties', 'qualities' or 'components' of the environment.

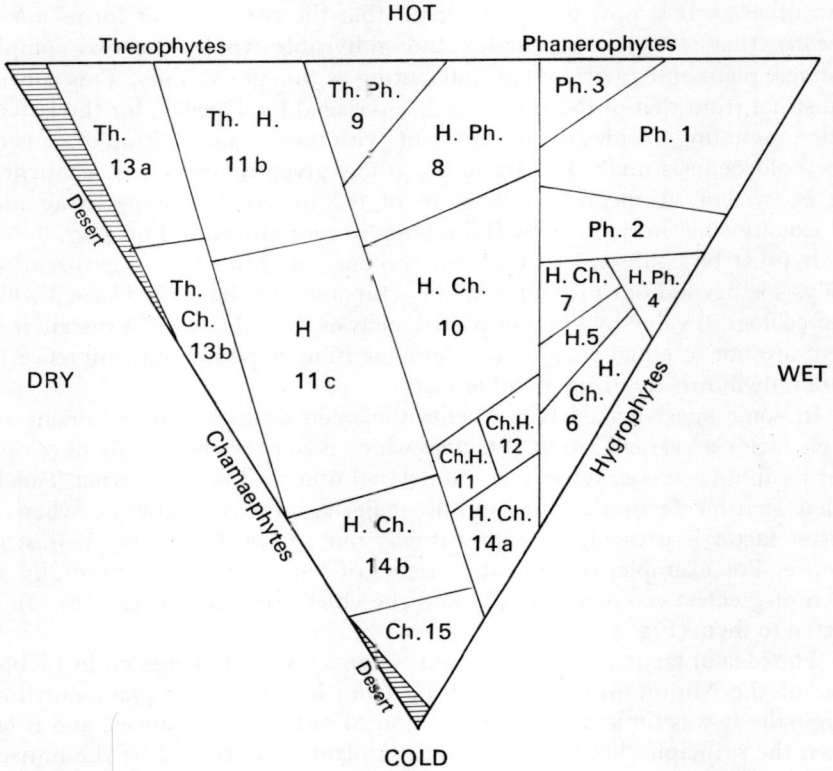

FIG. 4.18. The Bioclimatic Diagram after Dansereau. The letters show the prevailing Raunkiaer life-forms. The figures refer to the vegetation classes of Schimper and von Faber. The striped areas are the dry and the cold deserts respectively.

every such change will be effective, since a small change may still leave the factor in a limiting position. This is important for there is a tendency to think that any change in a factor assumed to be limiting must necessarily be reflected in a change of response by the plant and that if this does not occur the factor is not truly limiting.

At first sight it would seem that the principle of limiting factors must greatly simplify our attempts to comprehend the effects of environment, but things are not so simple as that. In the first place, tolerances vary between the biotypes of species, so that the environmental range of the species as a whole is presumably dictated by the ranges of its most extreme biotypes, which are not necessarily always present. Tolerances also vary greatly during the lifetime of each individual.

This means that the environmental reactions of a given species in a community will never be uniform. Biotypes are, we know, selected by the environment, so that the biotypes of a species in one geographical area may have different tolerances from those in another area and the species may thus appear to have differing reactions. There is no specific range of tolerance common to all factors, a wide tolerance for one may accompany a narrow tolerance for another. Furthermore, if a species is living in a situation which is far from optimum in respect of one important factor, its tolerances for other factors may be reduced, in other words it becomes sensitized to comparatively small changes. A species which has wide tolerances for many factors will usually tend to be widely distributed and will show comparatively little dependence on environmental changes (though it may be sensitive to competition) throughout the greater part of its area, but, near the periphery of the area, its tolerances become much more restricted and it can be more readily extinguished. Any marked climatic change will therefore always be signalized by changes in the distribution area of many species, either by increase or by decrease.

Decreased area may result in the breaking up of a continuous area into several disconnected areas, a condition called *polytopy*. There are many striking examples of discontinuous occurrence, which we shall speak of in the Section on Plant Geography (p. 3771). There are several possible origins of polytopy, of which the break-up of an originally continuous area is only one, but it seems applicable in many cases where the separation of the existing areas is not too great.

Another type of factor has been called a 'trigger factor', i.e. one which sets off a chain of reactions which destroy a previous balance and inaugurate a large-scale alteration both of conditions and of population. Such factors as fire, invasion by an aggresive competitor or the introduction of grazing animals are among the examples of such triggers. Their effects in many parts of the world have been profound and long-lasting, and it is not possible to predict with any certainty what the results will be. It is for this reason that the introduction of plants or animals to new areas should always be regarded as a potential danger and carefully controlled. There are too many unhappy instances of the disasters that can otherwise result.

The impact of the environment on a young plant changes continually throughout its life. Some of the changes are cumulative, some like the seasons, are cyclic. With these cyclic changes the development of the plant must keep step if it is to live to reproduce successfully. The gene controlled rhythm of the plant's development must coincide with the seasonal cycles or it will not continue. Native plants have arrived at such an adjustment through natural selection, but introduced plants may often fail in this respect. The vine, the Maize plant and the Soya Bean can all grow well in our British summer, but they are adjusted to a longer growing period than our climate allows, hence they can only form seed in years with an abnormally mild autumn.*

The existence of such a harmony of cycles in development and climate makes clear that meteorological averages for climatic factors have little meaning ecologically. It is not the mean, but the amplitude of climatic fluctuations which really matters to the plant. A great many species are excluded from areas which are apparently suitable because of the occasional occurrence of extremes, especially of cold, which they cannot tolerate. There are many

* See also under Photoperiodism in Vol. III, p. 3197.

'borderline' plants in our gardens which illustrate this. To take one example, some species of *Eucalyptus* can survive British winters for long enough to grow into fair-sized trees, but they could never be used for afforestation or become naturalized here because the occasional severe frost inevitably cuts them down. The search for a frost-hardy biotype of the Soya Bean has likewise failed to discover a race which could be introduced with safety into British agriculture, in spite of its potential value as a fat-producer.

A very important aspect of plant environment relations is the existence of *compensation*. This involves the interaction of factors in such a way as to modify the natural tolerances of the plant. An outstanding example is compensation for differences of latitude by differences of exposure. A sheltered southerly exposure enables species to grow further north than they normally would. Conversely, a cool northerly exposure can extend the range further south. Altitude plays a part in this and high mountains, even in the tropics, may bear vegetation of a type only found at lower levels much nearer the Poles. A warm and well-drained soil, such as that over limestone, will sometimes extend the tolerance of species for low temperatures or high precipitation. In the Scottish Highlands the presence of the calcareous and fertile Lawers-Canlochan schist, can compensate for low precipitation and southerly exposure otherwise unfavourable for a group of arctic-alpine species. Compensation for a climate with a rainfall below the tolerance limit of a species may often be provided by an ample supply of ground-water. The narrow strips of gallery forest following river beds through semi-desert country provide a well-known and widely occurring example of this effect. A parallel effect of ground water compensation is seen in the occurrence of colonies of species belonging to the tall-grass prairie of North America (e.g. *Andropogon* spp.) in moist, sandy depressions in the much drier climate of the short-grass prairie (*Buchloë dactyloides*) further west.

After all that has been said about tolerances it comes as an anticlimax to admit that we really have very little data on the subject. Apart from common forest trees and crop plants, which have been carefully studied, such data as we have for the generality of plants are simply observational. It is experimentally verified information that is needed and which is wanting even for ecologically important species. Close studies of single species and their ecotypes, such as those being slowly gathered in the *Biological Flora of Britain*, are most valuable.

The concept of environmental 'factors' runs through the whole of Ecology, each such factor being regarded as an element of the environment which has a distinctive influence on the plant. This is, however, a purely ideal concept. No factor acts in isolation, the action of each one is a function of the holocoenosis. Experimental attempts to factorize the environment by studying the effects of variation in one factor while keeping others constant automatically produce an unnatural situation, at least from the ecological standpoint. Even a factor which has a predominating influence operates only as a part of the environmental complex and its predominance may be locally nullified. It is imperative that this should be considered in all experimental studies. The laboratory methods of pure plant physiology, while they provide essential data for the ecologist, do not afford a model for his methods, for his aims are different. He has to bear in mind that the primary purpose of Ecology is the understanding of natural conditions and for this purpose the natural environment is the only appropriate laboratory.

Ecologists have to accept the environment as they find it and try to determine the condition and behaviour of the plants within it. Even if we cannot grasp the full complexity of the holocoenosis, we can at least discover the natural incidence of some of its leading characteristics, which may give some guide to the reactions of the plants. The attempt to factorize the environment by the analysis of single characteristics in isolation is leading away from the true ends of the science, while artificial modification of the natural environment for experimental purposes is not Ecology but horticulture. To show that plants may grow differently if their environment is experimentally transformed means little ecologically since it is true of every plant. The core of the ecological problem is to find how the plants cope with what nature has provided for them.

We have already referred (p. 3497) to seasonal cycles in the environment, but there are other, geochemical cycles which are indispensable to plant life. These involve the circulation of necessary materials, in which absorption by the plant is one episode. Most of these cycles involve materials of which the supply could otherwise be exhausted. We have previously (Vol. III) described the cycles of Carbon dioxide and Nitrogen. Were it not for the release of CO_2 by respiration and by ocean water the plants' resources could have been exhausted ages ago. It has been calculated by Norman that to produce 100 bushels of Maize per acre the plants require about 20,000 *pounds* of CO_2, which represents the CO_2 content of 21,000 *tons* of air. Although free Nitrogen is available in unlimited quantities the assimilable compounds of Nitrogen are definitely restricted in amount. Water too is unrestricted but fresh water is not and the land drainage-evaporation-rainfall cycle is a vital necessity. Phosphorus is a relatively uncommon element and it is also the subject of a cycle starting with soluble phosphates in the soil, through plants and animals and back through decaying organisms and excreta, by the action of phosphatizing bacteria, to soluble phosphates again. In fact most soil nutrients pass through a short cycle of absorption followed by return to the soil either through death and decay, leaf litter or leaching by rainfall from the leaves of trees (p. 3657).

All the natural cycles receive some extraneous input which helps to balance losses. Carbon dioxide from volcanoes adds to the atmospheric reservoir. Combined Nitrogen is increased by the action of Nitrogen fixing microorganisms. Volcanoes again exhale water vapour, and phosphates are increased by the erosion of phosphate-bearing rocks. Though nitrates are all soluble in the sea, some proportion finds its way into immobile sumps of deep water where it may be out of circulation indefinitely. There is also a considerable loss of Phosphorus to the sea. It is noteworthy that a Phosphorus deficiency is probably the most widespread defect of soils.

There is also, of course, the cycle of solar radiation energy, transformed by the green plant into energy of chemical combination and destined to be again released as the radiant energy of heat, either by burning or by respiration. Unlike the others this is an open cycle, for the liberated heat does not return to its source and, so far as we know, the sun's traffic with the earth is one way only.

Most of the ecosystems of the world were established before the appearance of mankind and his impact upon them has been largely disruptive, deflecting their resources to his own use. In earlier times the effects of this exploitation were negligible, but as populations grew the situation changed and dangers loomed which demanded collective action. As a result movements of conservation have grown to worldwide proportions, the objects being to prevent important ecosystems from disappearing and to preserve the action of natural cycles. The measures taken have been partly restrictive of exploitation and partly restorative, such as afforestation, conservation of water supplies and conservation of soils. The latter has been the most recent development, for it was only in the late nineteenth century that land clearances and the injudicious extension of arable agriculture exposed large areas of soil to erosion, with cumulative effects. Conservation measures have arisen from the belated realization that soil resources are virtually irreplaceable. The low standard of living in many countries has arisen through ignorant destruction of these resources or through the insufficiency of the methods of utilizing what existed. Industrial growth in developing countries cannot compensate for agricultural inefficiency for it is out of the soil that all life is drawn. The omnivorous grazing of the goat in some countries has done more to reduce human populations to poverty and decline than has ever been estimated. The agronomist is an applied ecologist and it is his knowledge of crop management and ecology, 'husbandry' in short, which can alone restore in many areas the prosperity which has been lost through mishandling. Why else has the good farmer been called the husbandman but that he protects, cares for and fertilizes his soil.

Every product of the soil impoverishes the soil. In a natural ecosystem the product dies and decays on the spot and its materials are returned to the soil. If man removes all or most of the product there is a depletion of soil resources which must be made good. If this is neglected the natural resources of the soil are said to be 'mined' and, like all mines, they inevitably become exhausted and the soil is ruined. The cynic may be inclined to gloat over this depiction of a natural ecosystem: we eat, we are eaten and we all decompose.

The longest-term cycle in nature is the erosion cycle, which involves the formation of sediments, their elevation and exposure to erosion and the transport of the eroded material to form new sediments. This is a downhill process, a loss of material from uplands to lowlands and thence to the sea. By our standards it is very slow, but very widespread and in times of geological quiescence enormous areas are levelled off by it to a peneplain. While this is going on many ecosystems are built up on the slowly eroding rock-soils and the cycling of materials within these systems has great importance in delaying the loss of essential elements and controlling the rate of their escape to lowlands where they may not be needed biologically or are simply lost to the sea.

The *productivity* of an ecosystem is a measure of the *rate* at which organismal substance is being produced. As a rate it is measured by increase of dry weight per unit area per unit time, the area being generally one square metre and the time one day or one year. This is clearly different from *yield*, which is a static measure of the total dry weight of vegetation remaining at the close of the growing season after all natural losses, or, in the case of long period growths, such as trees, at the maturity of the principal species. Yield depends on a subjective decision as to time, and is a human rather than an objective ecological quantity. Its chief value is in application to more or less homogeneous crops, such as grass or timber and it would be very difficult if not impossible to assess in a heterogeneous natural population.

Productivity, on the other hand, is related to a definite objective value, namely, the net rate of photosynthetic assimilation by the plants. This is, apart from a relatively small import of materials by dust-fall or by animals, the primary source of the mass of organisms forming the system, so long as it remains in a steady state. In theory this is simple but in practice there are great difficulties in estimating by sampling methods the assimilation rate and the respiration losses in a mixed terrestrial community. Aquatic communities lend themselves more readily to experimental methods. The only attempts that have been partially successful have been by measuring CO_2 consumption in large bell-jars placed over the vegetation, as compared with similar estimations under a blackened jar, with the aid of an infrared gas analyser which speeds up the operation. Failing this a method of periodic harvesting and estimation of dry-weight production is the simplest, approximate resource.

The question of productivity naturally raises the concept of the *energy flow* through an ecosystem in a stable state. Solar radiation is the primary input. What becomes of it? Plants under natural conditions fix about 2 per cent of the light energy they absorb, but only about half of this is absorbed by chlorophyll and becomes effective in assimilation. Under water the ratio is still smaller and in the sea the average has been estimated to be as low as 0·18 per cent. The assimilating plants are the primary producers, which includes green plants at all levels from microscopic algae and flagellates, in plankton or on the soil surface, upwards. If the plants formed a pure community their assimilated substance would pass directly to the agents of decomposition, protozoa, fungi and finally bacteria, whose numbers are very great, but whose total mass compared with that of the plants is very small. This implies a large dissipation of energy in respiration, estimated by Russell and Russell at 5×10^6 kilocalories per acre per year in a highly organic medium.

Under normal conditions the ecosystem includes many species of animals, among whom a food-chain exists of eaters and eaten, based upon the herbivores which eat the plants and are eaten in their turn by the lowest rank of carnivores. At each stage of consumption there is necessarily a loss of energy by degradation into heat which is lost by radiation. The series can thus be represented by a pyramid or pyramids in which each stage is represented by either the collective weight, the numbers of individuals or by the energy content of their respective masses. Each stage is less than the stage below it, the top carnivores being necessarily the fewest. The food-chain series is not, how-ever, a rigid organization, for the higher levels do not restrict themselves to prey in the rank immediately below, but may, in times of shortage, eat lowly grubs and insects or anything else that is edible, even including berries or other plant material. Food-chains in extreme climates such as the Arctic tend to be short and easily disrupted by environmental changes, since there are no food alternatives and if one item is depleted all suffer. In southern and more favourable environments the food-chains are longer and may interlock to form a web, since many alternatives are present. There is no necessary gradient of size in a food-chain, though one often exists. Herbivores are of all sizes, up to the elephant, and the larger ones have sometimes no predator but man. The biome, the unit which includes both plant and animal populations, may be wider than the plant association, since allied associations, for example, various tree associations of the northern coniferous forests (probably phases of a continuum) are associated with the same animal population. Apart from the energy dissipated by plant and animal respiration, there is a large amount locked up temporarily in the increasing populations, but all eventually passes to the microbial decomposers in the soil, by whom is finally dissipated the energy of solar radiation which was the foundation of the whole system. This is too variable a quantity in different situations to be given even an average figure. One estimate, in the State of Michigan, was 1884×10^6 kilocalories per acre per annum. An Alfalfa crop in a similar latitude weighed 1785 pounds per acre representing an energy content of 14.9×10^6 kilocalories, which agrees reasonably well with the estimate of 1 per cent of the radiant energy supply, allowing for respiration losses. The time lapse between assimilation and release of energy is known as the turnover period and depends on the lifetime of the various elements of the food-chain. It is shortest among the bacteria, where even a large increase of activity results in a comparatively slight increase in numbers. This refers to turnover of energy. Turnover of materials, as it affects the plants, is more complex and is quite different in soil and in water environments. We shall have more to say about it later (p. 3657).

The balances achieved by organisms with the supply of essential nutrients, involving the cyclic turnover of supplies, are often delicate and can easily be upset by interference. In a balanced ecosystem there are, however, mechanisms, still imperfectly understood, which tend to maintain its balance against disturbance, a situation described as *homeostasis*. The regulation of growth and the cyclical storage and release of nutrients form at least elements in this maintenance of stability which has a survival value for all the organisms concerned. Despite the competition it entails, the community is the essential 'home' of the plant, to the maintenance of which its whole life is adjusted just as surely as it is for a human citizen.* The true aim of conservation is to allow these natural balances and cycles untrammelled play. If the system is al-

---

* There are exceptions. Some plants can become community predators, 'outlaws', whose unchecked expansion may wreck a community. The author has seen an area of rain forest literally smothered by the overgrowth of *Calamus*.

ready stable it must fend off extraneous disturbances, especially those resulting from human actions, but natural development should not be hindered
nor any attempt made to impose artificial stability. Conservation means protection, it does not mean regimentation.

A plant without an environment is inconceivable, but there is a two-way
action between them, for in some cases the environment would not exist
without the plant. This effect of the *reaction* of the plant on the environment
takes a multitude of forms and it is of fundamental importance. Sometimes
it is an individual or a species reaction, more often and more obviously it is
effected by the whole community. Some examples are familiar. A forest
creates its own special environment. It results from the mere existence of the
community and it affects all its members. Sand-living grasses accumulate
blown sand and build up sand-dunes, which they stabilize with their rhizome
systems. Salt-marsh communities protect the marsh from erosion and collect
silt which builds up the ground to the highest tide level, where a successional
community takes over. In the soil there may be many effects, for example the
increase of humus, with correlative effects on the water holding capacity of
the soil, on drainage and aeration and on the micro-flora of saprophytic
organisms. Roots may break up rocks and start the process of soil building.
They may deplete water reserves and lower the water table or, conversely,
they may block drainage channels and raise the water table. They may impoverish the reserves of some essential nutrient, or they may, like so many
Leguminosae, enrich the Nitrogen reserves. Individual species, like Sphagna,
may change the pH of the soil and thus alter the solubilities of nutrient salts or
they may, like *Encelia*, excrete a growth-suppressing toxin which keeps away
competitors. In short, there are a thousand ways on which the plants affect
their communal environment and much normal succession is facilitated by the
changes thus brought about.

Many of the changes we have outlined are favourable, but not all, and
some reactions may be directly inimical. For example, in *Pinus* forests in
Sweden the accumulation of highly acid humus on the ground may go so far as
to suppress the germination of the pine seeds and hence put a stop to regeneration. To get over this, recourse is had to controlled burning of the humus,
which exposes the mineral soil again but is not sufficiently intense to kill the
trees.

As the method of factorizing the environment for investigation is admittedly limited and only partially satisfactory, Clements pioneered the idea
of using plants themselves as integrators, whereby the effect of the environment as a whole might be ideally estimated. He called such indicator plants
phytometers. Three methods were proposed by Clements. The first was to use
plants of species selected for their wide range and adaptability, that is plants
of extended tolerances. Important crop plants were also used. These species
were cultivated under controlled conditions and their reactions recorded,
especially for transpiration, growth and photosynthesis. They were then
taken, planted in containers, to the site of investigation and their functioning
measured regularly over a period in conjunction with a battery of recording

instruments. As these plants were in their own containers it is obvious that only climatic factors were being considered. The second method was to establish a series of permanent quadrats, each with a population as nearly as possible identical with the others. These were visited regularly and the progress of growth recorded. The third was that of transplants, soil blocks with the plants growing on them being taken from a single community and transferred to several different habitats.

The first method has not been widely used because of certain limitations and difficulties. First, it is necessary to use a variety of test plants, theoretically the more the better, and it is essential to establish the genetic identity of all plants used belonging to the same species, no light task. Second, a sufficient number of plants must be used to enable their modal response to be calculated and to eliminate individual eccentricities. Third, results in different regions are not so easily correlated as are instrument readings, while, lastly, the transport and, in populated countries, the guarding of the battery of test plants present practical problems. However, as a means of testing local variations of climate the method has attractive possibilities. The second and third methods are closer to nature and have been frequently used.

An extension of the phytometer idea is the use of plants in nature as indicators of soil and climatic conditions. This is, in fact, a much older idea and must have presented itself very early to the minds of primitive agriculturists, who learned to distinguish the probable value of soils from the plants naturally growing on them. There are thus indicator communities as well as indicator species. The latter are chiefly plants with limited tolerances and a restricted choice of environment. This is closely bound up with vicarism (p. 3348) where two closely related species replace each other on different soils. We have previously mentioned *Galium hercynicum* as an indicator of acid soils and *Galium sylvestre* of calcareous soils. *Picris echioides* and *P. hieraciodes* are both calcicolous, but the former inhabits pure chalky soils, the latter the dolomitized limestones of the Jurassic Öolite.

Life-forms in the Raunkiaer sense are, broadly speaking, valuable climatic indicators and Raunkiaer used his 'biological spectra' in this way to characterize climate types. The dominant life-forms of the climax vegetation are clearly related to the climate, as Schimper showed. On a smaller scale, habitat-forms can be indicators, especially where abnormal habitat form is developed in response to an abnormal habitat. Their indicator value is implicit in Warming's divisions of hydrophytes, mesophytes and xerophytes, with their subdivisions of halophytes, lithophytes, etc.

The community as an indicator is a complex of the preceding factors and as Clements says, it is a complete scale on which all the indications of the habitat are written and can be deciphered by analysis. This, however, requires a knowledge of the indicator value of different species for different factors. Factor indicators are very numerous, indeed almost every plant of limited tolerances is in some measure one. Only a few instances from the *British Flora* can be cited here. *Juncus effusus* as an indicator of waterlogged soil; *Suaeda maritima*, saline soil; *Clematis vitalba*, calcareous soil; *Calluna vulgaris*, peaty soil; *Salix herbacea*, an alpine climate; *Rumex acetosella*, acid soil. Indicators of lighting conditions we shall deal with later in speaking of Wiesner's 'light ration' of plants (p. 3550).

Despite the ideal of the holocoenosis, it has been customary to consider the various influences of the environment, for convenience, under separate headings: sub-aerial, edaphic, biotic and also aquatic. Each group possesses an extensive literature and only a limited consideration is possible in a general textbook, though enough, we hope, to indicate the main features.

CHAPTER XLVI

THE SUB-AERIAL ENVIRONMENT

THE two most important properties of a climate from the vegetative standpoint (or indeed from any standpoint) are the *temperature* and the *humidity*. The latter is chiefly conditioned by the rainfall, its amount and its distribution in time. Apart from its climatic effects rainfall is the most important source of the soil water and hence its effects are paramount both above and below ground. Other influences which are of importance are *illumination*, which comprises the quantity and quality of solar radiation, the plant's main source of energy and the prime cause of temperature variations, and *exposure*, which includes both radiation and wind, with consequent extremes of temperature. Thus we can see that all these influences form together a complex system in their effects on the plant. In trying to consider their individual significance we must bear in mind that any variation in one of them has, apart from direct effects, a train of consequences with respect to other factors, which it may be difficult to trace. We shall try, as far as possible, to emphasize the direct effects about which generalizations may be permitted. The interaction of factors makes the deduction of causality for any one factor a danger to be borne in mind.

In estimating the importance of any factor we are faced with the difficulty of deciding which of the various possible ways of measuring it is the most significant. Temperature, for example, can be represented by the annual mean, the monthly means or minima, or the diurnal range, or in several other ways. Generally a decision requires a knowledge of the biology of the plants concerned but some positive indication can be obtained by calculating the regression of some important effect, such as growth rate, with the variation of the factor as measured.

Climate, in the broad sense, is the determinant in the geographical distribution of species and hence in the distribution of vegetation types, but it is also a very important influence in the local topographical distribution of species and individuals, though in this respect its influence is shared and may sometimes be overridden by soil properties. The respective importance of climate and soil depend on the tolerances of particular plants. The ecological importance of strictly local variations of climate was first made prominent by Kraus in 1911, who showed that critical differences of climatic factors, such as temperatures and humidity, might occur within distances of a few feet. We have to think, not of the climate in regional terms which interest us as human beings, but of the plant climate as it affects the plant itself on the exact site of its occurrence, now unknown as the *microclimate*. Boyko has maintained that the tolerances of the plant which limit its success topographi-

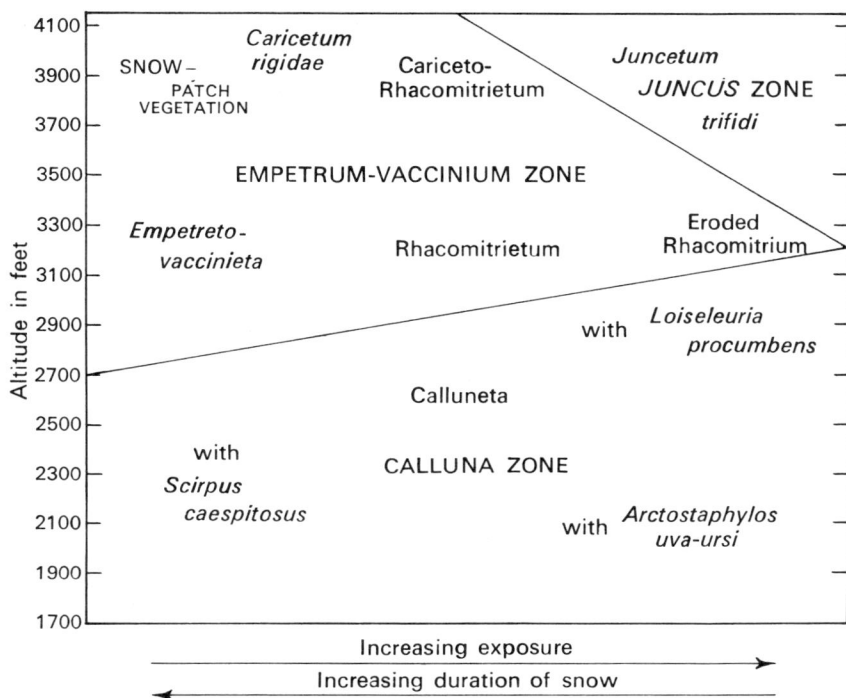

FIG. 4. 19. Diagrammatic representation of the interaction of altitude, exposure and duration of snow cover on the distribution of montane vegetation in the Cairngorm Mountains in Scotland. (*After Watt and Jones.*)

cally to certain microclimates are also operative in determining its geographical range. This principle, which is strictly applicable to ecotypes rather than to whole species containing ecotypes with various tolerances, he has crystallized as the Geo-ecological Law of Distribution, 'the topographical distribution is a function of the general geographical distribution', a version, if one may be permitted the comment, of the ancient hermetic maxim 'as above, so below'. Acting on this principle, he explored the possibilities of deduction in several directions, for example, to deduce the topographical (micro-) distribution from the geographical (macro-) distribution, or, with due caution and consideration of topographic and other local factors, the macro-distribution from the micro-distribution. Another form of the argument, from macro-distribution or micro-distribution to ecological requirements, is one already much used by foresters, gardeners and other cultivators in the selection of species for planting and of the sites in which to plant them.

Boyko gave some examples of the working of his principles in deducing ecological amplitudes, or tolerances, from distributional data, that is the use of species as indicators of climate. He took the road from Jerusalem to Jericho, which passes from a Mediterranean climate, with more than 330 mm of rain annually, through an Irano-Turanian zone, with 180–330 mm, to the Saharo-

Sindian region with less than 130 mm. In the Mediterranean zone *Ononis natrix* occurs on all slopes indifferently, in the Irano-Turanian zone it is limited, in diminishing abundance, to northerly slopes, and disappears in the Saharo-Sindian zone. Conversely, *Limonium thouini* is absent from the first zone, occurs, with increasing abundance, on southerly slopes in the

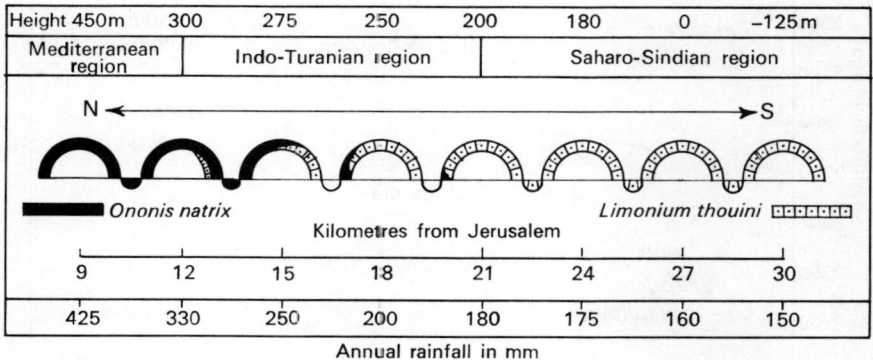

FIG. 4.20. Variation in the degree of exposure to insolation of a Mediterranean species, *Ononis natrix* and of a Saharo-Sindian species, *Limonium thouini*, along a traverse from Jerusalem to Jericho. The latter species appears at first only on south slopes but eventually replaces the *Ononis* on all exposures. (*After Boyko.*)

second zone and on all slopes in the third zone (Fig. 4.20). These facts are correlated with what he calls the IE factor (Insolation-Exposure). Solar radiation over this limited area is fairly uniform but by measurements of the solar radiation received per annum at different angles of slope and direction of exposure he was able to show the respective tolerances of these two species to insolation and how these tolerances change with decreasing rainfall. Thus the range of tolerance of *Ononis natrix* for insolation fell off rapidly with increasing aridity, while that of *Limonium thouini* increased with increasing aridity (Fig. 4.21).

The distributor of species as an indicator of climate is illustrated, even on a relatively small scale, in Britain. H. C. Watson in 1846 pointed out that certain species, which he called 'Germanic', were confined to eastern counties, while others, the 'Atlantic' species, were limited to the west, corresponding respectively to a more continental and a more oceanic type of climate, the latter being milder and more humid. The difference of climate is not very great, but it is reinforced by the prevalence of dry, sandy soils in the Suffolk area which is the chief centre of the Germanic species, such as *Veronica spicata* s.sp. *spicata* and *Artemisia campestris*. Humidity is of importance in this division and some species which occur in both areas show a marked difference of moisture tolerances in the east and the west. Thus the campanulacean *Wahlenbergia hederacea* grows in open grassland in the humid west, but only in boggy places and wet woodlands in the drier east.

Other species are more sensitive to temperature differences. The Atlantic

| Mediterranean | | | Irano-Turanian | Saharo-Sindian | | |

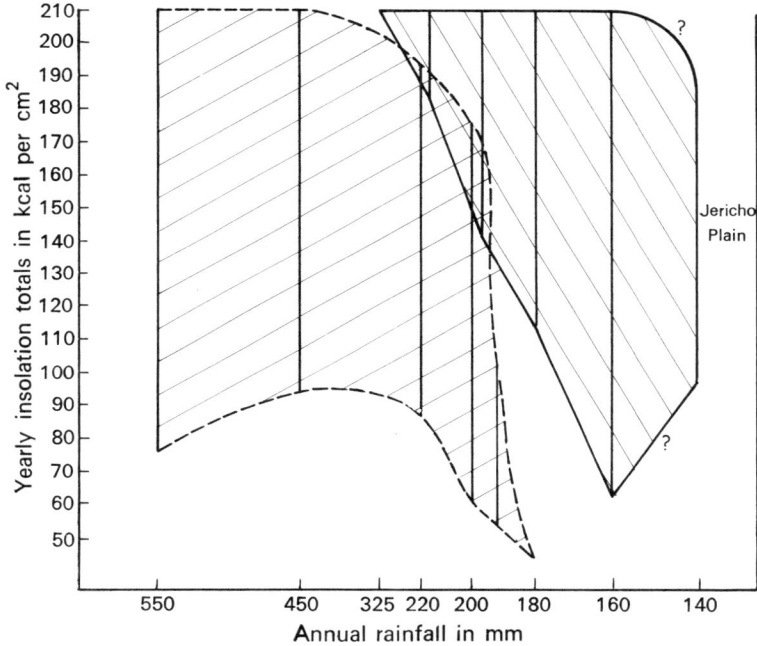

FIG. 4.21. Changes in the tolerance ranges with regard to insolation of *Ononis natrix* (Mediterranean) and *Limonium thouini* (Saharo-Sindian) in the traverse between the Mediterranean climate of Jerusalem and the Saharo-Sindian climate of the Jericho valley. (*After* Boyko.)

species *Rubia peregrina* is limited both in Britain and in Europe by the January isotherm of 4°C. As Salisbury points out, the plant begins to form its new shoots in January. *Ruscus aculeaius* on the other hand is more influenced by temperature at its flowering period and in Britain its area is practically confined by the mean maximum isotherm of 9°C for March, for at or beyond this limit it only sets seed in exceptionally favourable years. As Salisbury has emphasized, however, the coincidence of a distribution with some climatic feature is not proof of a causal connection, it only establishes a probability, and may well be influenced by edaphic factors as well as climatic.

It is rare for a climatic distribution boundary to be sharply marked, and in many cases it can only be observed statistically from a diminution of frequency (in the colloquial not the technical sense) and the absence or rarity of sexual reproduction beyond the boundary. Many plants can flourish vegetatively beyond their climatic boundaries, if they are protected from serious competition, but two things prevent their permanent establishment in such situations; first, their tolerances become so restricted that they are unable to compete on equal terms with the plants which are, so to speak, at home in the area; second, they cannot or only rarely form seed. In this respect plants which have an active means of vegetative propagation are at an advantage for they are to some extent relieved of both the above drawbacks. This is illustrated by *Ficaria verna*, which, in the northern part of its British range sets little or no seed and relies almost entirely on reproduction by bulbils.

Perennials have an advantage over annual plants in many climates with only a short growing season, especially in arctic and high alpine climates. Their leaf-initials, and often their flower initials, can be prepared beforehand and they are thus in a position to make full use of the short season, while autogamy or apomixis may afford compensation for a relative scarcity of pollinating insects.

An exception to this generalization is seen in semi-deserts with a short rainy season, during which the climate is so highly favourable that rapid growth is possible. Here the perennial population may be very sparse, consisting of a few specialized shrubs, but there may be an enormous population of ephemeral annuals, subsisting as seeds for the greater part of the year, but with the advent of the rains covering the desert, almost overnight, with a magic carpet of brilliant flowers. The shelter provided by the perennial shrubs may be of value in fostering the germination of seeds, just as clumps of trees or shelter-belts planted in treeless areas foster invasion by species which are unable to colonize the open, unprotected space.

Another case in which extremes appear to meet is in the adoption of the evergreen habit, for it predominates both in tropical rain forests and in the sub-arctic. The former environment permits uninterrupted growth throughout the year, with no resting period. The short summer in the sub-arctic also favours evergreens, which need lose no time in reclothing themselves with foliage when spring arrives. Deep frosts and drying winds, however, require that the leaves shall be xeromorphic. The niche is filled by the conifers of the great northern forests, or *taïga*, which answer both requirements. By travelling northwards, on almost any continental longitude we pass from rich, mixed broad-leaved forests to northern broad-leaved forests of fewer species and more homogeneous character. From these we pass to coniferous forests, occupying the great northern land masses, and finally meet a boundary zone of dwarf scrub, before trees disappear entirely and we enter the open heathlands of the arctic tundra.

Increasing altitude simulates the effects of higher latitude and a similar zonation according to height is shown by the vegetation. From the valleys we pass upwards through broad-leaved forests, of whatever type is indigenous to the lowlands of the area, then into coniferous forest and finally, through a scrub zone on to alpine grassland, rich in perennial herbs. The outline is the same but the parallelism does not hold true in details, since the alpine species and the arctic species may be quite different, though physiognomy and life-forms may be closely similar. In Britain, within a relatively short compass, it is notable that some species which are high alpines in Wales or northern England, are found at sea level in the far north or Scotland, e.g. *Saxifraga oppositifolia* and *Saussurea alpina*.

The adaptability of some plants is sometimes due to the natural selection of variant ecotypes, but it can also be brought about directly. There is no evergreen ecotype of *Quercus robur* in Europe, but in Jamaica it has assumed an evergreen habit. Trees of northern species raised in the southern hemisphere soon 'learn' after an initial period of uncertainty, to adapt themselves to a different seasonal rhythm. Others may also become evergreen, or, like *Salix*, remain deciduous but contract their defoliate period to a couple of weeks.

This contrasts with the conservatism of other species such as the deciduous trees of equatorial forests, where defoliation has no ecological meaning. It has been surmised that these may be survivors of an earlier xerothermic period before rain-forest conditions were established. If this were so it would imply a very high degree of conservatism with regard to defoliation, combined with wide amplitude of tolerances with regard to many other changed factors.

An important feature of any climate is the length of the frost-free period, which defines the growing season for perennials and the available life period for annuals. Here topography has a compensating influence. Anything which provides shelter from light frosts will prolong the growing period and may be decisive for species which have a critical length of growing period lying near the annual climatic mean of frost-free days. The influence of an oceanic climate along a coastal belt causes the prevalence of higher mean temperatures than occur inland and may eliminate frost as a factor altogether. The northerly range of southern species is usually more extensive along the sea-littoral than elsewhere. Salisbury has pointed out that 34 per cent of the vascular plants in Britain which are confined to the sea coast are of southern distribution, some definitely Mediterranean in origin, while among inland plants the percentage of southern species is only 7.5.

Not only strictly littoral plants are affected by an oceanic climate. *Quercus robur* extends to 63°N on the coast of Norway, while in continental Eurasia its limit is several degrees further south. The example of *Q. petraea* is even more striking, for it attains to about 61°N on the Norwegian coast, but in Russia does not occur north of the Caucasus.

By contrast the altitudinal limit of trees is much lower in oceanic than in continental areas. This is generally attributed to the inimical effect of strong winds on trees. The tree limit in Britain is about 457 metres (1500 feet), while in the more sheltered area of the Black Forest it rises to 1450 metres. Many oceanic islands are completely treeless, and the treelessness of the huge pampas area of southern Argentina has been attributed to the force of the winds which sweep across it. The sheltered valleys of the Andes at the same latitude are richly forested, while on the mountainous west coast of Patagonia and

Tierra del Fuego there are forests of *Nothofagus* at sea level, although these areas are much further south. The benefit of shelter from wind can be seen on many mountains in Britain, where sheltered gullies may be wooded to over 609 metres (2000 feet). Exposure at high elevation on mountains sets an upward limit to tree growth everywhere, but the shelter which the mountain

FIG. 4.22. The Franz-Joseph glacier in Westland, New Zealand. The glacier comes down to sea level and the influence of the coastal climate is shown by the subtropical vegetation, contrasted with the Southern Alps behind.

mass provides from wind allows this limit to reach higher levels on the side away from the prevailing winds, and there is a positive correlation between the size of a mountain mass and the prevailing tree limit or timberline within its area. Probably the highest tree limit in the world is among the Himalayas, where it may reach 3660 metres, only 240 metres below the snow line. That it is wind exposure rather than cold which sets the limit is evidenced by the fact that some of the coldest spots on earth, in Siberia, where January temperatures may sink to nearly $-70°C$ lie within the forest region.

Attempts have been made to express the correlation of important climatic factors by some method of integration. Amann proposed an *Index of Hygrothermy* to integrate precipitation and temperature. The Index is:

$$\frac{PT}{t_H - t_C}$$

$P =$ The annual precipitation in cm.

$T =$ The mean annual temperature in degrees C.

$t_H =$ Mean temperatures of the hottest month of the year.

$t_C =$ Mean temperature of the coldest month.

Walter and Lieth have created and applied a diagrammatic expression of the characteristics of a climate, of great value geographically, which we shall illustrate later, in the section on Plant Geography (see p. 3831).

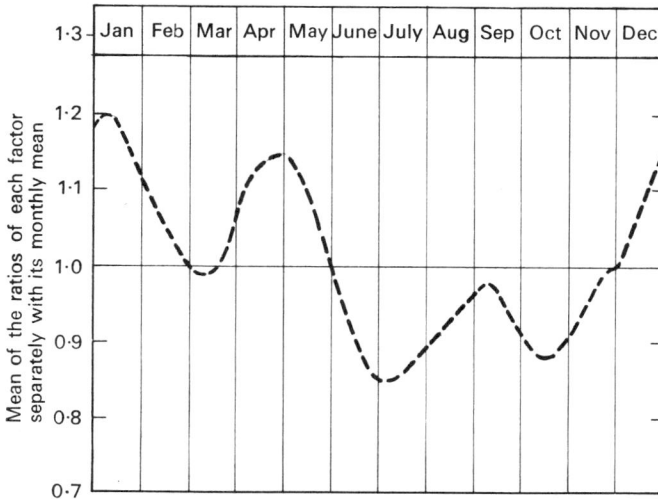

FIG. 4.23. Annual variation in the consensus of favourability factors in the climate of Rio de Janeiro. For explanation see in text.

In 1919 the present author suggested a method of constructing a 'favourability curve' of the variations of climate throughout the year, which he found useful. Four factors were employed, which it was assumed would increase plant growth, other things being equal. These were; temperature, rainfall, relative humidity and number of clear days, i.e. days in which cloud-cover was less than 50 per cent. Dividing the year into weeks the ratio of the weekly mean to the annual mean was taken for each factor separately and the mean of the four ratios plotted. The resultant curve for the neighbourhood of Rio de Janeiro is given in Fig. 4.23. This shows a considerable seasonal variation for a tropical area, but not surprisingly great for a mountainous district not far from the southern tropical limit.

Temperature as a climatic factor

Schimper, in his great book on Plant Geography, places temperature second to humidity because he considered that the physiology of water supply was better understood than that of temperature and that evidences of adaptation to variations of water supply were clear while adaptation to temperature variations was not. Today, however, with increased information, we must take a rather different view and recognize that temperature and humidity are equally important as climatic components and that temperature has probably a broader effect on the world's vegetation than humidity.

It is important, at the start, to distinguish between temperature and heat, for the two are not synonymous. Temperature is a measure of intensity, physically a measure of the amount of molecular agitation which expresses heat. Heat is measured in calories. The gram-calorie is the amount of heat required to raise 1 g of water from 15°C to 16°C. For practical purposes the kilocalorie is often substituted, which is the heat required to raise one kilogram of water through the same temperature interval. The calorie is therefore a measure of quantity, while temperature is a linear measure for which any suitable scale may be used. Fahrenheit, somewhat arbitrarily, took the coldest freezing-mixture known to him as his starting point. Celsius adopted the universal constant of the melting point of ice for the zero of his centigrade scale, which corresponds to 32°F. What is called 'absolute temperature' is reckoned from −273°C, at which point molecular movements cease, or in other words, there is no heat at all. This is the zero of the K or Kelvin scale.

Both heat and temperature have their ecological importance. Primarily, solar radiation is the only important source of energy, including heat, for the effects of the earth's internal heat are negligible so far as the biosphere, the zone of living organisms, is concerned. This radiation produces changes of temperature in objects which absorb it, but in green plants it also supplies energy for metabolic processes, so that only a fraction of the energy supply appears in raised temperature. The amount of energy received in this way at the surface of the earth varies greatly with the angular height of the sun above the horizon, which implies a daily variation everywhere and also a seasonal variation which depends on latitude. It also varies with the altitude of the ground, for rather more than 50 per cent of the incident radiation, is absorbed by the atmosphere, which is warmed by it. As the greater part of this radiation is absorbed in the lower, denser atmosphere, it follows that mountains receive more radiation energy than lower regions. Absorption affects the ultraviolet and the short wavelengths more than the long waves, so that mountains also receive a higher proportion of the short wavelengths, which have effects on the growth of alpine plants.

The highest measurements of solar energy received were made on the

Peak of Teneriffe (3683 m) and on Mt. Whitney in California (4420 m). These agreed at 1·64 gcal/cm²/min. By extrapolation from measurement at different heights a mean value of 1·94 g cal/cm²/min has been arrived at as the

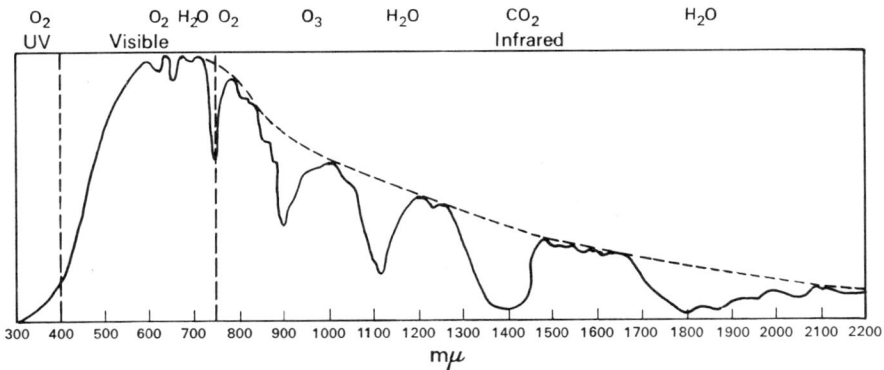

FIG. 4.24. Distribution of energy in the solar spectrum, showing the chief regions of absorption by Ozone, Oxygen, Carbon dioxide and water vapour in the atmosphere. (*After Langley*.)

amount of radiation entering the atmosphere. This is called the *solar constant*, which depends on the sun's activity and only varies over immensely long periods. (See under Ice Age, Vol. 3, p. 2572.) Dorno has calculated that 75 per cent of this energy reaches down to 1800 m. above sea level and less than half of this amount actually arrives at sea level itself. The amount absorbed by the air is not equal for all wavelengths (Fig. 4.24). Some 50 per cent of the energy of solar radiation is in the infrared, and this and the short waves (below 500 mμ) are largely absorbed. The extraterrestrial energy maximum in the solar radiation is at about 470 mμ, but at sea level the energy maximum is at about 650 mμ, in the yellow-red region of the spectrum, which is the region chiefly active in photosynthesis. The sun shines on us, in fact, through a yellow filter, and near the horizon, its beams come through an extra length of dense, lower air as through a red filter.

The energy absorbed by the atmosphere (which includes suspended dust and water droplets) is reradiated at infrared wavelengths as thermal radiation, partly back into space and partly downwards to the earth. Of the remainder, which reaches the earth, a part is lost by reflection back as light. The amount thus lost depends on the nature of the surface. Forests generally reflect less than 10 per cent, but grassland may reflect up to 32 per cent. This is a considerable loss, but it is much less than from some uncovered surfaces, e.g. sand 12–50 per cent, water up to 75 per cent, and fresh snow, 80–90 per cent. This is the light by which the earth is visible from the moon. It is expressed by astronomers as the 'albedo', i.e. the mean reflection as a percentage of that from a totally reflecting white surface.

The energy absorbed at the earth's surface is accounted for, first, by raising the temperature of the soil; second, by the latent heat of water evaporated; third, in a similar way by promoting transpiration in plants; fourth, about 1 per cent in photosynthesis; fifth, a small amount in raising the temperature of plants.

From the earth's surface there is a continuous reradiation of heat, the thermal loss. This, and the loss by reflection, have to be subtracted from the incoming solar radiation to get the *net radiation*, which is positive in the daytime, owing to the normal excess of solar radiation, and negative at night when solar radiation ceases, with the lowest negative value when the sky is clear. The net radiation is that available for raising the temperature of soil, the air near the ground and for the evaporation of water. In winter it may be continuously negative over some periods, leading to condensation instead of evaporation and eventually to freezing. In an arctic climate radiation may be continuously negative from September to April.

Apart from radiation there is also some exchange of heat between air and soil by convection, and conduction by air turbulence, which, in temperate climates, is only considerable during the summer months, but in a hot, dry climate, where there is little or no evaporation, it may be important all through the year.

This balance of gains and losses expresses the *energy budget* for any particular locality, by means of which the vegetation must live and which forms one of the most important sectors of the total environment. If we can, by experience, learn to interpret the signs given by the vegetation they provide the best and most sensitive indicators of climate. Severe climatic exposure provides perhaps the most easily recognizable signs in the growth of plants, either in prostrate mat-like forms (e.g. *Prunus spinosa*), or dwarf, close, cushion-like forms (e.g. *Calluna vulgaris, Ulex europaeus, Vaccinium myrtillus*).

It is probably true to say that, apart from the frequent accompaniment of drought, no situation is too hot for plant life in some form, and few, putting aside the effects of wind, are too cold. The latter point is illustrated by the fact that the lowest temperatures on record occur in the forest region of Siberia. Continuous cold, if it keeps the soil frozen, prevents growth simply through drought. It has been estimated that about one-fifth of the earth's land surface has permanently frozen soil (permafrost) only a thin surface layer of which thaws out to give roothold to small plants. Even in Antarctica exposed rocks may thaw out in the summer, but the erosive effect of the fierce ice-laden winds of winter inhibits the establishment of any but the lowliest of plants. Temporary extremes of cold, although decisive in the distribution of many species, are not in themselves totally prohibitive of plant life.

The energy income, measured in calories, varies with latitude, according to the sun's elevation, as is shown in Fig. 4.25, which shows also the seasonal variation. These figures refer to horizontal surfaces, but the angle of inclination of the surface also makes a great difference to the energy receipt.

A surface inclined towards the sun, so that its incident rays are normal to the surface will naturally receive the maximum energy and a surface similarly inclined away from the sun will get the minimum (Fig. 4.26). The receipt depends not only on the angle of the slope, but on its direction. Boyko gives figures which show the following values for slopes of 30° in different directions, near Jerusalem. If the annual energy receipt on a horizontal surface be taken as 100 then slopes of 30° facing south receive 120 units, SE/SW, 108; E/W 90; NE/NW 56; N 53. This is at a latitude of 31° 48′N, and at higher latitudes the differences become much greater, as the north facing slopes may receive no direct sunlight at all, even in summer. Rübel has estimated that in the Swiss Alps a southern slope may receive five and a half times as much insolation as a northern slope (Fig. 4.27). These differences in energy receipt,

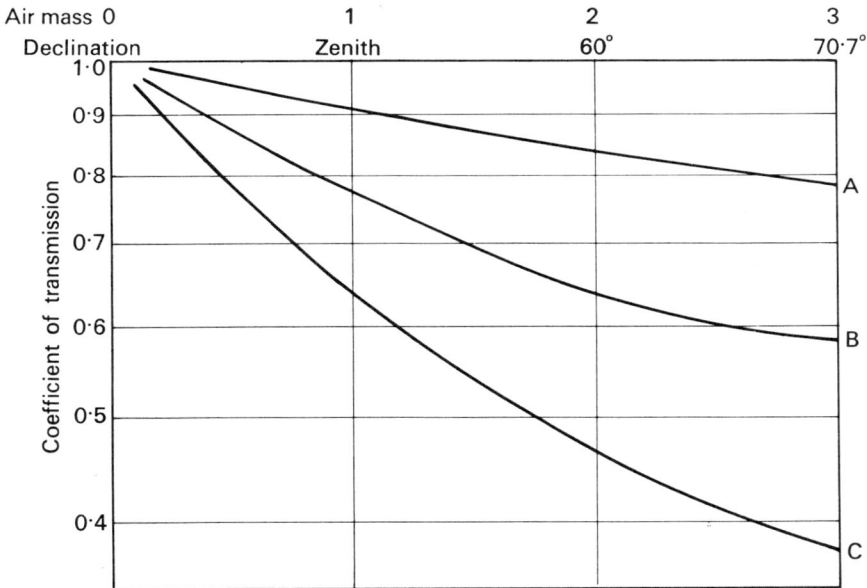

Fig. 4.25. Variation in the amount of solar radiation transmitted by the atmosphere at different heights and varying declination of the sun from the zenith, for places in mid-temperate latitudes in summer. The numbers 1, 2 and 3 represent the airmasses traversed by the solar rays. A. The Jungfraujoch, 3457 metres. B. Davos, 1600 metres. C. Washington D.C., 127 metres. The curves are extrapolated backward towards the super-atmospheric value. The amount of atmospheric absorption diminishes logarithmically with height. (*From Duggar.*) (See also p. 3543.)

apart from the supply of energy to the plants, have marked climatic effects on air temperatures, soil temperatures and the length of winter conditions, and in consequence there are often great differences in the vegetation on north and south slopes (Fig. 4.28). Even when the same communities exist on both slopes, they show considerable differences in the heights to which they reach. An example from the Italian Alps, quoted by Schröter, showed a difference of 100 metres between N and S slopes for the upper limit of *Pinus–Larix europaea* forest and a difference of 250 metres for the upper limit of alpine meadows.

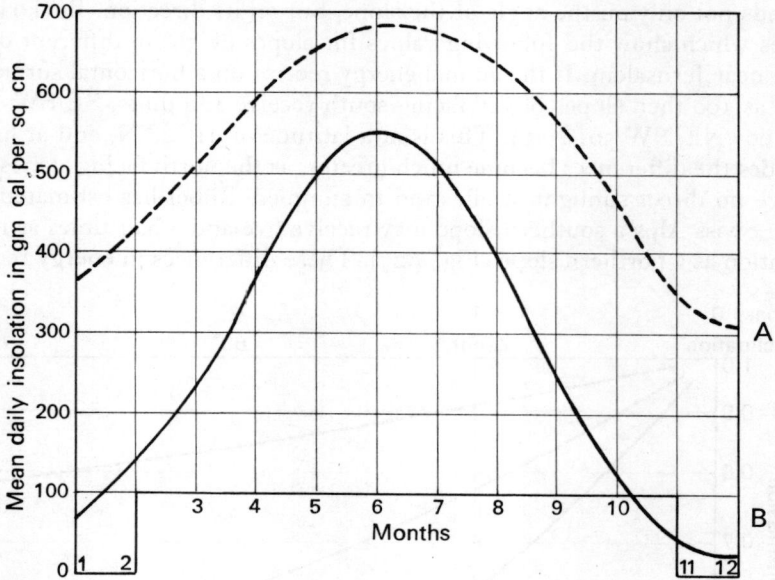

FIG. 4.26. Variation in the amount of insolation throughout the year on two surfaces, one horizontal (A) and the other (B) lying at an angle of 40° NNE. Drawn for data for Jerusalem lat. 31° 48′ N. The difference is least at mid-summer when the sun is highest. (*After Boyko*.)

Schröter gave a classic description of the effects of exposure in the Findelen valley in Canton Valais. On the slopes facing south, rye is cultivated up to 2100 metres, accompanied by a sparse turf of xerophilous grasses, while on the north slope, only about a kilometre distant as the crow flies, there is a dense pine forest accompanied by dwarf shrubs of the type of the arctic tundra, representing a difference equivalent to 30–40 degrees of latitude.

When the south exposure also happens to lie in the 'rain shadow' of the hill, so that it receives only a scanty rainfall compared with the north exposure, the difference in the conditions can be much exaggerated. For example, Gail found in Idaho that the mean relative humidity in summer was 22 per cent

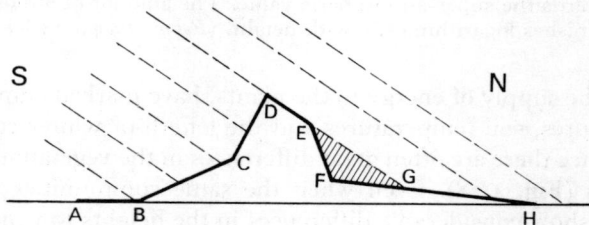

FIG. 4.27. Profile of an ideal mountain along a N–S axis, showing the differing degrees of insolation. C–D receives most, D–E the least, while E–G is in perpetual shadow. (*After de Martonne*.)

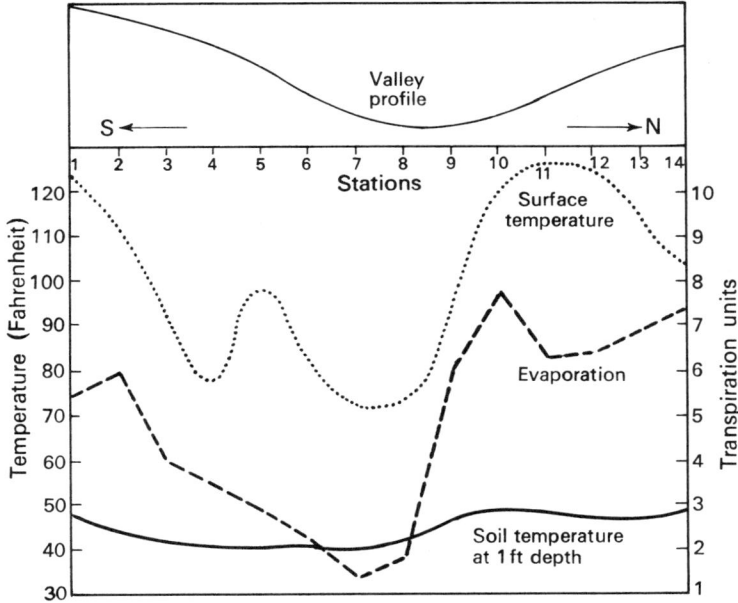

FIG. 4.28. Diagrammatic representation of the differences in some environmental factors between the two sides of a valley of which one side faces south and the other north (average inclination about 20°). (*After Bates from Braun Blanquet.*)

higher on the northern than on the southern slopes of a mountain. The northern side was covered with *Pseudotsuga taxifolia*, the southern side bore only the xerophilous *Pinus ponderosa*.

Many factors may operate to modify the difference of exposure, such as the relative angles of slope, the water-holding capacity of the soil, relative humidity of the air, rate of evaporation, the length of time that snow lies, the liability to avalanches or landslides, the character of the rock, etc. Nevertheless, the difference is always present in some degree and it is noteworthy as a generalization that, in central and northern Europe, species of Mediterranean affinities frequent southern exposures while, in the Mediterranean area, northern species are found on northern slopes.

Among its various effects the heating effect of solar radiation on the air and the soil are highly important; they are, however, very different. There is, of course, a great deal of quantitative variation due to differences of latitude, season, cloud cover, relative humidity and topography, but there is also a diunal variation which is common to most localities due to the intrinsic differences between air and soil.

During the night the surface of the soil loses heat at rates of from 0·12 to 0·15 g/cal/cm²/min. This heat is rapidly removed by convection in the air, with the result that in the early morning the soil surface is colder than the air immediately above it and this, in turn, is colder than the air at 1·0–1·5 metres above ground level. This heat loss is greatest in dry air and under a clear sky, which accounts for the tremendous drop in temperature at night which may occur in deserts. It also accounts for the fact that ground frost may occur without air frost.

The heat loss at night is reduced by the vegetation cover. The deeper the vegetation the smaller the loss. It is least in forests, which tend to minimize temperature changes both by day and night.

After sunrise conditions are reversed: the ground surface heats more quickly than the air and during the forenoon hours the surface is warmer than the air immediately above it and this air layer may be from 2·5–8°C warmer than the air at 1·5 metres. The difference may be even more in an arid climate. Towards evening an equilibrium is approached, when these temperature differences become minimal. Under strong insolation the temperature of the soil surface rises very high indeed, over 65°C having been observed in Nevada. The surface may be 20°C hotter than the air close to the ground. If there is any plant cover the plants may reach temperatures 10–15°C higher than the surrounding air, according to their absorptive power, which is greater for open grassland than it is for woody vegetation.

Below the soil surface temperature conditions are very different. While the surface heats in the morning more rapidly than the air above it, in the evening and at night this is reversed and the surface cools more quickly than the air. In the early afternoon the soil surface under full sun is hotter than the air, but this applies only to the uppermost 1 or 2 cm, below which the temperature gradient drops rapidly. Kraus measured a difference of 8°C between depths of 2 cm and 10 cm at Wurzburg, i.e. a gradient of 1°C per centimetre. Sinclair in Nevada, found a drop of 29°C between 0 and 2 cm depth and 15°C between 2 and 10 cm, but only 5°C between 10 and 20 cm. This gradient is subject to diurnal fluctuation down to a depth of 30 or even 50 cm, beyond which the direct influence of daily insolation does not extend. From this level downwards the temperature gradient is much less steep and the temperatures are subject only to long period variations, which form an annual cycle. The effects of seasonal differences of insolation are, however, felt to very considerable depths, as much as several metres in sub-tropical climates.

Dry soil has a low specific heat, but is a poor conductor, this is shown by the rapid surface heating and by the steep fall of temperature below the surface. The soil water, however, heats and cools more slowly than the soil materials. The annual cycle of temperature changes in the soil lags behind the seasonal changes at the surface and the deeper the level the greater the lag, so that at 3 metres depth the maximum may not be reached before September or the minimum before April. Moreover, the annual variation becomes markedly smaller with depth, though a small variation is still perceptible at a depth of 3–4 metres in the sub-soil. In Montana, with a hot summer and a very cold winter, the soil at 3 metres depth showed an annual variation of 7°C, while at 30 cm depth the variation was 20°C (Fig. 4.29). At this depth the soil was frozen for three months of the year, but at 1 metre the temperature barely reached 0°C and at 3 metres it did not fall below 3°C. There is thus a seasonal reversal of the temperature gradient in the soil, at least in temperature climates, the deep soil being warmer than the upper layers in winter and cooler in summer, with a change over in spring and again in autumn.

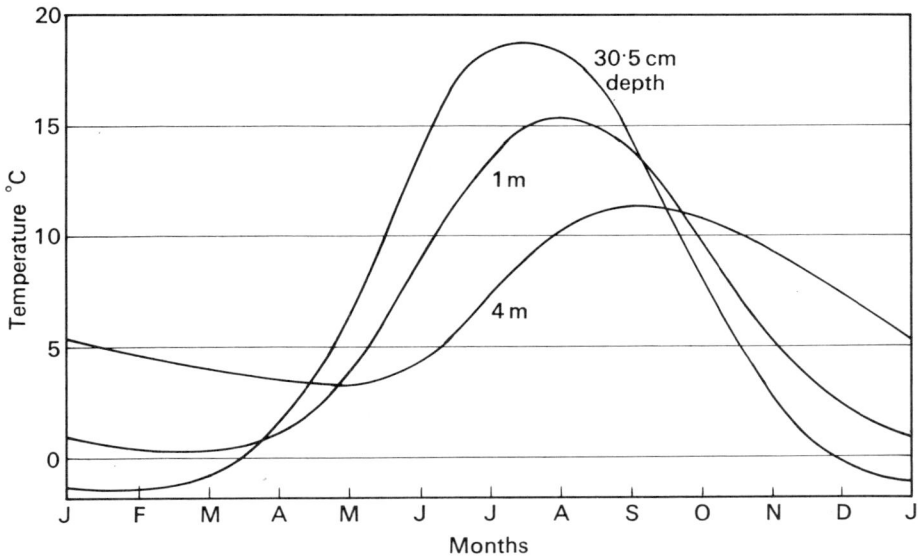

FIG. 4.29. Mean monthly temperatures at three levels in the soil, in Montana. The gradients become less with increased depth, show an increasing lag and are reversed between summer and winter. (*After Fitton and Brooks.*)

Prolonged freezing of the soil, which in arctic-alpine conditions may extend much deeper than 40 cm has a very restrictive effect on vegetation as it limits root penetration to the surface layers. On the other hand temporary freezing of the soil in many alpine situations is compensated by the retention of warmth in the deeper layers of the soil, which enables the deep roots of alpine plants to survive and even to function in the absorption of water during the cold weather. As the movement of heat in the soil is almost exclusively by conduction, dense soils lose heat more rapidly than loose, humus soils, which are therefore resistant to freezing.

The intense heating (70–80°) and the rapid and extensive changes of temperature at the soil surface which may occur in sub-tropical climates, are naturally most inimical to seedlings and may limit germination to a short season when these factors are at their least unfavourable.

The fact that the temperature coefficients of the three leading physiological processes of the green plant, photosynthesis, respiration and growth, are highest in the range 0–10°C is very favourable to plants of cold situations. Especially is this so for alpines, and accounts for the striking rapidity with which such plants spring into active growth and flowering as soon as the snow cover melts, or even before it quite disappears, as does, for example, *Soldanella alpina*, whose purple bells emerge through the last inch of snow.

In very cold climates snow is in fact a most valuable protection for plants, not only from extremes of cold, but even more importantly from the drying effects of sub-zero winds. A comparatively slight cover can have a markedly protective effect. There may be an increase in the daytime of 8°C between a 3 cm layer of snow and the air above it and the difference between the snow and the air at 1·5 metres may be greater still, but at night the coldest zone of air is that immediately above the snow.

Some protection against the effects of extreme cold is afforded by an increase of the osmotic value of the cell-sap after exposure. As measured by plasmolysis, this may amount to a doubling of the value, which lowers the freezing point of the fluid. Several other factors of cold resistance have been suggested, such as an excess production of pentosans, but the matter has not yet been fully investigated. The most effective protection is, however, the growth habit of the plant, either passing through the winter in seed form or perennating only by underground parts. Protection against overheating is chiefly by means of transpiration which consumes incident energy as the latent heat of vaporization. Its practical value appears in some tests on leaves growing in tropical sunshine. When transpiration was checked by vaselining the stomatal surfaces, many of the treated leaves turned brown and died within a short time.

An influential temperature factor, which is not always given its proper consideration, is that of cold air drainage. Cold air, being heavier than warm air, always tends to sink, and flows, like a liquid, towards lower levels. If this flow is checked by topographical features such as hollows or cul-de-sac valleys, the cold air piles up to great depths, with the coldest air at the bottom. Such 'frost-pockets' as they are called can suffer exceptionally low winter temperatures and they are rightly dreaded by professional cultivators. In these places the usual temperature gradient is reversed, the temperature increasing with height above the bottom level. A mountainside may thus provide a warmer winter habitat than the valley below it and this may be reflected in a reversed zonation of the vegetation, with the least hardy plants forming the highest zones. Temperature differences of well over 20°C may exist between the bottom and the upper margin of such frost hollows, and, even more damaging to vegetation, the frosty period at the bottom may be prolonged as late as June (Fig. 4.30).

Just as there is a downward temperature gradient in the soil, so there is an upward temperature gradient in the air. The mean figure for the decrease of temperature in still air is 5·5°C for each 100 metres of altitude which is independent of latitude, but is somewhat greater in summer than in winter, when it may drop to 1°C per 220 metres. The gradient is affected by topography, being somewhat steeper on slopes facing south than on north slopes. It is also affected by diurnal variations. In this respect there is a great difference between air and ground. With increasing height the heating power of the sun on the air diminishes with the decreasing density of the air, that is to say that less of the initial value of the solar radiation is lost by air absorption. This leaves a greater proportion available for heating the ground and the plants. The daily

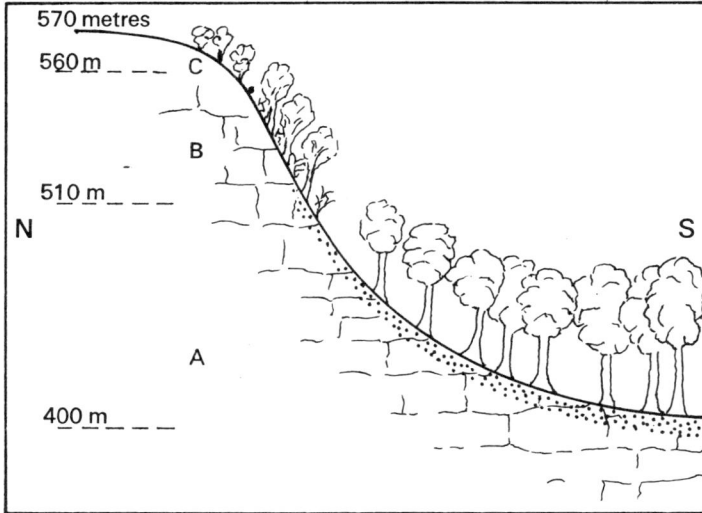

FIG. 4.30. Section through a frost-hollow, the ravine of Bugarach in the eastern Pyrenees. The hardiest species, *Fagus sylvatica*, (A) grows at the bottom, the less hardy *Quercus-Buxus* (B) on the upper slopes and the least hardy, *Quercus ilex* (C) at the top, a reversed sequence. (*After Braun Blanquet and Susplugas.*)

variation of air temperature is thus less as the height increases, though the average temperature is lower. The same is true of the annual variation and Hann has calculated that at 9500 metres seasonal effects disappear and the zone of perpetually low temperatures is reached.

At the level of the summit on Montblanc (4810 metres) only 5 per cent of the incident radiation has been removed by air absorption, while at the sea level it is diminished to 32 per cent. The heating effect on the ground at high levels is thus much greater than at low levels and, although the air may be thin and cold, both plants and soil are intensively warmed. The difference between sun and shade temperatures is greatly accentuated; at 3000 metres a difference of 33·5°C has been recorded. Reradiation at night is also more intense at high levels and may be as much as 0·20 langleys. Alpine plants in the sunny places therefore suffer a sharp difference between day and night temperatures, much greater than those in the shade, which were only at air temperature during the day. The heating effect of the insolation on the soil is considerably greater at high altitudes, just as the heating effect on the air is less. At 1811 metres height in the Engadine, Diem found the following annual mean temperatures, measured at midday over a period of five years: air temperature 1·79°C; soil temperature at 5 cm depth, 5·12°C; at 30 cm, 4·53°C; at 120 cm, 4·92°C. This deep-warmed soil is of the greatest value to alpine plants, and, as Schröter says, it shows the difference between the 'plant climate' and the purely 'meteorological climate'.

The difference between the mountain climate and the lower levels is greatest in spring, when the energy of insolation at alpine levels is taken up by the melting snow, rather than by the soil and it is least in autumn by which time all levels have been warmed up. Schröter instances that on the Rigi at 1800 metres the primrose (*Primula acaulis*) flowers six weeks later than at Zürich, but the autumn crocus (*Colchicum alpinum*) flowers simultaneously at both levels.

The alpine climate is thus very different from the arctic climate. Both have low annual air temperatures and both have prolonged snow cover, but there the resemblances end, so that there is little in common between their floras. In the mountains there is in summer intense illumination and well-marked soil heating, which persists to considerable depths. There are also sharp alterations of air temperature between night and day. Even in winter plants not covered by snow can be warmed by the sun into growth and flowering.

In the Arctic the weak illumination in summer is not compensated by its long continuity, and the ground is not warmed above winter temperatures except at the surface. Below a foot in depth it may remain permanently frozen. In the whole archipelago of Spitzbergen there are only 123 species of flowering plants, while 4·5 ha (about 11 acres) near the snow limit in Switzerland yielded 131 species, which shows how much more unfavourable are the arctic conditions.

Topographical features can upset general rules. Reflection from rocky mountainsides may raise the temperatures in a wide valley much above that which would be normal for its altitude. On the other hand a high plateau is much colder than the equivalent height on a mountainside. On a large scale, geographical features have also profoundly modifying effects. The proximity of the sea promotes a mild and moist climate, with only a small seasonal range, the oceanic type, in contrast to the high average temperatures, lower humidity and pronounced seasonal changes which characterize the continental type of climate. Within the oceanic zone the direction of ocean currents produces extensive modifications of climate. The effect of the Gulf Stream drift on north-western Europe is familiar, but there are many other examples. In South America the east coast is warmed by an equatorial current, the west coast is cooled by an antarctic current. In South Africa the same conditions obtain and the contrast between the cold Atlantic water and the warm water from the Indian Ocean can be observed even on the two sides of the Cape Peninsula, a few miles apart. The sea has a high specific heat and a drop of even 1°C in the sea temperature gives up enough heat to warm the supernatant air by 10°C.

Bodies of water, like the soil, are warmed by insolation at the surface. Water which is sufficiently shallow to be turned over by the wind will tend to be uniform in temperature, but deep water shows a downward gradient of decreasing temperature to a depth which depends on the effects of wind and wave action. At this level there is a narrow zone, the *thermocline*, at which the temperature rapidly drops to the level of that in the deep water and below this level it remains practically constant downwards. The surface water, above the thermocline, is called the *epilimnion* and the deep water the *hypo-*

limnion. This condition persists throughout the summer, but in winter the surface water cools until it becomes colder and heavier than the deep water; it then sinks to the bottom and displaces the deep water upwards. This annual turnover affects all bodies of water apart from the equatorial zone, though in the deepest seas it is too slow to be fully effective. It is of immense importance biologically in bringing to the surface reserves of nutrients stored in the deep water and preparing for the spring outburst of planktonic life.

From what has been said it will be obvious that the heat-status of plants is far from constant and that there is a continuous exchange with the environment. The plant gains heat from solar radiation and from convection in the air and conduction from the soil and neighbouring bodies. It loses heat by surface reflection, by convection in cold air and by conduction to cold soil and, most of all, by the latent heat of evaporation.

The effects of heat on plants and their reactions to it are specific characteristics or even to some extent individual and the individual reactions may change considerably during a lifetime. Thus, generalizations are few and, such as they are, have many exceptions. Ecotypes with different temperature tolerance are often found in species which have a wide range in latitude and individuals from the two extremes of the range may react quite differently.

Species may be classified as either *stenothermic*, having a narrow range of heat tolerance, or the opposite, *eurythermic*. Many marine Algae and Bryophyta are closely stenothermic, which limits their geographical range, while lichens and some wide-ranging Angiosperms, such as *Phragmites communis*, are remarkably eurythermic. In fact the geographical and topographical distribution of species is, in general, a useful guide to their status in this respect, but it is not the sole criterion since photoperiodism (Vol. 3, p. 3197) is also effective in limiting geographical range. Topographical circumstances must also be considered, for a plant occupying a situation with violent temperature changes between night and day must be regarded as eurythermic, although its geographical range may be restricted by some entirely different factor.

The normal temperature range for the physiological activities of plants is between $0°$ and $45°C$, but some hardy species can carry on photosynthesis at -3 or $-4°C$ and, although they may become inactive at $45°C$, many plants can survive much higher temperatures if they are not too prolonged. There are even some thermophilous algae whose whole range of activity lies above this point; they can survive at temperatures over $80°C$.

While it is true in a general way that increased temperature increases physiological activity, within the above range, the temperature coefficients of physiological processes (Vol. 3, p. 3177) differ from each other to some extent and the coefficient of any one process diminishes as the temperature rises, i.e. it is greatest in the range $0-10°C$. The most favourable temperature at any stage of the plant's life history depends, therefore, on the balance of the physiological activities at that stage. This is known as the *harmonic optimum* and the series of harmonic optima throughout the life history constitutes the *ecological optimum* which must correspond with external conditions or the plant cannot survive. The harmonic optima can vary markedly at different stages of development, particularly in plants of temperate climates and there may be phases in which low temperatures are essential, especially before the time of germination (see discussion on the maturation of bulbs, Vol. 3, p. 3179). Most temperate plants require lower temperatures at night, but equatorial plants are much less variable in their optima and can be cultivated at constant temperatures.

Sudden changes of temperature can be more injurious to plants than slow changes. This applies in both directions. Plants which have survived a night frost may be killed by rapid thawing in the morning if they are exposed to an early sun: this is the reason for the somewhat paradoxical method of preventing injury by spraying frozen plants with water; it delays thawing.

Susceptibility to cold injury varies at different points of the life cycle, seedlings being generally more susceptible than older plants, while plant organs also differ among themselves. Least susceptible are seeds and spores, some of which can survive immersion in liquid Hydrogen ($-270°C$) without apparent injury. For this reason some perennials become annuals in the coldest parts of their range, thus escaping dangerous cold. Stems are more resistant than roots and young leaves than old leaves, while floral parts are generally the least resistant. The cambium of the stem may survive when young, living xylem elements are injured. Such a zone of injured cells in the xylem may remain as a permanent feature called a 'frost ring', a serious defect of timber as it splits there very easily.

Some tropical plants are so sensitive that they succumb to temperatures below $10°C$, but well above freezing point, especially if exposure is prolonged. Most plants of cool climates, on the other hand, have a definite cold requirement, a period of low temperature at about $3-8°C$ in order to accomplish a rest period during which biochemical changes necessary for further development take place. This requirement prevents the growth in the tropics of many cultivated plants from temperate climates, except at high levels. If the cold period is curtailed the plants may pass into a condition of secondary dormancy, without growth, until the next cold season comes round (see under Dormancy and also Vernalization in Vol. 3, pp. 3181 and 3183).

In the dioecious species *Stratiotes aloides* the male and female plants have different temperature relationships. The geographical area of the species forms a belt across Europe from west to east. Salisbury has pointed out that south of latitude 52°N the male plant is almost exclusively found, while between 55° and 68°N the female alone occurs. There is an intermediate zone where both occur, and it is only in this region that seed formation is possible. Valentine, in comparing closely related pairs of species with differing ecological requirements (ecospecies), points to a similar separation of *Nuphar lutea* and *N. pumila*: the former is widespread in southern, lowland waters in Britain, while the latter is a local northern species. Salisbury, again, discussing the distribution of *Bartsia viscosa* in Britain, attributes its southern and western preference to the need for mild temperatures in September the month in which it ripens its seeds.

We may reiterate, in conclusion, that one may be easily deceived by coincidences into interpreting plant distributions as dependent on some temperature factor, without considering the fundamental question whether temperature is or could be truly a limiting factor in the given circumstances. Wide generalizations which seek to relate areas of distribution to such factors as annual mean temperatures or the number of frost-free days must be regarded as questionable.

Humidity as a climatic factor

Humidity of the atmosphere is quite a distinct factor from water supply, though both depend chiefly on precipitation. Water supply is a factor of the substratum and is considered later in connection with soils (p. 3619). The soil, it is true, contributes importantly to atmospheric humidity by evaporation, and, ultimately, precipitation means the return to the soil of moisture previously derived from it or from the sea. Roughly about one-third of the water precipitated on the ground is directly evaporated, the other two-thirds escape, either by surface drainage or through subterranean channels, and return eventually to the sea.

The downwardly percolating water raises the water content of the soil till it reaches its *field capacity*, that is the amount of water it can hold against gravity, any excess of water moves further downwards to join the *ground water*, which is drawn downwards and laterally by gravity till it finds an outlet in springs, rivers, lakes, etc., from which its flow seawards is free. The level of this ground water depends on local circumstances and is therefore variable both in place and time. It is called the *water-table* and is represented approximately by the standing level of water in a well. It always exists, even beneath deserts, though it may lie too deep to be of direct value to plants. As this water may travel great distances underground a dry area may have a plentiful supply of ground water derived from some higher, rainy area.

It has been calculated that the earth's atmosphere holds, at any time, only the equivalent of twenty-five millimetres of rainfall, about ten days supply. It is therefore evident that the cosmic circulation of water by drainage, evaporation and precipitation must be both free and rapid. It is, however, by no means uniform, as different climates testify, and while some places like the Khasi Hills in India have an almost perpetual rainfall, others, such as the northern parts of Chile get practically none at all.

Before considering these remarkable differences we must say something about the water vapour held invisibly in the air. The total amount which the air can hold is a function of temperature, increasing steeply as the temperature rises. The actual weight of vapour held by unit volume of air is called the *absolute humidity*. It is measured by the *vapour pressure*, that is to say the partial pressure of water vapour in the air as expressed in millimetres of mercury. Between 0° and 10°C this increases from 4·6 mm to 9·2 mm, a rise of 4·6 mm, but between 50° and 60°C it rises from 92·5 mm to 149·3 mm, a difference of 56·8 mm. These are the saturation values. The amount by which the actual vapour pressure falls short of the saturation value at a given temperature (of which there are published tables) is expressed in millimetres of mercury as the *saturation deficit*. This is an important factor as it is a measure of the *evaporating power* of the air, since it tells us how much more water the air at that temperature is capable of absorbing by diffusion. Practically, the evaporating power is increased by wind so that it may fluctuate independently of the saturation deficit. Only in still air would the two correspond.

The only direct method of measuring evaporating power is by measurement of the water lost from an evaporating surface of known area in a given time. The standard of reference is a free horizontal surface of pure water, but its use is hedged about with too many experimental difficulties to make it a

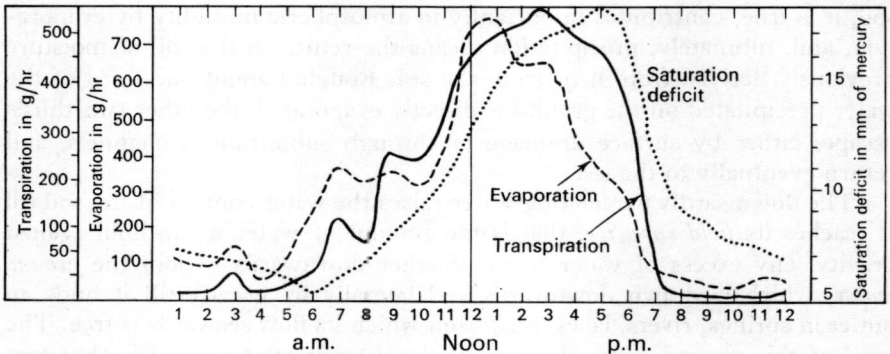

FIG. 4.31. Daily variation of the transpiration rate compared with evaporation and the saturation deficit in the atmosphere. Taken on a clear day at Akron, Colorado in August. (*After Briggs and Shantz.*)

practicable field standard and recourse must be had to solid evaporating surfaces, filter paper, linen or porous porcelain (i.e. the Livingston atmometers) which have been calibrated against the free-water values. The Livingston types can be purchased already calibrated. Each instrument must have its own coefficient to convert its readings into free-surface values.

The evaporative power of the air is not only related directly to the rate of water loss from the soil, but in a more general way to the rate of transpiration. The latter is, however, importantly affected by the action of light on the stomata, which is unconnected with evaporation, so that the two quantities do not march in step. Transpiration increases in the morning hours more rapidly than does the saturation deficit, the latter only reaching its peak in the later afternoon, some time after transpiration has begun to decline, while evaporative power, aided by wind, may be maximal earlier than either of the other two quantities (Fig. 4.31).

One of the most commonly used measures of atmospheric humidity, which is easily ascertained, is the *relative humidity*, which is a figure of the percentage saturation of the air. This again depends on temperature, as do the saturation values; thus at two different temperatures the same percentage relative humidity may occur when the saturation deficits, and hence the amount of water that air is able to absorb, are quite different. Daubenmire quotes the example that an RH of 60 per cent at a temperature of $15\cdot6°C$ implies a saturation deficit of only $5\cdot24$ mm, but at $26\cdot7°C$ the same RH implies a deficit of $10\cdot38$ mm, almost twice as great.*

Relative humidity is measured with paired thermometers, the bulb of one being wrapped in linen or cotton material which is kept wet. Air is drawn across both bulbs or else the thermometers are mounted so that they can be rotated rapidly through the air. The evaporation of water from the wet bulb lowers its temperature and from the difference between this reading and the dry-bulb reading the RH can be found by reference to tables.

When air is cooled, its absorptive capacity for water diminishes rapidly, with a corresponding fall in the saturation deficit, and if cooling continues saturation point will be reached, without any change in the absolute amount of water in the air. The temperature at which a given body of air reaches saturation is called its *dew point* and any further cooling results in the surplus of water vapour being eliminated from the vapour phase as liquid droplets, with some release of latent heat to the air, which tends to counteract further cooling. It is for this reason that a dew point above 0°C is supposed to prevent air frost.

The eliminated water takes different forms according to circumstances. If the air has been cooled by rising, the water forms clouds, whose height above the ground depends on the degree of saturation of the rising air. Electrical forces may cause the condensation of the cloud droplets until they are heavy enough to overcome the friction of the air, at which point they fall as rain. If air is cooled at or near the land surface, condensation takes the form of fog; if a layer of air is cooled by contact with a cold surface which has lost heat by radiation, then condensation occurs as dew.

A few plants can absorb water directly from the vapour phase if the RH is high. Among these are lichens and some mosses and probably also some epiphytes in very humid air. Fog droplets condense readily on solid surfaces and this fog-drip may be an important source of moisture for some plants. The western coasts of America in the warmer zone have a generally low rainfall, but they are persistently foggy and the fog-drip provides moisture sufficient to support a mesophytic vegetation which would be otherwise impossible.** The same is true of the cloud belt around many high mountains, where the clouds are in contact with the ground and the amount of drip and high humidity favour a type of 'moss forest' in which the trees are thickly draped with bryophytes and lichens. High humidity may be favourable to higher plants by limiting the amount of water loss by transpiration, but it is

* Since RH is the ratio of existing vapour pressure to the saturation vapour pressure, then p the actual vapour pressure at a given RH is given by $RH = p/p_{sat}$; p sat can be found for any temperature from published tables.

** In northern Chile efforts are being made to condense fog by mechanical means to supplement the very scanty water supply.

also favourable to the spread of fungi and bacteria, and continuous humidity encourages the attacks of parasitic fungi and may herald epidemics. The continuous high humidity under snow cover even provides the conditions suitable for some fungal parasites, particularly those of young conifers in northern forests.

Humidity is obviously a very variable quantity. There are diurnal and seasonal rhythms dependent on the local climate. Vegetation affects it in many ways; e.g. by transpiration, by screening evaporation from the soil, by shading, by mutual coverage and by protection from wind. Humidity below the main foliage level is consistently greater than just above it, especially in forests, and in the dense undergrowth of tropical rain forests the RH may remain little below saturation point for many hours continuously. In deserts, on the other hand, it may drop below 10 per cent at midday, which may cause cracking of the skin and other symptoms of acute human discomfort. Humidity also decreases with the height above the ground. This follows from the fact that atmospheric humidity, in the absence of rain, is derived from upward evaporation from soil and vegetation or from water. There is some return under special circumstances, the deposition of dew, for example, or the intake of air into the soil under reduced barometric pressure, but these exceptions hardly detract from the general rule that upward evaporation is continuous. The layer of air next to the ground is supplied in this way with moisture, the further upward movement of which is due to eddy movements in the air (Fig. 4.32).

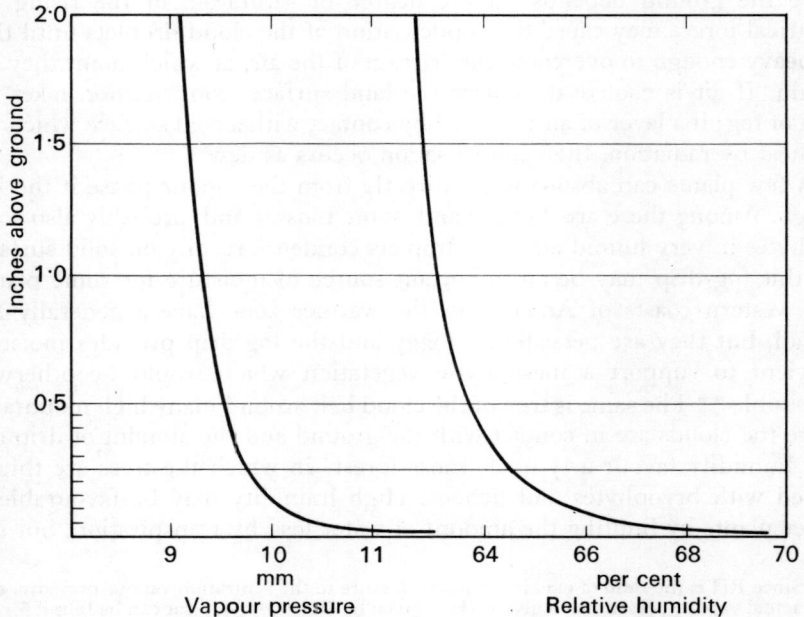

FIG. 4.32. Variation of the vapour pressure and the relative humidity close to the ground. Daily averages for seven days in July in Finland. (*After Rossi, from Geiger,* 'Climate near the Ground'.)

Geiger speaks of a 'wet type' relationship where the layer of air next to the ground is the most humid, and a 'dry type', which occurs mostly at night, when the lowest stratum of air loses moisture by condensation and there is an inversion, the higher air being moister than the lower. This inversion disintegrates at sunrise when evaporation from the ground begins again. This type of distribution is typical of moist and relatively cold climates. Under continental conditions and particularly in hot and dry climates, the high temperatures at noon dry out the surface of the soil and the RH is lowest near the ground, so that the midday phase is 'dry' and the night 'wet'.

The distribution of rainfall is influenced both by geographical and topographical factors. As the greater part of atmospheric moisture comes from the sea it is generally true that coastal areas tend to have a moister climate and more frequent rain than inland, especially inland continental areas in which precipitation is not only lighter, but is more markedly seasonal. At the heart of each large continental mass lies an area of drought which is often desert or semi-desert.

Over each of the polar ice fields there is a more or less permanent area of high atmospheric pressure from which air flows out radially to the temperate zones, becoming warmer and more humid as it moves over the sea. Vortices form around the edge of these polar air masses, centred on low pressure areas, the resulting winds spiralling upwards, becoming cooled and delivering rain. These are the cyclones, of which belts exist both north and south of the equator. Cyclone centres may persist for some time and they travel latitudinally along well-marked tracks bringing rain, strong winds and changeable weather with them as they follow one another.

In warmer climates and in continental areas on the other hand, rising air and therefore rain is usually due to the differential heating of land areas according to the season. If the disturbance is sufficiently strong, moist air will be drawn in from the direction of the sea and rain tends to be seasonal. The great monsoon region of southern Asia is governed by the seasonal northward and southward movement of the sun's path, with a consequent shift in the zone of maximum heating effect, the winds blowing alternately from NE and SW. In India the NE monsoon is dry, since the wind comes across China, the SW monsoon brings the rain from the Indian Ocean. In the East Indies, on the other hand, both monsoons bring rain, since both come over the ocean.

Fig. 4.33 shows three well-marked types of rainfall distribution. There is the variable Atlantic cyclonic type; the continental type with summer rains; and the Mediterranean type with spring and autumn rains, but with a very dry summer. Further south the two rainy periods merge into one period in late autumn and winter, becoming progressively scantier as desert conditions are approached. The activity of vegetation is completely dependent on such short rainy seasons. Most of the year is spent in non-growing dormancy and the arrival of the rains is the signal for a sudden and short-lived outburst of growth and flowering.

In North America the rain zones, from oceanic through continental to

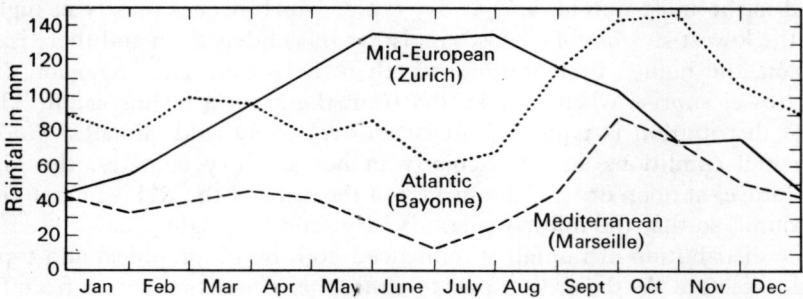

FIG. 4.33. Three contrasting types of annual rainfall distribution. (*From Braun Blanquet.*)

Mediterranean types, are aligned between NE and SW rather than due north and south, the driest areas being the south-western deserts. South Africa lies to the north of the southern cyclonic storm belt, which is between latitudes 40° and 50°S (the Roaring Forties). Its climate is therefore chiefly of the Mediterranean type, shading from SW to NE into desert. In all these areas the approach towards the equator leads into the tropical zone of heavy and regular rainfall due to the worldwide belt of rising air over the heat equator.

The geographical pattern of rain climates is subject to widespread disturbance by local topographical or, more importantly, orographical factors. It has been said that 'mountains alone divide, seas unite', and certainly mountains often divide climates as surely as they divide human societies. Any feature which deflects an air current upwards, thus cooling it, can determine rain, but the result depends, first, on how humid the air is and, second, on the height of the obstruction. The more humid the air the less elevation and cooling will be needed to bring it to saturation point. The higher the mountains the more complete will be their effect in causing precipitation.

A low range of hills may not force the air high enough to cause precipitation unless the air is very humid, their effect will not therefore be very marked, though they may have a somewhat higher rainfall on the side of the prevailing wind, they are more likely to produce mist. A range of high mountains, on the other hand, may cause almost total precipitation on the side of the rising winds. On their leeward side the air, warmed by its descent from the heights, will be ready to absorb moisture rather than precipitate it. This results in a dry belt on the leeward side, called the 'rain shadow', which may extend for a remarkable distance.

Examples of this division are numerous and notable; they are found wherever high mountains face humid winds. The Cascade mountains in the States of Washington and Oregon have a rainfall of 2032–2540 mm annually on their western slopes, which drops to about 254 mm on the eastern side. To the east of them there stretches a zone of semi-desert scrub, covering the whole of the upper basin of the Columbia River. Over the watershed, on the Pacific slope, one plunges at once into a rich and luxuriant forest of conifers.

The difference is exaggerated by the Japan Ocean current which here approaches the coast bringing with it warmed and highly humid air. Further south the Sierra Nevada continues the same effect, separating the dry state of Nevada from the rich forest on the California side, which includes the giant *Sequoiadendron*. On the other hand the low coastal range in California has little effect on the rainfall, and on its inland side dry conditions obtain until the foothills of the Sierra are reached. Irrigation has made this dry zone fertile.

The southern Andes separate the forests of the Valdivia coast from the relatively dry pampas which extends right across the narrow part of the continent. This mountain range however, is so divided by passes that some of the moist wind escapes precipitation and the forest extends through the range and some rainfall is available for the grasslands beyond. Only north of latitude 34°S, where the west winds become dry, is there a strip of desert in the rain shadow, around Mendoza. The western Atlas mountains in Morocco, also divide the *Cedrus-Quercus* forest on the windward, western slopes from a sun-baked semi-desert to the east, with no trees and only clumps of *Anabasis* among the stones. The Himalayas play a similar part with the SW monsoon, but here the picture is complicated by the great height of the Tibetan plateau on the northern, leeward side, so that there is no descent of the air after passing the mountain ridge and dry conditions extend for hundreds of miles.

On the windward side of the mountains rainfall increases up to about 3000 m, but beyond this height it tends to diminish, partly due to the depletion of atmospheric moisture already achieved and partly due to the escape of air over the passes, without further depletion. Deep valleys on the windward side may receive abnormally high rainfall due to the air rising all along their sides.

The seasonal distribution of rain is as important to plants as is its amount. Equal annual amounts of rain differently distributed through the year will have quite different results in vegetation. The prevailing temperature also determines the effectiveness on the vegetation. A long rainless period when the ground is covered by snow is not so unfavourable to plants as a long rainless period under a hot sun. In the former case the soil is saturated with water as soon as the temperature rises in spring. In the latter case the soil is so dried that a considerable amount of rain must fall before it has absorbed enough to bring it up to its field capacity. Delicate herbs and grasses may flourish under a comparatively rainless winter, but only tough xerophytes can withstand a summer drought.

Forests generally require a high annual rainfall and are comparatively indifferent to periods of drought, provided that enough rain has fallen to provide a reservoir of water in the subsoil, which their deep roots can tap. Some trees can even send roots 9–12 m down to tap the permanent level of ground water (sometimes called 'phreatophytes'), which enables them to grow in a dry climate that would not otherwise support tree growth. Many forest trees require a high relative humidity in their growing season and the maintenance of this depends on temperature. Thus, a rainfall of 762 mm per annum which would support forest growth in England would be totally inadequate to support tropical forests, which require upwards of 2032 mm per annum to maintain the RH at 80 per cent.

The relatively shallow roots of grasslands require a rainfall that is evenly distributed through the year, but it need not be as great as that for forests. If there is any irregularity in the rainfall the effects will be quite different if the deficient season comes in winter or in summer. Two grassland areas with the same annual rainfall will show quite a different population of species if they differ in the above respect. Any area with an annual rainfall of less than 254 mm will be semi-desert and will carry nothing but xerophytic plants, however the rain is distributed. Steppe is intermediate in character between true grassland and semi-desert. It is subject to dry periods, especially in summer, and may have very cold, snowy winters. For these reasons it contains a higher percentage of xerophytic herbs and grasses than true grassland.

Apart from its seasonal distribution the way the rain falls is important. As every cultivator knows, occasional heavy rain is more effective in supplying water to the soil than even a constant succession of light showers which never penetrate deeply and are quickly evaporated. Conversely, very heavy rain following drought may be more erosive and destructive than beneficial to the soil, as it is not quickly absorbed and is largely lost as flood run-off, carrying away quantities of silt.

The nature of the vegetation covering the ground has naturally a considerable effect upon the results of precipitation. The denser the covering the greater the proportion of the rainfall which is intercepted by the plants and, although a small quantity of this may be directly utilized by foliar absorption, most of it is re-evaporated without reaching the soil, so that its effects are largely atmospheric.

The present writer made some measurements of these factors in 1912. The vegetation was dense rain forest in southern Brazil. The water content of the soil in the absorbing zone of roots was determined immediately before rain was expected and immediately it had ceased. The amount of water represented by the rise in the water content of the soil was compared with the amount measured by a rain gauge outside the forest. In this way it was shown that, on the average, only 2 per cent of the rainfall became effective for supplying the roots. This was called the *rainfall efficiency* of the rain forest. With so low an efficiency it becomes plain why rain forest requires so high a rainfall for its economy. Grasslands, on the other hand, may have relatively high efficiencies, increased in their case by their caespitose habit of growth which funnels down much of the intercepted rain towards the roots. Grasslands can flourish with a much lower annual rainfall than forests provided it is evenly distributed through the seasons, and it may well be that the rainfall efficiency of different vegetation types is one of the factors determining the physiognomical character of the prevailing vegetation in relation to the annual rainfall available.

Forests generally have only low efficiencies when they are in leaf, but Geiger showed, on the contrary, that clearings in a forest, when the ratio of the diameter of the clearing to the height of the surrounding trees was more

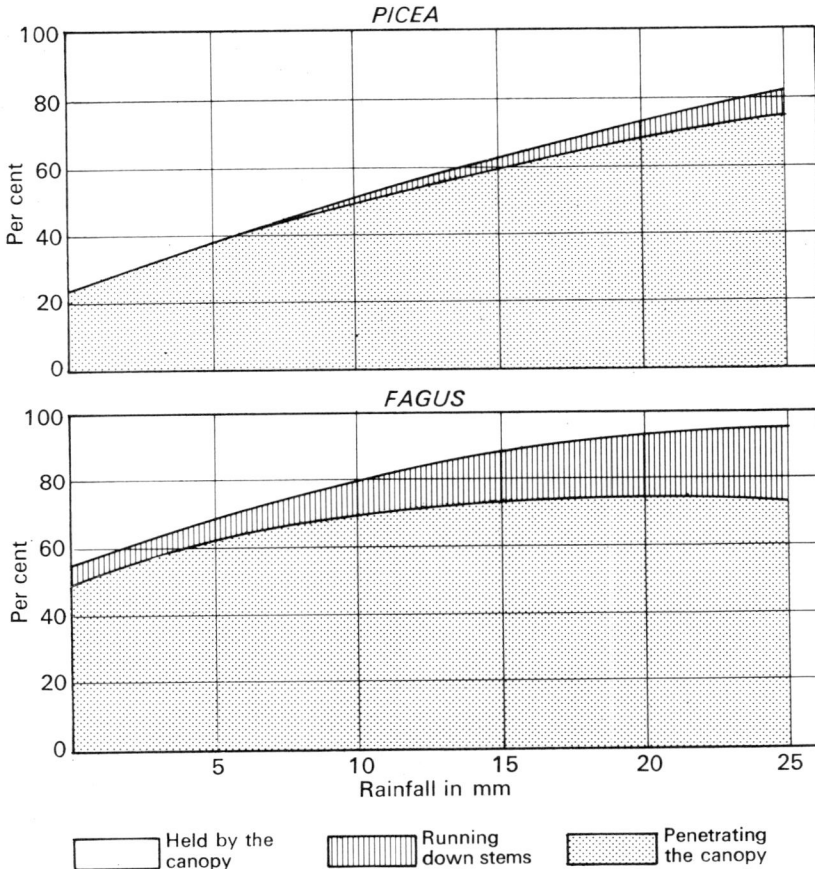

FIG. 4.34. Rainfall disposal in two contrasting types of woodland, *Picea* and *Fagus*. (*After Geiger.*)

than 1·40, caught more rain than the open country, with a maximum excess of 5 per cent when the ratio was 1·47.

Snow has many effects, both beneficial and otherwise, which are quite different from those of rain. When snow melts on ground that is not frozen it adds gently and beneficially to the soil moisture, but a sudden general thaw in spring may result in disastrous flooding. Among mountains with a perpetual snow cover there is also a tendency to diurnal floodings in the summer afternoons, as the snow melted by the sun fills up the drainage streams. The water reserves of the melting snow are of particular value in promoting the spring flora.

Also beneficial is the protection that snow affords against drought and cold winds in winter, and, on the contrary, against water saturation in winter which is so inimical to alpine plants grown in lowland gardens. The danger of drought to exposed plants is due to the frozen soil, strong dry winds and intense sunshine, especially on mountains. Minor benefits result from the raising of the air temperature by reflection of sunshine and the addition to the soil of atmospheric dust caught by or precipitated with the snow.

Prejudicial effects of snow are first, mechanical damage to trees caused by weight of snow. Trees may also be permanently stunted by being buried under snow for long periods when young. While buried they are also liable to the attacks of certain fungal parasites which find the snow a favourable medium.* Second, comes the shortening of the growing season due to any delay in the disappearance of the snow cover. Third, the promotion of *solifluction*, that is the downward drift of semi-liquid surface soil due to over-saturation with water in spring.

The shortening of the growing season is particularly severe on mountains and in high latitudes. Various estimates put the curtailment of the snow-free period at an average of $11\frac{1}{2}$ days for every 100 m increase in altitude. On southern slopes it may be less, about 10 days per 100 m. In the Swiss Alps the average length of the open period at 1800 m (approximately the tree limit) over a period of 16 years, was 5 months, with an amplitude of variation, on the northern slopes, of only 3 weeks in either direction. At 2400 m the period is reduced to $3\frac{1}{2}$ months. Within these narrow limits must be compressed the whole cycle of growth and reproduction for the year. Obviously preparation for immediate and copious flowering is vital to plants with so short a time at their disposal and growth must be limited to very small annual increments. With progressive shortening of the growing season tree-growth either becomes impossible or is limited to prostrate growth forms which are protected in winter by the snow covering. Indeed some conifers show the approximate depth of winter snow by the dense growth of prostrate branches which surround their bases.

Of great ecological interest is the effect of snow banks or snow patches which lie at certain spots on mountains for exceptionally long periods, usually always at the same places each spring. From such places the normal surrounding vegetation is excluded and they are colonized chiefly by bryophytes which can be satisfied with a growth period of $1\frac{1}{2}$–3 months and can, exceptionally, survive a whole year under snow.

Among the pioneers of these snow-patch soils in Europe and Britain are *Anthelia juratzkana* and *Pleurcclada albescens*, followed by *Polytrichum sexangulare* which often becomes dominant. If the open season is somewhat longer, up to four months, some Angiosperms may also appear, principally *Salix herbacea*, *Gnaphalium supinum* and *Cerastium cerastioides*, but all are extremely small. Characteristic of these places is the development, below the *Polytrichum* carpet, of a gley soil (see p. 3599) due to its prolonged saturation with water.

The opposite condition to soil saturation is, of course, drought, which may be more or less permanent, due to annual rainfalls of less than 254 mm,

---

* Such as *Herpotrichia nigra*, which covers conifer leaves with a felt of brown mycelium and, in northern lands, species of *Fomes* and *Trametes pini*.

or it may be seasonal, due to uneven distribution of rainfall throughout the year. Drought may thus be a serious factor, even alongside a fairly high annual rainfall that is limited to a short 'wet season'. Permanent drought means desert conditions with vegetation extremely sparse or absent. Seasonal drought usually means semi-desert, which bears an open but not always sparse vegetation of specialized xerophytes.

From the point of view of drought, plants may be divided into 'drought endurers' and 'drought evaders'. The former class is made up of phanerophytes and chamaephytes whose morphology and anatomy permit economical control of water loss, sometimes combined with internal water storage. The drought evaders are those plants with a short vegetative season, either geophytes which retreat to underground structures, or therophytes, short-lived annuals and ephemerals which pass the dry season as dormant seeds. The former are true xerophytes, what one might call 'professional' xerophytes, the class of evaders we may call 'amateur' xerophytes, since they are often native to mesophytic conditions and their presence in semi-desert is, in a sense, an intrusion, despite which the open space they find there enables them often to flourish with amazing abundance. There is, however, an intermediate or 'semi-professional' class, mostly annuals, which combine with their drought-evading growth-habit some degree of xerophytic adaptation, so that they find the semi-desert their most congenial home.

Drought is most conspicuous in the sub-tropics, in which zone the true deserts and semi-deserts form two world-encircling belts, north and south of the tropics. This means that drought is normally associated with high temperatures and powerful insolation, two factors which greatly encourage evaporation and intensify the difficulties of the plants.

While the very low rainfall of deserts is usually reliable, for it is due to stable geographical factors, seasonal rainfall which depends on meterological factors is unstable and unreliable. It shows a tendency to irregular periodic and unpredictable variations, which may be calamitous for both plant and human populations. Even the monsoons may fail and, affecting as they do one of the most densely populated regions of the world, their failure may spell famine and death on a terrible scale. In some parts of North America historical records tell of successions of almost rainless years, recurring at long intervals, which turn the country affected into a dust bowl. Although such events are only periodic, one such period may have ruinous effects on agriculture and have profound social repercussions.

Plants vary greatly in the relationship of their water economy to growth, that is to say in their water requirements or the ratio between water consumed and dry weight produced. Knowledge of this physiological character is of great importance in the technique of dry-land farming and we have discussed it in Vol. 3, p. 2898.

Under desert conditions, the density of the vegetation is directly correlated with the rainfall. At low rainfall levels the cover is sparse and the ground is almost fully occupied by the centrifugal spread of roots which tends to produce regularity of dispersion. At higher levels there is a tendency to aggre gation. Walter has maintained that the variations of density with rainfall are such as to ensure that each plant receives as much water per unit of transpiring surface as in a humid climate.

Irrigation is of the first importance in the utilization of dry lands. Semi-desert soils, which have not been subject to intensive leaching by rain are often potentially rich and only require water to make them highly productive. In any dry country the contrast between irrigated and non-irrigated land is extremely striking. The state of California, to quote only one example, could never have achieved its present wealth without vast expenditure on extensive irrigation systems.

Just as different species have different water requirements, so do different phases in the lifetime of an individual have distinct requirements, and this must be taken into account in the arrangement of irrigation. Generally, it is the most active growth period which has the highest water requirement, while germination and immediate post-germination stages and likewise the seed setting and maturing stages are less demanding and may be injured by excessive watering. Rice is a conspicuous example. Seeds are sown in nurseries which are not flooded. The seedlings are planted out in flooded land and maintained thus until the seed is formed, at which stage the fields are drained and remain dry until harvest.

We have previously (p. 3530) referred to the stratification of humidity in the air. This is markedly affected by the vegetation cover, especially in the first few centimetres above the ground level. For example, Stocker observed the following values in a meadow at Freiburg in Switzerland, at 11 a.m. on a calm day in July with an air temperature of 29°C. Relative humidity: at 1 m above ground, 57 per cent; at 13 cm, 78 per cent; at 2 cm (in the grass), 96 per cent. This shows how moisture can be trapped among the leaves of dense growth. Closely similar figures have been found by Martini and Teubner in grassland near Hamburg, namely RH 56 per cent in the open air and 90 per cent in grass 10 cm high, measured at noon. Complete saturation of the air among plants, even in dense growth, is rarely if ever achieved, for evaporation depends on the temperature of the evaporating surface, not on that of the air and the fact of evaporation cools the surfaces and reduces evaporation which tends to prevents saturation point being reached. Here again, dry lands are exceptional. Among the sparse growth of semi-desert plants there is no rise of humidity to be observed. There is little evaporation from the soil surface and the heated air rising from contact with the soil effectively prevents accumulation of moisture among the leaves and branches.

Forests also trap moisture in their interior spaces, though not to the same extent as dense grass or herb cover. Averages over 11 years for forests in Switzerland showed the following comparative figures of RH:

Larix decidua	Fagus sylvatica	Picea excelsa
69·5	78·9	85·5
+ 4·1	+ 3·6	+ 9·9

The lower figures are the averages of increased humidity inside the forest as compared with the open air.

The forest atmosphere also shows a humidity gradient with height above the ground. Figures given by Braun Blanquet showed that in a *Quercus ilex* woodland in the Riviera climate, the saturation deficit in the field layer was 9·5 mm, and that in the shrub layer at 1½ m height was 12 mm, at the time of daily maximum.

Humidity is one of the most influential factors governing the geographical distribution of plants, and this we shall consider in more detail in the next chapter. Precipitation depends on evaporation, which consumes one-quarter of the total solar energy received by the earth. As the greater part of this evaporation is from the sea it is natural that precipitation is more plentiful in regions near the sea coast than in continental areas, but there is also a general gradation by latitude, the highest figures being at the equator (annual average 1956 mm) and falling off in both directions polewards. Diminution of the rainfall with latitude is much more marked in the great land-masses of the northern hemisphere than it is southwards where ocean predominates. There is also, in both hemispheres, a minimum zone at about 30° which marks the zone of deserts.

The influence of any factor is greatest when it is at its relative minimum, that is when its relative influence is greatest. In such circumstances the effect of one factor may be decisive in determining vegetational boundaries and the water factor has widespread geographical effects of this kind, since the irregularities of rainfall create many areas where the water supply falls to minimal values. The vegetational boundaries between forest and steppe, desert or tundra, and on the mountains, are generally due to this factor, but on the other hand the geographical boundaries between deciduous and coniferous forests are not governed by the moisture factor but generally by temperature differences. In the broad zonations of the world it is the interplay of moisture and temperature that clearly provides the key to understanding, but its application in detail may need difficult investigation of all the possible factors.

Light and solar radiation as ecological factors

Sunlight is the primal source of energy for plant life, but this energy flux is not confined to the wavelengths of the visible spectrum, i.e. that which we normally call 'light', but is distributed over a wide span of wavelengths in the electromagnetic spectrum, and it is therefore the source of additional and highly important thermal radiation. Approximately 50 per cent of the energy in the solar spectrum lies in the infrared.

The received value for the 'solar constant' is 1·94 cal/cm^2/mm. This is the energy flux received by a surface normal to the sun's rays at the upper surface of the earth's atmosphere. It only changes, if at all, over long periods of time. The average annual value for the northern hemisphere allowing for differences of latitude and season is, however, much less, only 0·485 cal.

The measurement of visible radiation is usually based on the *lux*, the standard metre-candle, or sometimes on the *foot-candle* (10·764 lux). One lux produces a light-flux or quantity of light of 1 *lumen* per square metre and a foot-candle gives 1 lumen per square foot.

To enable comparisons to be made throughout the spectrum, it is preferable to use the energy content of the radiation, which can be applied to any wavelength. The basis for this is the gram-calorie (p. 3514) and 1 gcal/cm^2/min is the unit of radiant energy flux or *langley*. One langley represents a radiation intensity of about 100,000 lux and the value of the Bunsen-Roscoe unit used by Wiesner in his researches is about 88,000 lux (see p. 3550).

While the value of the solar constant is about 2 langleys, clear noonday radiation at midsummer at sea level lies usually between 1 and 1·3 langleys with a maximum of 1·5. In the solar spectrum the ultraviolet, despite its photochemical activity, only contains from 0·1 to 5·0 per cent of the total energy. Of the rest, 41–45 per cent is in the visible spectrum and 50–58 per cent in the infrared.

The stream of solar energy does not reach the surface of the earth without complex interference by the atmosphere due to molecular absorption, dust in suspension and clouds. Gates* gives the following figures for the disposal of the incident radiation. If the average value 0·485 cal be taken as 100 units, the clouds dispose of 52 per cent. Of this they absorb 10 units, reflect 25 back to space and transmit 17 to the ground. From a clear sky 24 units reach the ground directly along with 6 units of diffuse light from the sky itself. This represents a positive increment of 47 units at ground level, less than half of the incident radiation above the atmosphere, or 0·228 cal/cm^2/min.

The chief countervailing factor is re-radiation from the earth, which acts approximately as a 'black body'. The properties of a black body in radiation are, first, that it gives maximum radiation for all wavelengths which can be emitted at the temperature of the body, the total energy thus radiated being proportional to the fourth power of the absolute temperature. Second, the wavelength of the maximum of the emitted energy multiplied by the absolute temperature is a constant. For the earth's surface at a temperature of 288°K (abs.), i.e. 15°C, the emitted energy maximum is at about 1000 mμ or 1μ. This is well into the infrared, so that all this re-radiation lies in the thermal region of the spectrum. This does not take account of visible light reflected from the earth's surface, which varies greatly according to the 'albedo' or comparative reflecting power of the surface (p. 3514). As the reflectivity of green leaves in the visible spectrum does not appear to rise above 15 per cent, it is a relatively minor factor in the earth radiation, whereas the reflectivity in the infrared region, at about 1μ is as much as 50 per cent.** The fate of the visible reradiation can be regarded as following the same lines as the thermal radiation. The latter passes, directly to outer space, 10 units radiated out of 119, the residue, 109 units, being absorbed by the atmosphere. The radiation thus absorbed by the atmosphere is re-radiated in the proportion of 105 units back to earth to 56 units to space. There is thus a to-and-fro passage of energy between earth and atmosphere, involving, if there were no compensating input, a gradual net loss to space.

* GATES. *Energy Exchange in the Biosphere.* Harper and Row. New York. 1962.
** This is why vegetation appears white in photographs taken by infrared.

Apart from radiation, the chief output from the earth and vegetation is by evaporation which consumes 23 per cent of the input units as latent heat. The water vapour rises and at some level in the atmosphere condenses, with the liberation of an equivalent amount of heat to the atmosphere.

There is also the direct loss of heat to wind, partly by convection and partly by conduction, heat which is again passed to the general atmosphere. In both the latter processes some part of the lost heat is restored to the earth by atmospheric radiation.

Molecular absorption in the atmosphere is more complex and is qualitative as well as quantitative. The three chief absorbents are ozone, oxygen and water. Their *absorbent* effect is different from the *scattering* effect of molecules, which is a complex factor depending on the index of refraction, the density of the air and varying inversely with the fourth power of the wavelength.

Molecular absorption is due to molecular or atomic resonance at definite wave frequencies and, hence, is highly selective. Ozone is present throughout the atmosphere. At the ground surface it only amounts to an average of 0·4 ppm, but the amount increases upwards and reaches its maximum of 6 ppm in a layer of the stratosphere at 25 kilometres above the earth. This layer absorbs strongly in the ultraviolet below 400 mμ, so that only a small proportion of the ultraviolet in the solar spectrum reaches the earth and none at all below 290mμ. This absorption is shared to some extent by Oxygen and their combined action in this respect keeps the earth habitable by protecting it from the destructive effects of the short-wave radiation on organic compounds. Ozone has also a strong absorption band in the infrared at 9600 mμ. Absorption by Carbon dioxide is all in the infrared, the region of thermal radiation, where its absorption is almost total at 4300 mμ and in the region between 14,000 and 16,000 mμ.

It is this powerful thermal absorption which has led to the suggestion that certain periods of apparently high temperature in geological history, especially the upper Carboniferous and the Eocene Periods, might have been due to the increase of CO_2 in the atmosphere acting as a blanket to restrict thermal radiation from the earth. The theory is attractive, but it is still doubtful whether changes of the CO_2 concentration in air of the required magnitude are possible. At present local fluctuations in the CO_2 content of the air are small and the general atmospheric concentration is practically stable at 0·03 per cent. (Argon is 0·93 per cent, thirty times as much.)

Water vapour is much the most variable component of the atmosphere, varying between 0·1 and 1·0 per cent by volume. Its absorption is also all in the thermal infrared and is correspondingly variable. There is a very strong absorption at 6300 mμ and many absorption bands in the far infrared from 18,000 mμ to 23,000 mμ (23 microns) where the solar spectrum ends.*

The great importance of these absorptions is that the radiation from the ground at 1000 mμ falls in the region absorbed by ozone, while the atmospheric radiation at an average temperature of 263°K ($-$10°C) falls principally into the zones absorbed by water and Carbon dioxide respectively and is

* See Fig. 4.24 on p. 3515.

therefore held back from escaping into space. Were this not so the climate of the earth would approximate to that of the moon, unbearably hot by day and insupportably cold by night.

The short-wave absorption of ozone and Oxygen also protects us from the occasional floods of ionizing and X-rays emitted by solar flares, which could have devastating biological effects.

The effect of water vapour is readily perceived at night in the difference between a clear, dry night and one that is overcast by cloud. We feel warmer in the latter conditions because we are receiving downward thermal radiation from the atmosphere, although in fact there may be little difference in the ambient air temperature, and what we receive the plants also receive.

Dissipation of light by fog is not selective, it is a process of diffuse reflection which is independent of wavelength.

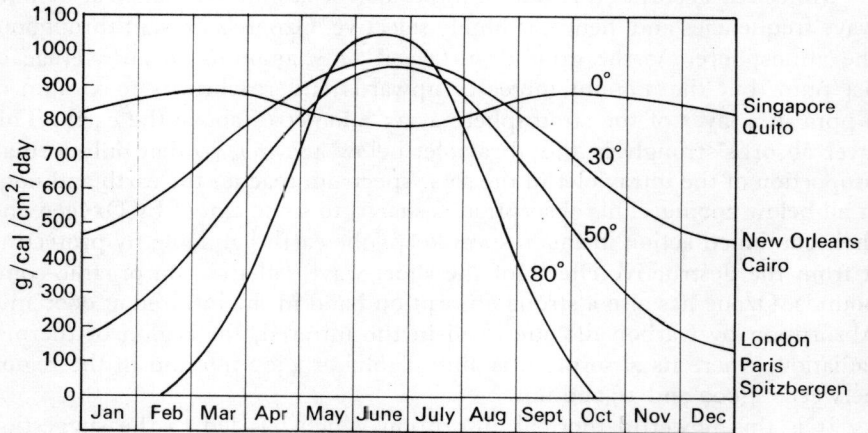

FIG. 4.35. Variation of the solar radiation received daily on a horizontal surface at different latitudes throughout the annual cycle. Solar constant taken at 1·94 cal/cm²/min. (*After Gates.*)

Geographical and topographical factors must obviously modify the preceeding generalizations. The daily total of solar radiation received on a horizontal surface external to the atmosphere varies greatly both with the latitude and the season of the year. Figure 4.35 shows the extent of his variation for latitudes between 0° and 80°N. The equatorial zone, 0–10°N shows comparatively little variation throughout the year, with consistently high values, averaging 400 cal/cm²/day, but with an actual diminution during the months May–July, when the sun's path is at its most northerly. At the other extreme, Spitzbergen, at latitude 80°N is entirely without solar radiation during November, December and January, but in the continuous daylight of June it reaches the highest daily total of all, 1050 cal/cm². A comparison of Miami (26°N) with Fairbanks, Alaska (64·5°N) is instructive. On 21 June Miami was receiving a daily average of 514 cal/cm² while Fairbanks had 522, but on 20 December Miami was still getting 283, while Fairbanks had no more than 4 cal. The annual totals were Miami 156,177 and Fairbanks 75,647.

There is an hourly variation at all latitudes due to the changing elevation of the sun, and this also affects the seasonal values at latitudes outside the equatorial zone, in which the midday sun does not vary more than $23\frac{1}{2}°$ from the vertical. This variation is due to two factors, first, that as the sun declines towards the horizon its rays strike the earth's surface more and more obliquely, so that each unit of the incident flux is spread over a greater surface area. Second, that the oblique rays traverse an increasing length of path through the air. When the sun is vertical the air mass penetrated by its rays, at a pressure of 760 mm, is called air mass 1.

At 60° from zenith the air mass = 2.
At 70·7° from zenith the air mass = 3.
At 78·7° from zenith the air mass = 5.
At 83·2° from zenith the air mass = 8.
At 84·7° from zenith the air mass = 10.
At 89·2° from zenith the air mass = 29.

The air mass is approximately equal to the secant of the sun's zenith angle.

These figures show how steeply the atmospheric path increases as the sun nears the horizon. This not only decreases the ratio of direct to diffuse light, but decreases differentially the amount of the short-wave component in the spectrum.*

The diminution of the sun's direct radiation as its zenith angle increases also affects the ratio of the direct radiation to the diffused radiation received from the sky as a whole. The ratio is highest at the equator, with a vertical sun. It is lowest in the Arctic where the direct lighting at certain times may become almost zero and the diffuse light is all important. The ratio varies from 1 to about 6, but for temperate latitudes at midday in summer it is approximately 4.

At any given latitude there will be a certain angle of the sun at which the direct insolation and the diffuse light are equal, but this depends so much on local atmospheric conditions that the point cannot be generalized. In clear summer weather at Vienna (lat. 48°N) the angle is approximately 57°, which is roughly air mass 2. A cloudy sky will increase the proportionate amount of the diffuse light. Thus, Wiesner records that at noon in May in Vienna (48° 10′N), with a clear sky, the diffuse sky light amounted to 36·5 per cent of the total radiation on a horizontal surface, but that with a fully cloudy sky the proportion rose to 44 per cent. In Washington, USA, at a lower latitude (39°N) and at noon in July with a clear sky the diffuse light was only 12 per cent of the total radiation on a horizontal surface, but at 8.0 a.m. and again at 4.30 p.m. it rose to 20 per cent, i.e. the ratio of direct to diffuse light was 4. As the atmosphere re-radiates at the same wavelengths that it absorbs, it follows that diffuse light is somewhat richer in short-wave radiation, 290–300 mμ, than is direct sunlight.

* So the sun looks redder.

Topographically, the two factors most affecting the amount and quality of radiation are the slope of the ground and the height above sea level. The former we have already described (p. 3517) The effect of the latter is principally due to decreased atmospheric absorption, by reason of diminished air density and dust content and sometimes decreased water vapour content. Primarily these conditions increase the intensity of the total radiation. At sea level the intensity at midday in summer is of the order of 100,000 to 150,000 lux, but at a height of 2000 m intensities of nearly 200,000 lux have been recorded, the maximum value, at the upper limit of the atmosphere, being 212,000 lux (2·0 cal/cm²/min). The balance of radiation is also changed: the proportion of direct insolation to diffuse lighting becoming greater as the amount of atmospheric scattering becomes less. The content of ultraviolet and of thermal wavelengths, which are normally absorbed by the atmosphere, is also increased and the considerable ultraviolet content at high levels is particularly noticeable in its biological effects.

At high levels the amount of ground radiation is not greatly changed, except under snow, from which, owing to its high albedo, reflection predominates over radiation, while the ground below is protected from radiation losses and remains warmer than uncovered ground. Little of this returned energy is absorbed by the thin atmosphere and is lost to space, this greater loss having to be set against the greater intensity of incident radiation. The sun therefore has little effect in warming the atmosphere, but produces a greater thermal load on any absorbing surface.

This difference is clearly seen in the comparison of sun and shade temperatures at different heights. At Diavolazza (2980 m) the sun temperature was 59·3°C as measured by a black bulb thermometer, the shade temperature only 6°C. At Pontresina (1800 m) the sun temperature of 44°C was matched by a shade temperature of 26·5°C, but at Whitby, at sea level a sun temperature of 37·8°C corresponded to a shade temperature of 32·7°C. The combination of an intense radiation load on an absorbing surface, with cold air temperatures at high levels is apparent. The radiant heat available to the plants compensates for the lower air temperatures and alpine vegetation, as Schröter says, gives no impression of being stunted by unfavourable conditions. Indeed it has been observed that rye (*Secale*) at 2100 m ripens faster than in the valleys.

If the sky is clear and the air is dry the greater loss by re-radiation at night may more than balance the increased insolation by day, so that at certain times and places the net account of radiation may be negative. The high mountain climate is one of extremes, but not wholly unfavourable.

At lower elevations, where clouds predominate, conditions will be quite different, the greater humidity and high thermal absorption by cloud imply decreased incident radiation and decreased re-radiation losses, so that the thermal exchanges are at a low but fairly uniform level.

Net radiation, that is the balance between gains and losses is a useful gauge of climate. It is positive almost everywhere except during winter in the Arctic and Antarctic, when during some months there may be a net loss. It is highest over the oceans owing to their more extensive cloud cover and the highest values are over the Indian Ocean, 120 kcal/cm²/annum; near the Persian Gulf it even reaches 140 kcal. Over land areas the value is generally below 100. For Britain the value lies between 30 and 40; the USA, 40–60; India 80–100; Russia 10–40. The long days of the northern summer tend to equalize values at that time. In the month of June Britain has a monthly gain of 8–10 kcal/cm², which is the same as that for Central America and India in the same period. Russia also has an average of 8–10 for the month. It is possible, as has been shown by Budyko, to characterize a climate graphically by comparing the net radiation, month by month, with the losses by convection, measured by the average wind velocities, and losses by evaporation measured by the ratio of precipitation to evaporation. Equatorial climates, as thus expressed, are equable, with high evaporation and low convection losses. Desert climates have relatively low evaporation losses or none; temperate climates show a maximum of all three quantities in summer as also do arctic climates, but with this difference, that in the Arctic all three values fall to zero or below for about five months of winter, while the summer maxima only cover two months.

The stream of radiated thermal energy, downwards or upwards, is practically never less during the twenty-four-hour period than the downward stream of solar radiation during the day and is often more. The visible solar radiation is, of course, essential for photosynthesis, but from the point of view of energy supply the infrared thermal radiation is at least quantitatively its equal.

We can realize how much ecological significance the thermal radiation has if we consider what it means in the case of a horizontal leaf exposed to full sunshine. The upper surface is receiving the total solar radiation, direct and diffuse, and the infra-red thermal radiation. The lower surface is receiving the upward thermal radiation from the ground and the sunlight reflected from the ground, the amount of which depends on the albedo of the surface.

Gates gives the following illustrative figures for an average case of a leaf over a grassy surface on a clear summer day.

Air temperature 30°C.
Ground surface temperature 40°C.
Albedo (reflectivity) of grass 7 per cent.
Vapour pressure (millibars) 36·1.
Energy flux in cal/cm²/min.
Solar (sky): 1·00. *Solar (reflected upwards)*: 0·07.
Downwards (thermal): 0·556. *Upwards (thermal)*: 0·792.
Net flux 0·694; total flux 2·418.
Absorbed by leaf per 2 cm² (1 on each surface), less loss by reflection, 2·210.

The thermal effect of such an energy absorption is given by the equation

$$\frac{\Delta T}{t} = \frac{Q}{t} \frac{A}{Mc}$$

where $\Delta T/t =$ rate of temperature change, $Q =$ radiation flux, $A/M =$ mass of the leaf per unit area, taken as 0.02 g/cm^2, $c =$ specific heat of the leaf, taken as 0.88.

Then, with a load of 2.0 cal cm^2/min, the rate of temperature change would be $114°C$ per min: as this is impossible it is obvious that there must be an efficient dissipation system. If the leaf radiates thermally as a black body this will dispose of 95 per cent of the load, but even this loss leaves enough energy to raise the leaf temperature well above permissible limits. The maximum recorded leaf temperatures are about 10–$12°C$ above air temperature.

The other means of heat dispersal are evaporation and conduction by wind. Leaving conduction on one side for the moment, it is possible to calculate the amount of transpiration required to dissipate the remaining 5 per cent of the incident energy and keep the leaf temperature at only $10°C$ above the air temperature. This amounts to 0.081 g of water per 2 cm^2/hr (4.05 g/dm^2) if the leaf temperature is $40°C$, which is rather a high figure though not impossibly high. It is, however, well above the average for mesophytes and suggests that conduction by wind must generally play a part in heat dissipation.

Consideration of the effect of the thermal load nevertheless throws a new light on the importance of transpiration, which is altogether obscured if only the visible solar radiation is considered, as has usually been the case; it can hardly be dismissed as a necessary evil, as has sometimes been done.

The figures help to explain why transpiration in desert plants is generally at a higher level than might be expected from xerophytes, and that, if transpiration in such plants has to be restricted, their chief safeguards against destruction by heat must be either increased mass or restricted area or both. Even so, tolerance of high internal temperatures becomes important, and temperatures of $50°C$ have been recorded from succulent tissues.

Convection with stationary air is not very effective as an agency of heat dissipation, since it is only proportional to the fourth root of the temperature difference, but conduction by moving air is much more efficient as it varies linearly with the temperature difference and the rate of transfer is proportional to the square root of the wind velocity.

Calculation of the heat-dispersive power of wind over a leaf is a complex matter with a considerable number of parameters, some of which are not easily evaluated.* One important factor is the breadth of the leaf in the direction of wind flow, for cooling is differential, being greatest at the leading edge and diminishing progressively across the surface. One example will show how influential this factor is. With a wind velocity of 10 mph (447 cm/sec) and a temperature difference of $5°C$ between leaf and air, the heat loss from a leaf 1 cm in diameter is 0.600 cal/cm^2/min, while from a leaf 5 cm across it is only 0.270 cal, little more than one-third the amount. It follows that plants with many small leaves will lose heat more rapidly than plants with fewer, large leaves. The latter are therefore more likely to show high internal tem-

* For full details see chapter 4 of the book by D. M. Gates, cited on p. 3540, which is strongly recommended to physical ecologists.

peratures unless they transpire actively. Objects of a cylindrical shape, such as small branches or needle leaves respond very quickly to heat exchange. At 1 mm diameter, with a wind velocity of 10 mph (447 cm/sec) and a temperature difference of 5°C the heat loss will be 1·095 cal/cm^2/min or about 1·9 times that from a leaf 1 cm across.

Plants have other means of protection against heat as well as these. The leaves of many sun plants have polished surfaces which increase reflection, or they may produce screens, such as multiple epidermes, or anthocyan pigments which reflect chiefly the red and infra-red wavelengths. Another form of protection consists in the angle at which the leaf is held, any departure from the horizontal leading to a reduced radiation load. The extreme in this respect is the vertical position of leaves, as in many Monocotyledons and in the so-called 'compass plants', in which the leaves are twisted into the north–south line, with their edges uppermost, thus offering the minimum exposure to the midday sun. Among alpine plants the prevalence of small and divided leaves is to be noticed as facilitating heat loss under intense irradiation.

The influence of clouds on illumination is quantitative rather than qualitative, for the diffusion of light by water droplets is independent of wavelength. The influence is also extremely and rapidly variable, ranging from the effect of a single cloud momentarily occluding direct sunlight to a totally overcast sky. The extent of cloudiness is usually expressed in tenths, from 0/10 for a clear sky to 10/10 for one entirely covered. Usually the effect is that of diminishing the light, but it should not be overlooked that scattered clouds, especially cumulus clouds, reflect a great deal of sunlight and can markedly increase the proportion of diffuse to direct illumination.

The annual extent of cloudiness is so closely linked to the climatic factors of temperature and humidity that it has a certain stability for any given region and, like other climatic characters, can be mapped. The maximum annual cloudiness lies in the two zones of 60°N and 60°S latitude. There is a secondary maximum at the equator, while between these zones the warm temperate and sub-tropical zones are regions of relative cloudlessness. This does not, however, mean that they always receive intense illumination, for the lack of clouds means lack of rain and consequently a dust content in the air varying from high to very high; a dust storm may indeed produce a complete black out. Arid regions are naturally associated with unclouded skies, and the largest area of this kind is that which stretches from the Sahara eastwards across Arabia into Iran and Central Asia. Other such areas include the south-western United States, the northern part of Chile, south-west Africa and central Australia.

The annual extent of cloudiness can best be represented by the number of days recorded without sunshine. In most parts of the world, regardless of latitude, the number of sunless days is greater in winter than in summer, but at high altitudes the reverse may be true, the high mountain climate is sunnier in winter than in summer. Schröter quotes figures for the sunshine between 11 a.m. and 1 p.m. at Sonnblick (3100 m), which were: December, 47 hours; June, 27 hours. This compared with Vienna, where the figures were: December, 21 hours; August, 66·5 hours. Averaged over the year, however, there is not much difference. At Zürich (473 m) the ten year average of sunshine was 43 per cent of the possible hours and at Säntis (2500 m) the average was 42 per cent of the possible.

A formula was put forward by Ångström for calculating the relationship between the amount of radiation received by a horizontal surface from a cloudless sky and that received from an overcast sky:

$$Q_s = Q_o[a + (1 \cdot 0 - a)S]$$

Q_o is the radiation received from a cloudless sky and Q_s that from the sky overcast. S is the prevailing percentage of the possible hours of sunshine and a is a constant.

Kimball and Hand adopt the value 0·22 for a, the equation thus becoming

$$Q_s = Q_o[0 \cdot 22 + 0 \cdot 78S]$$

Figure 4.36 shows the recorded curves for cloudless and overcast sky for the year in Washington DC. As would be expected the difference in gram-calories is maximal in midsummer (240 cal) and least in mid-winter (90 cal) though the ratio remains about the same (1 : 1·5).

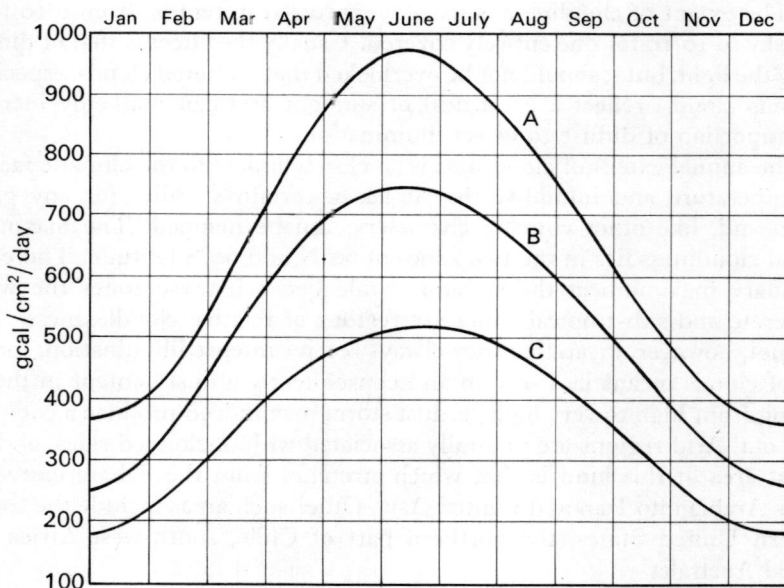

FIG. 4.36. Annual cycle of the solar and sky radiation received on a horizontal surface at the latitude (38° 56″ N) of Washington D.C. (height
127 m). A. Radiation outside the atmosphere. B. Radiation received on
a cloudless day. C. Radiation received on a day of 'average cloudiness'. In the last case the curve is smoothed from weekly averages
which show variations of 30–40 per cent. (*After Kimball and Hand,
from Duggar*, 'Biological Effects of Radiation'.)

It is now well known that not only the relative intensity, but also the daily length of the illumination period has a profound effect on the maturation and the reproductive activity of plants. This influence, known as the photoperiod, has already been discussed from the physiological viewpoint in Vol. 3, pp. 3197 to 3206. As the natural day-length varies with latitude and also cyclically through the year in extra-tropical latitudes, there is good reason to suppose that it has an important influence on the spatial distribution of species.

Within the tropics the day-length does not vary much from twelve hours throughout the year. With increasing latitude the seasonal difference between summer days and winter days with respect to length increases towards the polar extreme of twenty-four hours light in summer and nil in winter. Plants vary much in their response to these differences of day-length. Some species are neutral, but many can be classed as either short-day or long-day plants. These may be roughly described as those which will not flower under illumination of more than twelve hours and those which will not flower if the illumination is less than twelve hours. For obvious reasons tropical plants generally belong to the short-day category, while long-day plants are excluded from the tropics for the same reasons. Short-day plants can find a home in temperate latitudes, where they form the population of spring-flowering and autumn-flowering plants, utilizing the short days of those seasons, while long-day plants flower during the long days of summer.

It has often been suggested that this difference of habit is a determining influence in the geographical distribution of species. This assumes that the habit is constitutional, that is, genotypic and that the distribution has been acquired to suit it, but the reverse may be true and the remarkable adaptability of cultivated plants to changed environments rather supports that view. Roses, for example, will flower in the tropics, though they do not flower very freely, and they are long-day plants. That the optimum photoperiod may be an adaptive character is also borne out by the fact that plants of the same species, taken from the northern and the southern limits of their natural range, can show marked differences in their photoperiodic reaction. The same is true of cultivars from different areas, especially those of wheat.

In plants growing at the same latitude the length of the vegetative or maturation phase preceding flowering seems to be controlled chiefly by temperature, and this may be why short-day southern plants do not succeed in reaching flowering maturity in temperate climates. Vegetative growth, indeed, is controlled by quite different factors from flowering, and a plant may grow vigorously under day-lengths where it does not flower; only if it can reproduce vegetatively can it survive as a wild plant in such conditions. With cultivated plants, however, the difference may be economically profitable. The tobacco plant, *Nicotiana tabacum* reaches great vegetative vigour in the climate of Virginia (around 38°N), but flowering and seed-setting is much more prolific in the extreme southern States at about 30°N.

A prolonged vegetative phase under long days may have other ecological effects. Plants of high latitudes close their season of vegetative growth while the summer days are still long, but southern plants will continue active growth until the frost comes, which kills them before flowering. Frost hardiness, therefore, which is a controlling influence on northward spread, depends in part on the plants' reaction to long days.

Variations in the intensity of illumination have only a slight influence on photoperiod. It is not the 'light-sum' but the period which is effective. Nevertheless, intensity cannot be ignored, since flowering and seed-setting have a higher optimum light intensity than vegetative growth.

This raises the question of the condition of plants in shade. It is a matter of observation that plants show marked differences in their preference for exposed or for shady habitats. These differences were first looked at quantitatively by Julius Wiesner, who published his results as *Der Lichtgenuss der Pflanzen* in 1907. *Lichtgenuss* has usually been rendered in English as 'light ration', which is not strictly correct, since *genuss* really means the 'use' or 'enjoyment' of something in a legal sense. However, to say the 'light enjoyment' of plants is neither very clear nor very graceful.

What Wiesner did was to measure the range of light intensities in which natural species habitually grew, expressed as fractions of the full light intensity in the open. He only had at his disposal a somewhat primitive photometric method, depending on the darkening of a standard sensitive paper to match a standard tint. The time required was measured in seconds and the light intensity expressed as the reciprocal of this time, which gave Bunsen-Roscoe units (1 sec. time = 1 unit), equivalent to about 88,000 lux. The method is purely comparative, difficult to standardize, and the experimental errors so considerable, that to calculate reciprocals to three places of decimals, as Wiesner did, was meaningless. Intensities in the various habitats were expressed as fractions of the full prevailing daylight and the range stated as lying between the maximum and minimum fractions. In the case of trees of the forest canopy, the light range is measured from the minimum inside the crown of the tree as this gives a measure of the extent to which their seedlings can survive in shaded conditions among undergrowth, that is their capacity to regenerate in the community where they are growing. This may very well change with time, as during the lifetime of a mature tree the undergrowth may develop or change to such a degree that tree regeneration is stopped and thus the stability of the community is threatened. Trees with open canopies, such as *Larix decidua* (1/5) and *Betula verrucosa* (1/7–1/9) have a high minimum as shown by the fractions. *Pinus sylvestris*, though evergreen, is also light-demanding (minimum 1/9–1/11). *Quercus pedunculata* is less so (1/26), while *Fagus sylvatica* is one of the least demanding (1/60–1/80). The higher minimum in the latter case was taken in close canopy, the lower figure from an isolated garden tree.

It has been found in a number of cases that the minimum light intensity for trees and shrubs is lower for those growing in good soil than for those which are on a poor soil and less vigorous. As we have mentioned on p. 355 the compensation point, which may be taken as an index of the absolute light minimum, varies greatly among forest trees. It is as high as 30·6 per cent of full sunlight in *Pinus ponderosa* and lowest in *Fagus grandifolia* and *Acer saccharum*, among North American species. This has some significance with regard to succession, as those trees with the greater shade tolerance will tend to replace others with less. In discussing the continuum analysis as applied by Cottam and Curtis in Wisconsin (p. 3430), it was mentioned that the two species, *Fagus grandifolia* and *Acer saccharum*, are graded as having the highest continuum index in that they are characteristic of the last phase in woodland development. In Britain, *Fagus sylvatica* has in many places suppressed and replaced the light demanding *Fraxinus excelsior* and has itself been invaded and partially replaced by the even more shade tolerant *Taxus baccata*. In central Europe *Fagus* is the dominant tree at lower levels and *Picea excelsa* in the mountains. Lee has also shown in North America that the order of woodland types from the highest to the lowest light demand is also, by and large, their order of ecological succession.

The light minimum of species increases with latitude, which may very well be related to the lesser intensity of direct sunlight at higher latitudes. Wiesner recorded that the minimum for *Acer platanoides* at Vienna (48°N) was 1/55, but that at Tromsö in northern Norway (nearly 70°N) it was only 1/5.

The light conditions in woodlands have attracted much attention since they affect both the ground flora and the regeneration of trees. Deciduous woodlands change greatly with the unfolding of the leaves. In the pre-vernal aspect the branches allow from 40–70 per cent of the prevailing light to reach the ground, consequently it is this period, which in Britain lasts till the end of

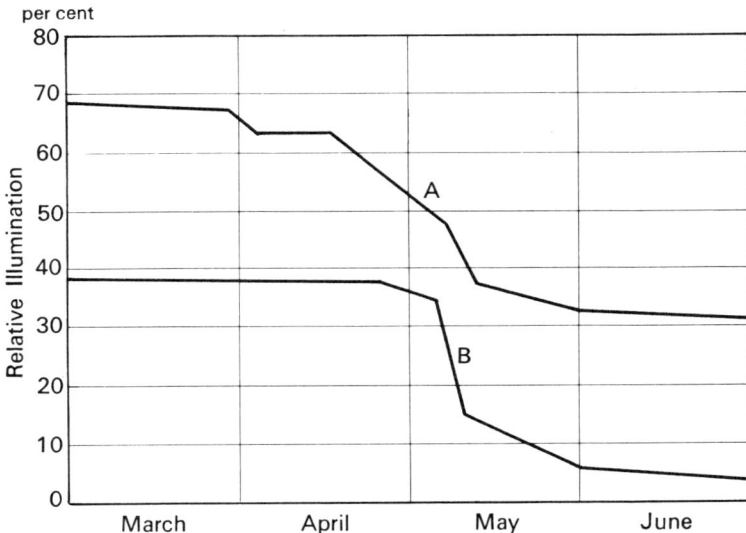

FIG. 4.37. Decreasing illumination in woodlands due to leaf-growth in spring. Lighting measured relative to the external lighting at the same time. A. *Betula* woodland. B. *Fagus* woodland. (*After Hueck.*)

April, that the most abundant plants of the field-layer flower, though they may continue assimilation throughout the summer, under greatly increased shade.

These plants of forest undergrowth have always been looked upon as typical shade plants, many of them growing at very low light minima, and this applies *a fortiori* to the sparse field layer in tropical rain forests. Species growing below deciduous trees can, however, have a very wide light range, from the spring months to midsummer. They sometimes also flourish in other situations, where they are not shaded, and the question arises: are there any true shade plants and if so what are their special qualities?

Blackman and Rutter investigated this question with reference to a common and characteristic woodland shade plant, *Scilla non-scripta*. They used two criteria, the net assimilation rate and the leaf-ratio, which is the ratio of leaf area to plant weight, both of which quantities show a linear relationship to the logarithm of light intensity. The 'relative growth rate' is the product of the two values.

They found that *Scilla* was not a true shade plant and that the relative growth rate did not reach its maximum with light intensities of less than 0·6 of daylight. The woodland preference of the species they attribute to, first, its protection from trampling damage by grazing animals, which is serious in grassland and, second, the restriction of competition by grasses. In grassland, grazing by rabbits could be favourable, as the grasses are eaten preferentially and there is no trampling damage. A high Nitrogen level in woodland soils was inimical because it increased the growth of grasses.

Several biotypes of the species were found which differed in their light relationships and these differences were traced to differences in the leaf–area ratio. In comparison with other shade species of woodlands and some sun-loving species they found that the sun and shade species did not differ significantly in either net assimilation rate or in their compensation points, but in shade species the optimal growth rate was attained at lower light intensities than in sun species. This they attributed to an increased leaf–area ratio at lower light intensities which they interpreted as characteristic of true *sciophils* (shade plants) whereas in *heliophils* (sun plants) the converse was the case.

Lundegardh, repeating F. F. Blackman's experiments on the effect of increasing light intensity on CO_2 assimilation, found a marked difference between sun and shade plants. At low light intensities the rate of assimilation was practically the same for both types of plant, but *Oxalis acetosella*, taken as a true shade plant reached its maximum at 1/10 of the full lighting and its curve thereafter was horizontal, whereas the curve with a sun plant, *Tropaeolum speciosum*, continued to rise up to practically full light intensity and only gradually tended towards the horizontal. Some shade plants, e.g. *Stellaria nemorum*, had their maximal assimilation rate at even lower levels than *Oxalis*.

Wiesner pointed out that *Hepatica triloba* is found in deeper shade under *Fagus* than under *Pinus*, attributable to the pre-vernal months of high light intensities under beech while its foliage is developing. This points to a character of shade-enduring rather than shade-loving and it is probable that the moisture factor is at least as important as the light factor in most shade plants.

Evergreen trees generally cast a deep shade, often deep enough to exclude all but a few flowering plants from the ground. At ground level in a Mediterranean wood of *Pinus pinea* the light at ground level does not rise much above 3000 lux and in a wood of *Laurus nobilis* it is even less. Against these low levels may be set the influence of moving 'sun flecks', filtering down through the canopy, to which the present author drew attention in 1919 in tropical rain forest; these provide short spells of relatively high intensity lighting, which in the daily aggregate are of not inconsiderable importance. Thus a plant at ground level in a rain forest may normally receive no more than 1/140 of the external daylight, but a sun fleck can raise it for a short period to 1/8.

Plants with a high demand for moisture, such as the comparatively unprotected Bryophytes and ground Algae are generally sciophilous and have unusual capacities for growth in shade. Mosses can endure lighting of only 1/2000 of the full daylight and Algae, like *Pleurococcus*, even less. Some ferns also, like *Adiantum capillus-veneris*, also endure extreme shade, provided that constant moisture is available, and such plants can colonize the sides of caves to a remarkable distance from the entrance. Where caves are illuminated artificially, the lamps encourage quite a rich growth of lowly plants around them, although, compared with daylight, their output may be very low.

Lichens, on the whole, are among the most sun-loving (or drought-enduring) of all plant groups, though some corticolous forms, like *Lobaria*, develop most luxuriantly in moist shade. Fungi are commonly supposed to be independent of light and certainly many can grow in darkness, but even some of them require light for reproduction, though low intensities suffice. Among these are notably the Saprolegniaceae and also some Agarics.

In *Pilobolus*, the sporangiophores (which are phototropic) require light for development and for spore-shedding. Lighting under water is quite a special problem, on account of the variable loss by surface reflection and by turbidity, and also because of the differential absorption of different wavelengths by the water itself. Water varies greatly in its transparency. This is generally tested with a Secchi disc, a circular white plate about 30 cm in diameter, suspended horizontally and lowered into the water to the point of disappearance. The limit of effective light penetration is reckoned as roughly twice this depth (since the light reflected from the disc has traversed the water twice); more accurate measures with photoelectric appliances show that small amounts of light penetrate much deeper than this. Schmidt estimated that in clear sea water the penetration of light varied with depth as follows: at 10 cm, 54·9 per cent; at 1 m, 35·8 per cent; at 100 m, 1·4 per cent. In the freshwater Bodensee the light at 50 m had fallen to only 0·0006 per cent, while in Esthwaite Water in the English Lake District Pearsall found that the light at 1 m, had fallen to 30 per cent and at 6 m was only 3 per cent. *Nitella flexilis* grew to depths of 3·5 m where the light intensity was about 5 per cent. Fresh water is often more turbid than the sea due to the amount of plankton growth and sediment in suspension, which both absorb and scatter light. One of the clearest waters in the world is that of Crater Lake in Oregon, in which attached Algae grow as deep as 120 m where there is still 0·5 per cent of the full light.

The amount of light lost by reflection at the water surface is not great. With the sun at zenith angles of c–60° the loss is only 2–6 per cent. At 70° it rises to 13 per cent, and it is only with the sun at low angles that the loss reaches the order of 30 per cent. Disturbance of the surface does not make a great difference, but a covering of foam introduces a considerable amount of total reflection from the bubbles and is a serious restriction. Diffuse light, coming as it does from all angles will be similarly affected to direct sunlight.

The differential absorption of wavelengths by water is important. (See also p. 3712.) Absorption is more or less proportional to wavelength, the blue-violet rays penetrating much more deeply than the red. Pure water has itself a blue tint due to this differential absorption. Quantitative estimates of relative absorption are naturally as variable as are other qualities of natural waters and the following are only quoted as examples.

> Hüfner found that chemically pure water in a layer 10 m thick passed (shown as percentages of incident light): red, 2·0; yellow, 32; blue-violet, 75.
> In the freshwater Bodensee, Oberdörfer found at 1 m depth:
> red, 69·3; blue, 74·4; violet, 81·2; at 10 m depth the remaining light showed: red, 2·5; blue, 5·1; violet, 12·4 per cent of the incident intensity.
> For sea water Ewald and Green in clear Mediterranean water found at 5 m: red, 3.4 per cent. At 100 m the red rays no longer darkened a red-sensitive plate, but the blue-violet could still be photographically detected at 1500 m.
> Green in the North Sea found at 10 m: red, 2·7; green, 16·6; blue-violet, 80 per cent. At 100 m red had disappeared, but there were still: green, 0·03 and blue-violet, 1 per cent.
> Traces of blue-violet light have been photochemically detected in the Mediterranean at a depth of 550 m, but these traces of short-wave illumination can hardly be of much significance to photosynthetic plants. Plankton organisms with chlorophyll have been dredged in the Adriatic from 1200 m, but they can hardly have been photosynthesizing at that depth and the depth level of planktonts is liable to sudden and rapid changes due to water currents.

While light is undoubtedly an important influence on the growth of submerged plants, which often show a well-marked zonation, it is not the only determining influence. Temperature, oxygen supply and dissolved substances must also be given due weight (p. 3703). That the plankton itself has a marked effect on light penetration is shown by the often observed fact that the transparency of water is greater in winter than in summer. In the Bodensee the whole-light penetration at 1 m depth in summer was 69·0–72·5 per cent and in winter 82·1–83·0. In Lake Geneva the Secchi disc depth of disappearance was 6·8 m in summer and 14·6 m in winter. There are many observations which show that the transparency of natural waters is inversely proportional to the bulk of plankton they contain; so generally, is the colour. Clear waters, poor in plankton, are blue (cf. Schutt's famous observation 'Blue is the desert colour of the sea') or, if very deep, black, as may often be seen at sea; plankton-rich waters tend to green, while brown water may be due to a diatom flora, but more usually to humus.

The adjustment of submerged plant life to depth is associated with varying optima in the relationship of CO_2 assimilation to illumination. Like shade plants on land, the plants which grow deepest have low optima, above which an increase in illumination does not increase, and may decrease, the rate of assimilation. Here, as elsewhere, however, populations often contain differing biotypes, and it has been found that individuals from shallower positions have higher illumination optima than those of the same species from deeper portions of their range, just as individuals among land plants taken from high altitudes possess greater hereditary frost resistance than those from lower levels.

The ecological influences of the air

The atmosphere is obviously the vehicle for the operation of the major climatic influences, such as illumination, moisture, and temperature, but it has some particular influences of its own apart from being a universal stock of Oxygen and Carbon dioxide. Its composition is too nearly uniform to have any differentiating effects ecologically, but local adulteration or pollution of its purity may have considerable effect.

Atmospheric pollution is generally of industrial or, at least, of human origin; the discharge of volcanic vapours is too local and too sporadic to be seriously considered. Industrial pollution on the other hand may affect large areas more or less permanently and at the moment seems to be inseparable from advancing civilization. It takes two forms, solid and gaseous. The former is principally soot, finely divided Carbon with varying amounts of hydrocarbons, which is highly adsorbent of gases, both natural and artificial. Soot deposits on foliage have thus both a physical and a chemical action, which may be lethal through causing foliage to shed. In the Hunslet area of Leeds in Yorkshire the soot-fall has been estimated* at 1 pound avoirdupois (450 g) per square yard per annum and almost the only plant species which survive its effects are the curiously assorted pair, *Sambucus nigra* and *Iris germanica*, so familiar in neglected gardens in Britain. Apart from its direct effect upon leaves the soot-fall also increases the acidity of the soil, which we shall speak about below.

Industrial fumes are also in part solid material; fumes from metallurgical smelters can carry enough of the metals, such as Copper or Zinc, to be absolutely lethal to surrounding vegetation. The Copper Hills in Georgia USA are a famous example of copper-fume poisoning which denuded a large area. Bare hillsides around Swansea, South Wales, also carry their own evidence of metallurgical pollution. As this represented an uneconomic waste of metal, refined methods (or industrial failure) have largely eliminated this particular source of pollution.

In Great Britain some thirty gases have been listed as atmospheric pollutants, mostly hydrocarbons, but including aldehydes. The most prevalent and

* In the 1920s. Much of this pollution has now (1970) been reduced.

most damaging is Sulphur dioxide, which causes blackening or bleaching and death of leaves. Evergreens suffer most, from the condensation of fog or dew carrying the SO_2 in solution. Species vary greatly in their susceptibility and so do the effects of weather conditions. The threshold value for visible lesions is about 0·5 parts per million, but some trees can withstand 60 ppm. With low concentrations there may be no visible lesions, but injury is shown by a marked decrease in yield, as much as 45 per cent of dry weight. Sulphur dioxide pollution decreases in the spring and after a summer minimum rises again in autumn, which points to domestic fires as a principal source.

The effect on the soil is shown by increasing acidity, down to pH 4·0 or less, caused by the formation of sulphurous acid in the soil and its oxidation to sulphuric acid. This increases losses by leaching, notably of Calcium, and it inhibits bacterial nitrification thereby decreasing the fertility of the soil to a point where it becomes practically barren.

Variations in atmospheric pressure, such as occur naturally, are not known to have any direct effect upon plants, but they do have an effect, the dimensions of which are not fully known, in promoting gaseous exchange between air and soil. Observations in mines suggest that the effect may be very far-reaching and that it is active in at least helping to maintain the equilibrium between soil air and the atmosphere. There is a tendency to belittle the importance of mass flow of air between soil and atmosphere and to emphasize the role of gaseous diffusion, but the rate of diffusion under unit pressure gradient is a linear function of pore space, whereas the rate of mass flow in the same conditions is proportional to a higher power of the pore space. This means that for a given soil under a constant pressure gradient mass flow is more rapid than gaseous diffusion. If diffusion alone were active in gaseous exchange it would be difficult to understand the rather wide diurnal fluctuations that occur in the composition of the soil air. Pressure gradients can arise from temperature differences as well as from barometric pressure, but temperature differences can only effect a very shallow layer of the soil. (See also p. 3587.)

In the soil, Oxygen is being consumed biologically and Carbon dioxide produced, yet the soil air remains close to the atmosphere in percentage composition, with rather more CO_2 and less Oxygen (CO_2 0·25/0·03, O_2 20·6/20·96) though the sum of their partial pressures is only slightly less than that in the atmosphere, which suggests that gas exchange, either by thermal diffusion or mass flow must be pretty efficient.

Of the various effects of air on plants it is, however, its movement which has the most obvious influence on growth. Wind has both physiological and morphological effects. The former involve the promotion of heat loss and water loss, the latter involve deformation of growth.

We have already (p. 3546) discussed the effect of wind on heat loss from leaves; the escape of water vapour is similarly increased. In perfectly still air vapour escapes from the leaf surface by thermal diffusion, proceeding outwards indefinitely. Movement of the air changes this immediately. At low air speeds the flow over the leaf is laminar (stream-line flow). Against the leaf surface is a thin zone of motionless air and it is only through this thin layer that molecular diffusion persists. Passing outwards from this there is a succession of zones with increasing speed of movement until the velocity of the free air is reached. The path of diffusion is thus greatly reduced and in all moving zones water vapour is carried off by the much more rapid process of conduction. As the wind speed increases, the thickness of the motionless layer diminishes but only proportionally to the square root of the wind velocity, so that the curve of the rate of transpiration against wind speed rises steeply at first but later flattens out; in other words light winds have a greater proportional effect on water loss than have strong winds, for purely physical reasons and quite apart from such secondary phenomena as stomatal closure. Roughness of leaf surface or a covering of hairs introduces a new parameter, the drag force at the surface, which lies between 1 and 10 dynes/cm^2, and has the effect of increasing the thickness of the motionless layer and thus reducing the rate of removal of water vapour.

Quite apart from the movement of air there is the very important factor of the gradient of vapour pressure between the interior of the leaf and the outer air, the effect of which we have fully discussed in Vol. 3, p. 2863.

Truly laminar flow of air in nature is limited to low velocities and the immediate neighbourhood of the leaf surface. The mass movement of air through vegetation will almost always be turbulent, that is will consist of a system of eddies, which also modifies the rate of vapour transference. Thinking in terms of vegetation rather than of the single leaf, the situation can be thus expressed:

$$E(x) = -D(z)\rho \left(\frac{\delta e}{\delta_2} \right)$$

$E(x) =$ Vertical flux of water vapour at height z, in g/cm^2/sec.

$D(z) =$ Proportionality factor representing 'eddy conductivity' in cm^2/sec., which is a function of wind velocity.

$\rho =$ Density of water vapour.

$e =$ Water vapour concentration in the air at height z in grams per gram of air.

To estimate the vapour flux from a plant community it is not enough to measure gross quantities such as wind velocity and humidity. These factors must be related to the nature of the vegetation as expressed by its effect upon eddying wind movement.*

The physical structure of the leaf naturally affects its reaction to wind, and species may differ considerably in their degree of sensitiveness. *Betula* leaves showed an increase of 160 per cent in water loss with winds of only 0·5 m/sec (1·2 mph), but did not respond to increases of velocity above this. *Helianthus annuus* continued to show increased water loss up to wind velocities of 4·0 m/sec (9 mph).

---

* Fuller details are given by RITER and ROBINSON (1951) *Quart. Journal of the Royal Meteorol. Soc.*, **77**, p. 375; and RIDER (1954) *Phil. Trans. Roy. Soc.* A, **246**, p. 481.

Other things being equal the wind velocity increases with height above ground level; Stevenson and Hillmann give the following comparative figures:

	Stevenson			Hillman		
Height (m)	0·05	0·5	8·0	12·0	16·0	32·0
Velocity (m/sec)	1·04	1·95	2·43	3·29	4·86	5·54

Hillmann's figures are yearly averages and with increasing altitude the yearly average increases, the amount depending a great deal on local topography. Thus at Sonnblick in the Alps (3100 m) the yearly average was 7·5 m/sec, while on the isolated Mount Washington USA (1928 m), the yearly average was 15·0 m/sec.

The desiccating effect of high wind averages is a powerful influence on the growth of alpine plants and in other exposed places such as sea coasts and in the Arctic. The habitual prostrate or cushion forms of species in such localities is not due so much to mechanical trimming of the shoots as to the effect of dry winter winds on any plants which are not protected by snow cover. Such winds, at times when the soil is frozen, can be exceedingly dangerous to plants at any level and this influence is predominant in fixing the Arctic tree limits. On the open arctic tundra no shoots which rise above the level of winter snow can survive the drying effects of the wind.

This does not apply, of course, to exposed sea coasts where the deformation of trees and shrubs is very remarkable. Buds and young shoots on the windward side are killed, partly by desiccation and partly by salt spray carried on the wind. Growth is all to leeward, that is away from the sea, which gives the appearance of having been bent by the force of the wind. The unilateral growth often causes the trunk of the tree to bend and there is a notable tree of *Fraxinus* at Rhosilli in South Wales which is completely horizontal and covers a farmyard like a roof. The same effect can often be seen in alpine trees, becoming increasingly marked with altitude and resulting, towards the limits of tree growth, in what is called in Switzerland *Krummholz* or twisted wood. If the mountains are snow-covered in winter, the branching below the level of the snow cover may be quite symmetrical and that above snow level one-sided. Mechanical damage just above the snow surface is usually due to the blast of small ice particles carried by strong winds.

Wind erosion of soils is a significant feature of many dry climates. Dust storms can carry away and redistribute thousands of tons of surface soil. The very extensive deposits of loess, a deep, fine-grained and stoneless soil, which are widespread in the areas of prairie and pampas and in western China and in Australia are simply age-long accumulations of airborne dust. How massive such deposits can be is shown by the fact that between the tenth century and the nineteenth century the floor of the Roman Forum became buried under more than thirty feet of dust.

A similar formative-destructive influence of wind is also shown in sand-dune formation both on sea coasts and in deserts. Sand, carried by strong winds, is trapped, either by tufts of large grasses, such as *Ammophila* or *Elymus*, or else builds up against permanent obstacles. Where living grasses are concerned they help to bind the sand progressively until small hills, up to thirty metres high are formed. Stabilized in this way the sand may then be colonized by mat forming mosses and many other small plants, followed either by grassland or by woody plants such as *Salix repens, Hippophaë rhamnoides* and *Calluna vulgaris*, forming a sandy scrub or sandy heath. If a dune is not bound by vegetation, as is usually the case in deserts, the dunes will wander with the wind and they assume a characteristic crescent shape, with the horns pointing away from the wind.

Analogous sequences on sandy shores are worldwide, with different plant species involved. In the tropics the binding of the sand is generally brought about by *Ipomoea pes-caprae* or *Ramirea maritima* assisted by spreading leguminosae, *Canavalia, Vigna* and *Clitoria*, followed by shrubby *Pandanus*, and *Scaevola frutescens (Koenigii)* and developing into a sandy scrub of *Barringtonia, Terminalia, Clusia, Calophyllum* and many other genera, according to the part of the tropics concerned, and passing over inland into forest.

A secondary, but none the less important, action of wind is the transport of pollen, spores and small fruits or seeds, which we have discussed in Vol. 2 as an agency of pollination and dispersal.

All the climatic influences form an interlocking network in their effects upon plants, and to discuss them as if they were separable is an artificial simplification for purposes of study, which is justifiable only so long as we realize the true position. To realize it and to comprehend it are, however, two different things. The ecological problem is one with an indefinite number of related variables, and our human limitations make it almost inevitable that we have to be satisfied with partial solutions, in the hope that these may build up into a generalized scheme that is not beyond our grasp.

Climate is the most potent influence in determining the world types of vegetation. It has direct effects upon the plants, which we have hitherto been discussing, but it has also its indirect effects in determining soil types. Tansley sums it up thus: climate plus parent rock determine the soil; climate plus available species determine the vegetation and these two, soil and vegetation, must harmonize under the sovereignty of climate if a stable equilibrium is to be achieved. Stability, as we have recognized, can only be relative to our time-scale, yet to us it is real, and recognizable as the end term of a finite series, a series tending, like all organismal processes, towards states of increasing probability in the mathematical sense. The series progresses by steps which we recognize in plant succession, or, in Clements's term, the *sere*, the motive force of which is *biotic*, active both in competition and in biotic reactions on the physical environment. We must now take a closer look at the biotic environment.

CHAPTER XLVII

THE BIOTIC ENVIRONMENT

COMPETITION between plants in a community is chiefly along four lines. In the first place comes competition for space, without which no growth is possible. Second and third, come the competition for water and for light, sometimes one and sometimes the other being the more important factor. Fourth, there is competition for mineral nutrients, which is not usually so acute, as the supplies available are not often critically low. Competitive power is an expression of the life-habit of the plant. Rate of growth, leaf spread, wide tolerances, stature and longevity are all positive factors in competition, but equally important is the plant's 'vitality' by which we imply its vigour in spreading, especially by vegetative means and its reproductive capacity, which may be summed up as *aggressiveness*, the plant's power of occupying and holding space. Vitality is to the plant what strength is to the animal. A plant of relatively low vitality may succeed in growing and reproducing, but it cannot realize its full competitive potential. (See also p. 3364.)

Controlled competition cultures offer a very interesting field for observation: some conducted by Clements gave striking results, but these were cultures of the same species. Mixed cultures can be even more interesting though more difficult to interpret. Some experiments on mixed seedling competition carried out by Willis in 1912 led him to believe that survival was a matter of pure chance, for the margin of competitive ability at that stage was too small, as between species, to be decisive in the selection of survivors. This deserves further investigation for the success or failure of a species in a community is probably often decided at an early stage.

Clements principally used *Helianthus annuus* and *Triticum vulgare* in his experiments. The Sunflowers were sown in plots with intervals between plants of 2, 4, 8, 16, 32 and 64 in. The supply of radiant energy was roughly proportional to the square of the distance between them and the amount of water and soluble nutrients was roughly proportional to the cube of that distance. Stem diameter in the most crowded plants averaged 0·2 in., and in the most widely spaced 2 in. The respective areas of assimilatory tissue per plant were 39 cm^2 to 2750 cm^2. The number of seeds per flower-head was 15 as against 1803.

Plots of wheat were sown at the normal rate of sowing (in Minnesota), at about half this quantity, and at twice and at four times the normal rate. These showed developmental features parallel to those of the sunflower plots, but with the significant difference that the weight of seed produced was greatest in the plots sown at twice the normal rate. That crowding may actually increase the collective reproductive capacity is an unexpected and important fact, not without human implications.

Another point of comparison brought out by Clements was that the daily period of stomatal opening was decreased by crowding, thus reducing the period of assimilatory activity to the plant's disadvantage.

Observations on prairie grasses by Clements brought out the advantage of stature by showing that where grazing was excluded, the tall Blue Stems (*Andropogon* sp.) always supplanted the short Buffalo Grass (*Buchloë* sp.).

The importance of tenacious occupancy of the soil was also brought out by attempts made by Clements to establish marginal forest trees in prairie areas. So complete was occupancy of the soil by the grasses that in spite of experimental assistance the trees all died in a few years.

Grasses show themselves to be formidable competitors in mixed competition cultures, and where the climate favours their establishment the successful invasion of a grassland is extremely difficult, largely because of the difficulties attending seedling establishment in the dense growth and in penetration of the soil closely occupied by the grass roots. In the initial stages of colonization broad-leaved weeds may hold their place against grasses by virtue of their greater rate of growth, but in spite of their prolific production of seeds they generally leave no second generation of successors. Even biennial or perennial weeds can only successfully invade a grassland where the grasses are handicapped by mowing or by grazing.

The xerothermic conditions obtaining in the post-glacial Boreal period enabled the prairie grasslands to extend eastwards in the USA driving a broad wedge, the 'prairie peninsula', into the eastern deciduous forest region. The succeeding period, which was cooler and wetter, allowed the forest to recapture much of this territory, but extensive prairie openings in the forest still remain as testimony to the power of the grasses to resist invasion by tree seedlings.

The process of succession depends on an unstable equilibrium in competition between species, which resolves itself in the displacement of some species and the extermination of others by secondary species, with very often a change in the predominant growth-form. The successful species may be either newcomers or they may be species already present which are selectively favoured by some change of environmental conditions, either due to external causes, or to internal changes brought about in the habitat by the reactions of the plants themselves. An example of an external cause might be the leaching of lime from the soil, which can allow the supersession of a calciphile community by a heath flora. An internal cause might be the increased humus content of the soil, changing its water-holding capacity or its pH. Succession is often said to involve the extermination of the plants of one phase by those of the next, but this is by no means always the case. A species formerly dominant may be reduced to a subsidiary position, but if it has wide tolerances for physical conditions it may survive in reduced numbers even up to the final phase.

Competition does not cease when a stable vegetation becomes established, but the competition between the limited number of species involved is also stabilized. The condition of stability is that the species are, by selection,

limited to those which are evenly matched or those which have a special
ecological niche, not menaced by the others. In this latter category we would
include climbers and epiphytes, both angiospermic and cryptogamic. The
overall composition of, for example, a natural forest, does not markedly
change with time. Its relative constancy is the expression of a stable harmony
of the species collectively with each other and individually with the prevailing
physical environment. Natural catastrophes, extensive wind-throw or fire,
may upset the equilibrium temporarily and may act selectively on species to
give survivors a temporary advantage. The survival of *Sequoiadendron**
giganteum (*washingtonianum*) to such an immense age may be attributed to
the fire-resistant quality of its very thick bark, but it did not, in spite of this
advantage, displace the other forest species. There may be temporary disturbance, but the equilibrium is eventually re-established, unless conditions have
permanently changed. (See also Mt. Rainier forest, p. 3392.)

The early stages in the colonization of an area by plants are largely dominated by chance, the chance, that is to say, of arrival in the area. Further progress in succession depends on a sorting out by competition of the species
which are less harmonized with the environment and their replacement or
subordination by those which are better harmonized. The chance assortment
of biotypes with which colonization began is thus replaced by a group in
which the range of tolerances and environmental demands are more or less
similar, the beginning of a natural association. This does not necessarily mean
a diminution of competition between them, on the contrary it may become
stricter, but it will tend to run along more limited lines and the margin of
competitive advantage between species will become narrower. This results in
a situation where the elimination of one species by another becomes increasingly improbable, though fluctuations in the population may often occur,
generally associated with varying seasonal conditions. The suggestion is often
made that the progress of succession involves a more complete usage of the
resources of the habitat, but this is open to question. That the species represented in later phases make a more efficient use of the resources is probably
true within the limits of their requirements, but these may leave much that is
unexploited.

The ranges of tolerance of the species finally associated together must at
least overlap, and for most their optima will be closely related, but the extent
of the different specific ranges may be very different. Those with narrow
ranges will probably be 'exclusive' to the one association, in Braun Blanquet's
sense, while those species with wider ranges, with a 'reserve' of tolerances,
will show this by their appearance in other habitats and other associations.
Many species have such reserves which enable them to survive in very
various environments. Under cultivation they may appear almost cosmopolitan, though two restrictions, their reproductive requirements and their

---

* May we suggest that when this species was separated from *Sequoia*, with which it had
been united, it should have reverted to the former generic name of *Wellingtonia*. *Sequoiadendron* is a tautonym.

lack of competitive strength, may hold them to very restricted natural habitats. Where their natural competitors or predators, insect or fungal, are absent, such species may be able to settle and flourish in many areas where they are introduced. They may even find environments to which they are better suited than their native one, or they may be able to tap resources not exploited by the indigenous plants, as is witnessed by the almost explosive expansion of some species brought to new countries (p. 3377).

As all factors interact, a plant's range of tolerance for one factor may be modified by a change in another. We have seen (p. 3551) how the light demand of a species varies with latitude. Similarly, in the USA, *Poa pratensis* in Kentucky shows a wide tolerance for variations of soil moisture, but its extension southwards is restricted by higher temperatures, under which its range of moisture tolerance becomes very limited. On the other hand *Cynodon dactylon*, Bermuda Grass, has a wide range of moisture tolerance under high temperatures, but this becomes very restricted under cooler northern conditions. Such differences are very common.

Plants with unusual tolerances may gain advantage in being able to exploit environments which few other plants can utilize. This is the case with most desert plants, sand-dune and salt-marsh plants, plants of waterlogged soils, soils with either very low or very high pH or soils poisoned by certain metals, especially Zinc, Lead and Copper. The plants capable of living in such conditions may be either species peculiarly adapted to them or they may be specialized ecotypes of widespread species. The latter is usually the case in poisonous soils.

Every biotic environment begins with colonization, which comprises migration, arrival or invasion, and ecesis, Clements's handy term for the process of establishment.

The power of migration means the power of leaving the parent area and in general it depends on the production of seeds, spores, gemmae, etc., collectively called *propagules* or *disseminules*, in numbers in excess of the natural wastage, which is often enormous. The wonderful variety of the means of dispersal of the propagules makes it a fascinating subject of study (see Vol. 2), but all we need note here is that the efficiency and the range of dispersal are very different in different species, so that succession in arrival is inevitable from the start. The first arrivals belong to species with light, easily transportable seeds, the great majority carried by wind and for the most part annuals, which are absolutely dependent on efficient dispersal for their survival. Migration is thus a selective process. Trees are usually slow migrants, partly because of their comparatively long immaturity and partly from their limited range of dispersal. Ridley estimates the average range of dispersal of the wind-borne fruits of Dipterocarpaceae as only thirty yards. Exceptions are *Salix*, *Betula* and *Populus* which therefore are frequent pioneers of woodland conditions. Their small seeds also escape the attention of birds, mice, squirrels, etc., which sometimes destroy the entire output of heavier-seeded trees. Successful migration is thus largely a matter of reproductive capacity. Even after arrival, competition, at least in the later stages of succession, can eliminate all but a small percentage of the beginners and successful invasion, like pathogenic infection, largely depends on mass action.

Salisbury in his book *The Reproductive Capacity of Plants* (1942) gives very interesting tables of the comparative weights of seeds, which show clearly that there is a positive correlation between the average weights of seeds and the stages of succession from an open habitat to woodland. If the seed-weights for each type of associes in the succession are arranged in classes, the upper limit of each class being twice that of the class below, then the modal point for the curve of each habitat separately, rises steeply towards the woodland end. As heavier seeds carry a larger supply of nutrients the figures suggest that seedlings from heavy seeds have a better chance of survival in closed and well-developed vegetation than the smaller seeds which predominate among the early colonists. Another factor coming into play as succession advances is increasing shade. A comparison of a selection of meadow species with ground herbs from woodland again shows that the average seed-weight for 22 species of meadow grasses and herbs was only 2 mg, while the average for 27 species of woodland herbs was 14 mg. The average seed-weight of 19 British woodland trees (excluding *Castanca sativa* but including *Corylus*) was 0·48 g. Another factor active in this connection is the sensitivity of seed germination to light. Although few facts are available it is noteworthy that the seeds of two species which colonize open ground or cleared areas, *Verbascum thapsus* and *Digitalis purpurea*, are light positive, while the seeds of *Actaea spicata* and *Leucojum vernum*, which commonly occur among shady herbage, are light negative.

Annual plants rarely have the stature or the food reserves required for the production and dispersal of large seeds. They rely on the production of large numbers of small seeds, thus overcoming the disadvantage of the greater rate of mortality which attends those species in which the small seedlings are more liable to suppression than those from larger seeds with a considerable reserve of food. On the other hand, remembering Harper's observations on germination in relation to the character of the soil surface, the small seed stands a better chance of finding a suitable niche for germination on open ground than does a large seed. The 'plan' of the small-seeded annual is one for escaping competition, that of the larger-seeded phanerophyte is one for overcoming competition.

With regard to seed-output, as distinct from seed-weight, Salisbury was able to show, on the basis of very extensive observations, that among British plants species normally found in habitats that are intermittently available, e.g. cleared woodland, exposed mud and shingle, have by far the largest output followed in descending order by species of permanently open habitats, semi-open habitats, closed but unshaded and lastly, closed and shaded habitats. The lowest category is physiologically significant in that flowering and seed production generally require higher temperatures and greater illumination than does vegetative growth. The shaded conditions that depress seed production will therefore promote vegetative propagation in species which possess the necessary structures.

Severe competition also reduces seed-output somewhat drastically and as vegetation develops successionally and competition increases there is an increasing number of perennials present, especially perennials with vegetative propagation. Frequently such species have a very small seed output or none, relying, at least in some areas, almost wholly upon vegetative spreading. This is also characteristic of species growing under extreme arctic or alpine conditions.

Vegetative spreading above soil level confers two distinct advantages. It may be either close and slow, holding space very effectively and denying entry to possible invaders, or it may be rapid and extensive, owing to the absence of resistance to runners on the soil surface. Underground rhizomes must overcome the resistance of the soil and differences of soil texture will control their average spread. In loose, sandy soils they may play quite an important part in the occupation of bare areas as, for example, do the sand-binders of sandy shores. In heavy soils their spreading may not be so rapid, but it confers the power of penetrating and colonizing closed communities, in which invasion by seeds may stand small chance of success. The growth of branched rhizomes in loose surface soil may combine both advantages, as conspicuously shown by *Urtica dioica*, in which the solid and rapid advance against formidable competitors, combined with its tall stature, is only limited by physical factors.

In the absence of any directive force, migration by one means or another is usually radial from the source, but unilateral agencies sometimes intervene. Such agencies are: a prevailing wind, the force of gravity on steep slopes, the downwash of heavy rain or water-carriage by streams. Any of these may direct seed migrations or may at times carry whole plants or portions of them to new habitats. Such directional forces may decide not only the migration of individuals but the advance of a whole community, especially at or near an ecotone where the balance between two communities may be unstable.

We can seldom trace migration in progress, what we see and what we judge it by is its product, invasion of some area, which becomes visible when young plants of the migrating species appear. For these to appear there has to be a process of *ecesis*, or accommodation to the new environment. Unless this is successful migration is valueless. In ecesis germination is the first obstacle and here the numbers and weight of the seeds may be decisive factors ('mass action') as we have seen above. After germination the seedling is exposed to many dangers; from predators, both animal and fungal, from physical conditions and from competition either with established plants or among the seedlings themselves. Germination is often long delayed or irregular or spasmodic. Some species produce dimorphic seeds, differing in size or shape and with differing times of germination. Such irregularities may be of positive benefit in ecesis by multiplying the chances of avoiding some lethal phase of conditions, such as drought, and thus increasing the probability of some seedlings surviving. Even under good conditions, however, the seedling mortality is very heavy and the decision of survival or non-survival for the individual is probably largely accidental and not strictly a matter of Darwinian natural selection. However this may be, in view of the enormous output of seeds by many species, the extent of mortality is a very important factor in maintaining the balance of vegetation. When we contrast the comparatively sparse occurrence of a species like *Verbascum thapus* with its output of over 100,000 seeds from a mature plant we get some realization of how small the chances of survival are for any single seed.

The ecesis and the lives and deaths of generations of pioneer plants on a new area, start the process of the amelioration of the habitat which permits invasion by newcomers and thus promotes, at an increasing *tempo*, the sequence of stages in a succession.

There are two main elements involved in this, so far as the plants are concerned; reaction and co-action. The term 'co-action' is preferable to 'co-operation' for it is more neutral; the latter term has overtones of purpose and common will, which are out of place here.

Reaction (see also p. 3385) implies the effects which a plant community has on its environment through the lives and deaths of its individual members. The relationship between environment and plants is mutual, each effects the other. Reactions on the soil are most marked in the early stages of succession, those on micro-climates are greater in the later stages, when large phanerophytes become predominant.

The most fundamental reaction is in the actual formation of a soil as a medium for growth. This may happen in various ways: first, by the breakdown of rock into particles either by chemical or mechanical action. Second, by the accumulation of plant remains. Third, by the formation of mineral deposits extracted from solution by water plants. Fourth, by the collection and binding of drifting materials, which we have already referred to in connection with maritime sands (p. 3559).

These primary events are almost entirely the work of plants, but in later reactions animals also play a part including the micro-fauna (and flora) of the soil, by mixing, aerating and manuring the developing soil.

Further reactions on the soil may take some or all of the following forms:

1. Reducing erosion by wind or water. Creeping plants are often introduced by man for this express purpose where erosion is serious.
2. Addition of humus to the soil. This has several side effects. It enriches the soil, it improves the soil structure by the formation of soil crumbs; this also improves soil drainage and aeration; it increases the water-holding capacity of the soil, to the benefit of the plants and to the reduction of the surface run-off of water; it changes the pH of the soil; it encourages the growth of micro-organisms in the soil, including nitrifying bacteria.
3. Changing the water content of the soil. While the addition of humus raises the *water-holding* capacity, the plants may cause a reduction in the total *water-content* through absorption and transpiration. In the

contrary direction, *Sphagnum* may absorb and hold such large quantities of water that natural drainage is blocked and the area becomes water-logged or flooded.

4. The circulation of nutrients, absorbed from the soil and subsequently returned to the soil, either by surface leaching by rain, decay and humi-fication in the soil or removal by grazing animals and later return else-where in droppings.

5. The excretion of acids or toxic substances from the roots. We have already discussed the evidence about this on p. 3369.

Reactions on the micro-climates are mostly modifications of rain-efficiency (p. 3534), humidity and illumination, due to increasing closeness of growth and the incoming of woody plants and trees. The growth of shrubs and trees may so reduce the lighting as to cause the disappearance of many herbaceous species, lessening competition for the seedlings of the woody plants themselves but introducing conditions of shade at ground level which the seedlings must tolerate until they are of full stature to reach the sunlight. This latter factor introduces a condition of competition between the woody plants themselves. The first comers are usually species relatively intolerant of shade and forming an open canopy, but they are liable to displacement by a sequence of species with increasing density of canopy, with which goes greater plasticity in leaf structure and the concomitant ability to grow in shade during their early stages as seedlings. Such sequences among forest trees are best exemplified in countries with a rich flora of tree species, but even in Britain the suppression of *Fraxinus* by *Fagus* and of the latter by *Taxus*, on a basis of shade tolerance, is well attested. In addition to the above there are also effects upon the rate of evaporation, due to wind-screening, and reducing the range of temperature changes.

Co-action reaches its highest level in symbiosis, but is present to some extent in all plant communities where, in the broadest sense, it includes all interaction between plants, whether destructive, as in the case of parasitism, competitive or co-operative. The reality of competition does not rule out some degree of co-operation, however involuntary it may be. In this sense there is a bond of co-operation between the species of a community with regard to their reactions on the soil environment. There must indeed be a harmony in their environmental reactions, which must be either coincident or complementary, otherwise they would not form a community. It is true that the reactions of a pioneer community usually prepare the way for their supersession by another set of species, but among these, in their turn, there must be a co-operative harmony with respect to the environment. Mutuality, in the form of mutual protection afforded by the massing of plants together, is inevitable, protection against wind, against excessive illumination, against excessive transpiration and sometimes against browsing animals, as where thickets of *Ulex europaeus* shelter much finer, flowering specimens of grasses than can be found in the exposed pasture land. Yapp has pointed out that well-marked contour levels above the ground occur in wind-swept habitats, caused by the massing of shoots and leaves at those levels which means that the leaves must benefit from their protective association.

The degrees and conditions of co-action between plant species are extremely varied and complex. The study of interspecific association (p. 3406) is largely the attempt to unravel and understand such co-action, which may involve whole groups of species both plants and animals. Sometimes the reasons for it seem simple. For example, tussock grasses often leave vacant spaces between tussocks, providing a home for smaller grasses and Bryophyta, which find competition too strong for them in a sward of creeping grasses. Nodule-bearing Leguminosae enrich the Nitrogen reserves of the soil and a positive association between *Trifolium repens* and the nitrophilous *Lolium perenne* has been recorded. Some examples of specific associations are based upon a common range of tolerance for some exceptional condition in the environment. In other cases two species appear to complement each other in ways not fully understood, so that each develops better in association than singly.

Woodhead introduced the idea of complementary association, between species which exploit the environment in different ways, thus minimizing competition. His classic case was the association in Britain of *Holcus mollis*, *Pteridium aquilinum* and *Scilla non-scripta*, the root systems of which utilize different levels of the soil, descending in the order given. Assuming a common range of tolerance for physical conditions among species thus associated, co-action is more important than competition in governing their relative abundance and distribution.

The influences at work in the co-action of members of a community are often subtle and complex and have not been intensively investigated, but there can be little doubt that it is an element in community building and in the development of pattern in the community. Instead of considering the natural plant association as defined by the three elements: flora plus tolerances plus competition, we should add co-action as a fourth element in the integration of a simple aggregation of plants into an association. The consideration of co-action as well as competition may also aid us in understanding the basis of the categories of social importance of species in a community, which Brockmann-Jerosch first recognized as: constants, accessories and accidentals.

What Clements describes as 'disoperation', that is disadvantageous or destructive co-action between species of plants or between plants and animals may often have a constructive side effect. Competition between two species may handicap both, but this may promote the success of some of their associates. Selection by grazing animals may change the whole composition of a grassland if carried on for several years.

If co-action be accepted as a reality it forms an argument against Gleason's individualistic theory of the plant community (p. 3406). This regards the community as no more than a chance aggregation of species selected by the nature of the environment, the only relationship between them being one of juxtaposition. Such a view reduces the biotic environment if not to nullity to a very subordinate position. Admittedly the initial aggregation of species on a new site is largely accidental and each species owes its arrival to its individual powers of migration, but though this may be so, it does not follow that the survival of the species through subsequent developments on the area retains this accidental and individualistic character. To accept this is to assume that the character of a mature community was decided by the conditions obtaining at its initiation, which could only be the case if all the species involved had equal rates of migration and arrived together and that all had similar growth forms and longevity, propositions which are manifestly untrue. The only truly independent or individualistic plant is the isolated plant.

Competition is not the only influence at work in a biotic community. As the community develops many relationships of dependence are formed, in which the establishment and survival of an incoming species depends on conditions created by some previous inhabitant. This may not involve competition. The benefit is entirely one-sided and the earlier plant may be unaffected by the arrival of the later comer. As Yapp points out, there is no competition between the woodland trees and the moss upon the ground, but the latter is dependent on the former for the conditions of its life. Of course dependence may later develop into competition. The cuckoo in the nest is no stranger to the plant world. Shrubs, for example, may create a nursery for the seedlings of the trees which will eventually overwhelm them.

Yapp also distinguishes what he calls 'priority' from true competition. Priority means that an organism is so situated that it can intercept and retain all it requires of some necessity before other organisms can get any. The obvious case is the interception of light by the tallest plants, though this hardly meets Yapp's definition, since all receive their quota of light simultaneously. Better perhaps is the instance of the deep roots of trees having priority of access to the reserves of soil water or that of the interception of the Oxygen supply to the soil by the carpet of lowly plants in the ground flora. Here there is a relationship between species which is neither truly competition or co-action, but it is nevertheless definitely an element of a biotic environment.

The living populations of the soil are in a special category, for they may be regarded either as part of the biotic or of the edaphic environment. They are, however, constituents of the biocoenose, the living aggregate of the community, and it is as such that we shall view them. (See also p. 3679.)

Foremost are the mycorrhizal fungi of the soil, the association of which with the roots of higher plants appears to play an important part in the nutrition of these plants. This is not the place to discuss the physiology of mycotrophy, but its significance as a biotic element is shown by the large and increasing number of species in which fungal associations have been detected. Certain angiospermic families have long been known as conspicuous in this connection, e.g. Pinaceae and Amentiferae (ectotrophic mycorrhiza) and Orchidaceae, Ericaceae and Compositae (endotrophic mycorrhiza). The fungi concerned are usually Hymenomycetes or Phycomycetes. The relationship in many cases is obligate. In most Orchidaceae germination cannot proceed until the embryo has been infected by the appropriate fungus. For this reason orchid seeds are usually sown into agar cultures of the fungus. Formerly seeds were sown on soil in which the parent orchid had grown, a procedure intended to secure seedling infection. Extreme cases are those species which have lost their chlorophyll (e.g. *Neottia*) and have become saprophytic in association with a root-fungus. Many species, perhaps the majority, have only an accessory or facultative relationship to the fungus, which apparently confers benefits by the exchange of materials between fungus and root, although the higher plant can survive without them. The term 'infection' is appropriate for the connection between the two partners as it appears to be a relationship of 'balanced parasitism' rather than a true symbiosis. Some of the fungi involved, e.g. *Armillaria mellea*, *Phoma*, etc., are aggressive parasites and are capable of destroying the young plant if it is weakened by unfavourable conditions (see also p. 3693).

The activity of the soil bacteria in the Nitrogen cycle is well-known and is of prime importance in the maintenance of soil fertility. It has also been shown that the bacterial flora on the surfaces of roots and in the soil in their immediate vicinity contains large numbers of organisms producing phosphatase, which hydrolyses organic Phosphorus compounds, liberating inorganic phosphate, though to what extent the higher plants benefit from this has not yet been proved. (See 'Rhizosphere Effect', p. 3691.)

A large field of study in biotic relationships lies in the genetical make-up and relationships of the populations of plants which constitute communities, a study generally called 'genecology'. Few populations of species, except perhaps small populations growing under extreme conditions, are genetically homogeneous and the biotypes present can hybridize freely thus increasing the biological potentialities of the species and its range of 'variability'. Even ecotypes, that is biotypes with disparate ecological preferences, which are normally segregated by ecological barriers to crossing, can hybridize freely, if they happen to meet, though such hybrids are not likely to be successful unless a 'hybrid environment' happens to be available where a previous ecological barrier has disappeared.

Anderson has drawn attention to the genetic consequences of 'hybridization' of the environment, that is to the opening up, either by man or by some other natural catastrophe, of fresh areas which can be colonized by populations which had been previously segregated ecologically, which facilitates inter-crossing, with the production of 'hybrid swarms'.

Hybrids between natural 'species', which should be, according to traditional ideas of the species, impracticable, have been recorded so often, on unimpeachable evidence, that new definitions of species have been required. Thus the 'coenospecies' is recognized as an aggregate which is in fact cut off by genetical barriers from gene exchange with other coenospecies. Such groups, however, may be made up of 'ecospecies', between which a limited exchange of genes is possible, though the hybrids are often reduced in fertility. The barrier to exchange of genes in this case is usually ecological separation. If the ecospecies differ morphologically to an extent which has earned them separate specific names, we have the case where 'species' may form hybrids, if they meet.

Species hybridism seems to be commoner in some countries than in others. Cockayne and Allan have recorded large numbers from the New Zealand flora, some of which, at least, are of proved fertility. The alleged polymorphism of many New Zealand species and the apparent occurrence of linking intermediate varieties between species are both put down to hybridism. Over forty cases are recorded of pairs of species which have normally distinct ranges, either geographically or topographically, but which meet at some locality (e.g. *Ranunculus lyallii* and *R. buchanani*; *Olearia colensoi* and *O. angustifolia*). In their typical areas the species are relatively invariable, but where they meet hybrid and intermediate forms abound. Significantly, Cockayne remarks that forest areas which have been cleared or burned provide the most plentiful harvest of hybrids.

These cases would seem to be explicable as those of ecospecies, taxonomically distinct but still interfertile when brought together though normally separated by ecological barriers.

The extremely interesting observations by Thoday of the genus *Passerina* (Thymelaeaceae) in South Africa also seem to show a collection of ecospecies so well separated by differentiation of habitats, and in part by geographical range, that he does not mention any occurrence of hybrids between them. It would be interesting to test their capacities for gene exchange.

When two ecospecies meet and cross, the F_1 generation is liable to back-crossing with one or other of the parent species, which are likely to be present in superior numbers. Repeated back-crossing may swamp the hybrid, whose offspring will revert to the appearance of one of the parent species. Nevertheless new genes have been introduced and some of these at least may remain effective to modify the character of the offspring so that a population of variant individuals is produced among which there are varying degrees of resemblance to one or other of the parents. The balance will incline towards one parent more than the other if the environment is more favourable to that species, for the variant population will be subject to environmental selection. This infiltration of genes from one ecospecies into another is called *introgression* and it may occur extensively wherever two ecospecies which can cross-fertilize meet one another.

Anderson and Sax (1936), who first drew attention to this phenomenon in *Tradescantia*, compared populations where two species met, with pure populations of the separate species. They used a tabulation based upon a number of chosen morphological characters. The species involved were *T. canaliculata* against *T. occidentalis* and *T. virginiana*. Six characters were used. Characters of *T. canaliculata* were scored zero, those of the other species were scored 1–3 according to their degree of development. Histograms based on the scores, compared with the parent species, showed the range of variation in the mixed population and also towards which parent the bias lay. This partial fusion of two specific genotypes is probably the basis of specific 'intermediates' and of the 'highly variable' species of the older taxonomists.

When a hybrid swarm has been subject to a long period of selection one or other of the types in the swarm may be the only survivor and attain a distinctive taxonomic rank, even specific rank. Sometimes introgression may produce a new character in the offspring, not present in either parent, which may be due to the bringing together of two genes formerly separated in the course of species evolution in the genus. If this were the case the 'new' character would really be the reappearance of a lost ancestral character.

Genetic hybrids which breed true and constitute new species, sometimes of great vigour and ecological importance, generally arise through one of two conditions; apomixis or polyploidy. *Apomixis*, i.e. the absence of sexual fertilization, has several cytological bases, but it is in any case a specialized form of asexual reproduction comparable to the more usual forms of vegetative reproduction. If a species has a vigorous and successful method of vegetative propagation the whole of a local population may form a *clone* of genetically identical individuals and if the species is self-incompatible sexual reproduction is thereby eliminated, unless chance pollen is introduced from another population outside the clone. *Lysimachia nummularia* is such a species.

Triploid hybrids, which are normally sterile, can propagate themselves vegetatively apparently indefinitely and thus come to rank as species. In the same way, apomixis or pseudogamy, which is a form of apomixis in which the stimulus of pollination is necessary, though no fertilization results, can perpetuate a chance hybridization (which may be only a once in a million chance) in the form of a constant population taxonomically indistinguishable from a species. The numerous 'critical' species of certain polymorphic genera, chiefly in the Rosaceae and Compositae, are known to have originated in this way. Some species, such as *Poa pratensis*, show a mixed condition, including euploid strains which reproduce sexually and a series of aneuploids with varying chromosome numbers, which are apomictic.

The new characteristics in a hybrid may enable it to exploit a novel environment or sometimes to usurp the territory of one of the parent species. The genetic processes by which such new types arise may therefore have considerable ecological importance.

Polyploidy is the name given to the condition of having an increased number of chromosomes. Each genus has a basic number of chromosomes, which is often that found in the species with the smallest number in its gametophytes. This number is, of course, doubled at sexual fertilization. There may be, however, other species in the genus with higher numbers. Suppose a species to have a haploid number of 8 and diploid 16. Another species may have haploid 16 and diploid 32. The latter is properly called a 'tetraploid', because the sporophytic nuclei contain not two but four sets of the basic chromosome number.

Polyploidy may exist in a genus to almost any degree. In many cases the numbers rise by cardinal multiples of the basic number, which is called a *euploid series*. In such series the even numbers are called balanced, because pairing of homologous chromosomes at meiosis is possible. The commonest cases are: diploid, tetraploid, hexaploid and octoploid; higher numbers are less frequent. Odd numbers: triploid, pentaploid, heptaploid, etc., are called 'unbalanced' because normal meiosis cannot take place; the number cannot be halved and some chromosomes have no homologous partner. Irregular departures from a euploid series are called *aneuploid*. Balanced euploids may be fully fertile but unbalanced euploids and aneuploids are either sterile or have very reduced fertility.

There are two conditions of polyploidy, according to its origin. It may result from an internal change, either the failure of meiosis resulting in functional gametes which are diploid, or less commonly, by the doubling of chromosomes in somatic mitoses. This is called *autopolyploidy*. Alternatively, it may arise through hybridization, in which case it is called *allopolyploidy*. If hybridization gives an unbalanced number of chromosomes, e.g. triploid, the offspring will probably be sterile, but this can be overcome by the auto-polyploidic doubling of the chromosome number, giving a balanced hexaploid, which is fully fertile and behaves as an independent species. Autopolyploidy can be induced by various artificial means, of which none is so successful as the application of the alkaloid colchicine. It has often been resorted to as a method of restoring fertility to valuable hybrids and because polyploidy is often accompanied by increased size of plants and larger flowers.

This short description of the nature of polyploidy is not irrelevant, because the morphological and physiological changes associated with polyploidy may have considerable influence on the plant's competitive ability and its environmental tolerances. Changes which have been observed, in some though not in all cases, include: increased water content, increased chlorophyll content, changes in leaf size, venation, increased stomatal size and reduced frequency, changes in hairiness and chemical changes in the production of anthocyanins and vitamins. A change which seems to be general is a slowing down of development and postponement of reproduction, which tends to transform annuals into biennials or perennials. Indeed this effect, together with the prevalence of polyploidy among perennials, has led to the theory that perennials, including woody plants, have evolved from annuals, in opposition to the anatomical theory of the reverse process.

There is general agreement that polyploid races in a species have different tolerances from the diploid races and that the range is generally broader. That they should be different would be readily concluded from the above list of changes in structure and habit, and it may well be that such changes of tolerance could lead to the development of new ecotypes or ecospecies. They flower usually later and for a longer period and this could facilitate hybridization by overlapping with the flowering period of other species within the same comparium or coenospecies.

Broader tolerances can lead to wider distribution and to entry into other communities. In particular polyploids have been widely credited with greater hardiness, that is, resistance to cold and other extreme conditions, so that they can grow where diploids cannot. Cyto-geographical surveys have led to the view that the percentage of polyploids in the flora increases with latitude. This was first noticed by Tackholm in 1922, who showed that high-polyploid species of *Rosa* extend further north in Eurasia than diploids or low-polyploids. Hagerup in 1928 observed a similar distribution among members of the Ericales, with the notable exception of *Vaccinium*, and put forward the general principle that polyploidy was associated with extreme climates.

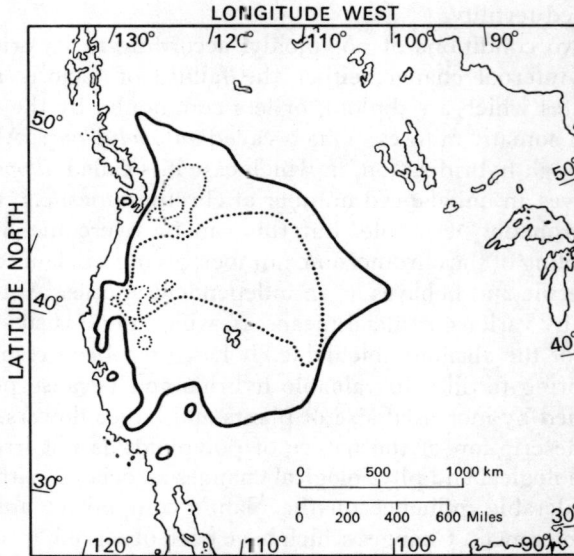

FIG. 4.38. The relative distribution of the polyploid and diploid species of *Crepis* in the western States of the USA. The diploids (dotted lines) are limited to small discrete areas and the polyploids (thick line) include a large continuous area. (*After Babcock and Stebbins.*)

Gustafsson in 1948 argued that the apparently high percentage of polyploids in arctic climates was only a secondary effect due to other causes, such as the absence of annuals (usually diploid) and the prevalence of perennials, in which Müntzing has found a higher percentage of polyploidy than among annuals, and of apomictic species and those relying on vegetative propagation, all features which, apart from polyploidy, would be favourable in the Arctic. Löve and Löve (1949) contravene these arguments and maintain that the increased percentages of polyploid species in arctic floras is due to physiological characters related to polyploidy *per se*, in particular to increased resistance to cold and drought. There is also some suggestion of a correlation between polyploidy and long-day photoperiodism, a necessity in the Arctic. There is certainly a parallelism, as the percentages of both polyploids and long-day plants both tend to increase with latitude. There is little information cytologically from the tropics but as there is a steady decrease in the percentage of polyploidy from 73·6 per cent in Spitzbergen to 37 per cent in Sicily, it may be assumed that in the tropics, where most species are necessarily short-day plants, the percentage is under 30. It has also been claimed that polyploidy can break down self-incompatibility, and self-fertilization is a prominent feature of arctic floras. While the geographical question is open to argument, there can be no doubt that polyploidy may have very important effects ecologically.

We now come, rather late in the day, to the inescapable question of the influence of animals as part of the biotic environment of the plant community. This is quite different from the matter of animal communities, though the two subjects interlock.

To most plant ecologists the animal inhabitants of a community are apt to be regarded as intruders with a chiefly destructive influence. Certainly in the majority of cases, the plants are the pioneers, making and developing an environment which the animals utilize. The opposite case is rare, except for human activities which constantly provide new habitats for plants. The scratched and overturned soil of a rabbit warren may offer an environment for ruderal plants, provided they are not on the rabbit menu. The lairs of animals, sheepfolds, etc., create the requisite conditions for a nitrophilous community, of which *Urtica* is the most prominent member. Bird-droppings on rocks may pave the way for lichen colonization and animal-droppings provide a home for fungi; but these are relatively minor events. Nor, of course, is the vegetation the only factor of the animals' environment. They too are influenced by physical factors, like the plants, and by the presence of absence of other animals.

The vegetation provides the matrix for the life of the animal communities and the composition, stature and structure of the vegetation may have a decisive influence on the composition and life histories of the animals living in it, since they are bound to the plants by their food chains as well as by many other requirements of shelter, breeding, etc., but the influence is bilateral since the animals as soil-makers, as pollinators, as agents of dispersal and as selective browsers have a controlling influence upon the constitution and the development of the vegetation, partly by direct action and partly by their reactions upon the physical environment. The mutual dependence of the two communities is so close that they cannot logically be separated. There are not two communities but in reality one, the biocoenose or biome, which is the living fraction of the ecosystem. While it might be argued that plants are independent beings which could live without animals, the fact remains that they do not do so. Where there are plants there are always animals.

There are in fact two aspects of the plant-animal relationship. In one they are united as a biotic community in which the two sectors must be, by *a priori* specification, coextensive. In the other aspect the animals are not part of the biome, but act from the outside as raiders and predators. Such influences may be temporarily profound, they may even deflect succession, but they are not inherent in the biotic community. Admittedly the distinction may be hard to draw in particular cases, but that is a matter for local investigation. The point at issue is simply that the animal population is neither wholly within the biome nor exclusively external to it but can be partly one and partly the other.

One good reason for the neglect of the concept of the biome is that few people are equipped to deal with all sides of it and naturally follow their own bias in studies. Botanists and Zoologists are alike in this. One of the few who have tried is John Phillips, who in 1931 gave an account of animal-plant relationships in the biotic communities of the Knysna forests on the south coast of Africa and of the East African plateau land in the former Tanganyika territory. These provide an interesting contrast, for although animals are abundant in both areas, in the forest the animals are mostly small compared with the plants while in the plateau savannah they are mostly large in comparison with the plants and a different balance obtains.

At Knysna experimental 'plots' were set up, consisting of four large, screened areas, walled and roofed with a light reed-thatch which did not keep out rain. These were adjusted to give four different light values internally, 0·5, 0·2, 0·06 and 0·01 of full daylight (1·0) corresponding to four stages of forest development, the last representing light conditions under the full forest canopy. In each of these shelters a population of evenly-matched young trees of the chief forest species was set in soil taken from the high forest. Growth of the young trees was very depressed in the two cultures with the most light, in which weed growth was strong, and also in the 0·01 daylight culture which represented the most shaded conditions of the high forest. It was best at 0·06 daylight, and for this and other reasons Phillips held that the cultures were equivalent to four successional stages of forest development. With regard to the animal populations, the big snail *Achatina*, an associate of the climax forest was only found in numbers in the two darkest cultures. Millipedes, which are characteristic of cultivated soils, were, on the contrary, only found under the lighter screens. The same was true of Orthoptera, which are not forest-living types. Hymenoptera, especially bees, were only found in the two lighter screens where abundant flowers of the weeds developed. The gall-forming Cynipidae, on the other hand, formed their galls on plants under the darker screens, true to their forest-living habits. Common frogs stuck to the lighter cultures, but tree-frogs favoured the darker ones. Lizards sometimes favoured the dark screens, despite their sun-loving proclivities, because of the many Diptera to be found there. Snakes, too, were sometimes drawn to the darker screens, hunting for the field mice living there, and avoiding the lighter screens. In short each habitat produced a particular community of plants and of animals related in some way to the condition of the vegetation, reacting in a way similar to that which they show in relation to stages of forest development, from herbaceous communities with *Virgilia capensis*, through evergreen scrub and evergreen bush.

Earthworms were most plentiful in the forest soils which contain 25–35 per cent moisture (on a dry weight basis) where there were 20,000–30,000 per acre. This type of forest is attractive to elephants whose raids are followed by wild pig or baboons. The soil thus gets rooted up and heavily manured, thus encouraging germination and at the same time dispersing quantities of seeds from the fruits consumed.

Among birds, the ground-living forest doves keep to the darker forest, but the bee-catcher prefers the lighter stages where bees and flowers are mostly found. The animals, as a whole, are undoubtedly attracted to and inhabit the areas where the aerial, soil and plant conditions provide best for their needs. There are certainly some persistent species, ranging from protozoa to mammals, which continue from stage to stage during the plant succession, while there are others which show marked preferences as the plants develop.

A striking example of a complex co-action at Knysna begins with the honey-bee. Fifty-two of the sixty-three more important species of trees and shrubs depend almost entirely on the honey-bee for their fruit and seed crops. The bees also depend on the flowering of the trees for their nutrition and survival. Their most potent foes are the bee-catching birds. These are kept down by the small carnivores *Herpestes* and *Zorilla* (the 'mouse-hounds') which, in their turn, are the prey of the wild cat and the leopard. The destruction of these bigger animals by man has brought about a great increase of the 'mouse-hounds', which have been forced to attack poultry. This has brought man against them. If they are reduced the birds will increase and the bees decrease, again helped by human honey raiders, leading to a heavy check to forest regeneration. It is reminiscent of Darwin's classic co-action chain: old maids–cats–field mice–humble bees–red clover.

Man has, of course, formed part of the biotic environment from the beginning. Perhaps his earliest impact, while he was still a hunter, was due to his trackways with their impacted soils and traffic pressure. Although their influence is obvious it has only recently been evaluated quantitatively by a team of investigators in the Scilly Islands (1971). Here are conditions of heavy tourist traffic, mostly pedestrian, over open country but with a number of defined and popular routes. Transects across popular routes show distribution gradients which are directly related to the traffic pressure.

The calculation of partial coefficients between traffic pressure and factors other than pressure, such as moisture content, pH, soil-water conductivity and soil depth enabled the effects of single factors to be successively eliminated to isolate the traffic effect. As a result significant negative correlations to traffic were found for three species while two species showed a positive traffic correlation, though not at a significant level.*

The effects of trampling pressure on sheep paths in chalk pasture were also observed by Thomas (1959) and Pering (1959). They found that *Phleum bertolinii* reacted positively to trampling, especially where the sward was broken. *Dactylis glomerata*, on the other hand, reacted negatively (Beddows, 1959).

Man has naturally had the profoundest effect upon the forests. By felling and burning he has started all manner of secondary successions which have changed both plant and animal populations. He has exterminated the buffalo and greatly reduced the elephants, the leopards and the various species of

---

* At the Bouche d'Erquy in 1909, when the local species of *Salicornia* on the marshes were being examined, the taxonomists found themselves obliged to recognize certain 'formae troddenupon' as characteristic of pathways!

buck. He has harried the bees for honey, has drawn them away from the forests with gardens of exotic flowers and has introduced weeds which invade the native vegetation and change its development; a formidable catalogue of non-beneficial actions.

On the East African savannahs we still have great concourses of grazing and browsing animals which survive man's efforts to destroy them. The grazers undoubtedly stimulate the growth of grasses, but the browsers on tree shoots tend to prevent regeneration and reduce trees to shrubby growth. Selective browsing and the attacks of termites on wood, limit the life of many species and prevent the establishment of others.

The soil is impacted by the battering of myriads of hooves and enriched with vast quantities of manure improving its fertility and water-holding capacity, an important matter in a dry region. Equally important is the effect of the hordes of burrowing rodents under the soil and the work of termites in soil aeration and the production of humus. The annual grassland fires not only act as a check on the animal populations, but they inhibit the completion of plant succession over many areas. Phillips cites the open woodland of *Berlinia–Brachystegia* which is held back by fires from developing into the scrub climax.

An account such as this, even if it does little more than indicate the co-action which may exist between plants and animals under natural conditions, does however show that the arguments in favour of a unified concept of the biome or biocoenose are very strong. Investigations of vegetation in the future should no more neglect the effects of the animal population than they should neglect the physical conditions as features of the ecosystem.

An ecosystem is a natural 'whole', an organic entity which is more than the sum of its parts considered separately. The holistic concept of Smuts should be part of all biological thinking.

CHAPTER XLVIII

THE EDAPHIC ENVIRONMENT

THE study of s ils has become a distinct branch of science, with an immense literature commensurate with its importance to mankind. The study of soil for its own sake, as an independent objective, is known as 'Pedology' and has reached a high degree of advancement. We shall not attempt here to follow it in its more recondite aspects; there are many other sources for such information for the benefit of the specialist. What we shall try to do is to present the soil as a vitally important element in the ecosystem. The contact of the individual plant with the soil is most intimate and no doubt every facet of its physics and chemistry affects the individual plant to some extent, but it is not with the physiology of the individual that we are concerned but rather with the soil as a background to the plant community and as a factor in its development.

We may begin by asking, 'What is soil?' Plants can utilize the most varied substrata as rooting media, but most of these are substitutes for soil, they serve as soils, sometimes quite satisfactorily, though they are not true soils. Such media are: shingle, sand, raw gravel, mud, rock crevices and peat.

A true soil is a complex system in which the principal components are: first, the parent material or mineral skeleton of the soil, consisting of the weathered and finely comminuted fragments of rock, and, second, organic matter, the decomposing remains of plants and animals which have lived on the area. Accessories in the system are water and gases, chiefly air, which circulate through the pores of the soil. Essential to fertility are the living organisms of the soil, its flora and fauna, ranging from burrowing rodents through earthworms down to bacteria and protozoa. For many of these soil is the normal habitat and they play a vital part in its development and maintenance. Sterile soil is still a usable rooting medium, but such a soil has no future and will sooner or later become exhausted unless artificially fertilized.

Soil is therefore not an independent natural system, it is the product of the interaction of many factors, the principal of which are: the native rocks, the climate, the topography and the living population, and the outcome of this co-action is so variable that an inclusive definition of soils which covers all the possibilities is almost impossible; there will always be exceptions.

For example, the definition we have presented above excludes peat, yet it is a very widespread and important surface deposit which serves the purposes of a soil but does not contain a skeleton of rocky material. Pure sand would also be excluded though many plants can live on it and as humus accumulates it shades off into a sandy soil so gradually that no dividing line can be drawn. Many other instances could be cited of what we may call 'partial soils' or *pedoids*, which are functionally soils though incomplete in the terms of our definition.

Pedogenesis and soil structure

We cannot think of a soil without thinking of the factors which have formed it and this process of soil development or *pedogenesis* should be our first line of inquiry. From this point of view we must add *time* to the other contributing factors mentioned above, since soil development is slow and the nature of the product will depend on its age. Zero time is that of the first exposure of the parent surface. Time is not in itself a constructive factor, but it influences the effects of all the other factors.

Jenny in 1941 summed up the process of soil formation symbolically in an equation:

$$s = f(cl, o, r, p, t., ..)$$

the soil being a function of the interaction of environmental climate (*cl*), organisms (*o*), the relief of the environment (*r*), parent material (*p*) and time (*t*). The dots indicate that the equation is open to possible additional factors.

It is no accident that the study of soils began in Russia and has been most keenly followed in the United States. Both are large countries, including a great range of climates and of geological structure and both have great agricultures dependent on the soil. In fact, the father of pedology was Dokuchayer who published his study of the Russian black earth in 1883. He clearly analysed the factors of soil formation, showed how their contributions to the process could be separately assessed, but that they all contributed by their variations to the variable final result. In short he laid down sound principles which are not in dispute. His work unfortunately was practically unknown in the West until Glinka's book on the geography of soils was published in German in 1914, an unpropitious date. It was not translated into English until 1927. In Britain the chief pioneers were G. W. Robinson and Sir John Russell and in the United States E. W. Hilgard, H. Jenny and J. S. Joffe.

Every part of the earth's surface, except perhaps that protected by the deep oceans, is subject to erosion by weathering, and landscape is the result of the interaction of its various factors. When a surface is exposed it becomes covered with a weathering complex or mantle which represents erosion in action, and soils constitute by far the greater part of this weathering mantle. As erosion is continuous there is no finality in soil formation; thus what is called a 'mature soil' is simply one which has lasted long enough for the weathering processes to have developed in it a characteristic system of horizontal layers, to be seen in section as the *soil profile*. Such soils may be of great age: 20,000 years is not too long for the development of the profile in some soils and some mature soils may be half a million years old, but from the geological standpoint they are all simply phases in the cycle of formation–development–erosion. Though erosion is continuous there may be a balance between it and formation so that the soil mantle remains continuous, though its constituent materials change. In other words, soil formation and rock erosion may be integrated into one process in equilibrium. The soil mantle may thus last as long as that particular phase of erosion persists. Alternatively it may be swept away by one of the agents of denudation: water, wind, gravity, or ice, and the underlying rock again exposed. Jahn (1954) has expressed the situation on an eroding slope (the usual situation) thus: D is the rate of soil formation, S the rate of removal by denuding forces, M the mass movement by gravity and A is the additional material coming from higher up the slope. Then if $D = S + M + A$ there is a balance of equilibrium. If D is high, soil will accumulate. He called this 'passive balance'. If D is low, then the condition of active balance prevails, or gradual removal of the soil until the rock is freshly exposed. The prevailing condition of balance will be different at different parts of the eroding area and we may distinguish the soils concerned by different adjectives. Thus *eluvial* soils are those in which removal overtops production; *colluvial* soils those in which there is more or less of an equilibrium between the accretion of eluvial materials and their removal, and *illuvial* soils, those in which eluvial materials come to rest and the soil accumulates. These three conditions, obtaining from the top to the bottom of a slope, form a *catena* of soils genetically related to each other, though their characteristics may be very different. The original, simple idea of the catena becomes much more difficult to apply where the slope is made up of geologically different beds, all eroding and contributing their materials to the soil mantle. The genetical connection of the soils is thus overlaid and, so to say, hybridized and the catena concept can only be applied with reservations. *Alluvial* soils, it should be said, are those deposited from water carriage, while *drift* soils, in formerly glaciated areas, are those deposited by the ice covering. *Aeolian* soils are those formed of wind-blown materials.

It used to be customary to distinguish between sedentary and transported soils. The former are those soils formed *in situ* by the weathering of the underlying rock, to which, therefore, they are intimately related; 'rock' is here used in the wide geological sense of any kind of mineral substratum, whether compacted or not. Transported soils or, more correctly, soil materials, obviously are quite unconnected with the underlying rock, their mineral constituents being derived from elsewhere. The distinction, though largely obliterated in the elaborate classifications by soil scientists, still has relevance to the ecologist.

A sedentary soil may be covered over by a transported soil and preserved 'fossil'. Such relic soils may be of great age and preserve a record of climatic conditions quite different from the present.

Weathering of a rock occurs in two ways. First, *mechanical*, the splitting of the rock into progressively smaller pieces either by the action of changes of temperature and of frost or by the wetting and drying of colloidal clay materials, which causes alternate swelling and shrinkage. Second, *chemical*, which involves the solution of minerals. In this process the action of carbonic acid dissolved in rain water is important. Some materials, such as carbonates, are thus dissolved fairly readily, as is testified by the vast caverns dissolved out of limestone rocks. Other materials, such as quartz, are relatively insoluble and therefore tend to accumulate in the residue of weathered rock. As this process of solution by rain water, called *leaching*, goes on perpetually, it may be said that everything eventually dissolves to some extent. There is not only solution involved, but also chemical decomposition attributable to the action of weak acids which replace kations with H_2, forming carbonates and silicic acid* by the breakdown of the hydrated metallic silicates which make up a considerable fraction of most rocks. These contain chiefly the kations Potassium, Aluminium, Magnesium and Iron, mostly as compound silicates. The K and Mg form hydroxides, the Al and Fe appear as sesquioxides. Ferric hydroxide is also formed by the oxidation of ferrous Iron in Ferro-magnesium silicates.

Clarke has estimated the proportions of the soil-forming minerals in the superficial crust of the earth as: Felspars, 57·8 per cent; Ferromagnesium silicates, 16·0 per cent; Quartz, 13·7 per cent; Micas, 3·6 per cent. Felspars (compound aluminium silicates) are thus the most important contributors to the weathering complex.

Weathering tends to reduce the amount of silicic acid in the rock. This is mostly removed by water in the form of the more soluble alkali silicates. A high proportion of silica is characteristic of soils of arid regions where this removal does not take place.

Leaching also removes some of the alkali and alkaline-earth kations from the rocks, which appear in solution as carbonates or bicarbonates in the drainage water, a loss which falls more heavily on Ca and Mg than on K or Na. Here again the extent of leaching depends on the climate and in dry areas the soils may be highly alkaline.

During weathering, materials may be also precipitated in the weathered complex. Calcium carbonate is frequently precipitated from bicarbonate in solution through loss of CO_2 from the solution. In some circumstances it may be so abundant that it acts as a cement uniting the weathered material into a friable calcareous mass, as in tufa around fens. It has also been pointed out that silicic acid sols carry a negative electric charge, while sesquioxide sols are electropositive. Electro-neutralization may therefore occur with mutual precipitation as a stable, electrically-neutral complex.

Soil maturation is a slow process. The beginning of visible podsolization was observed after 100 years in the soil under Lake Ragunda in Sweden, which was drained in 1796. The full development of a podsol may take from 1000 to 1500 years and podsols of greater age than this show no further increase in the thickness of the profile horizons. In other words the curve of maturation is asymptotic.

Loss of minerals by leaching can also, fortunately, be slow. A soil exposed to weathering for 300 years was found to have lost less than 10 per cent of the

---

* Silicon occurs in the soil solution as monosilicic acid mostly from the hydrolysis of silicates, in amounts up to 80 ppm in soils of low pH.

Calcium carbonate content of the parent material. In the heavy soil of the Dutch polders it needed 100 years before the soil became sufficiently aggregated to allow for free percolation and the removal of $CaCO_3$ to begin.

The breakdown of a rock by weathering gives rise to a mass of particles with a wide range of sizes (see p. 3612) along with rock fragments and fragments of constituent minerals. A very important fraction of this mixture is that called the *clay fraction* which consists of particles less than 0·002 mm in diameter. This limit has been adopted internationally and it is not a purely arbitrary boundary because below this limit there is a marked change in constitution and properties. These particles have a very low settling velocity in aqueous suspension and form, in fact, colloidal sols. They are the product of chemical, not physical, weathering and are the chemically reactive portion of the soil. Although generally referred to simply as 'clay' it consists of a mixture of minerals, of which *kaolinite* (china clay), $Al_4Si_4O_{10}(OH)_8$ or aluminium hydroxysilicate, is one of the most important, especially in tropical soils.

Other clay minerals, the formation and proportions of which depend on the climate and the pH of the soil water, are *Chlorite*, *Illite*, and *Montmorillonite*. In hot, wet and badly drained areas Kaolinite may decompose to form the valuable *Bauxite* ($Al_2O_32H_2O$) and Silica.

A clay particle has a porous, lattice-like structure with a large aggregate surface and it is very actively adsorbent. It readily forms clay-humus complexes with organic matter in the soil and these complexes are responsible both for holding basic ions by adsorption and for exchanging them for other ions in the soil solution. This power is called the *cation exchange capacity* of the soil clays. Measured as milliequivalents per 100 gm of clay at pH 7, it is a standard of the nutritional value of the soil, that is of its fertility. Kaolinite is the chief clay mineral formed under tropical conditions and as it does not adsorb Ferric ions the latter remain free and when oxidized give the characteristic red colour to tropical soils.

The action of the general factors in weathering, physical and chemical, is liable to extensive variation in accord with the type of climate and the parent material concerned. The results, as soil, are equally varied and the more they are studied the more complex their classification becomes.

Parent rocks may be grouped into four main classes: igneous, metamorphic, sedimentary and unconsolidated, all composed of aggregates of minerals formed from the elements: Oxygen, Silicon, Aluminium, Iron, Calcium, Sodium, Potassium, and Magnesium, in that order, with smaller amounts of Titanium, Hydrogen, Phosphorus and Manganese. Obviously such miscellaneous aggregates will weather in different ways and at different rates.

Igneous rocks are either of deep formation (plutonic), in which case they are coarsely crystalline, or they are of sub-surface formation (volcanic) and are finely crystalline or glassy in texture. They are further classified by their silica content; those with more than two-thirds (66 per cent) of Silica by weight, are called *acidic* (granite); with 52–66 per cent they are *intermediate* (diorite); with 45–52 per cent they are *basic* (dolerite); and with less than 45 per cent they are *ultrabasic*, i.e. largely ferro-magnesian hornblends (peri-odotite).

Metamorphic rocks are those which have been altered from their original condition by heat, pressure and chemical action, involving recrystallization. Thus from granite is formed gneiss; from dolerite, hornblend-schist; from clays, schists or slates; or from limestone, marble.

Sedimentary rocks include both those formed from sub-aqueous deposits and those formed by wind-blown materials (aeolian), e.g. the Triassic red sandstone. They are the products of former cycles of erosion and have formerly passed through prolonged weathering and sifting, so that they lose less mass by renewed weathering than do igneous rocks. They are of all degrees of consolidation according to their age and to the cementing material which has infiltrated into them secondarily, which may be Silica, Calcium carbonate or Ferric sesquioxide. The purest sediments are those formed by chemical precipitation under water, for example, chalk which is 95–100 per cent Calcium carbonate.

Unconsolidated materials include gravels, sands and clay beds, the products of recent erosion. The clays include the extensive glacial boulder clay, and the aeolian deposit of loess may also be included here, though it shades off into a soil. Minor categories are beds of volcanic ash and dust and rock screes.

The position of peat is anomalous. It is not a true soil nor is it a parent material of soils and it is best treated separately (p. 3643).

The breakdown of a rock under weathering depends largely on its chemical constitution, its mineral complexity and on the fineness or otherwise of its grain. Chemically, Ca minerals and Iron oxides weather easily. Cal-careous rocks dissolve almost completely leaving only a small non-calcareous residue which provides a thin soil. As the infiltration of water is the chief agent in weathering its chemical effect will depend on relative solubilities and on the relative mobilities of ions. The most important ions concerned show the following order of increasing mobility.

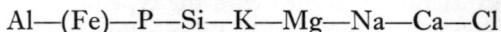

$$Al—(Fe)—P—Si—K—Mg—Na—Ca—Cl$$

The mobility of Fe is variable and depends on whether it is ferrous or ferric. Theoretically weathering would tend to remove the greater part of the more mobile elements, while the less mobile, especially Al and Fe, would show a relative accumulation, and in fact this is often true, and the soil may contain more of these elements than did the parent rock. In the course of soil development, however, the biological factor comes into play and worm action and absorption by plants may restore some of the more mobile ions from the subsoil to the surface soil which appears relatively rich in K, Ca and Mg. A similar enrichment of the surface may be caused by additions of aeolian dust.

Mineral complexity and fineness of grain increase weathering by pro-viding more points of attack for hydrolysis, hydration and oxidation.

In order to get an estimate of the relative effect of leaching in the removal of elements several ratios have been proposed. These avoid an error which arises where simple comparisons of the soil content of an element by weight are used, namely, the change of volume of the soil material near the surface by the incorporation of organic matter.

Aluminium, as the least mobile of the common elements in the soil, is used as the divisor in most of the proposed ratios.

$$\frac{S_iO_2}{Al_2O_3} = \text{the } Sa \text{ ratio of Marbut.}$$

$$\frac{S_iO_2}{Fe_2O_3} = \text{the } Sf \text{ ratio of Marbut.}$$

$$\frac{S_iO_2}{Al_2O_3 + Fe_2O_3} = \text{the silica-sesquioxide ratio.}$$

$$\frac{K_2O + Na_2O}{Al_2O_3} = \text{the } ba_1 \text{ value of Jenny.}$$

$$\frac{CaO + MgO}{Al_2O_3} = \text{the } ba_2 \text{ value of Jenny.}$$

Jenny further proposes, as an index of leaching, the ratio:

$$\beta = \frac{ba_1 \text{ of leached material}}{ba_1 \text{ of parent material}}.$$

The smaller the value of β the greater has been the relative leaching of the alkali metals with respect to the stationary Al_2O_3 value. He illustrates its value in the case of a weathered limestone. In this case, so great is the loss of volume by the removal of $CaCO_3$, that the amounts of Al, K and Na all appear to be greater in the soil than in the rock, but the small β value shows that in the course of weathering K and Na have in fact been drastically reduced.

The effect of weathering is to cover the exposed land-surface with a mantle of broken and disintegrated rock called the *rhegolith*, which can only be seen in its primitive nakedness in high alpine, arctic or desert areas, where the climate precludes plant life. If we now proceed to consider the sequence of processes of soil development in the rhegolith, it must be realized that this is only a pedagogical device. It would be naïve to conceive of weathering and soil development as consecutive in time, for wherever plant growth is possible weathering and soil formation are simultaneous and continuous, with close interactions the one on the other.

Soil formation involves one most important change, the addition of organic matter to the mineral soil, contributed by both microscopic and macroscopic organisms. This has far-reaching effects on the physical and chemical characteristics of the growing soil and it also paves the way for successional changes in the plant population. Among other effects it starts the bacterial cycle of Nitrogen fixation and nitrate production on which fertility largely depends. It also introduces the insect and worm populations to the soil, which are instrumental in the humification of plant organic matter and in its dispersal in the soil body.

The organic colloids not only affect the physical grain of the soil, building up fine particles into coarser aggregates, they also increase the water-holding capacity of the soil. The formation of the aggregate particles increases the pore-space in the soil and this leads to better aeration and better drainage.

This brings us to the two most important external agents at work on the soil: temperature and precipitation, coupled with the infiltration of rain water through the developing soil. These are obviously factors of the external climate and remind us that climate stands first in Jenny's 'fundamental equation' of soil formation (p. 3580). Jenny's factorial approach to soil formation has been much criticized, first on the ground that the various factors, treated as independent variables in the equation, are not in fact independent. Jenny himself admits that they are not, but believes that they can be sufficiently isolated to be treated as independent.

The second line of criticism points to the great difficulty of assigning quantitative values to the factors, involving a degree of uncertainty which quite destroys the value of the equation as a means of defining, let alone predicting, the character of the resultant soil. Those critics who claim that the equation never has been solved do not hesitate to add the rider that it never can be solved.

Qualitatively, however, a factorial approach is almost inevitable for in this way we can allow for the interaction of factors without attempting to evaluate them precisely. Furthermore, we have the soil itself as a given body for study and we can deduce from it the relative importance of the formative factors.

The early Russian soil scientists, surveying their extensive domain, showed that in the long run (where t, the time factor in the equation is large) climate has the most decisive effects on soil character. In two different climatic zones the same parent material will produce different soils, whereas soils in the same climatic zone eventually approach each other in character if the parent materials are not so different as to forbid it.

We can break down 'climate' for our purposes into moisture and temperature, the latter depending chiefly on radiation, which we have already treated in connection with vegetation (p. 3539 *et seq.*), but we must further consider their relationships. Radiation, as we have already pointed out, has a strongly heating effect upon the soil surface, depending both on the intensity of the radiation and on how much of it is absorbed. This is influenced by the colour of the soil, the darker it is the more radiation is absorbed by a bare surface. Light-coloured soils and sands are more highly reflective. Desert sands, for example, reflect 24–28 per cent of the radiation, as compared with zero for a truly black surface. A covering of grass will reduce the heating effect on the soil for grass is highly reflective, 14–37 per cent.* The actual rise of temperature caused by radiation will depend on the specific heat of the soil. Dry soils do not vary very greatly from one another in this respect; the specific heat of dry sand is only about 25 per cent greater than that of dry clay, but it varies markedly with the water content. The value for a saturated sand is more than doubled, while that for clay is more than trebled owing to its higher water content, for it is the high specific heat of water that makes the difference.† Thus dry soils are more readily warmed than wet soils and start spring growth earlier. The expression 'clay-cold' is literally true.

Under solar radiation the surface of the soil becomes hotter than the air above it. Heat is therefore given up to the air by conduction and convection and there is a steep lapse of temperature in the first two metres of air, which continues upwards with a rapidly lessening gradient. This promotes turbulent flow in the air. The temperature difference lessens during the afternoon and at night the gradient is reversed, the soil surface is colder than the air and may reach freezing point (ground frost) while the air above it is still comparatively warm. The downward lapse of temperature at night, which means that the colder, heavier air is below, does not promote vertical air movement and the flow of air tends to be stream line.

The penetration of heat downwards in the soil is slow, between 25–75 mm per day, unless heat is carried downwards by percolating water. The lower limit of daily changes of temperature is about 30 cm, below which daily differences reach vanishing point. The deeper one goes in the soil the smaller become the temperature variations and the greater the time-lag behind the prevailing air temperatures. Under grass the changes of temperature in the surface soil are reduced and even more so under forest, but the same relationships with air temperature and depth in the soil prevail, though on a reduced scale.

At a depth of one metre only seasonal changes of temperature occur and with a considerable lag. In Britain the soil at this depth reaches its highest temperature in late September–October and its lowest temperature in February–March. At this depth the seasonal variation is only about 9°C and at greater depths this decreases and finally disappears. At 30 cm depth the mean daily temperature follows closely that of the air though with a lag of several hours in attaining its maximum at about 8–10 p.m., while the minimum is from 8–10 a.m.‡

As with soil heating so with soil cooling, with particular reference to frost. Soil freezing is slow and the time taken varies linearly with the water content. It is a function of daily maximum and minimum temperatures, not of mean values. The depth of frost penetration depends on the soil covering. Under grass it only penetrates to half the depth that it reaches under bare ground

---

* The earth seen from space appears greenish-blue on this account, the blue being oceanic and atmospheric in origin.

† Average specific heat of soil-forming minerals (dry): 0·2 cals per gm. Water = 1.

‡ The facts about the distribution of temperatures in the soil were well known to Fourier and they played an important part in the advance of physical science since they formed the basis of his *Analytical Theory of Heat* (1822).

and under forest to only one-eighth the depth. Indeed under heavy forest litter the ground may escape freezing.

In regions with long and severe winters, at high latitudes and altitudes and in continental interiors, *permafrost* may form. This name is given to a layer, at varying depths, which remains permanently frozen. An illuvial humus-containing layer may accumulate above it. In summer the surface soil thaws to form an *active layer*. When winter comes the upper surface of the active layer freezes first and the still unfrozen layer below is caught between two ice-sheets and subjected to considerable pressures which create *cryoturbation*, i.e. folding, fracturing and flow, which are very characteristic of glacial conditions. The depth to which the soil thaws in the arctic summer varies from 30 cm to as much as 2 m in deep sands. There may thus be depth enough in some places for forest growth even over permafrost, though the shallower active layers are mostly covered with tundra vegetation, dominated by Ericaceae and dwarf *Salix*.

Frost causes the soil water to expand, resulting in *soil heaving*, most pronounced in soils rich in colloids. The pressure exerted is considerable and gardeners know well the need for firming the ground around plants after a frost; some plants may be pushed right out of the ground by this effect. In glacial or peri-glacial climates the heaving of the soil ejects loose stones centrifugally. Stones from adjacent areas collect along their boundaries with each other, forming *polygons* of stones surrounding areas of fine-grained soil. On slopes the polygons are drawn out by solifluction into long lines or *stone-stripes*, both well known features of arctic soils.

Frost has a disruptive effect on the soil minerals, releasing nutrients to solution. Solid freezing may also reduce the aggregation of soil particles, producing a finer-grained soil ('improving the tilth' as cultivators say) and if prolonged it puts a stop to bacterial action in decomposition, causing the accumulation of fibrous, semi-humified organic material. Frost effects are accentuated by repetition and it is not arctic regions with relatively few freezing-thawing repetitions which show its most disruptive effects, but rather central continental and alpine areas where the cycle may be repeated nearly a hundred times a year.*

The soil temperature has marked effects on microbial activity. Oxidative decomposition is suspended at temperatures approaching 0°C. The optimum growth temperature for most soil organisms is around 25°C. The chemical activities of bacteria may, however, increase up to 35–40°C, while nitrification is most rapid at 35°C. Significant activity of the latter process is therefore confined to the upper few centimetres of soil.

Temperature also affects the rate of water percolation downwards through the soil, due chiefly to variation in the degree of hydration of soil colloids and to changes in the surface-tension of the water.

Precipitation is disposed of in three ways: by evaporation, by surface

---

* The rocky detritus which is widespread on mountains is the result of the disruption caused by the constant repetition of the freezing-thawing cycle.

run-off and by infiltration into the soil. The first we shall treat last, bringing precipitation into relationship with temperature.

The relative amounts disposed of by the two last methods depend on the surface slope, the nature of the soil and its condition, the intensity of the rainfall and on the plant cover.

Infiltration is more or less inversely proportional to the intensity of the rainfall. Very heavy rain, of the order of 51 mm per 24 hours, has considerable erosive power, but has little chance of penetration. Rates of penetration decrease with higher water content and the first fraction of a heavy rainfall tends to block the soil by breaking up the aggregate particles and washing the fine material down into the pore-spaces of the soil. The nature of the soil is also a controlling factor, wet clay being practically impermeable and wet sand the most permeable. On a wet surface the force of impact of the raindrops scatters soil particles for surprising distances, as can be seen on the undersides of leaves after a storm, and on a slope this will cause downward translocation.

When there is an excess of rainfall over infiltration the excess will either form a standing sheet over the surface, in hollows or on flat areas, which protects against erosion, or it will move under gravity as a thin sheet of water on the soil surface, which is readily channelled by the surface topography and may then become very actively erosive. The thawing of snow cover may also produce temporary flood conditions which can cause serious erosion and the melting of the glaciers at the end of the Ice Age caused late-glacial erosion on a truly massive scale.

Erosion is probably most active in arid and semi-arid areas, where vegetation is sparse and rainfall tends to come in occasional heavy storms. Little of this water penetrates and the run-off causes gully formation and the conveyance of large quantities of solids to the river channels, which receive little ground-water but are rapidly filled by flood-water.* Where ground-water exists in arid regions it is either as a deep-lying stationary reservoir which has been slowly accumulated over a very long period, or as an underground flow from some neighbouring humid area, particularly from mountains.

Vegetation usually mitigates erosion and the planting of ground cover, such as the creeping *Pueraria thunbergiana* (Kudsu) has often been used effectively as a check.

Grassland, though it does not greatly diminish the amount of water reaching the soil, holds the soil very firmly with the mass of its fibrous roots and is highly protective both against the impact of rain and erosion by wind. Forests may cut off a large proportion of the rainfall from reaching the ground (p. 3534) and are, to that extent, protective, but they are not so protective against the effect of run-off on steep slopes, where it may often be seen that the soil has been washed away around the bases of the trees. Erosion on an extensive scale is, however, exceptional in humid climates, where really ravaging rain-storms are uncommon. Light rains mostly infiltrate directly

---

* This is proving to be a danger to dam-lakes. Dams are usually built in dry areas and the silt brought down by incoming rivers causes serious silting up of the lake water.

into the soil unless it is already saturated or its surface compacted by drought or traffic.

Prevailing rainfall and prevailing temperatures interact in controlling the loss of water to the atmosphere by evaporation, or more accurately, *evapotranspiration*, since the plant cover is also active in promoting loss of water. In regions with the rainfall maximum in summer a large proportion is absorbed and transpired by the plants. Where the rainfall maximum is in winter, less is used by the plants and a larger surplus infiltrates the soil. Leaching of the soil is thus more severe in such climates.

As a measure of the surplus of water available to the soil Transeau in 1905 used the ratio of precipitation to evaporation, the P/E index. He used figures for evaporation from free water surfaces, which require a calibration factor of 0·6 for bare soil and of 0·8 for grass cover. By mapping P/E values in the USA Transeau showed that zones of equal effective moisture were quite different from simple rainfall zones. This does not evaluate the proportion of infiltration to run-off in the effective moisture, which is greatly affected by local topography and by the intensity of the rainfall and is therefore a purely local and temporary factor. The P/E index itself is a feature of the macroclimate and its value is chiefly geographic in the delimitation of climatic soil zones.

As the evaporation from a free water surface is difficult to measure accurately, especially in the field, de Martonne has proposed an index of aridity in which temperature is substituted for evaporation, $p/(T+10)$, where P = the annual precipitation in millimetres and T = the mean annual temperature in degrees centigrade. This, like the P/E ratio, depicts the macroclimate and is chiefly of geographical interest.

In broad outline a P/E index of less than unity implies an arid climate and an index above unity a humid climate, and the range can be divided into series from a ratio of less than 0·2, which is very arid, to values above 2·9, which is extremely moist. As evapotranspiration is much affected by temperature, on which the saturation values of the air depend, such a series of index values has been called the hydrothermal range, since in the Arctic an annual precipitation of 500 mm will yield a P/E index of more than 2·0, which would only be reached in tropical temperatures with a precipitation of 3000 mm.

Voloboyer in 1956 represented this diagrammatically (Fig. 4.39) showing that the chief types of soil had characteristic hydrothermal ranges. Where these ranges overlap, the type of soil developed must depend on other factors such as topography or parent material As the P/E ratio defines the amount of water made available to infiltrate the soil, it also designates the extent to which infiltration and leaching may affect the soil and consequently the type of soil most likely to develop.

Glinka (1914) who first attempted to classify soils according to climate recognized that there were exceptions to the close relationships of the two qualities and he called soils which are climatically determined *ectodynamomorphous* and those which are controlled by other circumstances *endodynamorphous*. Nevertheless, he recognized that the difference is not absolute

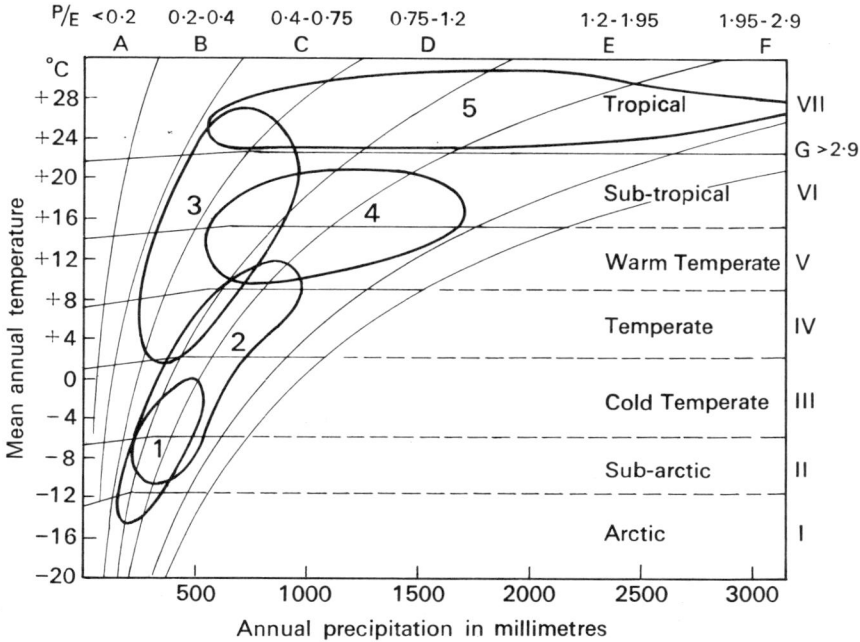

FIG. 4.39. Voloboyer's system of hydrothermal ranges for various soil types. The enclosed areas represent the ranges of the following: 1. Tundra soils, 2. Podsols, 3. Chernozems, 4. Brown earths, 5. Ferrallitic soils (tropical red earths). (*Modified from Bunting*, 'The Geography of Soils'. *Hutchinson.*)

and that, despite the effects to topography and parent material, the soils of the same climatic zone have certain common features typical for that zone. This primary relationship to climate has been generally recognized and appears in the classification of soils that we shall speak of below (p. 3601). The dependence of soil-forming processes on the precipitation-temperature relationship reaches its extremes in regions where the soil is frozen all through the winter and oppositely, in sub-tropical regions where it is dry all the summer. Either condition, while it lasts, reduces chemical change in the soil almost to zero and soil development is limited to one season.

To avoid the difficult measurement of evapotranspiration, Mayer in 1926 proposed another ratio, the NS Quotient:

$$\frac{\text{Precipitation (in millimetres)}}{\text{Saturation deficit of air (in equivalent mms. of Mercury)}}$$

Evaporation is a more or less constant function of the saturation deficit, and the NS values therefore parallel those of the P/E ratio and have been extensively used in climatic mapping. When temperature and NS quotients are used as climatic parameters they enable variations of any other factor to be combined in three-factor diagrams which can be very instructive in expressing soil variations over large areas.

Meyer himself used the quotient, in a range from 0–4000, as the basis for delimiting soil-climate regions. Although he did not use temperature in this classification it is obvious that increasing values of the quotient lead from dry/hot areas to cold/wet areas. The following are some of his values:

SOIL-CLIMATE ZONES	ANNUAL NS QUOTIENT
Deserts and semi-deserts	0–100
Mediterranean region	50–200
European Brown Earth region	275–400
Atlantic Heath region	375–1000
Baltic—Scandinavian region	300–1200
Tundra region	Over 400
High Alpine region	1000–4000

The process, or rather processes, of soil development are extremely complex and extremely variable and naturally their consideration has given rise to many divergences of opinion and to differing theoretical approaches which are frequently related to the experience and geographical location of the worker. One is strongly reminded of the similar theoretical divergences which have attended the consideration of natural vegetation. Indeed, the similarity is significant, for the soil is an integral part of the ecosystem and vegetation plays an important and sometimes dominating part in the development of the soil in which it grows. To attempt to isolate the soil from organic influences and consider it as a purely independent body would lead us to completely unnatural conclusions.

We begin with the parent material, which is commonly said to be the rock; but even this simple statement is misleading. From the very beginning organisms enter into the history of the material and the true parent material of soil is the rhegolith plus organic materials incorporated in it. Again, how far are we justified in thinking of soil development as a continuous, open-ended, process? Can it not be, as some maintain, subject to cycles? Many soils are extremely old and our information about their beginnings is necessarily vague. They have certainly existed through many minor and some major changes of climate. In the course of time they have also received accretions of new materials, partly from aerial dust-fall and rain-wash and partly from changed vegetation. They are therefore *polygenic*. Most soils are in this category, truly *monogenic* soils are exceptional.

The weathering process, which starts on the exposed surface of 'rock', is chiefly through the agencies of water and changes of temperature, i.e. the main components of climate. It is not a simple process and, apart from purely physical effects, it comprises solution, hydrolysis, ionic exchanges, oxidation, reduction and carbonation. These processes continue in the disintegrated residue and in the young soil which it becomes, but as the materials change so do their effects and special processes are set up which are distinctive of soils and may be called truly *pedogenic* (p. 3580), though no hard and fast line can profitably be drawn between weathering and pedogenesis. It is not the processes themselves but the material on which they work which is slowly altered, partly through the continued action of the weathering factors and partly through the appearance of additional organic material.

As the movement of water is predominantly downwards the fundamental effect in the young soil is the development of horizontal zones or *horizons*, which distinguish it from the undifferentiated rhegolith. In humid climates these horizons are basically three: The eluvial horizon, A, from which materials are being leached; the illuvial horizon, B, in which leached materials tend to accumulate; and, C, an underlying horizon of subsoil of rock, partially weathered and broken up, which is usually derived from the parent rock, or, in the case of alluvial or other transported soils may be an unrelated substratum. We shall deal later (p. 3599) with the finer subdivisions and variations of the horizons. The result, however, is the establishment in the soil of a characteristic and highly significant *profile*, which reveals the history of the soil's development and on which the discrimination of soil types is largely based.

The profile is principally due to the percolation of water, the effects of which depend on (*a*) the rainfall climate as represented by the P/E or NS indices, (*b*) the chemical nature of the soil, (*c*) the porosity of the soil, which controls the amount and the rate of water percolation.

The rate of infiltration of water into soils is a function of their clay content. Heavy, undrained clays are almost impermeable and may never, in consequence, develop a characteristic profile. At the other extreme, percolation in a deep sand may reach a rate of 15 mm/hr. The degree of previous saturation of the soil is also effective, a soil already heavily charged with water being naturally less absorbent than a moderately dry soil. A thoroughly dry soil, on the other hand, may also be resistant to penetration, since the influx of water is held back by the swelling of the soil colloids. The mechanical force of very heavy rain hinders penetration, as it breaks down the soil particles and chokes the drainage channels in the soil with fine silt. A soil in this saturated, impervious condition is said to be 'puddled'. In a heavy loam the rate of penetration may lie between 1·5 and 2·4 cm/day and a slow rate like this enables a greater proportion of the water to be withdrawn by plants and transpired, thus reducing the effects of percolation on the soil. Local variations, either in the amount of percolating water, due to topographical features, or in the nature of the vegetation, may so operate, even in a small area with a uniform climate and the same parent material, that a mosaic of profile types may be produced, quite deflecting the soil from the climatic 'norm', which, as Marbut insisted, can best be observed under gently undulating surfaces with uniform vegetation. Naturally marked variations of parent material may show exceptionally wide departures from the climatic 'norm', such as the appearance of a semi-arid soil in a humid climate.

The young soil, at the beginning of its development, is *azonal*, that is to say it has no profile and is not yet related to its climatic zone. There are three general types: *lithosols* on solid rocks; *regosols* on unconsolidated materials; and *alluvial*, transported materials. This young soil is the product of physical and chemical weathering and its mineral constituents are more or less disrupted according to their degree of resistance. In all but the most extreme climates there will be some addition of organic matter.

Quartz is the most resistant of the common minerals to chemical weathering and it therefore forms an inert residue in the soil skeleton. Physically it may be disrupted to any extent, from gravel to fine clay. Next come the Feldspars, among which there is some variation. Orthoclase (Potassium Aluminium silicate) is more resistant than either plagioclase (Calcium–Sodium Aluminium silicate) or albite (Sodium Aluminium silicate) which weather easily. The Micas also differ. Muscovite (white mica—Potassium Aluminium hydroxysilicate) is resistant to chemical weathering, but is easily comminuted physically. Biotite (black mica) is a complex Potassium–Magnesium–Iron–Aluminium hydroxysilicate and also contains Fluorine. It weathers easily and in soils of more advanced or intense weathering little or none is left. Among the Ferromagnesian silicates the pyroxenes are more resistant than the hornblends. Minerals of secondary origin are: calcite, dolomite, gypsum and haematite and the clay minerals (p. 3583). It is possible therefore by mineralogical analysis to gauge the extent of weathering which has occurred and also, within one climatic zone, the relative ages of soils.

Apart from accretions of organic materials and of material from the atmosphere, which may often be of great importance, the history of a soil perfused by water in a humid climate is one of loss, either of total loss to drainage water or of redistribution from higher to lower horizons accompanied by chemical recombinations. Plants counteract this loss to a large extent by the absorption of inorganic ions from the soil and their delayed return during decay. Polynov has pointed out that the actual order of migration of elements from the soil, as evidenced by the composition of drainage waters, does not follow the theoretical order of their ionic mobilities but, for the basic kations, follows the order: Ca—Na—Mg—K, a divergence which he attributes to the action of living organisms, either directly or indirectly through the adsorptive action of soil colloids. The process of loss can be, in any case, extremely slow. From marine clays, elevated above sea level 2000–3000 years ago, two of the most mobile ions, Na and Cl are still being actively leached out.

The downward movement of water in a soil is controlled by gravity and conditioned by capillary forces. Near the surface water is subject to loss by evaporation, but such losses diminish with increasing depth and eventually a level is reached at which the soil is completely saturated. This level is called the *water-table*. If a hole is dug to below this depth, water will stand in the hole up to this level. It is subject to seasonal variations, rising during the rainy season and falling during a dry season, but these variations are not great and the mean level is a more or less permanent feature of any locality, depending on the nature of the soil and the influence of topography on drainage.* Saturated soil cannot hold all its water content, so water below the table level escapes slowly, flowing in whatever direction gravity leads it and often following definite drainage channels. It may eventually emerge at the surface as a spring, at some lower level, or it may collect in a hollow to form a marsh or a

* City development and human drainage can, however, cause a dramatic drop of the water-table.

lake or it may continue underground to sea level, where it may often be seen issuing as a sheet of fresh water on the shore. This drainage water is an extract of the soil from which it is derived and its composition is a valuable guide to the effects of leaching in the soil.

The influence of the water-table upon the soil above it depends on its depth below the surface. This depth is greatest where drainage is best, it therefore increases as a slope is ascended and is least in valley bottoms. If it lies within the range of capillary rise it may contribute substantially to the evaporable surface water, especially in semi-arid areas, where the upward flow of water may predominate over the downward flow. When the water-table is as low as three metres below the surface it may be assumed that it has little or no effect upon the soil in the zone of free percolation. Where the table level is close to the surface it promotes acidity in the upper soil, with pH values down to 3·5. This interferes with humus formation, so that the surface is covered with a layer of highly acid vegetable litter. The ratio of silica/alumina is also higher in these badly drained soils. The water-table level tends to follow the surface on steep slopes, approaching the surface only near the bottom of the slope, as the rate of lateral drainage decreases towards the bottom. Some of the drainage water escapes to the surface through rock-channels higher up, but the greater part issues only at the base of the slope, where it saturates the soil to form marshes.

Soil formation, as we have already mentioned, is a slow process. The initial weathering of the rock may take hundreds of years, the time depending on the nature of the rock and the supply of moisture, while on unconsolidated materials; loess, alluvium or volcanic ash, the beginning of soil development may be recognizable in under 50 years. The development of the profile horizons may take very variable periods according to climate and material. Dating with any reasonable accuracy is only possible when historical or geological data allow for the estimation of a beginning, zero time. Organic material can usually be dated by Carbon-14 content. In the Arctic the humic surface layer may be 2000 years old, but in temperate climates surface humus may be no more than 100 years old.

The concept of soil 'maturity' is far too uncertain to be a guide regarding age. Every element of the soil behaves differently, and while a position of equilibrium may be reached with regard to changes of one soil element it will not apply to others. Percolation of water is the chief agent of soil development and so long as this continues no finality can be reached for the soil as a whole, although it may be reached partially by the complete leaching out of some soil element, such as $CaCO_3$. We can adopt Marbut's criterion of the development of visible horizoning in the soil as an index of maturity, but chemical analysis shows that horizoning is well under way long before it becomes visible and even after horizons are visible they may continue development in depth for an indefinite period, certainly for many thousands of years. Decalcification seems to be the most rapid process. Salisbury reported that it was complete in old sand-dunes after 280 years and Tamm reported that in the soils left by the draining of Lake Ragunda in northern Sweden in 1796, decalcification was complete down to 64 cm under mixed forest in a period of 116 years, while visible horizoning was scarcely perceptible in the same period.

Strongly leached profiles of a shallow character can develop quickly in temperature climates, about 400 years in northern England and about 1000 years in Sweden, but the extension of this leaching in depth may be very slow and soils several thousand years older show but little change. There is some evidence that a few centuries at a time of climatic change, such as the end of the Late-Glacial period, may have more effect in developing a soil than ten times that period under steady conditions.

The *development* of a soil is not to be confused with the *accumulation* of soil. The greater the amount of soluble or displaceable material in the matrix, the less will be the residue left as soil. Thus the residual soils left by the solution of calcareous rocks are always thin, decalcified and with a high clay content. Accumulation may be the result of many factors: abundant vegetation, aerial accretion or the downwash of aggraded material,* for example. The deepest soils are probably those on the mineralogically rich loess deposits.

The processes of soil development may be classified as: oxidation, solution, hydrolysis, ionic exchange, carbonation and chelation. Oxidation may be both inorganic and organic and is to be understood, in the wider significance of the term, as dehydrogenation, which implies that the oxidation of any substance is coupled to the reduction of another. Only in submerged or water-logged soils deprived of Oxygen do reducing conditions prevail, ferrous compounds taking the place of ferric. Hydration occurs as a physical process, causing swelling of minerals or colloids by the absorption of water, which can originate great pressures.

Carbonation is a potent agent of solution. Carbon dioxide and water form H_2CO_3 carbonic acid, which dissociates to give H^+ and HCO_3^-, the bicarbonate ion. This produces soluble bicarbonates from $CaCO_3$ in the first place and later from the metallic ions in silicates and oxides. The rate of $CaCO_3$ solution depends on the equilibrium between the water and the gaseous CO_2 available, the solubility coefficient of CO_2 in water being approximately unity. Soil air is richer in CO_2 than the outer air, the highest percentage (about 1 per cent) being in the organic layers at the surface. The soil water is thus more highly charged with CO_2 than river or lake water, but if soil water containing Calcium bicarbonate in solution comes into contact with air containing less CO_2 than that with which it is in equilibrium, the bicarbonate will dissociate, giving up CO_2 to the air, and Calcium carbonate will be redeposited, a process of great importance in some soils, as we shall see later. The removal of $CaCO_3$ starts from the surface downwards and causes a fall in pH. Further down in the soil, however, where the Calcium has not yet been leached, the reaction may be alkaline. As a result, in many leached soils the pH increases downwards and there may be at some level a definite acid/alkaline boundary or front.

Chelation is a process which originates in organic materials. The chelating agents are amino-acids or else ring-compounds, which readily exchange an H

* That archaeologists have to dig for their finds shows the extent to which soil accumulation occurs.

ion for a metallic ion. The minerals affected most are the sesquioxides of Fe and Al, which are decomposed, the metal ions being taken up by the chelating agent and held very firmly in solution often to a much greater degree than their normal solubilities would allow. Although stable in acid solutions the chelating agents percolating downwards pass into zones of increasing pH with the result that they decompose with the release of the metal ion at that depth. This is a very potent means for the leaching of the sesquioxides and the redeposition at depth is one of the most important agents in the formation of soil horizons.

Clay particles have a coarse (on the molecular scale) lattice structure and are somewhat like minute sponges, possessing a relatively large surface, the aggregate of which for the whole clay content of the soil is immense. The internationally agreed upper limit for the size of particles included in the clay fraction is 0·002 mm (2 μ), but clay particles as small as 10 mμ have been isolated. The particles are thus within the size range of colloids and the clay fraction behaves colloidally. It enters into a loose union with organic colloids in the humus to form a colloidal complex or clay-humus complex which can be regarded as a distinct entity in the soil. It should be remembered that clay minerals (p. 3583) are not original constituents of the soil, but are secondary products of the weathering process. Among the colloidal properties of clay is a conspicuous power of imbibing water, with consequent swelling and an equally marked shrinkage on drying, accompanied by vertical cracking into polygonal columns. It is also highly cohesive,* and in small amounts is responsible for forming the aggregates of mineral particles called the 'crumb' of the soil. In larger quantities it may bind the whole soil into an impervious mass only too well-known to cultivators.

The clay particle carries negative electric charges and these attract the H^+ ends of the polar water molecules to the surface of the particle, so that the water forming a film around the particles is molecularly orientated and is held very firmly. Further from the surface the attraction lessens, but entirely free water molecules do not occur within a distance of 0·1 μ from the particle surface. Other kations which are present in this water layer or which migrate into it are also attracted to the particle by virtue of their positive charges and may displace the H^+ ions and remain attached. The clay particles thus become richly charged with adsorbed kations, especially Ca, Mg, K, Na, Al as well as H, with Ca greatly predominating. Ions attached to the outer surfaces of the clay particle can be readily replaced by other kations, but some ions, particularly Mg, are attached inside the clay lattice and are not normally replaceable but can be released by subjecting the clay to fine grinding. Replaced ions are released to the soil solution and the process is called *ionic exchange* or *base exchange*.

The amount of bases available for exchange is called the *base exchange capacity* of the soil and is expressed in milliequivalents per 100 g of soil at pH 7·0. Hissink found the following values as the average of 25 Dutch clay soils: Ca 30·0; Mg 5·0; K 0·8; Na 2·5.

* Which makes pottery possible.

When the negative charges on the clay are fully satisfied by kations the particle becomes electrically neutral and the soil is said to be *base saturated*. If a soil is leached with a dilute acid the adsorbed kations can be progressively removed and replaced by H^+ ions. If this removal is complete the clay is then said to be an acid clay and the soil is fully *desaturated*. This occurs in nature by leaching with acid rain water, i.e. water in equilibrium with the CO_2 of the air, at a pH of 5·7. If there is no natural supply of $CaCO_3$ in the soil, the adsorbed bases are released to solution by replacement with H^+ and the clay is steadily depleted of nutrient and, in the absence of any natural or artificial accretion of basic substances, becomes acid.

Ions are adsorbed in the following order of strength: H, Ca, Mg, K, Na. The replacement of Hydrogen is thus the least probable, though it does happen, giving rise to *exchange acidity* in the soil solution.

The phenomena of base exchange were first revealed in two famous agricultural papers by H. S. Thompson in 1850 and by J. T. Way in 1852. The occasion for their researches was the introduction of soluble manures and the fear that these would be immediately washed out by rain. Thompson showed that NH_4 from Ammonium sulphate was fixed and Ca liberated. Way demonstrated chemical equivalence in fixation and liberation and identified the absorbing complex.

Base exchange also occurs to a limited extent between mineral particles which are in contact and also in surface contacts between roots and mineral particles, but the clay fraction is far and away the most chemically active fraction of the soil and it is from its exchange activity that most of the mineral nutrients of the soil are derived and it is on it that the fertility of the soil chiefly depends. Without clay the parent material would remain nothing but a disintegrated and barren ash-heap.

We have hitherto been treating soil development processes from the standpoint of humid climates in which leaching by rain predominates, but there are also large regions where this is not the case. Any region in which the P/E ratio is less than unity can be regarded as arid, because evaporation from the soil is greater than the precipitation it receives. This means that the soil either becomes desert or any available water is raised by capillary forces from below. In such soils the predominating movement of water is upwards, not downwards and any leaching that occurs is only occasional and frequently does not extend far into the soil.

Soil classification

Marbut suggested two useful terms for these opposite types of soil. He called the arid soils *pedocals* because they were characterized by a definite Calcium carbonate deposit horizon in their profiles, and the humid, leached soils he called *pedalfers* because they are characterized by a deposit horizon of Al and Fe sesquioxides. In the United States he showed that pedocals were characteristic of the western States and pedalfers of the eastern half of the country.

The soil profile demonstrates the history of the soil and it is the natural basis of soil classification. Soils in which a profile has not fully developed, either due to lack of time or to erosion are called *immature*. Erosion, if severe, may remove upper horizons altogether, leaving only a partial or *truncated* profile. Truncation may occur on the upper part of slopes or where there has been deforestation, exposing the soil to heavy rains. In the latter case the re-establishment of a vegetation cover may lead to the formation of a new profile, above the old truncated relict soil.

Taking the leached soil of a humid climate first, we find that the mechanical action of rain may have caused a *texture profile* to develop. Clay and fine silt are mechanically washed out of the A horizon, which becomes light and porous, while the B horizon receives these materials and becomes dense and heavy. Soils with a good crumb structure are very slow to develop texture profiling.

A fully developed profile in a humid soil will show the following horizons:

L   The superficial litter.

A_0 (or F) Fibrous layer containing roots.

A_1 (or H) Layer of amorphous humus.

A_2 The main eluvial layer, from which humus, sesquioxides and exchangeable bases are being removed. If the upper part is humus-stained it may be separated as Al_2.

B_1 The first illuvial layer in which humus is precipitated.

B_2 Second illuvial layer, in which Al and Fe sesquioxides accumulate. It is markedly brown-coloured with Iron.

C   The partially weathered rock or subsoil.

CR Unaltered rock or substrate. If this layer has no genetic connection with the soil it may be classed as a D horizon.

Waterlogged substrates under marshy soils are designated as G, for 'gley', typically a greenish-grey clay with mottled Iron stains. These are best developed with a fluctuating water-table where there is an alternation of oxidizing and reducing conditions. Under such conditions the oxidation of ferrous compounds coming from below the water-table may cause a considerable accumulation of ferric hydroxide, known as 'bog Iron ore'.

A soil profile may be characterized by the use of the letters, either by themselves, to show which horizons are present, or with measurements added. Letters in brackets (B), denote horizons weakly developed.

The removal of materials from the A_2 layer makes it lighter in colour than B and this process of eluviation/illuviation is called *podsolization*. When it reaches its extreme, under acid conditions, the A_2 horizon is completely bleached and forms a conspicuous ash-coloured or white horizon. This is the fully developed *podsol* from which the process is named. The word is Russian and means 'ash-soil', a very descriptive term. We shall have more to say about podsols later in discussing soil types (p. 3604).

We have mentioned above the sub-surface accumulation of ferric hydroxide in waterlogged soils with a fluctuating water-table. Similar zones of deposition in which colloidal materials are irreversibly precipitated are of wide occurrence in terrestrial soils. The most widespread type is dominated by ferric hydroxide which acts as cement indurating the materials into a hard layer practically impervious to the roots of plants and known as '*pan*' or '*ortstein*'. Iron pan is conspicuous by its colour, but in many cases there is also humus and alumina present and the pan may be blackish in colour. It forms in connection with podsolization and is often a factor restricting the growth of trees on heathy areas unless it is broken up by deep ploughing. The depth at which it forms depends on the depth of podsolization and may be anything from a few centimetres to a metre or so. The mechanism of this precipitation is not certain and may be different in different places, but it seems generally to be associated with some impedence of drainage and when it has consolidated the pan itself cuts off drainage very effectively, which leads to lateral leaching and the appearance of yellow sesquioxide deposits in drains and ditches.

It is notable that a pan or crust of ferric and alumina sesquioxides may form on the surface of tropical lateritic soils in areas where there is a limited rainy season and a long dry period. Such crusts are not typical of the equatorial zone, with a constant rainfall.

The opposite of these conditions prevails among pedocals of low rainfall areas with free drainage. The more humid type of these soils form the group called *tschernosems* (or more commonly, *chernozems*), the Russian name given to the black-earth soils of SE Europe. These are generally formed on loess though they are not limited to such parent material and in Germany chernozems have been formed both on boulder-clay and glacial sands.

The climatic conditions of chernozem formation are a light annual rainfall, mostly in the winter and often largely as snow, with a hot, dry summer during which evaporation gradually dries out the soil. Active plant growth is thus limited to spring and early summer. There is little leaching and the movement of water is predominantly towards the surface. The upper layers therefore remain base-saturated and there is no illuvial B layer, only A and C. The outstanding feature of these soils is the deep A layer, from a half to one metre thick, which is practically black with humus. Actually the proportion of humus is not large, 8–10 per cent is usual and 16 per cent about the maximum. The blackness of the humus seems to be related to the higher summer temperatures and the neutral or alkaline reaction, especially the latter, since other sub-tropical soils which are not base-saturated may contain greater amounts of humus with hardly any humic colour. Another character associated with their high base-status is their permanent 'crumb' structure, which makes them light to tillage.

Below the A layer lies the lighter-coloured C layer and at or near the junction occurs the zone of Calcium carbonate concretions, which is the unfailing mark of the pedocal soil. The depth at which this layer is formed depends on the rainfall, being nearer the surface in the more arid regions. In Colorado, with 254 mm annual rainfall, the lime horizon is less than 254 mm deep, but in following a line of increasing rainfall eastwards to Missouri there is a steady descent of the lime horizon, at a rate of about 64 mm for each additional 25 mm of rain, till with 1016 mm of rain it drops to a level of about 2032 mm. With higher precipitation it rapidly drops out of existence. The close relationship to precipitation suggests that the level of the horizon marks the lower limit of normal water percolation. Although there is no leaching of silica or sesquioxides in pedocals there may be enough percolation to wash out easily soluble salts and a partial leaching of Calcium carbonate from the surface layers.

The recognition and description of different types of soil and attempts to classify them along various lines have developed into a subject of astonishing complexity and have spawned a bewildering vocabulary of terms in Russian, Greek, American and, indeed, as Kubiena says, 'in any language and with any meaning'.

Soils may legitimately be classified from various points of view either practical or theoretical. The agriculturist views them from the point of view of his own interests and would classify them by texture, water-holding power and nutrient reserves. The soil technologist would classify them by their mechanical properties. The soil scientist may view them morphologically, on the basis of their profiles, or genetically with reference to the formative processes that have controlled their development, or by a combination of these criteria. The geographer is chiefly interested in distribution, but before he can map soils he too must classify them in some way.

When Dokuchayev wrote, at the beginning of the century, he recognized only twelve soil types for the whole world, now there are more than sixty in his own east European area alone. The intensive investigation of any area adds to the number of distinguishable types and sub-types, which acquire local names and the relationship of which to any general classification is often obscure. Each major author prefers his own classification and his own nomenclature and has his own followers. There is as yet no internationally agreed system and we are reminded of the historical troubles which all evolving classifications have passed through.

The boldest attempt at universality is that of the United States Department of Agriculture,* whose 'Comprehensive System' had reached by 1950 its 'Seventh Approximation', the latest tentative, not yet final. Unfortunately the system carries with it a system of agglutinative nomenclature which aims at conveying a complete morphological definition of the soil in question, but which results in compound terms of prodigious uncouthness. It would seem that the same end would be better served by symbolic formulae.

If we agree that the character of a soil is causally linked to climate we can recognize three main *orders* of soils: *zonal*, *intrazonal* and *azonal* (Sibirtsev, 1899). *Zonal soils* are matured soils, distributed over large areas with the same general types of climate, with characters which are primarily determined by the climate and only secondarily by the parent material. *Intrazonal soils* are local variants in which some accident of topography or parent material deflects the soil away from the prevailing climate type. *Azonal soils* are soils of a special type which may occur in any zone. They lack any definite profile,

---

* *Division of Soil Survey*, Beltsville, Maryland.

either by reason of immaturity or sometimes because of the nature of the parent material.

Zonal soils, as noted above, are mature. Differentiation of horizons and the development of a profile seem to be fairly rapid in freshly exposed material, but the process slows down and eventually change becomes so slow that an apparent equilibrium is reached between accumulation and destruction and the soil changes but little over long periods of time. This is the condition we call maturity, in which climatic factors have reached their full effect.

Characteristically zonal soils are not infrequently found in areas other than their normal zone and some overlap of zones must be recognized. Thus red-brown lateritic soils, which are characteristic of humid, tropical climates, occur in temperate conditions in Virginia and Oregon in the USA and as far north as the Yangtze Valley in China. Exceptions such as these need to be accounted for. In the case of these temperate soils the dark igneous rocks which are their parents lend themselves readily to the formation of such soils, which are probably the result of temperate climatic forces acting for an immensely long time, possibly since the early Pleistocene, during which there have been marked variations of climate and vegetation.

A. Zonal soils are divided in *sub-orders* and these into *groups*. The US Division of Soil Survey in 1949 suggested six sub-orders and corresponding groups.

Sub-orders	*Groups*
1. Soils of the cold zone.	Tundra, arctic and sub-arctic brown soils.
2. Light-coloured soils of arid regions.	Desert soils—red, brown and grey (Sierozem).
3. Dark coloured soils of semi-arid, sub-humid and humid grasslands.	Black earths (Chernozem); prairie soils (Brunizem); chestnut soils (Kastanozem).*
4. Soils of the forest–grassland transition.	Degraded Chernozem; non-calcareous brown soils.
5. Light-coloured, podsolized soils of forest regions.	Podsols; brown, grey and red-yellow podsolic soils.
6. Lateritic soils of warm-temperate and tropical forest regions.	Reddish and yellowish-brown lateritic soils.

B. Intrazonal soils are similarly divided, according to the environment factor chiefly responsible for their special characteristics.

Sub-orders	*Groups*
1. Halomorphic soils (saline and alkaline) in arid regions with impeded drainage, and in littoral regions.	Solonchak; Solonetz; Soloti.

* The term 'chestnut soil' has been the source of some misunderstanding. The humus layer of these soils is greyish-brown and the comparison intended by the name is with the bark of *Castanea*, not the fruit.

2. Hydromorphic soils of marshes and swamps. Ground-water soils.

Meadow soils (humic-gley soils); bog soils; half-bog soils; planosols.

3. Calcimorphic soils, calcareous.

Brown forest loams; Rendzinas; terra rossa (Krasnozem); ando soils (?)

C. *Azonal soils*

Lithosols; rankers.
Regosols: dune soils; loess soils.
Alluvial soils.

The Great Soil Groups

The synopsis above requires some further explanation of the nature of the groups mentioned in it. We will take them in order.

Zonal Soils

Tundra soils

These soils are dominated by cold and moisture. They are of several types, largely influenced by topography, but in all of them development is minimal, and they have the common character of overlying a zone of permafrost (p. 3588). The disturbing effect of freezing causes distortion and mixing (cryoturbation, p. 3588). There is little leaching and no clear horizons. Most are acid and have a surface layer of partially decomposed peat below which is a viscous bluish-grey layer, often gleyed with ochreous streaks. The best developed arctic soils are the arctic brown soils, formed on better drained ridges and terraces, where the A_2 layer is brown and acid, overlying a yellow A_3, below which is a brown, stony C which is neutral or alkaline.

Desert soils

These are always arid and may often, in addition, be saline or alkaline, in which case they should be classed as halomorphic. They are generally sandy, with the coarsest sand at the top. The whole surface may be stony ('reg'), generally due to wind erosion. Some desert soils in Australia and the USA are red, but the most widespread types are grey (sierozem). These are structurally undeveloped soils with a high Ca content and sometimes a calcareous pan (*caliche*). Crusts of Calcium carbonate and of gypsum are frequent. These may be sub-surface formations laid bare by wind erosion.

Black Earths

These soils, known as chernozems, have already been discussed (p. 3600). They shade off, with decreasing rainfall, into *chestnut soils* which differ from the black earths chiefly in the lower organic content in A_1, which is brownish-grey rather than black. Calcium carbonate may accumulate in A_1 and the concretionary zone is nearer the surface than in true chernozems.* Gypsum may occur in A_2.

Soils related to chernozems are the so-called 'black cotton-soils', which are variant types of the same group. One is widespread in South India and is known as 'regur'. It is a tropical black earth. Although it receives a higher rainfall than is typical for chernozem there is also a much higher temperature and intense evaporation, which impedes leaching and retains base-saturation. Other black cotton-soils occur in East Africa and in the sub-tropics in Australia, Spain and elsewhere. The soil of the Argentine pampas should probably be classed with them.

Prairie Soils

These are soils of permanent grasslands or steppe, widespread in the middle-west of the USA. They are well leached, but there is no zone of $CaCO_3$ accumulation and, although their climate is humid, they retain a high base-status. There is no break up of the clay-complex releasing silicic acid and sesquioxides and consequently no illuvial B layer. The deep A horizon is dark coloured and rich in humus, shading downwards into a brown zone and then into the parent material, which is frequently loess.

Degraded chernozems

Otherwise known as 'brown and grey forest-soils', these are formed in Europe in areas where forest has apparently invaded steppe; the original chernozem of the steppe period undergoing marked changes. These involve a decrease of organic matter, the breakdown of the typical crumb structure and the disappearance of $CaCO_3$ from the profile. The colour varies from brown to grey with increasing leaching and there is definite eluviation of sesquioxides into the B horizon. These soils are transitional between chernozems and true podsols, into which they pass under increased leaching.

Podsols

These also have been mentioned before (p. 3599). They are typically cold-temperate, formed under coniferous forest and in western Europe under heaths. Podsolization is the product of intense and prolonged leaching, and it occurs in all degrees of intensity, from slight to complete.

The outstanding character of the true podsol is the presence, below the surface humus, of a bleached layer with little or no sesquioxide content, usually greyish but in some sandy soils quite white (silver sand). This is sharply demarcated from the illuvial B horizon, in which is a zone deeply stained with ferric sesquioxide, often overlaid by a black zone of humus accumulation, though in less extreme podsols this may be absent. These zones are often indurated by Iron or humus cementing the material into hard pan.

The more extreme development of podsols is usually on sandy material

---

* In all soils of the chernozem type the depth at which concretions form is directly related to the annual rainfall.

with a low base-status originally, supporting a heath vegetation or conifers. The plant litter is strongly acid and decomposition is almost entirely fungal. This produces a thin acid peat which remains sharply separated from the mineral soil below, but its highly acid products greatly intensify the effects of leaching below it.

On clays and loams, with a higher base-status, podsolization is much slower and less marked. The eluviated layer is shallow and there may be little or no difference in colour. This is due to the greater stability of the clay complex under base-saturated conditions. It is only under acid surface material that the clay complex becomes desaturated and unstable and breaks up with release of Fe and Al and the formation of silicic acid. Apart from the molecular eluviation there can also be a certain amount of mechanical down-wash of clay into the B horizon and the two processes have not always been clearly distinguished from each other.

Brown Forest Soils

These are the typical soils of temperate deciduous forests and of lands from which such forests have been cleared. These brown earths occur in many parts of the world, with considerable variation. They are neutral or only slightly acid and there is no tendency to form peat, as the leaf litter contains considerable reserves of bases which are returned to the soil. Apart from this the profile may be leached free from carbonates, which are found only in the C horizon. Free ferric sesquioxide is present and gives the characteristic colour to these soils, but there is no downward eluviation and the silica sesquioxide ratio of the clay complex shows no marked change downwards. The humus is well distributed among the mineral components by earthworms.

From the zonal point of view Britain is a brown-earth region and as the greater part of the country was originally under deciduous forest most of the agricultural soils were probably derived from forest brown-earths. Centuries of disturbance, liming and manuring have altered them considerably from their primal state and they must now be classed as artificial. In Scotland, however, where the primitive forests were largely coniferous, podsolization was extensive. In all areas where deciduous forest has been replaced by heath or by conifer plantation the degenerative change of podsolization sets in, with progressive base-desaturation, dependent on the amount of rainfall. Naturally in a country as geologically complex as Britain there are many intrazonal and azonal soils to be found, and in some districts they may exceed in area the zonal brown earth.

Lateritic Soils

These are typically the soils of warm temperate and especially of tropical forest regions. They are predominately reddish, sometimes yellow to brownish but so predominant is the red colour that to most people red earths and the tropics are inevitably associated. The reddening process of *rubefaction* occurs after the soils have been leached free of carbonate and is due to free ferric hydroxide.

Despite the prevailing red colour, tropical soils are by no means uniform and their classification is still a matter of argument. Two factors are of general influence. One is the intensity of weathering due to high temperatures and high rainfall. Weathering has also gone on for much longer than in northern soils owing to the absence of any glacial period and may extend deeper than in temperate zones. The other factor is the rapid decomposition of organic matter, leaving these soils much poorer in humus than temperate soils. Exceptions are those of soils under savannah vegetation, where the intervention of a dry season helps to conserve humus from destruction, or under forest or grassland at high altitudes where deep organic layers may form.

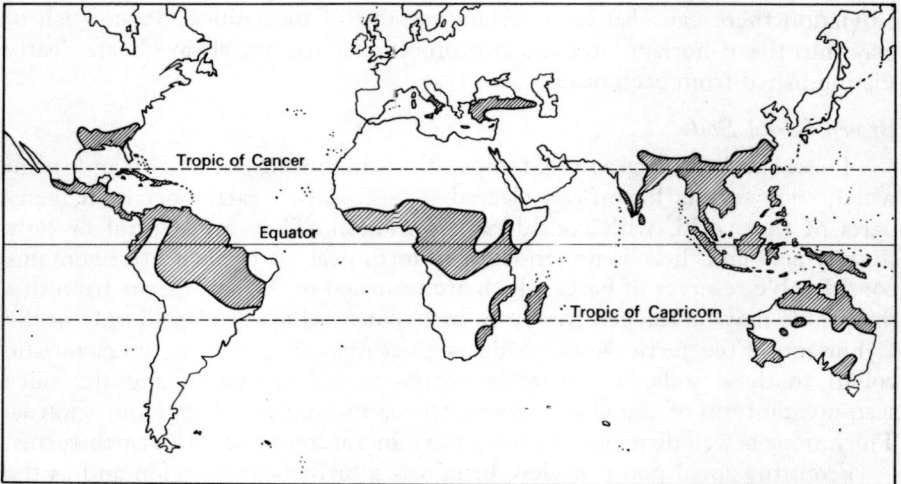

FIG. 4.40. The distribution of laterites and lateritic soils, mostly tropical and subtropical. Rain forest tends to deter hardening, clearance accelerates the completion of the process. (*After McNeil in* 'The Scientific American'.)

The term 'laterite' for red tropical soils has been used quite indiscriminately. The name was introduced in 1807 by Buchanan for a vesicular earthy mass, consisting chiefly of ferric oxides, with or without silica, and hardening on exposure to the consistency of brick (Latin: *later*, a brick) which can be used for building. True laterite is not in itself a soil, it occurs as distinct physical layers only in some soils. The iron-stained soils which do not fit into the category of laterite have been given the name of *ferralites*. Tropical red earths in general have been called *latosols* and the weathering process they have undergone is called *laterization*.

Robinson points out that the red tropical soils are the product of the primary weathering of crystalline rocks, whereas the northern temperate soils are largely secondary and made over from materials already partly weathered. He maintains that there is no essential difference between the product of primary weathering of crystalline rocks, where it can be found, in Britain, and the younger red earths of the tropics. The greater intensity of the red colour in the tropics he puts down to the prevalence in the tropical soils of less hydrated forms of ferric oxide, such as turgite and haematite, which are deeper in colour than the limonitic oxides of temperate soils.

The depth to which weathering may reach poses a difficulty in distinguishing soil from parent material. If unaltered rock is treated as the parent then the overburden of 'soil' may be 20 metres or more in depth, with considerable variations of composition and weathering in the profile. If, however, only the fully weathered surface layer is to be considered as true soil then we usually have only a shallow layer to consider, the main difficulty being that it has seldom been so considered.

The uppermost layers are greatly leached and may appear pale or greyish owing to removal of ferric hydroxide. The base/alumina ratio is very low indicating an almost complete removal of bases, while the silica/alumina ratio is also low, not only at the surface, but throughout the mass, showing a large removal of silica and relative accumulation of alumina, which may be much greater near the surface than lower down. The ratio increases with depth but seems to be always below unity. The accumulation of Al may be great enough to form bauxite beds, valuable as sources of Aluminium.

Although the foregoing is a description of laterite, it appears to be very difficult to define it in terms of composition or indeed of origin, which is almost certainly not uniform. Apart from the physical property of hardening into brick on drying, the accumulation of ferric hydroxides is one of the outstanding characters. This enrichment may be either endogenous or it may be exogenous, that is, illuvial, due to lateral seepage or to capillary enrichment with Fe from below in places where the seasonal variation of the water-table is considerable, especially on flat ground or in depressions. This can only occur where there is a pronounced dry season, not under rain-forest conditions, and laterites, especially at high levels, may be 'fossil', that is to say they are the product of climatic or topographic conditions which no longer exist.

During the process of leaching in high rainfall areas the relatively soluble colloidal silica and silicates of Aluminium are washed downwards. Owing to the absence of electrolytes or humus in the soil water they are not readily precipitated and their amounts increase downwards. The relatively insoluble oxides of Fe and Al stay behind, resulting as we have said, in a low silica/sesquioxide ratio, and the lower the ratio falls the more mature (older) is the soil. This is the opposite of podsolization where the sesquioxides are eluviated, but not the silica or silicates which are readily precipitated by humus materials in solution.

The laterite may be regarded as being in many cases the end product of these reactions, consisting almost entirely of Iron and Aluminium hydroxides with some residual quartz. This result is not usually reached under forest, but the tropical red earths may be generally regarded as soils subject to laterization. They are always acid in reaction.

Not all parent rocks can give rise to a lateritic soil. Sandstones and acid granite are resistant to desilicification. The latter give rise to a stable kaolinitic soil, not liable to hydrolysis and often red with illuvial Iron. Sedimentary clays and sand are also exceptions and sands may show typical podsolization, even in the tropics. These aberrant types would all be classed as intra-zonal.

In dealing with tropical soils we have to reckon with much more intense erosion than in temperate climates, so that truncated soils are common, and also recognize that the colluvial movement of soil materials by flooding is more extensive, often producing mixed or compound profiles with layers of different ages and origins.

Intrazonal Soils

These are more or less local soils which owe their special characteristics to certain factors of the environment or to the parent material and whose rela-tionships to climate and vegetation are secondary. The *first* groups are saline or alkali soils, called *halomorphic soils*.

Solonchak or white alkali soils, are soils saturated with Sodium salts usually from ground-water rich in Na, where the water-table is near the surface, usually in the vicinity of inland seas and salt lakes. The classic area is north of the Caspian on ground originally covered by that shrinking sea. Another principal area is around Great Salt Lake in USA in the area covered by the post-glacial Lake Bonneville, now shrunken to the existing Salt Lake. In both these areas the water level in the soil has fallen and the salts in the soil have been concentrated under an arid climate. The term 'white alkali' refers to the white crust of salts which forms on the soil under drought conditions.

Paradoxically the resalination of soils previously leached free of salt may take place as the result of abundant irrigation, which may raise the level of saline ground-water until it comes within reach of the top soil.

Solonetz. Black alkali soils. The term was originally applied to soils having a special structure, a laminated or platy surface horizon overlying a deep layer of columnar or prismatic structure with rounded tops to the columns. This is characteristic of an alkaline condition, but soils with this structure may occur which have little or no exchangeable Sodium so that 'alkali soil' is more specific than the Russian name.

These soils are completely deflocculated, that is they are in the single grain state, with maximum compactness and minimum pore space. The level of such soils is therefore usually depressed and black alkaline solutions of humus may collect on the surface, which gives them their name.

The outstanding chemical feature is the presence of Sodium carbonate, though this is a secondary rather than a primary feature. In a saline solonchak the principal salt in the colloidal complex is Sodium chloride and the reaction is neutral or only slightly alkaline. While an excess of NaCl persists the clay colloids remain flocculated, but if, through leaching or by other agency, the excess of Sodium chloride falls low enough to permit the hydrolysis of the Sodium-saturated clay colloids the released Sodium may form Sodium car-bonate by reaction either with Calcium bicarbonate in the soil solution or with the CO_2 of the air. The reaction then becomes strongly alkaline and the colloids are deflocculated. In other words a solonchak has become a solonetz.

Soloti (solod or solath). These are degraded saline soils which have been leached of Sodium in the absence of Calcium carbonate. Instead of a Calcium clay being formed by exchange with the released Sodium, an acid Hydrogen clay is formed which readily breaks up into silicic acid and sesquioxides, forming a complex which can be podsolized. The sesquioxides are leached out and move downwards and a bleached horizon is left, rich in silica.

As these three types of halomorphic soil are closely and often genetically related they tend to occur together as a complex or even as a catena around a salt basin. Differences of compactness and differential erosion of the soloti give rise to a pattern of micro-relief on the surface and afford a very interesting study in their relation to vegetation.

Salt-marsh soils and mangrove swamp soils are not generally considered as in the terrestrial category, though if they were they would probably be most suitably treated as varieties of hydromorphic soil. When marine soils are reclaimed from the sea, however, as has been done on a large scale in the Dutch polders, their transformation into agricultural soils involves a prolonged period of leaching by rain and top-dressing with lime to ensure the replacement of Sodium by Calcium in the colloidal clay.

The *second* category of intrazonal soils, are the *hydromorphic soils*, those dominated by ground water, the soils of marshes and swamps. These are all soils of impeded drainage with the water-table rising to or near the surface.

Meadow Soils (*Wiesenboden*). The soils known by this rather inappropriate name are soils occurring under similar climatic conditions to the podsols, but with a water-table near the surface. They are usually of topographical origin and commonly occur in areas of depression in regions of podsolic soil. Like the latter they are associated with cold, wet climates and often with mountain areas. They may also occur below the superficial peat as the mineral substratum of fen and swamp soils.

The essential characteristics are, first, a superficial humus layer of variable thickness which is sharply marked off from the sub-surface horizon; second, a pale sub-surface layer which resembles the A_2 layer of a podsol but without its acid reaction and, third, below water level the *gley-horizon*, pale-grey, bluish or olive green in colour, with rusty mottlings or streaks of ferric hydroxide. (See also p. 3599.) This is the diagnostic character and constitutes the gley. It results from the fact that the sub-aqueous layer is usually clay-enriched and often constitutes a plastic clay, which is anaerobic and reducing in reaction, with Iron in the ferrous state. When the water-table drops, air penetrates looser patches or along micro-channels and local oxidation and deposition of ferric Iron occurs. Manganese dioxide may also be formed.

The extent of humus accumulation at the soil surface is very variable and particularly where drainage is badly impeded the increase may reach a point at which a peat soil is formed. Peat formation usually starts on a podsol foundation, and if a layer of pan forms, with ground water held above it, the podsol may be gleyed, a *gley-podsol*. The growth of peat above this produces a *peat podsol* ('half-bog' and bog gleys). These pass over continuously into true *organic soils* or peat beds, in which the proportion of mineral matter is generally well below 50 per cent and may be so low that it represents no more than wind-borne dust.

Peat is formed under two distinct conditions. The first is as *blanket peat* which, as the name suggests, may cover a wide area regardless of topography and is generally based on glacial drift. This is a truly climatic formation of cold temperate regions with high rainfall and low evaporation. It should therefore find a place among the zonal soils. It is somewhat base-deficient and acid, which inhibits bacterial humification, and in texture it is loose and of low density. Blanket peat is less acid and has a higher base content than high-moor peat (see below), a condition which has been attributed to the action of sea spray.

The other condition is that of *basin peat*, where the topography allows water to accumulate. The formation of peat in such situations begins in lacustrine conditions and in many cases has been continuous since the end of glaciation. If the original lake water was rich in mineral bases the gradual filling up of the lake by vegetation gives rise to eutrophic fen with a peat covering which is neutral or alkaline in reaction and on which Ericaceae, *Molinia*, *Alnus* and *Betula* replace the water plants.

Where the water supply is base-deficient an acid peat develops known as *low moor* (German: *Niedermoor* or *Flachmoor*) with a different vegetation, in which *Sphagnum*, with great water-holding power, is prominent. At first the peat is still under the influence of ground water and is spoken of as *terrigenous*, but, if the rainfall is high, *Sphagnum* may build up the peat to a level above the reach of ground water, when the peat becomes dependent on rain alone and is called *ombrogenous*. As water is naturally more plentiful towards the middle of the basin the peat there will grow higher, producing a typical hump-backed profile which is called *high moor* (*Hochmoor*). These are purely descriptive names and have no reference to altitude. (See also p. 3645.)

Artificial drainage of peat beds allows the leaching of acids from the surface which then becomes humified, admitting mesophytic plants to colonize it. The subsequent disintegration of the peat reduces its mass and lowers its level by compaction. The drainage of the East Anglian fens has lowered the ground level extensively by many feet. Woodwalton Fen, which has been conserved, now stands up well above the surrounding country and has had to be diked-in to keep its fen character.

Planosols are essentially continental soils formed on horizontal plateau surfaces over loess, drift or alluvial substrata. They become water-logged in the wet season as drainage is slow and they have a hard pan of compacted clay washed down from the partially bleached A horizon. They are base saturated and of considerable depth.

The *third* category, are the *calcimorphic soils* in which the calcareous nature of the parent material is a dominant character.

Rendzinas. This name is given to a group of soils developed over limestone or chalk and containing free Calcium carbonate. They contain up to 12 per cent of organic matter and are generally dark coloured, but if the amount of carbonate is sufficiently great they may be either greyish or even whitish, as in some English chalk soils. They are never reddish. The A horizon may be up to 30 cm deep, but they are usually shallower. The reaction is slightly alkaline and sesquioxides are never mobilized, so no B horizon forms. The clay fraction is largely siliceous, with a high (5–6) silica/sesquioxide ratio. The superficial layer (A_1) may sometimes be quite black. English soils on the mesozoic Lias and Oolite appear to belong to the rendzinas.

Quite different are the red and brown limestone soils, typified by the Mediterranean *terra rossa*.

Red Limestone Soils (Terra rossa, Krasnozem). These are typical of the hard palaeozoic limestones, and they are best developed in the Mediterranean basin. In contrast to the rendzinas they are base-unsaturated, may be leached free from Calcium carbonate and have a high content of clay. The reaction may be acid.

These soils for the most part are shallow and lie directly over unaltered limestone which suggests that they are derived from the non-calcareous components of the limestone, but the view that they are simply weathered residues of the limestone is inadequate and the process of their formation is not clear. There would seem to have been an irreversible precipitation of sesquioxides liberated by degradation of the clay complex.

Soils closely similar to terra rossa cover the limestone in many parts of South Wales. Their silica/sesquioxide ratio is not more than 2, they are highly leached and usually slightly acid. They bear a heath vegetation, even close to the underlying limestone. Often they are intensely red.

The desaturated condition of the red soils suggests that they are more advanced in development than the rendzinas and may have developed from them, or they may be substituted for rendzinas under conditions of greater leaching and exposure. In either case one would expect to find intermediate grades and these may be represented by the so-called *terra fusca*.

Brown Calcareous Loams. Over the plateaux of the chalk in southern England the chalk is overlain by a brown-earth called the 'clay-with-flints', which is an apt description. It retains a high base-status though it has been leached free from carbonate and it supports natural oakwood or beechwood. It is generally regarded as an erosion residium of the chalk itself, and, if this be so, it implies the erosion of a very great thickness of chalk.

Brown calcimorphic loams may also form directly from limestone in warm climates with low rainfall or they may be the products of degradation of rendzinas with the formation of a B horizon. They are essentially an intermediate group, linked with terra fusca on the one hand and true rendzinas on the other.

The beechwood soils of England are rendzinas on well-drained slopes, but on plateau surfaces over chalk they tend towards a brown loam type, with a deep A_1 layer mixed with humus and a paler A_2 below. The B horizon is not well marked but appears as a reddish brown loam with a high, probably illuvial, clay content. These soils may become podsolized, with a thin, bleached layer below the highly acid layer of leaf humus.

Andosols. A name given to young soils on basic volcanic rocks. The A horizon is dark to blackish and is very fertile. They change with age, the A horizon bleaching brown to reddish and eventually assuming zonal characters, in the tropics passing into red tropical soils.

Azonal Soils

These are immature soils which are governed by the parent materials as they have not developed zonal profiles.

Lithosols

These are stony layers formed on slopes where erosion is active. They are coarse grained and gravelly and retain little water or organic matter. Thin lithosols on acid rocks have been called *rankers*. They may develop into podsols or rendzinas according to the nature of the parent rock.

Regosols

These are immature soils on loose substrata such as sand or loess. The thin grey soil which forms on old sand-dunes which have been covered with a closed vegetation is typical of regosols. The A horizon is highly organic, but leached and decalcified and no B horizon forms. The early stages of pedogenesis on loess are also regosols, with decalcified A horizon which is slightly acid, resembling a dune soil, but developing into zonal podsols or chernozem according to the climatic conditions.

Alluvial Soils

Local azonal types. They vary considerably with the nature of the alluvium, which may be a coarse gravelly material with a deep water-table, or sandy with a moderately deep water-table, or fine, silty material which is perennially moist. As all these are transported materials they may come from another climatic zone, notably in the deltas of long rivers. They are usually deep and base-rich and are agriculturally valuable. Peat may form on the wetter areas and in boggy basins.

Owing to the heterogenous materials deposited, the texture of an alluvium may vary in alternating layers giving an appearance of horizon development which is misleading. Flooding and down-wash may add successive deposits to an alluvial area and buried surfaces are not uncommon, while organic matter is often mixed throughout with the mineral soil.

Physical characteristics of soil

The physical characteristics of a soil, its texture and its behaviour with regard to water, depend largely on what is called its *mechanical analysis* which is a process of sorting out its mineral constituents into grades according to the size of grain.

The procedure has been standardized internationally and the following grades adopted:

Grade	Diameter limits in mm
Coarse sand	2·0–0·2
Fine sand	0·2–0·02
Silt	0·02–0·002
Clay	< 0·002

The successive limits are at equal logarithmic intervals.

The first step is to rub a weighed sample of the dry soil through a 2·0 mm sieve. All material retained by the sieve is put aside as 'gravel'. After sieving the sample is boiled with 6 per cent Hydrogen peroxide to destroy the humus present which would otherwise cause the soil particles to aggregate. Next it is treated with 0·2 N hydrochloric acid to remove carbonates and release bases from the clay particles. The sample is then washed and shaken up in 0·008 N Sodium hydroxide, with the object of breaking down all aggregates and reducing the soil to its primary particles.

The *coarse sand* fraction is sieved out by washing the dispersed sample through a sieve of '70 mesh', with apertures of approximately 0·2 mm. Beyond this point separation of the grades depends on their rates of sedimentation, in accordance with Stokes's Law, that the limiting velocity of a particle falling in a fluid is proportionate to the square of its diameter. *Fine sand* is estimated by placing the sieved sample in a large beaker filled with water to an exact depth of 10 cm, stirred and allowed to settle for 3 sec. The supernatant liquid is then poured off quickly into a storage vessel and the procedure repeated with fresh water until the supernatant liquid remains clear at the end of the settling period. The settled deposit is washed into an evaporating dish, dried and weighed.

The cloudy supernatant liquid is now poured into a tall glass cylinder and allowed to settle for 8 hours. Then a 20 ml pipette is lowered into it to a depth of exactly 10 cm and a sample quickly drawn off. This is transferred to a basin, dried and weighed. It represents the *clay*, i.e. particles which have not settled through as much as 10 cm in 8 hours. The weight of *silt* is found by difference when the combined weight of the other grades, expressed as percentages of the original sample are subtracted from 100. It is usual to express the composition in percentages of the weight of the mineral soil dried at 100°C. Humus and carbonates can also be estimated by the loss of material on treatment with Hydrogen peroxide and hydrochloric acid respectively, again based on the oven-dry weight of the original sample, less gravel.

Comparisons of soil can most readily be made by triangular co-ordinates, using clay, silt and the combined sand fractions as the poles of the triangle, the three fractions making up 100 per cent of the mineral soil. (See Fig. 4.41.)

While the boundaries separating the coarser grades are arbitrary and do not imply any abrupt change of character, this is not true of the clay fraction. The coarser grades are mostly original mineral fragments, but the clay is a secondary product of weathering and is colloidal. Like other colloids it holds water by imbibition and changes markedly in volume according to the amount of water it holds. At certain states of water content clay becomes plastic and cohesive, and it is this quality which is responsible for the building up of the

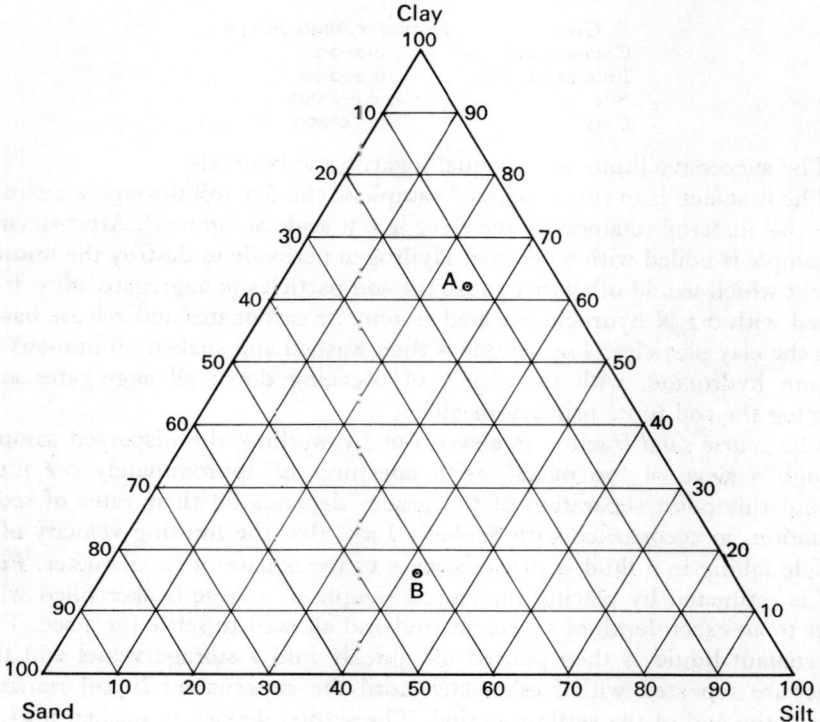

FIG. 4.41. Diagram of triangular co-ordinates for the graphical expression of the mechanical analysis of soils. A shows a heavy clay, sand 11·8 per cent, silt 26·0, clay 62·2. B is a medium sandy soil, sand 41·8 per cent, silt 42·4, clay 15·8.

particle aggregates in the soil, which we call 'crumb', a most important effect in opening up the soil to air and improving drainage, since fully dispersed soil material is practically impervious to both air and water.

Clay is also highly adsorptive of ions in solution, especially kations, and it is therefore the seat of that base exchange of which we have spoken, which is vitally important to the fertility of the soil.

The colloidal qualities of clay are shared by the organic matter of the soil, the decomposed plant and animal tissues, popularly called 'humus', the nature of which we shall discuss later (see p. 3633). From a functional point of view clay and humus are grouped together as the 'colloidal complex' of the soil, but, as the humus is more strongly colloidal than clay the properties of the complex will vary considerably according to their relative proportions in it. In some highly weathered tropical soils almost the whole complex consists of clay, while in deep ombrogenous peat the whole complex is organic.

Correlated with the state of aggregation of the soil is the extent of its *pore space*, which, in turn, is closely related to the soil aeration and its water-holding capacity (see p. 3619), and to the penetrability of the soil to roots. Furthermore, if aggregation is weak, the clay colloids may be carried downwards by leaching to form a clay-pan, which happens in some prairie soils. Pore space is not a fixed property of a soil, but varies with circumstances. Any impaction decreases it, frost (see 'Heaving' p. 3588) and cultivation increase it. It is measured by the difference between apparent density (weight/volume) and true density as measured by specific gravity methods, but single estimates can only be arbitrary. The mean specific gravity of soil minerals is about 2·65, but the apparent density in dry soil in its natural state is only about 1·4–1·8. The difference indicates that pore space is very considerable. In round figures it usually lies between 40 and 60 per cent of the soil volume.

Contrary to traditional belief the aggregate pore-space is greater in heavy, clay soils than in light, sandy soils. What makes the difference is that in the clay the pore-passages are all extremely small. As the flow resistance in a tubular passage is inversely proportional to the fourth power of its radius, the frictional resistance to mass movement of air or water is so much greater in the clay that the free circulation of either is almost inhibited.

A very important fact about pore space is, however, the difference between capillary pores, and larger, non-capillary pores. The former are more influential with regard to water-holding capacity, since they can hold water against gravity and the water they hold does not drain away. This is what determines (apart from colloid imbibition) the *field capacity* of the soil for water. The larger pores are more important for aeration, since after drainage they are practically filled with air. A low percentage of non-capillary pores, 12 per cent or less, creates a waterlogged soil which is unfavourable for most plants. The rate of filtration of water through an undisturbed soil is perhaps the best indication of the relative balance of pore spaces and is itself a quantity more directly related to ecological considerations than the estimation of pore space as such.

The *soil reaction* is a highly important soil factor. This implies the acid-base or H–OH balance as measured by the pH scale. This scale when applied to simple ionic solutions is logarithmic, but when related to soil properties there is evidence that the series should be treated as arithmetical. Soils are usually well buffered against pH changes, and in this action organic material plays a large part. Plant litter itself varies in reaction according to the species contributing to it and there is a high degree of correlation between the pH both of fresh leaves and of surface humus with the pH of the underlying soil, especially where the underlying soil is acid. Where the soil is alkaline the litter tends to be rather more acid than the soil.

It is often claimed that the litter from conifers tends to make the soil more acid than the litter from broad-leaved trees, but in fact there is a similar range of variation between species, in this respect, in both groups. Ovington and Madgwick made comparisons of 100 forest plots on a variety of soils with a number of unplanted plots on the same soils. Their figures show that the pH range under the conifer stands was greater than under the broad-leaved stands, but the two ranges overlap. Some conifers, e.g. *Chamaecyparis lawsoniana*, with leaves rich in Calcium, produce a more alkaline soil than the broad-leaved *Quercus* or *Fagus*. In nearly all cases there was considerable acidification of the litter during humification and that thereafter the pH, on acid soils, varied little with depth down to 60 cm, but on alkaline soils the pH rose in the A horizon to a constant figure of more than 7·0.

Because of this superficial acidification, and because leaching carries bases downwards, the pH of most soils tends to increase with depth.

The pH of soils is broadly correlated with their base status and particularly with the Calcium status. In soils at pH 8·3 the colloidal complex is usually Calcium saturated; above this level free Calcium carbonate is present and, in arid regions, often Sodium carbonate as well. At pH 3·5 the bases on the colloid particles have been practically wholly replaced by Hydrogen and consequently soils with a pH below this level are rare and generally contain free ionized acids.

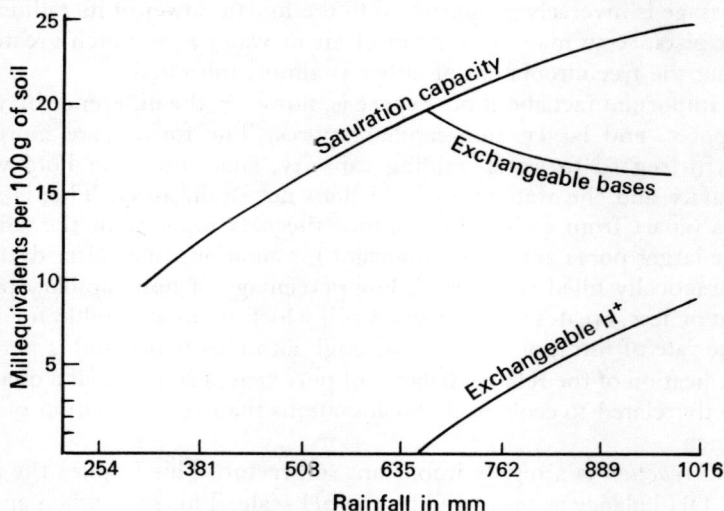

FIG. 4.42. Diagram showing the relation of rainfall to the saturation capacity, the exchangeable bases and the exchangeable Hydrogen in the surface layer of loess soils. (*After Jenny.*)

Acidification in the mineral soil is the result of the replacement of adsorbed bases by H ions (p. 3650). The total amount of the exchangeable adsorbed bases represents the *saturation capacity*, which is measured by replacing the bases with some neutral agent such as Barium acetate. The saturation capacity rises logarithmically with the rainfall, but at a certain rainfall value, depending on the nature of the soil, the amount of adsorbed bases begins to decline and the amount of adsorbed H ion to increase, as the latter replaces bases leached out. The degree of acidification is measured by the *saturation grade*, which represents the extent to which this process has proceeded. Expressing the quantities as milliequivalents per 100 g of oven dried soil, the saturation grade:

$$S = \frac{\text{Sum of the basic cations}}{\text{Saturation capacity}} \times 100.$$

A high saturation grade implies an excess of bases over Hydrogen ions and vice versa.

The saturation grade has chemical as well as physical significance. Calcium is one of the most conspicuous basic kations in soils, and plants react very differently to different saturation values of this ion. All higher plants appear to need Ca, it is one of the 'essential' elements, but there is a great range in the requirements or, indeed, tolerance of species with regard to it. While many species are comparatively indifferent, that is they have a wide range of Ca-tolerance, there are two groups with very pronounced proclivities in this respect, the *calcicole* and the *calcifuge*. The former can only survive or, at least, can only survive competition on calcareous soils, the latter only on acid soils. This physiological peculiarity, which deeply affects the distribution of such plants, has been the subject of a good deal of investigation.

The effect of a soil of unfavourable reaction appears to be definitely chemical and is independent of the physical nature of the soil. It is also clear that it is $CaCO_3$ not Ca ion as such which is important and that its effect is through its action on the soil pH. The alkaline condition has side effects on the solubility and availability of other ions which may be more influential than the presence of $CaCO_3$ itself. The latter is important though it may not be always the decisive factor. For example, Rorison showed that *Scabiosa columbaria*, a definite calcicole, would grow in water culture both at pH 4·8 and at 7·6, but the addition of Ca bases gave a positive improvement in growth. A high pH restricts the availability of Fe and of P, and Ca antagonizes the uptake of K. Calcifuge species, such as *Galium hercynicum* fail on calcareous soils, possibly due to ionic starvation, since the ash-content of P and K is greater on acid soils. Ionic deprivation may also occur on acid soils, since in such soils Al reacts with P, diminishing its availability. There is also the question of ionic toxicity. Alkalinity reduces the availability of Fe, but it also encourages the supply of N and it diminishes the availability of Mn and Al, both of which are toxic if their solubility is increased. At pH 4·8 Al is definitely toxic, but in organic soils it may be chelated, leaving Mn as the chief toxic agent.

The problem cannot be said to be wholly solved, but enough has been said to show that the determining effect of the presence of $CaCO_3$ in a soil is not entirely a matter of direct action, but may also have decisive effects on the availability or, alternatively, the toxicity of other ions.

At reactions round about pH 7·0 most important kations are sufficiently soluble to satisfy the needs of plants. The solubility of some, e.g. K, P, Ca and Mg, falls off with lower pH, while Fe and Mn fall off with raised pH. The supply of combined N falls off in both directions.

The soil pH is subject to several environmental factors of change. During the year it is influenced by the amount of rainfall, decreasing in rainy periods and increasing under drought. These variations are small in any one locality, little more than a single integer, but there is a marked difference between arid and humid climates. In the former the pH may rise to 9·0. There are also seasonal changes which are perceptible even at a depth of 1·5 m and not directly related to rainfall. These seasonal changes are greatest near the surface, the F and H layers of humification being always more acid than the mineral soil which begins at about 10 cm down. The acidity is greatest in winter and spring and diminishes in summer. It is probable that the seasonal changes are due, at least in part, to the varying activity of soil micro-organisms. Another modifying factor may be the rise of base-rich ground water towards the surface during periods of low rainfall and increased evaporation.

The range of amplitude of variations in soil pH is controlled by the extent of buffering which exists in soils of different types. Buffering substances are of many kinds, e.g. colloidal clay and humus, amphoteric electrolytes, weak acids and their salts with strong bases, or acid colloidal hydroxides of Iron and Aluminium. Humus soils are the most strongly buffered, sands the least, with, in consequence, the least constant pH. This applies in the same degree to the bleached layer in podsols.

Olsen in Denmark in 1923 showed that each species in meadows and woodlands had a characteristic pH range,* representing the amplitude of its tolerance. Observations in Britain, Switzerland and Norway have confirmed this. Salisbury tested the pH of the soil at places where *Pteridium aquilinum*, *Vaccinium myrtillus* and *Scilla non-scripta*, respectively, were found. Braun-Blanquet also tested the soil pH over 125 localities for *Carex curvula*, covering a great variety of soils. If the pH values are plotted against the number of localities in which these species occur, it was found that they could be fitted to a normal curve of variation. Both authors claimed that the mode of this curve shows the optimum pH for the species. This has been disputed by Emmett and Ashby who claim that this proceedure shows no more than that the variations of soil pH have a normal distribution. They maintain that it is essential to separate the distribution of pH over a whole area, regardless of the occurrence of the species, from the distribution by localities of the species itself. If this is done they claim that no special pH preference, within the specific tolerance limits, is shown by the species, or in other words, there is no optimum.

Changes of pH often play a part in the succession of vegetation. The plants themselves may change this soil factor, in either direction, among their other reactions on the environment, and then plants restricted by a narrow range of tolerance may be succeeded by species with different or with broader tolerances which can maintain greater vigour over a wider range of sites than can the earlier species.

A relatively high pH, implying a high base-status, has a great influence on the state of aggregation of the soil particles. Negatively charged primary colloid particles in clay show an electrostatic repulsion from each other and such unsaturated soils tend to remain finely particulate, i.e. highly dispersed and consequently impervious. When the charge on the particles is neutralized by adsorbed kations, their mutual repulsion is lost and they tend to flocculate gravitationally into compound particles, which loosens the soil structure. If

* An interesting contrast is afforded by *Deschampsia flexuosa* (pH 3·5–4·9, maximum frequency at 3·5–3·9) and *D. caespitosa* (pH 5·0–7·9, max at 5·5–6·4).

binding materials are available, humus, Calcium salts or sesquioxides, then the compound particles, which are only loosely associated, become further bound together much more firmly into aggregates, or 'crumbs', of very variable size, from 0·2–20 mm. The macro-structure of the soil is therefore an expression of its degree of saturation.

A similar flocculation of particles occurs when an amphoteric colloid is brought to a pH at which it is electrically neutral, its isoelectric point. This is the condition governing the deposition of illuvial sesquioxides in a leached soil. They are deposited and form a horizon in the soil at whatever level the pH corresponds to their isoelectric point.

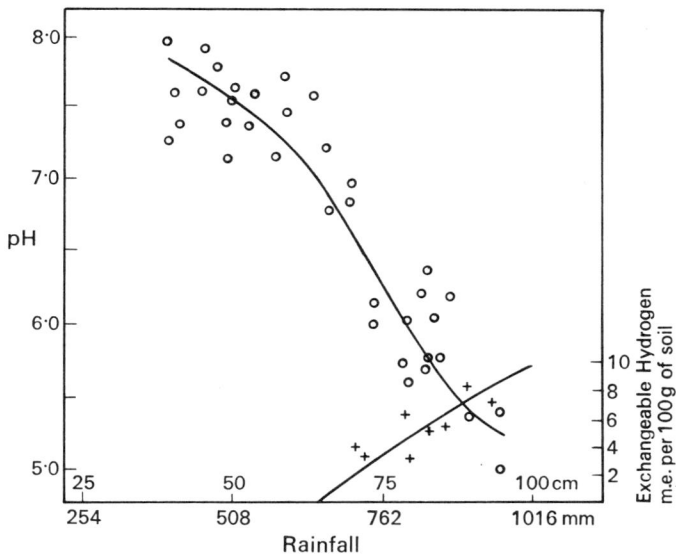

FIG. 4.43. Diagram showing the relation between rainfall and soil acidity in a loess soil. Circles mark pH values and crosses the amount of exchangeable Hydrogen. (*After Jenny.*)

Water in the soil

The condition and movement of water in soils depend upon a complex of physical and chemical factors which have called forth a great deal of investigation and produced an extensive literature.

A minor portion of soil water is held in the composition of hydrated salts and colloids. This is unavailable to the plant. The greater portion is held in two ways, both governed by surface forces, namely, imbibed by soil colloids, both organic and inorganic, or by capillary action on and between non-colloid particles. The way in which capillary water is held by a soil depends both on the physical structure of the soil and on how much water is present. Physical structure includes the relative proportions of different sizes of particles, the 'mechanical analysis', and on the amount of colloidal material, both organic and inorganic. It also involves the degree of aggregation of the particles into compound particles.

Water is held partly as film on the surfaces of mineral particles and partly in the spaces between them. The smaller the particles the greater is their aggregate surface. Other things being equal, a soil with a high proportion of fine particles will hold more water as surface films than a soil made up of coarse grains. Only a part of the water content is held thus, however. If water is abundant, much of it is contained in the pores and capillary channels of the soil. The water in the larger channels will drain away under the pull of gravity and be replaced by air but in the fine channels it will be held by capillary cohesion and will not drain.

The porosity of the soil depends on the packing of its constituents and a soil in which there is a wide range of particles of different sizes will pack more closely than one with particles all of one or of a few sizes, thus providing a greater proportion of narrow capillaries which can hold water against gravity.

The packing of the soil particles assumes special importance as the water content of the soil diminishes through drainage and evaporation. As the capillaries are emptied and the water replaced by air the remaining water is limited to the angles where particles are in contact with each other, in which the last water is held in 'wedges' in the minute tapering clefts between mineral grains. If there is only a point contact the water forms a ring around the point. Connected at first together to form an extensive web, these wedges are gradually isolated as the water content continues to fall. Another factor here comes into play, namely the surface tension of the water. Each wedge or ring has a concave, meniscus surface with a higher surface tension than a flat water surface and this tension is inversely proportional to the radius of curvature. As the wedges shrink their surface tension therefore increases steeply and their resistance to evaporation or absorption is consequently greater.* An 'air-dry' soil, i.e. one which is in equilibrium with the atmosphere, may thus retain an appreciable amount of moisture which can only be removed by oven-heating to over 100°C, the 'oven-dry' soil. This air-dry moisture is, however, too firmly held by physical forces to be available for absorption by the roots of most plants; it is 'unavailable'. Oven-dry soil picks up a small quantity of *hygroscopic water* when exposed to the air, but this again is unavailable. (See p. 3629.)

The various phases of water content in the soil are the following,

1. *Saturation*. All spaces filled with water and the colloids fully imbibed. The saturation capacity is a measurable characteristic, obtained by differential weighing between a dry and a saturated sample.
2. *Field Capacity*. The amount of water retained by a saturated soil after all free water has drained away.

---

* The water tension in the soil is a valuable index of the 'water-supplying power' of the soil, which is inversely related to the tension. It is measured by a tensiometer and is expressible as a negative pressure in millimetres of mercury.

3. What may perhaps be called the *loss phase*, during which evaporation (and root absorption) gradually reduce the water content to the limit of availability.

This important distinction between available and unavailable water is determined by the 'wilting point' of plants growing in a given soil. Soil at or near its field capacity is packed into an impervious container and a test plant is carefully planted in it. The surface of the soil is then sealed off so that all water loss from the soil must be through the plant. The test is then left until the plant shows permanent wilting, when the container is opened and the residual water estimated by oven-drying. This is the amount of the 'unavailable' water.

It might be supposed that the results would vary considerably with the species of plant used, but Briggs and Shantz in 1912 showed that this was not so and that, provided the rate of transpiration was limited to a very low level, the results were closely similar whatever species of plant was used. In other words, the unavailable water is a specific character of the soil, an expression of physical forces depending on its structure. A coarse-grained soil with high porosity and relatively few angles of contact between grains may retain as little as 1 per cent of its water content in an unavailable form. Fine-textured soils, on the other hand, have a very large aggregate surface on their particles; they pack closely, with many angles of contact and minute capillaries and, if they contain much clay or humus, they have a high imbibitional capacity. Due to frictional forces water moves very slowly in such soils and while they may have a lower saturation capacity than a more sandy soil, their field capacity is higher and the level of unavailable water is also higher and may be up to 40 per cent of the saturation value. As a fine-grained soil holds its water more tenaciously than a coarse sand does, it is more resistant to drought. Despite its high level of unavailability the water resources of a fine-textured soil will keep plants alive in a drought for much longer than will a sandy soil.

The latter have advantages in good aeration, good drainage and quick response to changes of temperature, but they dry out too easily and, for agricultural purposes, they need to be improved by 'marling', that is top-dressing with clay to give the balance of texture which constitutes 'loam'.

Several experimental values have been proposed and used at various times as measures of the water capacity of different soils in comparisons. One of these is the *moisture equivalent*, which is determined by submitting a sample 1 cm deep, saturated to its field capacity, to a centrifugal force of 1000 g. This removes all capillary water and leaves only the amount due to colloid imbibition and to the wedges in the angles of contact. It is an arbitrary quantity, since theoretically the amount of water extracted should increase progressively with the force applied. An extension of this procedure is to make a second estimation on a sample of the dry soil which has been saturated with xylol instead of water and determine the xylol equivalent. On the assumption that xylol is not imbibed by colloids, the difference between the two equivalents gives the amount of water which was held by the colloids alone.

Another quantity of value in comparisons is the *sticky point*. This is obtained by working a ball of wet soil between the hands until a point is reached at which the ball begins to pick up the soil adhering to the skin, i.e. becomes sticky. The soil thus treated has been reduced to its primary particles and these have been forced into their closest mechanical packing. The water content of the soil in this condition is regarded as representing the maximum imbibitional capacity of the soil colloids. It does not distinguish between organic and clay colloids, but if the soil is first treated with Hydrogen peroxide to remove humus colloids, the sticky point moisture bears a linear relation to the clay content.

Although it is convenient to distinguish between gravitational water and capillary water, the distinction is not sharp and depends among other factors on the freedom of drainage and the height of the water-table. The presence of the water-table within a metre or so of the surface has a definite effect in raising the moisture-holding capacity of the soil, while a water-table six metres down may be held to be without effect on conditions of free percolation in the surface layers. The field capacity is thus a more or less ideal value, useful for comparisons but with little relation to the constantly varying moisture conditions in nature. Moreover, it may differ considerably in different horizons of the soil profile, as must be evident from a consideration of their differing textures and colloid content. At the level of the water-table the soil is fully saturated with water and the diminishing water content from this level towards the surface represents the equilibrium between gravitational and capillary forces at each level.

A valuable measure of capillary force is the *capillary potential* introduced by Buckingham in 1907 and modified by Schofield in 1935. By analogy with electrical potential the capillary potential is the force determining the movement of water in the soil, though inversely, for in this case movement will tend to be from areas of low cp to those of high cp, the capillary potential being inversely related to the water content of the soil. Buckingham expressed the potential in terms of the height of the water column equivalent to the pressure difference required to produce movement of water from one portion of soil to another. The range of such differences is very great as between dry and wet soils, and Schofield proposed to use the logarithm of Buckingham's values as a convenient method of expressing this range. By analogy with pH he called these logarithmic values the *pF*.

Practically, estimations of pF may be made either by measuring the water retained by soil subjected to suction or by the water absorbed by a dry soil against a given pressure. Suction can produce pressure differences of up to one atmosphere and as this difference corresponds to a water column of 1000 cm, the corresponding pF would be log $1000 = 3 \cdot 0$. Higher values can only be obtained by adding excess pressure to the effects of suction. The important result was obtained that at corresponding values of pF the water retained by a wet soil and the water gained by a dry soil are not the same and the two sets of observations yield a pair of curves, related like hysteresis curves, with the curve for drying lying above the curve for wetting. The higher values of pF, verging towards the oven-dry condition, have been estimated from measures of the aqueous vapour-pressure of soils, which can be converted into terms of pF.

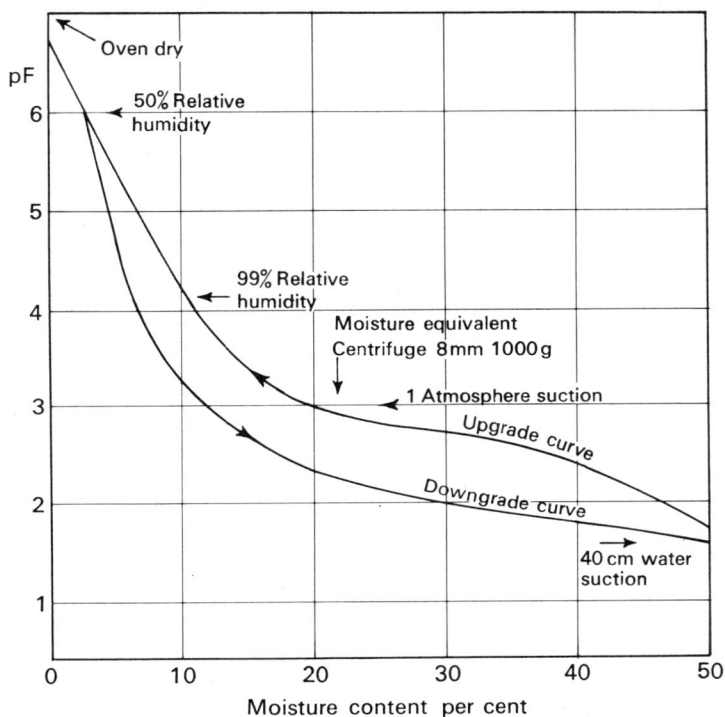

FIG. 4.44. Moisture and pF curves for the loam of the Rothamsted Grass Plots. (*After Schofield.*)

Capillary potential increases as the moisture content falls, and there is a corresponding increase in the power of the soil to retain water. The absorption of water by a plant's roots may be regarded as resulting from the superior pF of the root, but as the soil pF increases there will come a balance point at which it will no longer be possible for the root to withdraw water. Schofield showed that an average value for the pF of the soil at this point, the wilting point, is about 4·2.

The difference between the two pF curves, of a soil during wetting and one during drying, referred to above, means that at any given moisture content, say 20 per cent, a soil which is losing water will have a higher capillary potential than one which is gaining water. Conversely, two portions of soil can be at the same pF, although one, which is losing water, has a higher moisture content than the one which is gaining. As a soil gains moisture its pF drops, and vice versa. A dry soil in contact with a wet soil will draw water from the latter until capillary equilibrium is reached. At the conclusion of the movement the gaining soil will have reached its field capacity, and although the losing soil will still contain more water, any gains above this point by the gaining soil will be negligible. At this point of capillary equilibrium the pF will be approximately 3·0.

The principle of pF is basic to an understanding of water diffusion in the soil, as distinct from purely gravitational drainage or movements in the vapour phase which may be quite important with regard to the drying out of soils. (See p. 3626.)

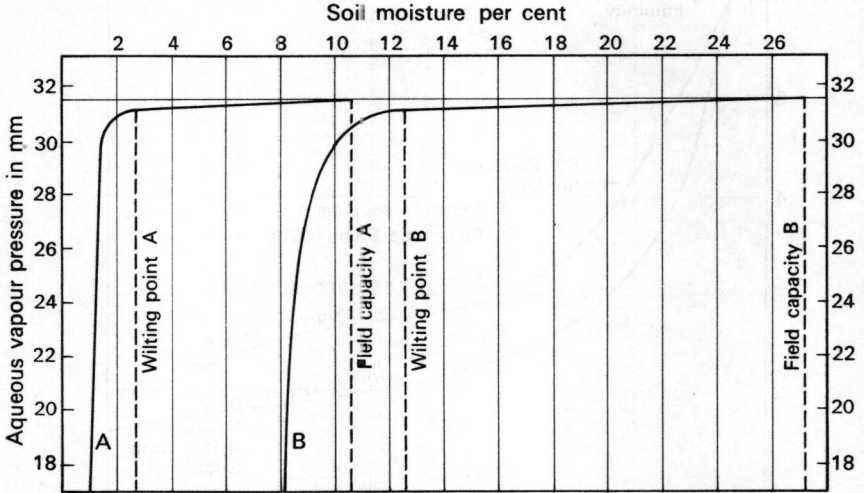

FIG. 4.45. Changes in the aqueous vapour pressure of soil water at different percentage moisture contents. A is a fine sandy loam, B is a clay loam. The vapour pressure scarcely changes with decreasing moisture until the wilting point is reached after which it drops rapidly. Note that the field capacity of A is lower than the wilting point of B. (*After Edlefson.*)

When water is added to a dry soil, by rain for example, the water percolates fairly freely downwards until an equilibrium distribution has been reached to a certain depth, depending on the amount of water available and the moisture capacity of the various horizons of the soil. Below this depth the moisture content falls off sharply. Subsequently, water is transferred by diffusion from the wet upper layer to the dry lower layer. The upper layer is drying by diffusion and evaporation, with a rising pF, the lower layer is wetting, with a falling pF. The rate of wetting of the dry layer drops with time, as the pF of wet and dry layers approximates at pF 3·0, when the upper layer will have uniformly reached its field capacity, at which point the dry soil, gaining water, is in equilibrium with the upper, wet, soil, though still at a lower water content. The final gradient of water content in the lower layer now falls much more gradually downwards and reaches a greater depth than at the beginning.

The sum of the forces with which the soil holds water is sometimes called the *moisture potential*, or better, the *specific free energy*. This is a measure of the work that has to be performed to remove water from the soil, i.e. to overcome the holding forces of the soil. It is expressed in ergs per gram, or, more usually translated into atmospheres (1 atm $= 1 \times 10^6$ erg). The zero datum level is taken as that of a free water surface.

The specific free energy has three major components. (1) Hydrostatic pressure (ultimately the gravitational field). (2) Adsorptive (including capillary) forces, which increase as the moisture content drops. (3) Osmotic potential, due to dissolved solutes. The total free energy is also sometimes called the *integrated water stress* of the soil, or the *water tension*. The last term is, however, misleading as it suggests that all the water in the soil is under a uniform tension. It has arisen because of the use in measurements of the water tension developed in a porous pot filled with water, in equilibrium with the soil water. In fact only a part of the water is actually in tension, due to the negative curvature of the air-water interfaces. Much of the water may be under positive hydrostatic pressure.

The free energy increases with decreasing water content until the point is reached where plants cannot withdraw water, the permanent wilting point. Below this point the wilted plant may continue to withdraw at a very low rate, because of the time-lag in water continuing to find its way to the root surfaces. This has given rise to the concept of a *wilting range* of water content, but the effect is only observed in plants growing in closed containers. Under field conditions there is effectively no further withdrawal of water below the wilting point. The field capacity and the wilting point are respectively the upper and lower limits of the available water stored in the soil.

The specific free energy of the soil water rises very slowly as the amount decreases below the field capacity. It only begins to rise steeply as the wilting point is approached. Over at least two-thirds of the available range of water content water is therefore equally available to the plant, despite the drop in actual water content. Thus, between a field capacity of 20 per cent water and a wilting point of 8 per cent, water is equally available down to a level of about 11 per cent. It has been contended, in opposition to this view, that increases of water tension of 1-4 atmospheres in the soil cause a definite impairment of both absorption and growth. There is no real contradiction here, since increased tensions of this order only occur as the wilting point is approached.

At the wilting point the free energy is about 16 atm, but Woodruff has shown that the free energy potentially available in the leaf by the absorption of solar radiation is far in excess of that required to extract water from the soil even at the wilting point and Gardner has suggested that the stoppage of absorption at this water level is due, not so much to the inability of the plant to absorb, as to the slowing down of the rate of water movement from the soil to the roots.

In an extended experiment by Veihmeyer in California a *Prunus* tree was allowed to exhaust the water of a large sealed container. When the wilting point was reached the water supply was again raised to field capacity. This was repeated some fifteen times between 1 May and 14 November. No variation in the rate of transpiration was observed at any level, the slope of the withdrawal curves was uniform throughout. With bean plants also the average transpira-

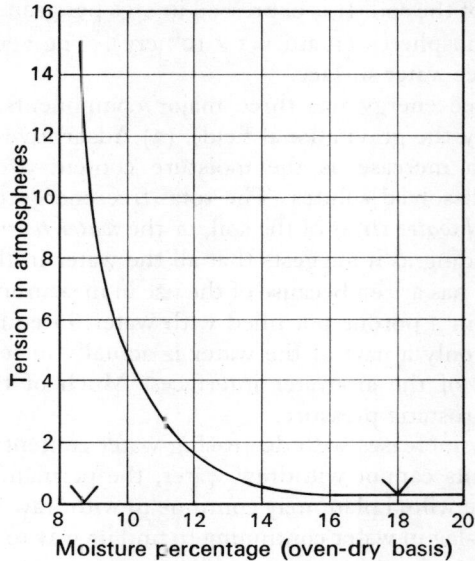

FIG. 4.46. The relationship between moisture content and water tension in the soil. Panoche loam USA. The arrows indicate the field capacity (18 per cent) and the wilting point (8·6 per cent) respectively. The water tension remains low to within 2 per cent of the wilting point. (*After Wadleigh in the 'U.S. Dept. of Agriculture Year Book'*, 1955.)

tion per 100 cm^2/hr/day of leaf surface showed no correlation with the soil water content.

As with transpiration, so with photosynthesis and growth, the bulk of the evidence supports the view that neither are affected adversely by a decreasing water content in the soil until wilting occurs.

Physical losses of water in the soil are attributable to downward percolation under gravity, to capillary action in any direction, and to evaporation upwards. The first mentioned can only occur when the content of the upper layers rises above field capacity, and as we have just seen, its downward range may be limited. The second movement is slow and also limited by changes in capillary potential. The third is the most important. Naturally evaporation is most active at and near the surface, but if the subjacent levels are warmer than the surface, as in winter, there will be a considerable distillation of vapour towards the surface. If the water-table is less than one metre below the surface a small amount of water may also reach the surface by capillary rise, but only if the surface layers are below their field capacity. Lateral capillary movement is also very limited.

The rate of evaporation follows Newton's law of cooling, i.e. it is constant over a wide range of moisture content, but falls off rapidly when the moisture content is reduced to the permanent wilting point and ceases when the soil reaches the hygroscopic limit. The amount of evaporation is greatly increased by vegetational covering, for the transpiration of the plants can draw water rapidly from the whole region penetrated by the roots.

All the factors affecting transpiration will then be active in determining the extent of water loss. On bare soil, however, temperature, humidity and wind are the most influential factors. Changes of temperature and perhaps too of barometric pressure, control the movement of air into and out of the soil, which will promote evaporation. Diurnal temperature changes may cause a turnover of 5 per cent or more in the soil air.

While the foregoing discussion reflects the fundamental principles governing the movements of soil water, it will be realized that in nature there are numerous complexities of conditions and situation which have greatly modifying effects. Conditions are rarely sufficiently stable for any equilibrium to be maintained or even reached and change is more or less continuous. It can be seen, for example, that downward movement of water in the soil is not a simple gravitational flow of free water and that, consequently, the amount of rainfall is a very imperfect guide to the extent of soil leaching which takes place. Only a careful study of the water draining from a soil can give reliable information about this.*

Study of percolating water gives some idea of the solution complex in the soil, but the picture is incomplete without considering the conditions of adsorption inside the soil, since the percolate contains substances released from adsorption. To sample the true *soil solution* as it exists in the soil is not easy. The simplest method is by downward displacement in a column of saturated soil, which drives out at the base of the column a solution which may be assumed to be in equilibrium with the adsorption complex of the soil. Further discussion of the nature of the soil solution we shall postpone until dealing with the mineral nutrient supply in the soil (p. 3649).

The soil mass consists of three components: solid (minerals plus humus), liquid (water), and gas (air), the air content (pore-volume) being the reciprocal of the water content. The latter, in a saturated soil, is made up of four components: free gravitational water, water held by imbibition, water held by capillarity and hygroscopic water. In a soil with a water content at field capacity (p. 3620) the first component is eliminated and the last component, though useful as a measure of the total internal surface of a soil, is too small and too strongly held to be ecologically effective. The water capacity of a soil is thus effectively determined by its capillary potential and by its imbibitional capacity. The capillary potential varies, in any given portion of soil, with the water content, but its value in different soils is determined by the relative proportions of particles in the different size classes, that is by the 'mechanical analysis' of the mineral particles. It can be shown that in a mineral soil the water capacity increases with increasing pore volume, that is to say with an increasing

* Large bodies of undisturbed soil are enclosed by impervious walls and provided with drains to catch the percolating water. Such a structure is called a *lysimeter*.

amount of inter-particle space, up to a maximum beyond which it rapidly drops, for the water in the larger spaces is chiefly free, gravitational water. The extent of capillary movement of water is also governed by particle size. In a pure sand it is maximal with particles between 0·2 and 0·5 mm. Although purely capillary rises of over one metre have been observed, the movement is extremely slow and its contribution to the supply of available water in the upper soil layers is negligible.

Water held by imbibition is a function of the colloid content of the soil, but this consists of two quite distinct entities, the organic (humus) colloids and the clay colloids, which have different water-holding capacities and occur in very varied proportions. There are weathered tropical soils which contain 90 per cent of colloidal clay, while peat from a raised bog may be 99 per cent organic colloids. Organic colloids have a much higher water capacity than clay colloids so that their relative proportions in a soil has an effect on the total water capacity.

An increase of the organic matter in the soil raises its water-holding power to some extent but has very little effect on the amount of available water, owing to its great retentive power. When one considers the small percentage by weight of organic matter in the soil, it is plain that to effect any considerable difference the amounts added would have to be very large. Table 9 gives the figures from an experiment by Veihmeyer (1938) which illustrate the effect of adding well-rotted barnyard manure to a soil.

TABLE 9

Soil		Manure Added, tons per acre			
		0	10	50	200
Fine sand	Organic matter	2·1	2·2	2·9	5·3
	Moisture equivalent	3·2	3·4	3·3	4·4
	Available percentage	2·2	2·2	1·9	2·1
	Wilting percentage	1·0	1·2	1·4	2·3
Silty loam	Organic matter	1·6	1·8	2·5	5·0
	Moisture equivalent	16·1	16·0	15·8	15·7
	Available percentage	8·6	8·5	8·2	7·7
	Wilting percentage	7·5	7·5	7·6	8·0

As the moisture equivalent may be taken to represent the field capacity, the available water is the difference between this figure and that for the wilting percentage. These percentages are calculated on a soil–weight basis, but it is generally accepted now that a soil–volume basis gives a truer picture of the soil–water relationships. In any case it is obvious that there has been no significant gain in available water in spite of the considerable gains in organic matter.

Even differences which appear quantitatively small, when translated onto a field scale may be vitally significant. A plot of the Broadbalk field at Rothamsted has received a dressing of 14 tons per acre of manure each year since 1843, while the next plot has received none. By 1943 the available water in the manured plot had only risen from the equivalent of 2·0 to 2·7 inches of water.

This gain of 0·7 ins. represents, however, 0·36 gallons or 3·6 pounds of water per square foot of soil, nearly all in the top 9 in., which is not inconsiderable.

Important though the water capacity of the soil is, its influence is relative to the operation of other variable or local factors. Too high a capacity involves the exclusion of air in a wet soil, producing the partial anaerobiosis and reducing conditions characteristic of bogs. Under an arid climate a soil with a high water capacity may be a definite disadvantage ecologically, since it can retain all the rain water in the upper layers, thus excluding the development of all deep-rooted trees and woody perennials.

The reciprocal relationship of water content and aeration implies that there may be a point of best balance or optimum water content with respect to any given plant community. In general terms, productivity in a mesophytic vegetation will rise with a rising water content up to about 60 per cent of the saturation value and begin to fall steeply at contents of over 80 per cent. With a water capacity which maintains the average water content at such high values, mesophytic vegetation will give way to bog or moor communities, while if the capacity is too low to maintain an average of 10 per cent only an arid community can thrive.

Furthermore the value of a high water capacity is subject to the proviso of availability to the plant. We have touched earlier (p. 3621) on the matter of availability and the existence of a lower limit of available water or 'wilting point'. Briggs and Shantz (1912) express this value as the *wilting coefficient*, that is the weight of water remaining in the soil expressed as a percentage of its dry weight.

That it is a quality of the soil as such was shown by the observations of Briggs and Shantz that the moisture equivalent of the soil (see p. 3621) divided by a factor of 1·84 ± 0·013 gave good agreement with the wilting point as determined by plant experiments.* Clements coined the term *holard* for the total water content, divided into the *chresard* (available) and *echard* (unavailable) fractions.

Although the wilting coefficient is a property of the soil its determination by tests with plants does involve some variation, as not all plants are equally sensitive to a water deficit in their tissues Knight has shown that in some plants the onset of permanent wilting may occur at a water deficit of only 1 per cent in the leaves, while other plants are resistant down to a deficit level of 8 per cent. To express these differences Briggs and Shantz used the *relative wilting coefficient*, that is the ratio between a single observation and the mean of all the experimental observations made on the same soil. Ratios above unity imply a higher wilting coefficient, that is an amount of water left in the soil which is above the average. The range of these ratios as found experimentally is, in fact, narrow. In a series of 1319 observations by Briggs and Shantz the range of relative wilting coefficients found was between 0·94 and 1·13 and was related more to the comparative development of the root system than to the xerophytic or hygrophytic qualities of the plants.

* The wilting coefficient is also related to the *hygroscopic coefficient*, being equivalent to the latter value divided by 0·68. The hygroscopic coefficient is the percentage of water acquired by a dry soil exposed to a water-saturated atmosphere at 20°C.

The ability to withstand a considerable water deficit in the tissues without permanent injury has a close bearing on the capacity of a plant for xerophytism, since such plants develop a very strong suction force without dependence on exceptionally high osmotic potentials. Water losses of more than 8 per cent are known to be frequent in desert and steppe plants. Thoday has observed deficits of 26–27 per cent in the inrolled leaves of *Passerina* in South Africa. A deficit of this order probably represents the critical saturation deficit at which permanent wilting occurs. In mesophytes it is usually below 25 per cent, but in true xerophytes it may be over 75 per cent. In mosses and some ferns, which possess the power of revival from desiccation, the water content may fall lower still without death resulting. Very few Angiosperms possess this power, but an exception is *Myrothamnus flabellifolia* (Rhodesia) which can recover from a deficit of 93 per cent.

While the transpiration rate of most xerophytes, reckoned *per unit of surface area*, is greater than the average for mesophytes, the former have consistently low values of the quotient *surface/water content*, and on a basis of fresh *weight* their transpiration is below that of mesophytes.

The concept of *physiological dryness* in a soil as distinct from physical dryness dates from Schimper's *Plant Geography*, and it was one of the foundations of his concept of edaphic as distinguished from climatic plant communities (see p. 3325). He cites as three causes of physiological dryness: abundance of soluble salts in the soil, abundance of humic acids, and low temperatures; all hindering root absorption even from wet soils. The idea has haunted ecological literature ever since, though it has been severely criticized. It drew support deductively from the apparently xeromorphic character of many bog and peat-living plants, e.g. Cyperaceae and Ericaceae. Wadleigh has defined the 'total soil moisture stress' as the sum of the hydrostatic water tension and the osmotic potential of the soil water, producing a diffusion pressure deficit which must be overcome before root absorption can take place. This is correct as far as it goes, but it does not allow for the diffusion of salts into the roots, which materially alters the situation and diminishes the importance of the soluble salts as a barrier to absorption. This, of course, can only occur where there is full liquid continuity between the soil solution and the root surface. Under conditions of rapid transpiration, however, a steep moisture gradient may develop in the immediate neighbourhood of the root so that liquid contact is interrupted and in such conditions, as Philip points out, we may speak of a true physiological drought.

The case of plants growing in highly humic soils is rather different. The salt content is too low to have important osmotic effects, and it seems more probable that any restriction of root absorption is due to the effects of humic substances absorbed by the roots (see p. 3643).

Some of the products of anaerobic decomposition in soils devoid of air are definitely poisonous to plants, e.g. Hydrogen sulphide and phosphine. Scheiner and colleagues isolated dihydroxystearic acid from bog soils, which is highly poisonous to plants even at concentrations of 1:20,000, while Lundegardh has shown that an acid medium reduces the permeability of root cells . All these factors tend to exclude most mesophytes from boggy soils and selectively limit the population to a few resistant species. The toxic substances are rapidly destroyed by aeration.

Consideration of the distribution of soil moisture in relation to rainfall and of its geographical effects on vegetation belongs rather to the following section of this book, but there are manifold local variations of water content in the soil which arise from edaphic causes unrelated to rainfall. Such are, diminished drainage and accumulation of humus, or increased drainage due to extensive root penetration of the deeper soil, or to some other cause. As most types of vegetation have a restricted tolerance in the range of soil moisture, any major change in this respect can effectively cause the disappearance of one vegetation type and its replacement by another, in a manner which is quite distinct from the normal evolutionary succession. The latter process is, however, sometimes correlated with changes of soil moisture arising from the reactions of the plants concerned, as in the increase of humus or in the raising of the soil level by the accumulation of silt or sand in a hydrophytic community.

Organic soil constituents

The chief feature which distinguishes a soil from weathered rock detritus is its content of organic matter, derived from the residues of the plants and animals which have lived on or in it and mixed with the mineral soil skeleton. The amount of organic matter is very variable, ranging from as little as 1 per cent to 99 per cent in pure peats. The lowest figures are for arid tropical soils, the highest for ombrogenous peats. Cultivated soils in moist, temperate climates average from 10–15 per cent, but higher values occur under uncultivated grasslands where the upper layers of the soil may contain up to about 40 per cent of organic matter.

The greater part of this matter is derived from plant residues, but a not inconsiderable quantity is contributed by the remains of the vast number of micro-organisms, bacteria, fungi and protozoa, which inhabit the soil, as well as by larger animals and their excreta, worms and insects in particular. It is, however, the decomposition products of green plants that give soil organic matter its chief characteristics.

Plants make two main contributions to the organic matter. One is from their roots, whose decayed remains permeate the soil throughout the zone of root penetration, often called the *rhizosphere*. Grass roots in particular are very abundant and relatively ephemeral and so make an important contribution to the high level of organic matter under grasslands. The other contribution is from 'litter' that is the masses of dead stems, flowers and especially leaves, which fall upon the soil surface and are only slowly incorporated into it. Trees contribute chiefly in this way, with their massive leaf-fall. Their roots are relatively resistant and long lived.

The downward incorporation of this surface material into the soil is largely the work of earthworms. Darwin estimated that in fifty years the whole of the top 25 cm of soil was brought up to the surface by worms. Some of the decomposing material is also carried downwards by drainage water (p. 3620). The latter process is also assisted by worm action in the soil, which increases the amount of large pore space, increasing aeration and drain-age. Dead worms are quite an important source of Nitrogen, as their dry weight is 54 per cent protein. Worm casts on the surface, which vary in amount between 2 and 40 tons per acre also show an increased exchangeability of kations and of Phosphorus, compared with the native soil. All in all, the worm population is a matter of some importance to the soil.

The downward movement of organic matter from the surface is normally very slow and there is a steady gradient of diminishing amounts downwards. The chief exception being in some heavily leached soils, e.g. podsols, where there is a zone of organic deposition in the B horizon. In highly acid soils the absence of earthworms precludes to a great extent the downward mixing process and the organic matter tends to remain as a superficial layer, clearly differentiated from the mineral soil below.

Some of the geologically older rocks contain organic substances or Carbon of organic origin, but most of these substances are stable hydrocarbons with a much higher Carbon content than recent plant remains, which are mostly relatively unstable.

The ultimate result of organic decomposition is to dissipate the substances concerned as Carbon dioxide and water with the release of other elements, N, P, S and kations in inorganic form. This complete disintegration is called *mineralization*. In some conditions it is a fairly rapid process which may be distinguished as *oxidative decomposition*. It is mainly due to microbiological activity and is favoured by free aeration, high temperatures and sufficient moisture. Surplus moisture, on the other hand, by limiting aeration, may inhibit the process. It is also favoured by freely available Calcium, especially Calcium carbonate. For all these reasons the process is most active at and near the soil surface and in hot, humid climates it leads to the rapid disappearance of organic matter from the soil. Cultivation also favours the process and cultivated soils, especially in the tropics, can be very low in organic content unless this is artificially supplied.

The decomposition of organic matter produces large quantities of Carbon dioxide in the soil and the rate at which it is produced is a measure of the rate of decomposition. This gas, in solution as carbonic acid or in combination as bicarbonate, is the most important agent in the weathering and leaching of the soil and in the release of adsorbed kations to solution. Its action may be supplemented by the formation of organic acids, but it remains the principal single agent in soil development.

Oxidative decomposition is destructive of organic matter, but it cannot be clearly separated from other processes in which the original components of the plant remains are transformed into other organic compounds which are relatively resistant to oxidation and therefore accumulate in the soil in the form of dark-coloured, amorphous material to which the name *humus* is properly applied. These processes are covered by the general term *humification*.

In humification, not only are the original components broken down, but chemical recombinations take place which build up new secondary substances, peculiar to humus and unrelated to the original substances.

Some authors make a distinction between the substances directly derived from plant or animal remains and those which are formed in humification and are peculiar to the soil, limiting the term 'humus substances' to the latter. In as much as both classes form one complex and many of the latter class are chemically variable and ill-defined the distinction seems to be more formal than practical.

Humification depends essentially on microbiological reactions and conditions which hinder or inhibit such reactions prevent humification. Thus, low temperatures, acid reaction or exclusion of air lead to the surface accumulation of plant debris in an almost unaltered condition, with the organic structure preserved. These accumulations form peat, in which organic structure is only very slowly lost as depth increases, through the action of fungi or anaerobic bacteria.

Anaerobic decomposition is also characteristic of submerged pond and lake soils, from which it releases methane and Hydrogen as well as Carbon dioxide and forms structureless lacustrine peat. It is a constant process in such soils, but in many bog soils it is intermittent, being limited to periods of flooding when the soil is saturated. A seasonal alteration of anaerobic and aerobic decomposition may give rise to a finely banded or 'varved' soil.

It is as well to emphasize here that the course that humification takes depends almost entirely on the prevailing circumstances and that a difference even in a single factor may cause the production of two very different soils. Climate, in a wide sense, temperature range, humidity range, pH, base-status, mineral texture, the presence or absence of salt or of sulphate are all controlling factors in determining the course of humification.

Highly siliceous soils are deficient both in soluble nutrients and in lime. They tend therefore to be more or less acid and to be colonized by plants which make only small demands on mineral nutrients and are antagonized by lime, the calcifuge plants, such as are most Ericaceae and Coniferae. Under such conditions humification is chiefly brought about by fungal mycelia; acidity increases and the acids further deplete by leaching the nutrient reserves of the soil. Earthworms are absent and instead of the humus being mixed with the mineral soil it remains as a superficial layer, which on dry heaths may be very thin, but in wet places accumulates as acid peat. It is under such conditions of acidity plus high rainfall that podsolization (p. 3604) occurs.

The processes of decomposition are also considerably affected by the constitution of the materials available. Micro-organisms preferably attack those remains which are richest in nutrients. As the leaf-fall from trees is perhaps the greatest element of surface litter in bulk, it is interesting to see what its chemical composition may be. The bulk of tree litter is notably large. Ovington, in a study of plantations of *Pinus sylvestris*, showed that the annual fall of litter from the trees, including leaves, twigs and cones, reaches a maximum at about the age of 20–23 years and thereafter continued at an average level of about 6000 kg/ha/year (dry weight). As he estimates the dry matter production at 17,000 kg/ha/year (including ground flora) this means that about one-third of the annual production was returned to the soil each year. Under heavily foliated deciduous trees the proportion is probably much greater.

The fate of the litter on the ground varies with the tree species. In general the deciduous leaves decay quickly and form only a thin layer on the surface, the leaves of Conifers are more resistant and form a thick carpet. Ovington found that under Scots Pine at the end of a 55 year period, 91 per cent of the total accumulation of litter had decomposed but that the remainder represented approximately the amount of litter contributed to the soil during the last 6 years.

The following figures in Table 10, after Ovington, show the percentages of the oven-dry weight represented by some of the most important elements in

TABLE 10

	K	Ca	Mg	P	SiO$_2$	C	N	C/N ratio
Quercus robur	1·20	1·03	0·22	0·25	0·69	56·15	2·91	
Castanea sativa	1·10	0·67	0·29	0·22	0·27	55·98	2·50	22·5/1
Fagus sylvatica	0·99	0·97	0·19	0·18	1·24	57·17	2·57	
Pinus sylvestris	0·56	0·62	0·09	0·13	0·19	63·71*	1·15*	
Larix decidua	0·48	0·61	0·27	0·22	1·26	56·39	2·16	40.42/1
Picea abies	0·46	0·98	0·11	0·12	1·03	59·12	1·40	
Abies grandis	0·46	0·91	0·16	0·11	0·19	62·77	1·46	

* Figures for *Pinus nigra*.

leaves of various tree species, all of which were of approximately the same (mature) age and were growing on the same soil.

Comparing hardwood with conifers, it will be seen that for Ca and SiO$_2$ there is no consistent difference, but that for K, Mg and N the hardwood leaves are richer than the conifers. There is also a small preponderance in their favour in P content, which is lessened by the abnormally high percentage of this element in *Larix*, which is characteristic of this, the only deciduous conifer included in the table. The C/N ratio is also much higher in conifers.

These constitutional differences in the leaves are associated with differences in the development of the type of humus formed and the morphology of the upper soil layers, described on p. 3599.

The keynote of the process of humification may be said to be conservation. Instead of a fairly rapid degradation of the original plant and animal materials to inorganic substances, a very complex system of chemical recombination takes place with the formation of new components, peculiar to the soil, which are relatively resistant to degradation. These may be collectively called *humin substances*. They accumulate in the soil and the total weight of such substances in the soil may represent the contributions of many generations of the living population. This chemical system is by no means stable, the equilibrium between decomposition and recomposition is dynamic and largely dependent on environmental conditions. At any given time breaking down and building up are going on simultaneously and every stage from unaltered plant materials to highly polymerized humins may be present. Under optimal conditions for micro-organismal activity: high temperature, neutral reaction, high humidity and aeration, the balance is well on the side of decomposition, the materials are quickly mineralized and little or no humus is formed. At the other end of the scale, where micro-organismal activity is inhibited, decomposition is so slow that almost unaltered plant remains accumulate on top of the mineral soil.

The mixture of humins is almost always dark-coloured, due mostly to the development of melanin, derived from protein break-down through the action of tyrosinase, and in part to dark-coloured quinones. The depth of colour varies, and it has been observed that it is blacker where the soil has a high base-status (see chernozem, p. 3603) which may give a false impression of the amount of humus present.

An important feature in humification is the interaction of the organic colloids with the mineral colloids. Adsorption of organic substances on clay particles has a great effect in stabilizing these substances and protecting them from decomposition so that mature humus is best conceived as a *clay-humus complex* rather than as a free body in the soil. It is the particles of this complex which are active, in turn, in adsorbing and holding the important kations which would otherwise be so easily leached from the soil. The clay-humus complex not only protects adsorbed kations, but also some at least of the organic constituents of the humus itself. For example, about 50 per cent of the Phosphorus and about 10 per cent of the Nitrogen in soils is in the form of nucleic acids which are protected against dephosphorylation by adsorption on clay minerals, especially bentonite (montmorillonite). Similarly it has been found that amino acids are protected by clay adsorption. C^{14} cellulose was fed to the soil and after 300 days it was found that 7 per cent of the original C^{14} was present in amino acid form, which represented 30 per cent of the C^{14} still remaining in the soil, the rest having been lost as CO_2. The addition of montmorillonite to the soil decreased this loss and increased the percentage retained, suggesting that the amino acids were protected by adsorption on the clay mineral.

The rate at which the dead material, vegetable or mineral, is broken down, depends not only on the environmental conditions but, as we have pointed out, on the character of the material itself, which varies greatly in its resistance to decay. Hardwood leaves decay more rapidly, on the average, than coniferous leaves, which create a deeper layer of litter in consequence. The silicified rachides of *Pteridium* are particularly resistant. Among woodland trees, *Betula* leaves on neutral sols are the fastest to decay. *Quercus* leaves, on either neutral or acid soils, are much slower and decay at about the same rate on both types of soil. *Betula* leaves lost nearly 90 per cent of their dry weight in six months, *Quercus* leaves only about 20–25 per cent. Such differences may also affect the type of humus resulting; the longer the period of decay the more acid tend to be the products.

The break-down processes are partly biological, by the agency of micro-organisms, and partly physico-chemical, the attack being primarily the hydro-lysis of the higher polymers and secondarily, oxidation processes, degrading the original insoluble substances to soluble monomers. The different rates at which this goes on is shown by figures in Table 11, from Bengtsson and Barthel, for materials added to the soil in measured quantities.

TABLE 11. PERCENTAGE OF ORIGINAL SUBSTANCE DEGRADED

Time	Neutral soil		Acid soil	
	Cellulose	Lignin	Cellulose	Lignin
After 1 year	95	36	79	13
After 4 years	100	63	89	37

About 25 per cent of the degraded cellulose is synthesized into bacterial body-substances.

The rapid mineralization of cellulose, chiefly by *Cytophaga* and other cellulose hydrolysing bacteria, compared with the relatively slow break-down of lignin, implies a rise in the percentage Carbon content of the residues.

The direct products of break-down are sometimes distinguished as non-humin substances, and humification, in the strict sense, consists of the for-mation from this plexus of products, of secondary polymers of increased stability, coloration and insolubility, which are the true humin substances peculiar to humus. The soil organic matter therefore is a dynamic mixture of break-down substances (10–20 per cent of the whole) and secondary humins in all stages of formation. Considering its chemical complexity and its dy-namic character it is small wonder that its chemical elucidation has provided a baffling problem.

The formation of humus is thus not the simple result of the break-down processes but involves secondary re-syntheses; these are accompanied by a reduction of the Oxygen content of the products and an increase in Carbon and Nitrogen. Indeed, the C/N ratio is one of the most useful ways of dis-tinguishing the humin substances formed.

Obviously those original substances which are the most resistant to break-down will contribute most to the organic complex which is the seat of humifi-cation and among such substances lignin figures prominently. It was at one time dubbed the mother-substance of humus and Waksman considered that an adsorption complex of lignin and proteins (adsorptively protected) was its most important constituent. This view was based upon the relative stability of the lignin complex under microbial attack. Not that it remains unchanged. Humus lignin is not identical with xylem lignin; it has, for example, a higher base-exchange capacity. How far this alteration goes in humification is an open question. There is evidence that in some conditions lignin can be com-pletely mineralized. Nor is the protein material in humus derived from plant proteins which are, as a rule, rapidly hydrolysed to amino acids, but from microbial proteins re-synthesized from the break-down products of the plants and animal bodies. This material differs from plant proteins in being resistant to de-nitrification.

Opinion is divided about acceptance of the Waksman scheme. That such adsorption complexes exist in humus is undeniable and also that proteins are involved in them, but how far they are partnered with lignin and how far with a mixture of aromatic polymers derived from lignin is uncertain. That such a complex figures largely in the composition of humus is shown by oxidation with dilute (1–2 per cent) Hydrogen peroxide, which does not attack cellulose. This agent dissolves 70–80 per cent of soil organic matter. This soluble and oxidizable fraction has a practically constant C/N ratio which argues a statistically constant composition, though not necessarily a simple one. The total process of humification must be regarded as a continuous series of steps leading from the initial materials among the non-humin products of decomposition, which are highly oxidized and of low energy content, towards end-products which are stable, energy-rich and of low Oxygen content. These changes are essentially micro-biological and depend on bacterial and fungal metabolism, in the absence of which they do not take place. The final long-term product is probably coal, but a genetic chain from lignin through humins to coal has not been established.

Though, as we have seen above, a large fraction of humins belongs apparently to a readily oxidizable entity, the humins as a whole present a complex of substances to which exact chemical definitions cannot be applied and which are separable only by empirical methods. They are also to some extent interchangeable, so that exact boundaries between one constituent and another cannot be drawn. Although defined compounds such as polyphenols and quinones can be isolated from humus it cannot be stated whether they are distinct parts of the humin complex or are break-down products from more complex assemblages, though the latter is probable.

The classic procedure for separation is extraction of the soil with Sodium hydroxide after treatment with dilute acid to remove carbonates and bases. The dark brown extract is then treated with an excess of hydrochloric acid. A dark brown precipitate is produced and a yellowish solution is left. The fraction precipitated by acid is called *humic acid* (or acids), the solution contains the unprecipitated *fulvic acids*.

Fulvic Acids

The crude solution obtained is always contaminated by varying amounts of non-humin decomposition products of relatively low molecular weights, from which it is possible to separate the high polymer fulvic acids by filtration through activated charcoal.

The fulvic acids, like the humic acids, are polymers of cyclic substances, both with or without Nitrogen, such as phenolic glycosides. They contain less Carbon (below 50 per cent) and more Oxygen than humic acids. The greater part of the Oxygen is in carboxyl groups, hence their relatively strong acid character and their ability to form salts in basic soils. In the free state, in acid soils, they are water soluble and very mobile and play an important part in podsolization. They are mildly reductant and form reversible redox systems with Fe ions as well as relatively insoluble compounds with Fe and Al hydroxides, which accumulate in the B horizon of podsols. In podsols they may form over 70 per cent of the humin substances, in brown earths less than 50 per cent and in black earths less than 20 per cent.

For a long time it has been customary to divide the fulvic acids into two categories, *crenic acid* and *apocrenic acid*, the latter being more easily oxidizable and more readily polymerized than the former. Polymerization with loss of Oxygen, leads over into the humic acid group, with which they stand in a reversible relationship. The differences between the two acids are, however, chiefly a matter of degree and the distinction has been dropped by many workers.

Humic Acids

These are stable high polymers with molecular weights running up to several thousand. Their large molecules are formed of polymerized and condensed ring systems, both isocyclic and heterocyclic, which are interlinked into a three-dimensional lattice. Physically they occupy an intermediate position between true solutes and colloidal hydrosols. They have an important kationic exchange capacity and are responsible for the formation of the adsorptive complexes with clay minerals. They form 'humate' salts. Those with alkali metals are water soluble, but those with Ca and Mg are insoluble, which implies that the humic acids are an important agent in the aggregation of soil constituents, as is well known to cultivators.

Three categories of humic acids are recognized, but they are imperfectly distinguished by their relative solubilities and precipitabilities. The first category is that of *hymatomelanic acid* (sometimes called *ulmic acid*. Ulmin is an insoluble form). These are soluble in alcohol and extractable from the acid precipitate of humus formed from an alkaline humus extract. The remainder, insoluble in alcohol, consists of a mixture of Brown humic acid and Grey humic acid. If redissolved in an alkali, they can be separated by the addition of an excess of an electrolyte. The Grey humin is precipitated, but not the Brown humin.

None of these substances is a true chemical entity. They form a series with a progressive increase in molecular weight, degree of polymerization, stability and insolubility, with the hymatomelanic acids at the lower end of the series and the Grey humins at the ultimate and most stable end.

Hymatomelanic acids are yellowish-red and are chiefly associated with decaying wood and indeed are not far removed, chemically from lignin itself. They have been called ligno-humic acids. They occur chiefly where aeration is restricted and are abundant in the humus type called *moder* (see below).

The Grey and Brown humic acids are really two groups of substances separable only by quantitative characteristics. For example, the Grey humins have a C content between 58 and 62 per cent, the Browns have a C content of less than 50 per cent. Under electrophoresis on paper the Browns move more rapidly than the Greys and are thus partially separable. The colour difference is not so great as the names imply, the Greys being rather blacker than the Browns. The Greys are found chiefly in base-rich soils and in the A horizon of podsols. The Browns are chiefly in acid soils and in the B horizons of podsols.

A biological and functional distinction can be drawn between two grades of humus, namely, into *persistent* (*dauer*) humus and *alimentary* (*nähr*) humus. The former title covers all the higher polymers which are the most durable constituents and the slowest to yield to chemical and microbial attack, such as lignin. This fraction is chiefly of importance for its water-holding and adsorptive capacities. In the latter respect it may have double or more the capacity of the mineral soil constituents. In sandy soils it increases the water-holding power of the soil, while in heavy clay soils, on the other hand, its aggregating properties increase the granulation of the soil and increase its aeration and permeability to water.

Some actinomycetes and fungi can attack even the stable polymers, but as a direct source of either C or N compounds for higher plants it plays probably only minor part.

The alimentary humus is that fraction which is freely used by micro-organisms as a source of C and N in nutrition. It is slowly but continuously mineralized and from it come the intermediate and soluble substances which are metabolized by plants of all classes. It has been estimated that 3000–6000 kg/ha are consumed every year in cultivated soils. By-products of this decomposition (apart from heat), which are of great importance, are CO_2 (see p. 3632) and many substances capable of forming chelates with mineral ions and preserving them from precipitation as insoluble compounds.

Chelation is largely the work of organic acids (though sucrose is also active), which are produced by microbial metabolism or may, in some plants, be excreted from roots. Their action is chiefly on Al and Fe, by which these ions are retained in solution and may become available for absorption by the dissociation of the chelate. Moreover these ions are thereby prevented from precipitating Phosphorus as insoluble phosphates and thus depleting the soil of available phosphate.

One of the most conspicuous deficiency symptoms in higher plants is *chlorosis*, which signifies the failure of chlorophyll formation, leaving the plant a prevailing yellow, carotenoid colour. Photosynthesis is thereby greatly reduced. The effect is usually due to an Iron deficiency in the leaves. As the absolute Iron content of soils is almost always sufficient it follows that the Iron is present in insoluble combination and the addition of inorganic Iron compounds to the soil is useless to combat the condition. Chlorosis is most common in soils with a high pH, associated with a high lime content.

Soluble chelates, owing to their high stability, retain the Iron in soluble form and prevent its precipitation so that they are available to combat chlorosis. A suitable Iron chelate is formed by Potassium ethylenediamine tetra-acetate, which has proved of value in horticultural practice.

There are two possibilities of action. Either the Iron is released by ionic exchange or the chelate is absorbed as such by the roots and the Iron released by a metabolic breakdown of the organic complex. The latter appears the more probable. Its molecular weight is below the maximum (800) for penetration of the root membranes and its low dissociation coefficient makes it unlikely that the Iron would be liberated without the breakdown of the complex. Furthermore, the dissociation of the complex increases with rising pH and in these circumstances the Iron is, in fact, released and precipitated as ferric hydroxide; the addition of the chelate to the soil thus failing in its purpose. Humic acids can also form soluble compounds with Iron and may play a part as natural soil-chelates keeping Iron available. In keeping with this is the fact that chlorosis rarely occurs on humus-rich soils with a low pH.

The total of organic matter in the soil is made up of humified material and the non-humified plant and animal remains. Humins can be estimated quantitatively by extraction with cold alkali and subsequent fractionation, as already described on p. 3637. For the estimation of the total organic matter there is no simple and accurate method. Resort may be made to complete proximate analysis, but this is too long a process for routine work. The classical method is to estimate the loss of weight on ignition of a sample to red heat, but in this there are several obvious errors, such as the formation of oxides, which increases the residual weight, and loss of combined water and of CO_2 from carbonates, which decrease the residium. Carbonates can be eliminated by pre-treatment with hydrochloric acid and the amount of combined water can be estimated by distillation from a sample, and these corrections increase the accuracy of the measurement, which is useful as a rough comparative guide. In soils with a very high humus content the method is indeed reasonably accurate, but the method now generally used is to estimate the amount of organic Carbon present by oxidizing it by either a dry or a wet combustion method and multiplying the result by a factor of 1·724 to give the corresponding weight of organic matter present. This conventional factor is based on the average Carbon content of the principal substances present, i.e. about 58 per cent. Jenny uses the total Nitrogen content as a similar index of the organic matter. In cold climates the Humus/Nitrogen ratio is about 12:1, in temperate climates it is 10:1 and in hot climates 9:1.

Under natural conditions the humus in the soil varies in character with the prevailing nature of the mineral skeleton and with the climatic conditions of the locality (see p. 3633). The type of vegetation is also a very influential factor. Temperature, soil moisture and the base status of the soil all serve to modify the biological activities in the soil, with corresponding effects on humification. The raising of all three factors within limits tends to promote biological action (and vice versa), with a corresponding effect on the rate of humification. High temperature and humidity in the tropics may, in fact, carry decomposition further and result in almost complete mineralization.

Microbiological action may be depressed either by low temperatures, low soil humidity, or conversely by excessive humidity with a concomitant deficiency of aeration and poor drainage, or, lastly by a low base status, associated with a low pH. Under favourable conditions the decomposition of litter is chiefly due to the actions of living organisms in great variety, from mice and worms downwards to bacteria (see p. 3679). Under depressing conditions, in cold, wet and base-deficient areas, decomposition is consequently slow and the result is a superficial layer of blackish humus containing a good deal of undecomposed litter and having an acid reaction. Bacteria are largely replaced by fungi, whose abundant mycelia ramify throughout the humifying material. This is called *raw humus* or *mor*. It is highly colloidal and has great powers of water retention. It holds back both water and Oxygen from the subjacent layers. Furthermore, its colloidal and acid character inhibits the precipitation of the hydroxides of Fe and Al and of Silica sols, so that these remain mobile, a very important factor in the development of podsols with their marked layering in the soil profile. A notable characteristic of mor is that the humus layer is sharply cut off from the mineral soil and rests upon a compacted and little disturbed surface of the latter. This is due to the absence of worm action which is chiefly responsible for mixing in neutral soils.

There is, however, a reciprocal dynamic equilibrium between the humification process and the local vegetation. The latter affects the humification factors in various ways, but above all through the chemical qualities of the litter. This has been arranged chemically by Hesselman into five categories, varying from those with an excess of acidic and a minimum of basic buffer substances, to those with an excess of the latter and a minimum of the former.

In the most acidic category is the litter from northern conifers and from *Calluna* and *Vaccinium*. Consequently it is under such vegetation that the best developed podsols and mor are found. Most of the deciduous forest trees yield litter which is only moderately acidic and rich in basic buffer substances, e.g. *Betula*, *Fagus*, *Salix*, and *Fraxinus*, while *Quercus*, *Larix* and *Acer* are rich in both acidic and basic substances. Most basic are *Carpinus* and *Ulmus*. Reciprocally, once mor formation is established, it tends to exclude all but the most acidic species of the higher plants and thus becomes self-perpetuating.

In contrast to mor the humus formed under temperate climates and in base-rich soils is neutral and adsorptivity saturated, with Ca ions prominent. This product is called *mild humus* or *mull*. In colour it varies from light to dark brown and it is intimately mixed with the mineral substratum, though slight eluvial zonation may occur. This constitutes the typical temperate brown-earths, characteristically deciduous forest soils, and their agricultural derivatives. Mull soils are productive, granular in texture and rich in micro-organic life. They can absorb large amounts of water without becoming gelled or 'puddled'.

Despite the opposite characters of mor and mull in their typical states they are connected by intermediates and a change in the vegetation may cause the change of one type into the other, even on the same soil. Similarly, Gorham has shown that local topography may influence the equilibrium in one direction or the other on one and the same soil. The dominant vegetation is sometimes decisive. Thus soils under *Fraxinus* are always mull soils while those under *Calluna* are almost always mor soils, but under *Fagus* either type may develop. It has been suggested that the development of mor under *Fagus* may be due to the growth of a subsurface stratum of fine roots which mechanically isolates the litter layers from the mineral soil and prevents its incorporation. This is also true of other plants which have a densely interlaced root-system. When mor develops under *Fagus* it prevents regeneration and the forest may be replaced by *Calluna* heath. Indeed the whole question of the relationships between root systems and soil development, especially microorganismal activities, deserves further investigation.

A chemical factor which has been found practically useful in distinguishing soil types is the Carbon/Nitrogen ratio. As might be expected this is highest in mor, from 20 to 50, while in mull it is below 20 and in lime-rich soil may be as low as 10.

A third type of humus, which unites some of the characteristics of each of the others, is called *moder*. Like mor it is acid, but it is also mineral rich like mull and transitions to true mull occur. Like mull it forms a comparatively deep layer, but it is largely colloidal, not granular and it remains unmixed with the substratum. Its C/N ratio is intermediate, round about 20. In well-drained areas moder and transitions towards mull may form considerable layers above the mineral soil, on which they have little influence. On calcareous soils this condition is called a *rendzina* and on siliceous soils, a *ranker*.

Foresters have adopted a terminology for the uppermost soil layers (corresponding to the A_0 and A_1 layers) which relates them to the process of humification. First comes the *L layer* or layer of free litter on the forest floor. Second, lies the *F layer*, or fermentation layer; this consists of the partly decayed and partly eaten fragments of litter, which are still recognizable. The colour is dark brown and the mass is interwoven with fungal mycelia and contains a considerable amount of insect droppings.

Both these layers are easily permeable to rain and lose soluble materials pretty rapidly, the loss of weight in one year varying from about 25 per cent (*Fagus*) to about 40 per cent (*Quercus*), the rate of loss being proportional to the mean summer temperature. These soluble materials include carbonic acid and several organic acids, which are highly important in leaching substances from the lower layers.

Third comes the *H layer*, the humification layer, in which the secondary synthesis of humins chiefly occurs and true humus is formed. The thickness of the *L layer* varies with the type of vegetation. It is greater under conifers than under hardwood trees, whose leaves decay rapidly. Under mor forming conditions F is usually thick (under *Pinus* and *Picea* it is the major layer of the three), H on the contrary, is thin and strongly leached by acids from F. Under mull-forming conditions, L is usually thin and F well developed, but H is the major layer, though it may be so well mixed with the mineral soil as to be hardly distinguishable as a separate layer. In moder soils F may be absent and is always inconspicuous compared with H. The H layer may sometimes be so compacted that it forms an impermeable bed and initiates bog or pond conditions.

Under the influence of von Liebig it was for long held as a principle that plants did not, or could not, absorb organic substances, but from 1930 onwards many workers have given contrary evidence and it now seems certain that products of humification, notably quinones and phenols are not only absorbed by the roots of autotropic plants, but that they may have powerful influences on the plant in regard to respiration, growth and the uptake of inorganic ions. Not only relatively simple substances but complex molecules may be absorbed, though the physiological mechanism involved is not clear. A molecular weight of about 800 appears to be the limit for root absorption. Below this limit substances of relatively low molecular weight; e.g. Penicillin (mw 330) may be translocated intact and appear in guttation drops. This has also been shown for thymoquinone and thymohydroquinone both of which can polymerize to brown 'humins' when oxidized. They cause accelerated germination in cereals, increased root growth and increased dry weight of seedlings. Oxyanthraquinones are active as growth stimulants and penta-oxyanthraquinone is seven times more active in this respect than auxin, but only with a supply of NO_3 Nitrogen. It is supposed that the quinone activates nitrate reduction and improves N nutrition.

Larger molecules, such as streptomycin (mw 581) may be absorbed, but are not transported, or at least not freely, and are probably broken down in the roots.

In line with the above observations are others which show the excretion of organic substances from the roots, which can affect other plants (see Alleopathy, p. 3369) and particularly the bacteria in the surrounding soil. There is here a reciprocal action, for the organic substances absorbed by the plant are the result of microbial action, while organic substances excreted by the plant encourage bacterial development, to such an extent that the bacterial population around roots can be 100 times as great as that outside the rooting zone (see p. 3691).

The forms of humus we have described above are all terrestrial, but a form which is only semi-terrestrial is that known as *peat*, which is characteristic of more or less waterlogged areas.

Peat is a growth indigenous to wet, cool, oceanic climates and in such conditions it reaches its greatest extent, although continental peat areas occur in Canada and Siberia and a few areas of true peat with high acidity, pH 3·0 or less, are known in the tropics. Peat has probably always been a feature of soil formation, but it is too easily eroded to have left fossil remains from earlier ages, unless coal be regarded as such. Interglacial peats are known and the existing growth of peat began in the Late Glacial and has continued intermittently ever since. Some deep basin peats, in fact, extend back in time to these early beginnings.

The essential condition for peat growth is waterlogging, which cuts off the air supply and reduces micro-biological action to an extent that permits the accumulation of partially decayed plant litter. It would therefore be classed as a *hydromorphic* soil. There are two main classes of peat: acid peat and fen peat. Acid peats are more plentiful and reach greater depths. They are formed over siliceous rocks with a very low content of soluble minerals. They have themselves a very low mineral content and a very high proportion of colloidal organic material, but in this respect it is possible to draw a fairly clear distinction between the basal layers, formed in contact with the ground water, which are described as *terrigenous* and the upper layers which are above the level of ground water and owe their water content solely to rain, which are called *ombrogenous*. The highly acidic character of this peat is strongly anti-bacterial and promotes its accumulation so that it may achieve depths of many metres.

Fen peat, on the other hand, is formed in contact with base-rich ground water, with which it is saturated, and is neutral or alkaline in reaction, with a pH of 5–8 as against a pH 4·5 or less in acid peat. In fen peat, therefore, waterlogging is the determining factor, not base deficiency or its concomitant, acidity. If fen peat grows above the water level its surface is leached by rain and it is possible for *Sphagnum* and other acidic plants to establish themselves to start a growth of acid peat over the fen. Not a few basin peat beds in typically acidic regions are found to rest on a foundation of fen peat.

Peat beds have been objects of interest to man from early times, partly as fuel and partly for the sake of the tree trunks embedded in them. A variety of popular names for waterlogged areas grew up, without precise definition, which the ecologist has inherited; such as bog, marsh, fen, swamp, mire, moss, moor.

Tansley has proposed to give these names a more exact connotation, to the great advantage of ecology. The following is his scheme:

Marsh. An area with a waterlogged mineral soil in which the water level is not normally above the soil surface.

Swamp (also *Mire*). Like a marsh, but with a summer water level which is normally above the surface. Commonly overgrown by *Phragmites*, *Scirpus* and other tall reeds and rushes and then called *Reed swamp*. May develop into marsh if the water level falls or the ground rises by periodic silting, and then reeds dwindle away.

Fen. Where the ground is waterlogged, but is not subject to periodic additions of mineral silt by flooding, an organic peaty soil develops, saturated with water rich in soluble minerals. If this layer develops above the ground water level it may be invaded by acidophilous species, which in time create a superincumbent acid peat. Unlike acid peat the fen peat can be transformed by drainage into a fertile agricultural soil; in East Anglia the formerly very extensive fens, fed by water from the chalk, have largely disappeared since the eighteenth century draining operations. The fen peat being very spongy, however, drainage causes it to shrink, the soil level subsiding over long periods at rates of from 26–77 mm/year. The fen reserve at Woodwalton in Cambridgeshire is now well raised above its drained surroundings and has had to be diked all round to hold in its water supply.

The surface of fen develops unevenly, with different effects on the vegetation of different levels. Where the surface stands at about the level of the winter water-table there often develops a type of woodland known as *fen carr*, dominated by water-loving trees such as *Salix*, *Alnus*, *Frangula* and *Rhamnus*, which may develop a rich undergrowth of marsh species and may, by the invasion of *Quercus*, pass over into typical Oak forest.

While the marsh and fen vegetation is classed as *eutrophic* expressive of the adequate supplies of minerals available, the various types of acid peat are classed as *oligotrophic* on account of their low content of mineral matter. The same terms are applied to the respective types of ground water.

Bog (*Moss*). Tansley used these as synonymous terms for acidic peat beds generally, however produced. The term *moor* he rejected, partly because of its general use in this country for high level areas of heathland and partly to avoid confusion with the similar German word *moor*, which is applied indifferently to any area of deep peats, whether acid or alkaline, at any level.

Tansley discriminated three main types of bog, topographically distinguishable. (See also p. 3610.)

Valley bog develops on the flat bottom of a valley or in depressions where water stagnates (basin peat). It may start either as fen or as bog, depending on the supply of mineral salts available, but it is typically less acid than other types of bog and supports fen plants as well as plants of acid peat.

Raised bog (German, *Hochmoor*) may develop from either valley bog or valley fen. Its chief characteristic is its convex surface, the better growth of *Sphagnum*, especially *S. imbricatum*, in the middle producing a hump from which the surface slopes gradually towards the edges. These bogs begin to form when the growth of peat has blocked the original water flow, thus spreading the flooded area. Some of the water finds its way round the edges of the peat, so that the bog is typically fringed by a margin of marsh, called by the Swedish name of *lagg*. These bogs are essentially topographical features and their extension is limited to the valley or basin in which they began. Their upper portions are raised far above ground water level and are purely ombrogenous with a low mineral content. If a section of such a bog is analysed, it is found that the change over from terrigenous to ombrogenous growth is marked by a simultaneous decrease in Si and Al. Other bases diminish much less and more gradually, because even rain water brings some soluble minerals with it.

Raised bogs present a certain danger of 'flowing', when the peat is supersaturated with water and becomes semi-liquid, flowing out catastrophically over the surrounding land.

Blanket bog is a term applied to peat growth which is not topographically limited but depends entirely on a very heavy rainfall. It covers the country indiscriminately like a blanket in places where heavy rainfall is combined with acidic rocks, notably in western Scotland and the west of Ireland. It does not reach the depths attained by basin peats and most of its growth has been relatively recent, as attested by archaeological remains found in it.

The post-glacial development of peat has been by no means uniform as is shown by the 'recurrence' layers, including the *grenzhorizont*, which mark old eroded surfaces where peat growth had stopped. The onset of the wet Sub-Atlantic period marked the most recent rejuvenation. Blanket peat

appears in zones Vlla and Vllb (see Vol. 3, p. 2602), and much of its development stems from the climatic deterioration which took place at about 600 BC.
Blanket peat was responsible for overwhelming the remains of the Sub-Boreal
forests, a process which continued until very recent times. At present there
would seem to be a recession in its activity, since it shows extensive erosion in
many places. Erosion causes dissection of the peat blanket by drainage channels
which cut down to the underlying rock, separating the peat into tabular
masses with bare sides and a mat of vegetation on top. This may well be a
natural and inevitable consequence of peat development on any drainage
slope. Water escapes at first along rock channels beneath the peat, which, by
the caving in and washing away of the overlying peat, gradually work their
way to the surface forming open gullies in which the water-flow follows its
natural tendency to cut back into its gathering ground. While this sequence of
events seems to prevail largely in wetter areas, there can be little doubt that
if the growth of peat is checked and its water content reduced, either by a
natural change of climate or by artificial drainage, the shrinking and cracking
of the peat will open drainage channels which lead to erosion and dissection
into a pattern very like that first described.

Valley bogs do not become dissected in this way, but are liable, as mentioned previously, to bursts or flows, when the mass of contained water turns
the peat into a fluid, the weight of which cannot be restrained by the surface
vegetation. Much of the peat then flows off, leaving a bare eroded surface
behind. The widespread draining of such bogs and the consequent establishment of a heathy vegetation, much firmer than the earlier *Sphagnum*, has
greatly reduced the liability to such outbursts.

An almost opposite set of circumstances, of which Pearsall has recorded
some examples, occur where accelerated erosion of higher slopes has brought
down mineral gravel or sand which has been incorporated into the peat surface and in some cases has actually buried and thus conserved peat beds under
a layer of stony gravel.

The flowing of liquified peat is a special case of soil-flow or *solifluction*
which chiefly affects mountain soils. Freezing of the abundant soil water
causes expansion or 'heaving' of the soil, which becomes very spongy, and
when it thaws the semi-liquid soil flows, even on very gentle slopes, forming a
succession of small terraces with the vegetation carried in front of them
leaving bare soil above. The heaving of the soil also tends to expel stones to
the surface and to produce the stone polygons or stone stripes which are a
conspicuous feature of arctic soils, the pattern assumed by the stones depending on the degree of slope.

Humification and the formation of a humus soil may also proceed under
water in lakes, ponds or slow rivers. The chief difference from terrestrial
humification is the prevalence of anaerobic decomposition, with the liberation
of methane and Hydrogen as well as Carbon dioxide. The character of the
product depends chiefly on two factors; the nature of the plant materials
available and the amount of mineral silt supplied by drainage from the land
surface. Here we can trace a series of conditions, beginning with the earlier,

submerged stages of fen peat formation, supplied abundantly with the remains of reed-swamp plants. It receives relatively small additions of silt but is well supplied with minerals in solution.

A widespread type is that called by the Scandinavian term *gyttja*, which we may simply call lake-mud, although the Swedish term originally applied to alluvial soils derived from lake deposits. Deposits which can be classed under this heading consist of a loose admixture of black humus with varying, but always considerable, amounts of mineral silt. An extreme type of deposit is that called *dy*, which is an amorphous, peaty substance precipitated from humins dispersed in water. It is very poor in Oxygen and in minerals and devoid of plant remains.

The study of submerged soils is of vast practical importance in connection with rice culture, on which so many millions of the least affluent human beings depend. The conditions in paddy-soils therefore figure largely in such studies.

The environment of a submerged soil is so different from that of terrestrial soils that quite different considerations apply. Perhaps the most obvious difference is the restriction of the Oxygen supply. The rate of Oxygen diffusion through water is only about $1/10,000$ of the rate in air. The same factor which limits the access of Oxygen to the soil also limits the escape of Carbon dioxide from it, so that there are two effects involved; shortage of Oxygen and excess of Carbon dioxide. Both these factors have direct effects on the plant life as well as on the soil. In the latter an excess of CO_2 promotes leaching and the deficiency of Oxygen promotes anaerobic decomposition, which decreases the CO_2 output and brings about the liberation of Hydrogen and methane, with sometimes considerable quantities of Nitrogen and Hydrogen sulphide and other noxious products.

More important because more widespread, is the indirect effect of changing the oxidation-reduction (redox) potential in the soil. On the oxidizing or reducing state of the soil depend several important chemical factors, for example, the balance between the oxidized and the reduced state of some substances, which may have serious consequences. The redox potential, measured with a bare platinum electrode in a wet but well-aerated soil may be of the order of $E_7 = 0.4-0.5$ volts. In a waterlogged soil it is < 0.3 V. In other words the waterlogged or submerged soil will be predominantly reducing. The critical potential for different redox pairs of ions varies, however, and as the potential of a soil rises with a lowered pH, the potential readings must be adjusted to neutrality, pH 7.0, for comparison. At this pH the critical potentials for the most influential pairs are as follows: nitrate/nitrite, $0.45-0.40$ V; nitrite/ammonia, $0.40-0.35$ V; ferric/ferrous iron, $0.30-0.20$ V; sulphate/sulphide, $0.10-0.06$ V (Mortimer). At the lower potential in each case the reduced member of the pair will prevail and at still lower potentials may be exclusively present. These changes depend on complex conditions and are partly controlled by micro-biological reactions and partly by the humus present acting catalytically. (See also p. 3729.)

The change from ferric to ferrous iron means a great increase of solubility of the Iron compounds. Ferrous ions thus become freely diffusible and are freely exchangeable with other kations in the soil colloids. The same conditions apply to Manganese. Thus the establishment of reducing conditions increases the base-exchange capacity of the soil, to the consequent enrichment of the water. The periodic flooding of rice-paddy soils has this important result, which is particularly valuable in view of the high iron requirement of the rice plant. The accumulation of ferrous and Manganese hydroxides also raises the pH of the flooded soil, which is beneficial to the action of anaerobic bacteria on the organic matter.

The reduction of sulphate yields principally Hydrogen sulphide and methyl sulphide, both of which are toxic and may be dangerous unless sufficient ferrous ions are present to convert them to ferrous sulphide. Where enough organic material is present (of the order of 20 per cent) to promote very active anaerobic bacterial action, the reducing zone of the soil becomes typically black with the accumulation of this ferrous sulphide.

For all practical purposes the series nitrate–nitrite–ammonia acts as a single chain and is readily reversible. Thus nitrite in the soil is essentially a transitory intermediate and only under special conditions can it accumulate to a toxic concentration. All three substances can, however, be reduced to gaseous Nitrogen by denitryfying bacteria, which is a dead loss to the soil unless it be fixed and recaptured by other bacteria or by blue-green algae, the latter being sometimes very active in this respect in tropical waters. Denitrification requires anaerobic, i.e. reducing, conditions and a low pH.

It will be realized from what has been said, that submerged soils are subject to more loss by leaching than terrestrial soils. Under a static pond the leached ions remain available to plants in the water, but if the soil has a downward drainage or if the water is periodically drained off, as in rice culture, the losses may be serious and the soil becomes degraded unless artificially replenished. Additionally there is the danger of the release of some product in toxic quantities; Iron, Manganese, nitrite, ammonium, Hydrogen sulphide, are all potentially dangerous in excess and the maintenance of a safe balance, especially in rice culture, requires considerable experience and local knowledge.

Under the general conditions of Oxygen deficiency, oxidizing conditions can occur only at the soil surface, where a very thin oxidized layer, which may be only a few millimetres thick, sometimes displays the rusty colour of ferric hydroxide. Below this is the deep anaerobic layer, bluish to black in colour. If the lake-mud overlies a permeable subsoil we may find a deep-lying oxidizing region, below the anaerobic zone and in such circumstances ferrous ions percolating downwards may be precipitated as ferric hydroxide to form a hardpan layer, analogous to that in a podsol. In the anaerobic layer there may also occur rusty streaks and patches of ferric hydroxide where local oxidation has taken place. These are frequently associated with roots, which can only function in an anaerobic medium by virtue of the Oxygen conveyed from above through internal channels. Oxygen or oxidants diffusing from the roots may produce a rusty ferric zone around them. (See gley, p. 3609.)

Without pretending to explore fully the very complex and distinct conditions in submerged soils, it will be evident, from what we have said, that the processes active in a submerged soil are in many respects opposite to those in a terrestrial soil. It must be remembered, however, that in a terrestrial soil, where the downward drainage is impeded, waterlogged and anaerobic conditions may obtain in the deeper layers which reproduce substantially the situation in submerged soils, although the upper layers of the soil remain well aerated. (See also p. 3675.)

Mineral nutrients in the soil

The *soil solution* is the term for all the water held in the soil against gravity, with all its content of dissolved electrolytes and gases, more or less in chemical equilibrium with the soil solids. All the elements of plant nutrition, with the exception of Carbon and some of the Oxygen and Hydrogen are absorbed from the soil solution.

Although most of the Nitrogen supply is derived from the air the higher plants can only absorb it as dissolved compounds from the soil solution.

Trace elements are required in such extremely small amounts that deficiencies of one or more of them are exceptional (see below) and the same is true of the major elements although variations in the amounts of the latter are very considerable. The sum of Ca and Mg together outweighs the others in all soils, but in those which are highly saline the Na and Cl together may run the others close. The soil solution is in most cases very dilute, the concentration for Ca, Mg, Na and K together averaging between 0·02 and 0·05 N, with lower values only in barren sands and higher values only in saline soils. The variable concentration of the soil solution has osmotic effects which we shall touch on later. Concentrations of kations in the ground water at water-table level may be considerably lower than those in the upper part of the soil.

The factors affecting the concentration of the soil solution are many. The most obvious factor is the water content. The concentrations of single ions varies inversely with the water content, though not always in strict proportionality, since changes in concentration may bring in side-effects on solubility and exchangeability. For example, the concentrations in solution of Ca and Mg in carbonate soils are affected by the partial pressure of CO_2 (and consequently by the pH) and also by the total salt concentrations. The balance of exchange of ions with those bound by the soil colloids will also have an effect on the concentration of ions in solution. Variation of solubility with general concentration has also been observed with phosphate anions. Additions of neutral salts to the soil, i.e. increased salt concentrations, decreases the solubility of phosphates if the pH is above the isoelectric point of the soil colloids (about pH 4·5), but below this pH the solubility of phosphate is increased, as at such levels anions from other salts are adsorptively exchanged for phosphate ions.

Mineral ions exist in the soil in several different conditions, which may be categorized thus: *readily available*, ions in solution or exchangeable from surface adsorption on particles of the clay-humus complex (p. 3635). *Slowly or partially available*: ions in combination in the organic matter. *Non-available*: ions held internally in the lattice of the colloid particles. These may become available if the particles are broken up, mechanically or by further weathering. Progressive organic decay and mineral weathering are indeed essential in wet, temperate climates to maintain the supply of available ions, as the adsorbed, exchangeable ions tend to be progressively replaced by H ions from the H_2CO_3 in the soil water. This leaching process, if unbalanced, would eventually lead to the soil becoming wholly acid and devoid of available mineral nutrients. This is also called base-desaturation, the opposite of the condition of base-saturation which is the ideal for soil fertility. (See also p. 3616.)

The ions differ in their representation in the above categories. Thus Cl and NO_3 are almost wholly in solution. On the other hand, K is largely held unavailable in the colloid lattice. The same is true, naturally, of the elements of the clay minerals which actually constitute the colloidal particles. In total, a large proportion of the elemental content of any soil is in non-exchangeable form and thus chemical analysis shows only the total mineral resources of the soil and is not evidence of its actual fertility.

The base-exchange (kation exchange) capacity of a soil may be measured by taking advantage of the high bonding-energy of the NH_4 ion and extracting the soil sample with ammonium acetate, which displaces all the absorbed basic ions. These appear in the filtrate as acetates and may be separately estimated, while the loss of NH_4 from the extracting solution gives the total exchange capacity. While this is a useful value it does not tell us anything about the actual equilibrium between the soil solution and the adsorption complex, which is governed by the activity of the ions.

In this respect the metallic kations generally follow the classic 'lyotropic series', which depends on the relative radius and hydration of the respective ions and their valency, increasing with the latter. Thus we find that at low concentrations the adsorptive energy of the ions increases in the order Na—K—Mg—Ca, the divalent ions being more strongly adsorbed than the monovalent, and this difference increases with dilution, that is with soil moisture, so that the ratio of $Ca+Mg/Na+K+H$ absorbed is greater the more soil water is present. At higher concentrations K behaves anomalously and moves up the series, which may be due to its entry into the colloid lattice. Heavy K manuring of a soil may therefore deplete a soil of exchangeable Ca.

The equilibrium also depends on the exchange capacity of the clay minerals present. Clays with a high exchange capacity, such as montmorillonite, adsorb a much higher proportion (22/1) of Ca, and other divalent or polyvalent ions, than of K; while kaolinite, with a low exchange capacity, has not the same affinity for Ca and the ratio Ca/K is only 5/1. Not only are the divalent ions taken up more readily but they are held more firmly by the clays with high exchange capacity, so that their ratios of release to the soil solution are in inverse order. That is to say that a kaolinite clay releases more Ca in relation to K than does a montmorillonite clay, thus making more available for root absorption.

These facts have a great influence on the power of a soil to retain its resources of mineral ions, since, unless they are adsorbed by the soil colloids, either mineral or organic, the ions are liable to be leached out and disappear in the drainage.

Release of ions to the soil solution may also occur by dissociation of simple compounds and this affects particularly the carbonates of Ca and Mg. The clay-silicates, on the other hand, only release their non-exchangeable reserves, chiefly of Ca and K, by weathering, which decreases particle size and increases solubility. Acid clays, i.e. clays with much exchangeable Hydrogen, may also themselves promote the weathering of the coarser particles. All the familiar leaching factors: acidity, rainfall and temperature contribute to weathering, though chemically the reactions in the particles are very complex and the clay-minerals differ greatly among themselves in regard to their susceptibility, felspars and micas being on the whole less stable than the zeolites such as kaolinite, which are, surprisingly, more stable than quartz. As weathering progresses the proportion of the more stable compounds in the clay increases and the release of non-exchangeable ions diminishes. Very prolonged weathering thus inevitably tends to eventual depletion of the soil's soluble reserves.

We have already touched in Vol. 3 (p. 2914) on the absorption of ions by direct contact between roots and clay particles, which is, in a sense, a form of weathering, since the root surface may be highly acid.

Like the clay colloids the organic material in the soil holds ions as either: (1) Soluble compounds, which contribute to the soil solution. (2) Ions adsorbed in exchangeable form. (3) As constituents of compounds derived from the living plants, e.g. phytin, nucleic acids, etc., or as secondary metal-organic complexes. Most of these are relatively stable and insoluble, so that ions from them only become slowly available; they are non-exchangeable.

The colloids of mor humus are predominantly negatively charged. In mull humus there is a balance of negatively and positively charged particles and the humus as a whole is therefore amphoteric. Positively charged particles will adsorb and hold anions, nitrate, sulphate, phosphate, chloride, etc., in exchangeable condition. Among the inorganic colloids this property is only paralleled by the positively charged ferric hydroxide and aluminium hydroxide which are largely responsible for the fixation of phosphate.

The acid reaction of mor humus is generally attributed to its preferential adsorption of basic ions, whereby H ions are released to solution. We should not, however, overlook the effect of the water-soluble fulvic acids, formed in humification, which probably also contribute to the acid reaction.

The concentration of ions on the adsorptive surfaces of the soil is much higher than in the soil solution, thus constituting a nutrient reserve which exchanges ions with the surrounding solution whenever the concentration equilibrium between them is disturbed. The adsorptive binding of ions of higher valency and lesser mobility acts, however, as a resistance to root absorption and the amounts absorbed by roots from soils of such ions as Ca and Fe are less than those absorbed from watery solutions of equal ionic content. Indeed, were it not that the adsorptive capacity of protoplasmic colloids is much greater than that of the soil colloids, it would be difficult for roots to overcome the competition of the latter for ions. However, we cannot here follow the ions into the root itself, interesting as are the problems involved. (See Vol. 3, p. 2899.)

Apart from the nutrient ions, many other substances are adsorbed by the soil colloids. The heavy metals are very strongly held, which applies to most of the trace-elements, which are thus kept down to very low concentrations in the soil solution. Gases and organic compounds may also be adsorbed, including antibiotics. One important aspect of this adsorption is the removal of germination inhibitors from seeds as a condition of their germination in the soil.

Two ions of particular importance in the soil deserve special mention for their ecological importance. These are the Ca kation and the PO_4 anion.

Calcium is in most soils, except some organic soils, the major ion in the soil solution and in the exchange complex. It is a constituent of many of the primary minerals of the weathered rock, especially the basic felspars, while in calcareous soils with a high pH various Calcium phosphates are present. The soils of arid regions, on the other hand, contain mostly free Calcium carbonate and Calcium Magnesium carbonate (dolomite), which are sufficiently abundant to influence the whole character of such soils and are the basis for their classification as pedocals (p. 3598). In temperate climates, rendzinas (chalk or öolite soils), and some fen peats may also have a high carbonate content. Lateritized soils of the wet tropics may, conversely, be very poor in Ca. The solution of the carbonates depends, as previously mentioned, on their conversion to soluble bicarbonates by the action of carbonic acid.

The dominating influence of Calcium carbonate in 'calcareous' soils is reflected in the plant populations. Cultivators have known for long that certain plants, e.g. most members of the Ericaceae, cannot be grown on calcareous soils, while ecological observation shows that the species-composition of populations on calcareous soils is very different from that on soils poor in Ca, and that many species show a marked preference 'for' or 'against' limey soils, which has given rise to such appellations as 'calcicole' and 'calcifuge' applied to them respectively. Opposite characters in this respect may appear between two genera of the same family or between two species of the same genus (e.g. *Galium hercynicum* and *G. sylvestre*, see p. 3617). Although many species appear to be indifferent in their soil preferences, Snaydon and Bradshaw have shown that this apparently wide ecological tolerance may really be due to the existence of differentiated ecospecies within a taxonomic species. Thus they found that the reactions of natural populations of *Festuca ovina*, growing on calcareous and on peaty soils respectively, differed as much in cultivation with regard to different levels of Ca in the soil as if they were distinct species, though there was no morphological difference between them and they were growing only a mile apart. Species with wide tolerances cannot therefore be accepted unconditionally as ecologically homogeneous, which is an important conclusion. Further, they found some degree of heterogeneity among the individuals of each population, which could be the basis of separatist selection. While the soil preference is thus very marked in many species, it is not possible to say, without population tests, which species are genuinely or wholly indifferent.

The question of calcifugy (see p. 3617) has been debated for at least a century. The question is whether the soil preference is a direct result of the Ca content, and if so, what is the nature of its effect on the plants? Calcium is necessary to all plants, but is an excess poisonous to some plants and not to others?

Opinions on these questions have been divided between those who regard the chemical effects of Calcium carbonate on the plants as the answer and those who prefer to regard the physical effects on the soil as more important. A modification of the chemical view is to refer the effects to indirect actions of the soil carbonates on the availability of other necessary ions, especially K and Fe. This latter view relies largely on the appearance of chlorosis in many non-calcicole plants (lime-chlorosis) in limey soils, which is attributed to the precipitation of Fe in insoluble form in soils with a high pH. This simple view, however, leaves unexplained why calcicole plants do not show chlorosis on limey soils, while some, but by no means all, calcifuge plants on the same soil become chlorotic.

Rovison (1963) has taken the view that it is Aluminium rather than Calcium which is the differentiating factor between the two types of soil. Using the markedly calcicolous *Scabiosa columbaria* he found that seedlings transferred to an acid soil died. Additions of $CaSO_4$ to the soil did not prevent this, but $Ca(OH)_2$ did, the Ca was not the critical factor but the pH. Below pH 5·0 Al is soluble, but above that pH level it is precipitated and its toxicity is eliminated. He backed this up with water-culture experiments which showed that ionic Al in acid solutions was toxic, but that the inhibition could be removed by raising the pH.

Snaydon and Bradshaw, on the other hand, maintain that the Ca level is important, quite apart from the toxicity of Al (and Mn) at a low pH. They point out that species differ widely in their Calcium requirements and that these are highly correlated with their soil preferences. The differences may depend on varying kationic exchange capacities in their roots. Calcifuge populations take up Ca preferentially to Na at low Ca/Na levels, but at high Ca levels they fail to prefer Potassium, which is antagonized by Calcium, or to obtain enough Iron, which is relatively insoluble at high pH, and may die from mineral deficiency.

Ramakrishna (1965) found three Indian ecotypes of *Euphorbia thymiflora* growing respectively on Gangetic alluvium, on a highly calcareous limestone soil and on a soil of intermediate lime content. In all cases germination was best on the natural soil of the ecotype. The Ca uptake in each case was approximately proportional to the lime content of the natural soil. Additional Ca decreased the yield of the ecotype on the alluvium.

Many calcicole plants show no positive lime-requirement in cultivation, provided the soil has the physical characteristics of calcareous soils, namely, high porosity and aeration, low water content and quick warming. Some calcifuge plants are quickly poisoned by water containing lime in solution, e.g. *Sphagnum* and many Ericaceae, but it is notable that this is only caused by $CaCO_3$ and $Ca(OH)_2$, which are alkaline and not by neutral Calcium salts, chloride and nitrate, for example, which suggests that the high pH of calcareous soils is the inimical factor.

Some calcifuge plants, on the other hand, which in nature show a marked preference for acid soils, are comparatively indifferent to soil reaction when cultivated in isolation, which points to competition as one of the deciding factors. For example, *Deschampsia flexuosa* in cultivation shows an optimum at pH 5·0, but grows in all soils between pH 3·0 and 7·5, whereas in natural populations it does not occur above pH 5·0 and its optimum is always well below that figure, even as low as pH 3·5.

We may conclude that calcifuge plants are not all physiologically unfit to grow on calcareous soils, but that they are there at a competitive disadvantage compared with plants which are not subject to that disadvantage. What that disadvantage may be is by no means clear and there may well be no general explanation, the effect may be specific.

It is perhaps worth considering whether Magnesium deficiency in a calcareous soil may be a factor inimical to calcifuge plants. Some species which are not found on pure limestone soils will nevertheless grow on dolomitized (magnesian) limestones. Further, calcifuge species can be induced to grow on limey soils by dosing the soil with Magnesium sulphate. Calcium can replace Mg on the surfaces of clay colloids and the soluble $MgCO_3$ is easily lost by leaching, while large amounts of Mg, exceeding the adsorbed Ca, are held unavailable in the colloid lattice structure and are only released on weathering or fine grinding of the particles. An Mg deficiency would also account for the appearance of chlorosis.

Priestley offered an interesting suggestion, that calcifuge plants (Ericaceae, in particular) contain super-normal amounts of fats. These could form insoluble Calcium soaps on calcareous soils and in the cell walls these would hinder the translocation of solutes in the tissues.

The habit of calciphily is to some extent dependent on climate, for it has been observed that many species which are constantly calcicole in northern Europe show no such preference around the Mediterranean. This may be explicable by reason of the preference of these plants for a dry, well-aerated soil which they find best on calcareous rocks, rather than owing to a direct action of Calcium. It may also be the result of soil reaction, since even the extremely calcifuge *Sphagnum* is more sensitive to OH ions than it is to Calcium, and it may be that the calciphile plants demand, conversely an alkaline pH rather than a calcareous soil.

There is a further question as to how much Ca is actually absorbed by plants on calcareous soils. Analyses by Iljin showed that the Calcium level in the plant is largely a specific matter, though on the average calcicole plants do absorb more Ca than the calcifuge, though specific differences can be so great that in some instances the reverse may be the case.

Phosphorus in the soil occurs either as inorganic phosphate or in organic combination in humus, especially in nucleic acids and phytin. The organic P may vary between 8 and 50 per cent of the total soil Phosphorus. The inorganic phosphate is very strongly adsorbed and only a small part remains exchangeable by neutral salts, so that the concentration of phosphate in the soil solution is normally very low. The amount of phosphate thus adsorbed increases with lowered pH and in acid soils the phosphate chiefly occurs as insoluble Fe and Al phosphates. In soils with higher pH Calcium phosphate predominates. Falling pH means greater ionization of the metallic hydroxides and of the Fe and Al groups in the silicates. Phosphate fixation is thus increased, partly by adsorption and partly by precipitation of the insoluble metal phosphates. At a pH level of 6·0–8·0 fixation and precipitation are minimal and phosphate availability is maximal as the sesquioxide-bound phosphate is changing to Calcium phosphate. Organic matter, especially some organic acids, such as citric, which may be formed during humification, increase the solubility of soil Phosphorus by chelating the Fe and Al and thus counteracting its fixation by these ions.

The whole history of phosphate in the soil is very complex and the balance between the various conditions of its occurrence is in a dynamic equilibrium, largely influenced by the pH, the amount of Ca available and by the silica/sesquioxide ratio. Phosphate solubility varies directly with the value of the latter, larger values implying a decrease of metal sesquioxides in the soil.

The *osmotic pressure* of the soil solution was put forward in 1910 by Gola as the basis for an ecological classification of soils, at a time when the study of colloids was beginning to make an impact on soil science. Since that date soil classification on climatic grounds has become so complex that Gola's simple views have been overlooked. Nonetheless, they deserve notice, for they do embody a factual and ecologically significant feature of soil character.

Gola divided soils primarily into two groups, *geloid* and *haloid* respectively. The geloid soils are those which have a high colloidal content and are poor in soluble salts, with a soil solution containing less than 0·5 parts per thousand of solutes. The haloid soils have a higher mineral content, are comparatively rich in soluble materials and have a soil solution of more than 0·5 parts per thousand.

In geloid soils the solute concentration is more or less constant or eustatic. In haloid soils it is liable to considerable changes; it is anastatic.

These two groups do, in fact, correspond to two easily recognizable and very different soil types from the ecological standpoint. The geloid soils are the acid, unsaturated and highly dispersed soils, the haloid soils are base-saturated, aggregated ('crumby') in structure and comparatively porous. Whatever classification of soils, based on pedogenesis, may be adopted, these two classes of soil are of opposite ecological character and are clearly reflected in their prevailing vegetation.

The solutes in soils are not all derived from the minerals of the parent rock. Quite surprisingly large quantities are supplied by rain. The figures shown in Table 12 are the average of monthly readings for two years taken at six stations in South and West Sweden (Tamm). Amounts in kg/ha/year.

The large values for Na and Cl may be due to proximity to the sea coast. Values for a station well inland were lower, Na 3·6 and Cl 5·6 respectively. The values for Na and Cl are not strongly correlated, indicating that more than one chloride is concerned. Annual fluctuations in the amounts of single ions present may be as high as 100 per cent so that averages between years are not significant, but the annual totals of all solutes, of the order of 30 kg/ha, are not negligible. Although Phosphorus was not included in the above analysis, some former observations by the present author showed a phosphate content comparable to that of sulphate given in Table 12.

TABLE 12

Precipitation	NH_4—N	NO_3—N	Ca	Na	K	Mg	Cl	S
530 mm	1·75	0 85	7·6	6·3	2·4	1·3	9·8	5·7

The figures for the solute content of rainwater have a bearing on the supply of solutes to ombrogenous peats, especially the upper zones of raised bogs which are out of touch with terrestrial sources of minerals.

A curious point, brought out by observations in Britain and Scandinavia, is that the uppermost zones of peat bogs are richer in minerals than those below, until the terrigenous basal layers are reached. In one case in Sweden, where pollen analysis enabled the peat zones to be dated, it was found that the zones formed between AD 1000 and 1942 had an average ash content of 2·14 per cent of the dry weight, whereas the average ash content for the earlier zones, back to 4000 BC was only 1·36 per cent. From AD 1000 onwards the ash content of each zone steadily increased and in the uppermost 10 cm it reached 5·30 per cent. This is taken to indicate an increasing supply of terrigenous dust in the air due to expanding human settlement and agriculture.

Heavy raindrops do not have time to come into equilibrium with the soluble matter in the air they pass through, so that their actual content of solutes is naturally fortuitous and variable. It is notable, for example, that fog and dew, formed near the ground, contain much larger amounts of ammonia than does rainwater. This may be ecologically important, as the estimated consumption of Nitrogen in peat formation exceeds the average supply from rain alone.

The extreme case of *Tillandsia usneoides* (Spanish Moss) growing epiphytically on telegraph wires in Florida (a common sight in the American tropics) shows dramatically how a plant can depend on rainfall for its nutrients. The *Tillandsia* contained 5·15 of ash per cent of the dry weight. One-quarter of this was SiO_2, followed by Fe and Ca, in that order, as the most abundant materials. Some solid dust may have contributed to this, but it was not microscopically visible. (See also notes on dust-fall on p. 3886.)

We have already spoken (p. 3502) of the mineral cycle between plant and soil whereby the mineral content of a soil under vegetation is replenished annually by the decomposition of plant litter on the soil surface. A study by Ovington of the balance sheet for minerals in plantations of *Pinus sylvestris* on glacial soils in East Anglia revealed some interesting features of the circulation. Both the trees and the ground flora were included in the analysis. The plantations ranged from 0–55 years old and the ground flora showed two phases, *Calluna* dominating during the early years, then dying out and being replaced, as the plantations were gradually cleared, by *Pteridium, Holcus mollis*, and *Rubus*. Up to the age of 11 years the ground flora contributed much more than the trees, both in respect of dry-matter production and minerals returned to the soil. This fell both absolutely and relatively, to a minimum between 20 and 23 years, the period of greatest production by the trees, when the contribution of the ground flora was only 5 per cent of the total. It then rose to about 22 per cent between 35 and 55 years, during which period the annual production by both trees and ground flora maintained a steady level. Not only litter was included in the analysis, but also the roots of harvested trees, left in the ground for slow decay, and allowance was made for the losses due to removal of the trunks of felled trees.

Each age group shows a surplus, very variable in amount, of production over return to the soil, except in the last group, where a small deficit can be attributed to the removal of harvested trees. The amount of dry matter returned to the soil varied between 24 and 98 per cent of the annual dry-matter production. The figures show little correlation with age except that the lowest figures are in the middle period when the ground flora is at its minimum.

Taking elements separately, the largest return was of Nitrogen. The ratio of the N in litter and roots to the annual uptake from the soil varied between 27 and 95 per cent and on two occasions was negative, i.e. in age group 11–14 years there was a net deficit of 11 per cent and in age group 20–23 (the most active growing period) a deficit of 16 per cent. In the same age groups there were also deficits of Calcium and Magnesium (11–14) and of Potassium and Phosphorus (20–23). In other words, during those periods there were net losses, covering four important elements, in the balance sheet of uptake against return, only partially due to the removal of harvested material. These were, of course, managed plantations. No natural community could support constant deficits, but even so it is evident that the percentage of the annual uptake returned to the soil can be surprisingly high. Next to Nitrogen in the amount returned to the soil were: Calcium, Potassium, Phosphorus, Magnesium and Sodium, in that order. Here again, the lowest percentage returns were in the middle-age period, showing how important a part is played by the ground flora and that it must be considered in estimating the economic balance-sheet of forest land.

Ovington's figures are based on the assumption that all the elements returned to the soil come from decomposed litter, but this is not strictly true. Quite considerable amounts of K, Ca, Na and Mg can be leached by rainfall from the living foliage, chiefly as bicarbonates, and returned directly to the soil. This has been estimated by comparing the mineral content of rain col-

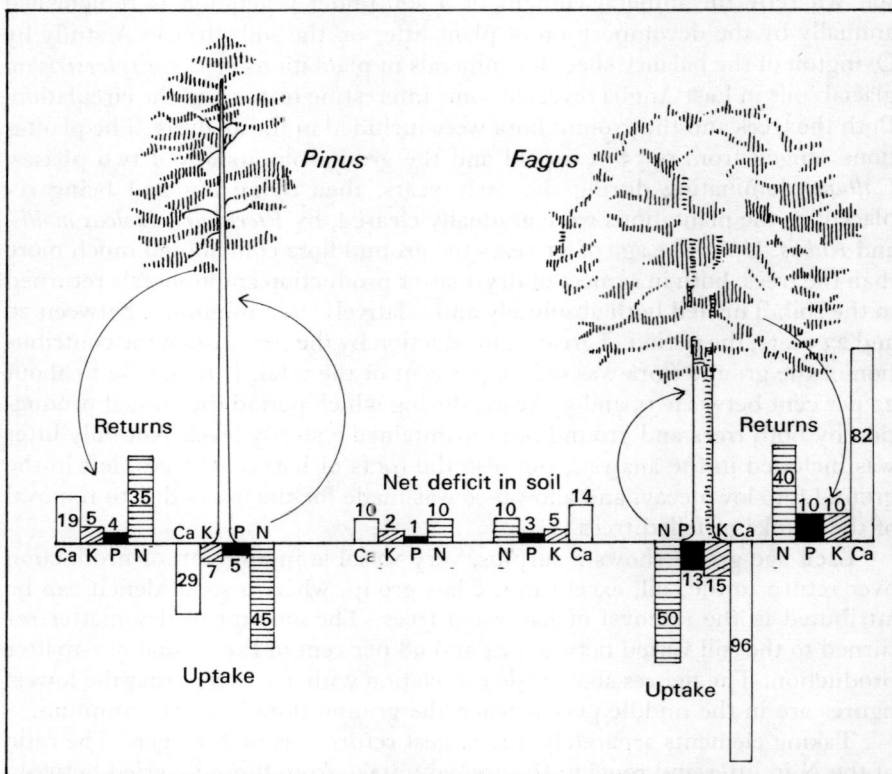

FIG. 4.47. The annual nutrient cycle in forest trees. The balance between uptake and returns to the soil in leaves, litter, etc. The amounts are shown in kilograms per hectare. (Pounds per acre, divide by 1·12.) (*After Dengler, from Feucht*, 'Der Wald als Lebensgemeinschaft')

lected under trees with that collected at the same time over open ground. Rain collected under deciduous trees in Sweden contained about ten times as much K as rain outside the forest, about three times as much Na and rather less than twice as much Ca. Under Conifers the K content was about 50 per cent less than under the deciduous trees, but there was no significant difference in Ca and Na. Potassium would thus seem to be the element most easily washed out from the leaves. Litter by itself is rich in Ca but poor in K, and it has been estimated that the return of K to the soil by precipitation is approximately equal to the amount returned by litter. It must be said, however, that the number of such analyses available is too small to serve for any sound generalizations on this subject.

Extensive measurements of the substances leached by rain from trees and returned to the soil were made by Carlisle, Brown and White in 1960 in a wood of *Quercus petraea* in the north of England. The mean annual rainfall is 171 cm and observations continued throughout the year by means of rain-gauges inside and outside the wood. The woodland canopy was fairly light and the total difference between rainfall and throughfall was only 21·18 cm or 13 per cent.

Table 13 shows the amounts given in kilograms per hectare per annum.

TABLE 13

Mineral	Rainfall	Throughfall	Throughfall and litter	Percentage due to Throughfall
N total	9·54	8·82	49·88	17·7
K	2·96	28·14	38·65	72·8
P	0·43	1·31	3·50	37·4
Ca	7·30	17·18	41·01	41·8
Mg	4·63	9·36	13·23	70·7
Na	35·34	55·55	57·22	97·1

The Nitrogen quota includes both inorganic and organic N in the proportions: *Rainfall* 6·28 to 3·26 and *Throughfall* 3·8 to 5·02 respectively. Inorganic N was taken up from the rainfall by the canopy in ten months out of twelve, including leafless months and organic N was also taken up in smaller amounts during four months, three of which were leafless months. Both could have been absorbed by the epiphytic microflora.

In addition to the net annual deficit of N in the throughfall, there was also a small deficit of P during the leafless months, again probably due to the moss and lichen flora. The large amount of Na in the rainfall reflects the proximity of the sea but the large addition in the throughfall is surprising. The largest amounts were in autumn when it might have been readily leached from dying leaves.

The total of organic substances leached was very considerable; 104·33 in the rainfall becoming 453·16 in the throughfall, an increase of roughly 340 per cent; the largest amounts being from June to August. In the latter month 70·7 per cent of the organic matter was carbohydrates, including glucose, fructose and melezitose, the latter probably from aphid honey-dew.

The rate of exchange of nutrients between plants and soil has been examined for tropical forest in Ghana by Greenland and Nye. The total annual return to the soil in litter, fallen timber, decay of roots and rainwash from foliage is 11 to 12 tons per acre or about 28 tons per hectare. The annual return of nutrient elements from fallen material is as follows, in kilograms per hectare: N 180, P 7·5, K 57, Ca 220, Mg 40·5. In rainwash from the trees the annual amounts, also in kg/ha/year are: N 9·5, P 2·5, K 170 (mostly as bicarbonate), Ca 22·5.

The amount of decomposition in the soil is virtually equal to the soil increment. Litter decomposes very rapidly. The initial rate is about 1·3 per cent per day, but this figure cannot be simply extrapolated since decomposition does not progress evenly but passes through variable stages governed by different groups of organisms. The rate of humus decomposition, on the other hand, is low, only about 4 per cent per annum but what proportion of the litter is humified is not known.

The principles involved in the circulation of mineral nutrients are fundamental to an understanding of vegetation–soil relationships. On base-saturated soil the litter will also be rich in minerals and its decay serves to increase the reserve capital of the soil. On a poor unsaturated soil the returns to the soil will be correspondingly small and, if mor humus forms, the organic acids it produces will tend to increase leaching and thus further impoverish the soil, so that the exchange between soil and plants dwindles away in time and the vegetation with it.

The tropical rain forests, however, stand on an entirely different, indeed, unique, footing. They apparently flourish in the poorest of soils, which contain a superfluity of Fe and Al sesquioxides, but from which practically all nutrients have been leached. This paradox has caused the ruin of many attempts to clear forested land for agriculture. The rain forest continues to exist in a delicate but unstable equilibrium, which has become possible owing to its great age and uninterrupted growth but which would be otherwise impossible. It is supported by a complex of factors which combine to make possible the most rapid circulation of nutrients through the cycle plant–soil–plant. It is not an exaggeration to say that such forests are feeding on themselves ('taking in each other's washing!') and that the soil provides little more than anchorage. Chief among the operative factors in this equilibrium are: continuous growth; continuous fall of litter; the rapid and complete mineralization of the contents of the litter; superficial root systems even of large trees and the widespread prevalence of mycorrhiza. If a tropical soil possesses even a minimal reserve of nutrients, it can, in time, regenerate a cleared forest, but on the majority of such soils, without any reserve, clearance destroys the equilibrium and the inevitable sequence is barren, stunted scrub or grass, coupled with accelerated soil erosion. It is to be understood that these remarks apply chiefly to the montane rain forests on ground that has a high relief. Rain forests on riverine flood-plains may be in quite a different situation and have a deep base-rich soil, but even so, the circulation of nutrients in them is exceptionally rapid and to a considerable extent superficial.

The productivity of an area in terms of vegetation can be estimated by either of two approaches; photosynthetic activity in the formation of dry matter, or, in relation to soil fertility. The two factor complexes are interdependent, since the efficiency of photosynthesis is influenced by soil fertility and is in great measure an expression of it.

The mean photosynthetic gain of a given type of vegetation is not easily estimated in the field without an extensive sampling of either dry-weight gains over a chosen time interval or of gas exchange values in foliage. Herbaceous vegetation may be treated by complete harvesting at the close of the active growing season. These values must be checked against the dry weight of sample crops taken at the beginning of the season and it must be remembered that if roots are included in this crop the total dry weight may be more than doubled. Clear cropping is hardly practical with forests. In the latter a practical estimate of the annual increment may be obtained by random sampling of tree girths, which can be checked by taking a boring into the wood of the trees sampled and measuring the width of the last annual ring or rings. This is best done at intervals of several years to minimize error. It is a matter of great economic importance in managed forests, as the greater part of the dry-weight production is stored in the wood.

The effects of soil nutrients on productivity are threefold. First come their direct physiological effects in the plant itself. Second there is the complex colloidal relationships in the soil which control their availability to the plant. Third is their effect on the physical properties of the soil: granulation, reaction, osmotic potential of the soil solution, etc. To unravel the network of events even for a single ion is extremely difficult. For the evaluation of their combined effects, the crop produced provides the only integrated answer.

As the expense of synthetic fertilizers is a considerable item of a cultivator's overheads, the relationship between crop weight and nutritive levels in the soil has been extensively investigated by agricultural chemists, to discover some method of judging the effectiveness of increasing the level of a nutrient in the soil to get the most favourable cost/production ratio and to avoid any ill-effects of over-fertilization.

One statement of relationships in this field, which may have an ecological bearing, is that called the 'relativity rule', a development of von Liebig's original Law of the Minimum. It runs thus: the relative activity of any factor is greater in proportion to its deficiency ('minimum') relative to the other effective factors. As the level of the factor rises so its relative activity decreases and becomes nil when it reaches its effective maximum.

As will be seen, this is a special case of the law of diminishing returns. It can be expressed more simply by saying that the effect of variations in any factor is greatest when it is in relative deficit, or, figuratively, a donation to a poor man is more valuable than the same donation to a rich man!

From 1909 onwards E. A. Mitscherlich endeavoured to give a precise formulation of the effects in crop production of the variation of a single factor, taking into consideration that all the other factors are also functional and each subject to the relativity rule. This he sought to achieve by reference to a theoretical maximum crop, produced when all factors are at their optimum level. The expression of what is called sometimes the 'Activity Law' and sometimes the 'Product Law' (*Ertragsgesetz*) was finally (after variations at the hands of Mitscherlich and Baule) embodied in the formula:

$$y = A(1 - e^{-cx})$$

where y is the increase of crop obtainable by x, the increase of a factor. A is the maximum possible crop, e is the base of natural logarithms and c is the activity constant of the particular factor. Mitscherlich maintains that this is a constant, independent alike of soil and plant and it is on this point that most controversy has centred and estimated values greatly varied.

Baule expands the equation to cover any number of factors simultaneously:

$$y = A_n(1 - e^{-cx}).(1 - e^{-c_1 x_1}).(1 - e^{-c_2 x_2})...(1 - e^{-c_n x_n})$$

which assumes that we know the number of operative factors, and their activity constants and that they can be varied at will.

This law appears to be only a mathematical model, one might almost say a platonic 'idea' of a law, based upon an extreme simplification of the plant–soil situation and strictly applicable only under rigidly defined conditions. Nevertheless, it roused great interest in Europe and gave rise to massive experimental tests, of great value, which seem to show that, however erroneous in some conditions, it does statistically give results approximating to the facts, provided that the elements of the equation are derived from experimental results and adjusted accordingly. As an effort to formulate the relativity rule quantitatively, it may have shortcomings but it also has merits.*

Serpentine and other special soils

There are a number of rocks and soils which provide special habitats by reason of their exceptional chemical constitution and the specialized vegetation associated with them is of great ecological interest. We have already spoken (p. 3652) of the problems connected with calcareous, i.e. high Calcium soils, as contrasted with soils of low pH.

Prominent among these special soils are: serpentine soils; salt soils, with an excess of Sodium chloride; and soils containing unusually large amounts of certain heavy metals such as Zinc, Copper, Lead or Selenium.

Serpentine is the name applied to a rock or the derived soil in which the chief mineral constituent is a hydrosilicate of Magnesium and Iron, formed from the silicate, Olivine, during weathering. The Olivine rock is called Peridotite. Chemically its main peculiarity is the abnormally high proportion of Magnesium in the resultant soil. The Ca/Mg ratio is always less than unity and characteristically may be less than 0·1, whereas in normal soils it varies from 1·0 to 10·0. The silica content is low and serpentine rocks are classed as 'ultra-basic'. There is also a marked shortage of other important substances, especially the nutrient anions NO_3, SO_4, Cl and PO_4 and the alkali metals. Similarly, there is a very low level of Molybdenum, which may occasion deficiency symptoms in the plants. Conversely serpentine has high levels of Chromium, Nickel and Cobalt, which are peculiar to this rock. As these metals are noxious to many plants and are absorbed and accumulated in the tissues because they are not used metabolically, they must be reckoned as of equal effect to the Magnesium excess in determining the special serpentine vegetation. In addition, the very high Fe content has a definitely limiting effect on the colonization of these soils, while the opposite shortage of Al militates against the formation of the important clay-minerals such as montmorillonite. It should be noted that it is the high ratio of Magnesium to Calcium which is important, not their absolute amounts, which vary greatly and in the case of Calcium are always low. There is thus no comparison between serpentine soils and dolomite soils, where both elements are abundant, and the vegetation of the two types of soil is quite different.

Characteristically, serpentine weathers at the surface into an unstable

* For a full critique of Mitscherlich see: G. E. BRIGGS (1928) 'Plant yield and the intensity of external factors'. *Annals Bot.*, **39**, p. 475.

debris, which tends to increase with time. The soil is poor in organic matter, 1·0 to 5·0 per cent. The heavy metal content is relatively high; Nickel 2000–3000 ppm. and Chromium 1500–13,500 ppm.

The low Ca content and the absence of sand produce a heavy, unaggregated soil structure and the poverty in strongly adsorptive clay-minerals allows the Ca to be leached out, so that weathering tends to change the originally relatively basic character to a relatively acid one, with a corresponding change of vegetation due to invasion by acidicolous species.

Wherever serpentine occurs, in whatever climate, the vegetation gives the same impression of limitation and toxic effect that is associated with soils rich in the heavy metals. In the northern hemisphere *Pinus* in various species, in the tropics *Casuarina*, and in the southern hemisphere *Araucaria* are the principal trees, all showing varying degrees of impoverishment. Trees are very thinly scattered, with a scrubby and stunted undergrowth and show a marked tendency to die out, even in favourable climates. Cultivation is not possible and the general aspect is barren.

There are, however, a number of species which are apparently unaffected by the nature of the soil and some which are true serpentine specialists and limited to such soils. The reasons for this endemism are very varied. Some species are limited to serpentine only in regions far from their centres of distribution, which might be attributed to 'avoidance' of competition in a relatively unfavourable environment. Other species are known to occur off serpentine only in poor, dry soils, on sandstones or marine sands. Serpentine areas sometimes open the way for species to enter zones where they could not otherwise colonize. Arctic species may form a serpentine tundra in what is otherwise forest country. Southern species may invade northwards, e.g. *Erica vagans* in south-west England, or alpine species may descend to lowlands. In all these cases it would appear that chemically tolerant taxa are taking advantage of the relatively light competition from the local flora. The status of these taxa varies from varieties of non-serpentine species to whole genera, e.g. *Halacsya* (Boraginaceae). These are morphologically distinct taxa, but it is highly probable that where no such distinction exists between serpentine and non-serpentine members of a species, the serpentine plants are definite ecotypes with a special serpentine tolerance (cf. calcifuge ecotypes, p. 3653).

The serpentine area of the Lizard in Cornwall is celebrated for the number of species otherwise very rare or unrepresented elsewhere in Britain. It was not glaciated and the serpentine rocks are largely overlaid by periglacial deposits apparently of loess character, so that the flora is somewhat mixed with both calcicole and calcifuge plants. Of these species very few are strictly serpentine plants. The most prominent of them, the southern *Erica vagans*, occurs also abundantly on the neighbouring gabbro and several others are examples of disjunct distribution, occurring on non-serpentine rocks at considerable distances from Cornwall. An exception is *Minuartia verna*, var. *gerardi* (Caryophyllaceae), which is a true serpentine variety and is adapted to a high Magnesium soil.

The families which seem to produce most of the serpentine species are: Caryophyllaceae, Gramineae, Leguminosae and among ferns the Polypodiaceae. In the latter family there are two species of *Asplenium* which have long been known as serpentine specialists; *A. adulterinum*, which shows a mixture of characters of *A. trichomanes* and *A. viride*, also *A. serpentini*, a close relative of *A. adiantum-nigrum*. It has been remarked that these two species are exceptional in showing remarkably luxuriant growth on serpentine, a fact also noted in the case of some non-specialist species, e.g. *Armeria maritima*. In connection with *Armeria* it is notable that maritime species seem to be more readily adaptable to serpentine than do mesophytic species. This has been attributed to the low Ca/Mg ratio prevailing in sea water.

Genetic studies have shown that the serpentine taxa investigated have all been closely related, as sub-species or varieties, to non-serpentine species. Indeed, the serpentine habit can arise in a single individual in a mass-culture of a non-specialist species, e.g. *Achillea borealis*. In 1951 Kruckeberg showed that in species which have serpentine and non-serpentine ecotypes the non-S types were markedly intolerant of serpentine soils, but, as we have mentioned, individual variants may occur which are S-tolerant and might be the basis for habitat-selected serpentine ecotypes. Kruckeberg suggested that physiologically this might mean the ability to absorb sufficient Ca without accumulating toxic amounts of Mg.

It is apparent that the number of serpentine endemics is a function of the richness and age of the surrounding flora. In the formerly glaciated northern countries the number of serpentine endemics is very small. Both in these parts and over the rest of the world there are indications that the serpentine areas have acted as refugia, the peculiar soil conditions keeping off many aggressive species and greatly reducing competition.

In Britain, the peninsula of the Lizard is the best known serpentine area, but several others are known in Scotland. The vegetation is a modified form of heathland and is part of the oceanic heath complex of western Europe. There is in fact a close similarity between the serpentine heath of the Lizard and the heaths of south-western France.

The species recognized as serpentine endemics in Scotland are: *Minuartia rubella*, *Arenaria norvegica*, *Cerastium nigrescens*, *Lychnis alpina* and *Asplenium viride*, all, except the last, members of the Caryophyllaceae.

Serpentine soils also occur in New Zealand where the sparse growth and stunted physiognomy contrast sharply with the prevailing beechwoods. There are several ecospecies endemic on these soils: *Myosotis munroi*, *Pimelea suteri* and *Nothothlaspi australe*; another ecotype is *Cassinia vauviliiersii* var. *serpentina*.

Myosotis munroi apparently succeeds in restricting its uptake of Magnesium, while *Nothothlaspi* tends to accumulate Calcium. *Pimelea suteri* absorbs Nickel and Chromium to high internal concentrations and appears to have a specific tolerance for them.

As in other parts of the world, there does not seem to be any single overriding serpentine factor, the serpentine denizens apparently responding specifically to one or other of the peculiarities of this complex ecosystem.

Increasing the Calcium content of a serpentine soil by additions reduces its toxicity and renders it less alien to extraneous species, but additions of Nitrogen increase the toxicity, presumably by increasing the solubility of kations as nitrates.

Rune, from a study of northern areas of serpentine, has summed up six general characters of their vegetation, which are also of general applicability.

1. The serpentine flora is relatively poor in individuals as well as species.
2. Wherever serpentine areas occur, the area includes taxa which are positively serpentine-tolerant, related as subspecies, vicarious species or varieties to local non-serpentine species.
3. Many of the plants on serpentine have disjunctive areas of distribution, their occurrences on serpentine being remote from their main centres of distribution.
4. Basicolous and acidocolous species often occur together. This is made possible by the combination of a low Ca content with a relatively high pH.
5. The serpentine flora has a relatively xerophytic character and is predominantly composed of heliophilous, sclerophyllous plants.
6. The serpentine flora is often dominated by certain families, e.g. Caryophyllaceae in northern Europe and eastern North America.

As regards the heavy metals in serpentine soils it is noteworthy that both Chromium and Nickel are absorbed by the plants, but principally the Nickel, even when the soil contains five times more Chromium. Moreover the Chromium taken up by the plants is chiefly found in the roots, especially their bark, and in the secondary wood of the stem, whereas the highest concentration of Nickel is in the foliage, where its toxic effects are naturally more evident. To these, however, Calcium additions are an antidote.

The amounts of Cr and Ni in serpentine soils varies greatly, both absolutely and relatively, so that they have no constant or unitary effect. Indeed, the same may be said of the serpentine characteristics as a whole and we may conclude that the preference for or tolerance of serpentine in certain species rests upon a variety of factors, not on a single overall conjunction.

The reactions of plants to soils which contain abnormal amounts of other heavy metals, such as Zinc, Copper, Lead, Selenium, Uranium, Vanadium, etc., are very similar to the reactions to serpentine, but tree growth on such soils is practically absent. Several widely dispersed species have physiological ecotypes which tolerate a heavy metal and give the impression of a specialization for that kind of soil because other ecotypes are excluded. In *Silene cucubalus* two distinct ecotypes have been found, one Copper tolerant, the other Zinc tolerant, the alternative metal being poisonous to each respectively. These ecotypes usually absorb and accumulate the metal in considerable amounts unless the absorption is antagonized by Calcium. A number of such metal specialists have been noted by prospectors as indicator plants, though it

FIG. 4.48. An outcrop of copper ore causing a wide break in bushland in N. Rhodesia. Photograph by Col. Rowland Fielding. (*From the* 'Journal of Careers', *by courtesy of the Editor.*)

does not follow that the metal interesting to miners is the decisive factor in their occurrence. For example, some Uranium beds also contain Selenium and it is by the occurrence of Selenium specialists that the beds can be recognized. Similar cases may be due to the presence or absence of trace-elements such as Molybdenum. Copper-tolerant plants may be influenced rather by the extreme acidity of the Copper soils, due to the formation of sulphuric acid, than by the Copper content. The indicator value of specialists which are only sub-tolerant or which are influenced by some accessory factor, may be sometimes due to visible defects or malformations, which are as conspicuous as the actual presence of a species.

Many plants accumulate Aluminium, and Chenery found that this was often related to pigmentation. Thus 87 per cent of plants with blue fruits are Al-accumulators. The well-known colour change from pink to blue in *Hydrangea* is due to the formation of a delphinidin-alumina lake which appears to be unique.

High levels of soluble Aluminium in the soil can be toxic, and this is particularly liable to be the case in acid podsolic soils below pH 5·5. At 40 ppm it inhibits the cultivation of alfalfa. Clarkson (1966) examined four species of *Agrostis* for Aluminium tolerance and found considerable specific differences. In order of decreasing tolerance were: *Agrostis setacea, canina, tenuis* and *stolonifera*. In water cultures *setacea* grew in concentrations of kationic Al as high as 1·6 mM, eight times as great as the tolerance limits of *A. stolonifera*. Chelated Al was non-toxic to all species and it was suggested that tolerance was due to the formation of stable complexes of Al with organic acids in the plant. The existence of such complexes in plants has been detected by several observers.

Selenium is readily absorbed by plants either as selenate or selenite, but more readily as selenate. Selenium is an essential element for animal livestock, though not for plants, and its deficiency in herbage has caused severe losses, especially in New Zealand. At levels as high as 50 ppm Se becomes toxic to plants, causing chlorosis. The amount absorbed appears to vary inversely with the amount of the clay fraction in the soil. Cruciferae, *Lolium perenne* and Leguminosae are the most active absorbers, in that order.

Uranium is another metal which is to some extent preferentially absorbed by plants. *Pinus sylvestris* accumulates it in the needles to a variable extent. On ordinary soils the median value is 0·48 µg/g of ash, but on the debris of an old Uranium mine in Cornwall oak leaves were found to contain as much as 160 µg/g of ash (Dean, 1966).

The number of species which accumulate unusual elements is very large and is not confined to specialists. Some widespread species accumulate a metal passively if it happens to be present. Thus *Ledum palustre* accumulates Copper even from soils with a very low Copper content, and *Orites excelsa* accumulates Aluminium. Nemeč even found that *Zea mais* accumulates Gold, perhaps as the soluble double salt with Potassium or Sodium chloride, and such species may be valuable as indicators.

Ecotypic populations of *Agrostis tenuis* have been found by Gregory and Bradshaw, around metal mines in north and central Wales, which showed separate and distinct tolerances for the metals Copper, Zinc, Nickel and Lead. Tolerance for Zinc and Nickel was combined, but each of the others was separate, thus three independent tolerances exist, which are genetic and are inherited through seed.

Absorption of the metals may be selective and not purely passive since it has been shown that the relative proportions of Lead isotopes in the tissues of a Lead-tolerant plant are not necessarily the same as their proportions in the soil.

The physiology of Zinc tolerance in *A. tenuis* has been investigated by Turner and Marshall (1971). They found that homogenates of the roots absorbed Zn freely and rapidly and that the major site of accumulation was in the cell walls, but that the proportion held by the walls was greater in tolerant than in non-tolerant plants. Other cell constituents only accumulated Zn to high levels in the absence of cell walls and it is suggested that the preferential concentration in the walls may be protective, as the mitochondria and ribosomes are probably the seat of toxicity. Absorption by the walls is linear to the concentration of Zn supplied. It is also independent of temperature and pH, which suggests an adsorptive exchange reaction. In non-tolerant plants a lower concentration in the cell walls is accompanied by higher concentrations in the sensitive cell organelles. A similar uptake of Zn has been found in excised barley roots, in algae and in the moss *Fontinalis*.

Premi and Cornfield examined the effect of some heavy metals on ammonification and nitrification in the soil. Copper increased ammonification at concentrations of 100 ppm, but depressed the process at 10,000 ppm. Zinc had similar effects, but Manganese and Chromium at low levels depressed ammonification and had no effect on nitrification.

The inheritance of the tolerant character shows that it is genetic in origin and has been selected from the general population, among which various degrees of tolerance occur sporadically.

Bradshaw and Gregory and others have shown that in non-tolerant populations of *A. tenuis* a small proportion of individuals show an exceptional tolerance for Lead. By selection from these strains complete tolerance can be achieved in about five generations. Bradshaw has succeeded in selecting a number of genetic strains of grasses highly tolerant for various heavy metals, which can be utilized for the sowing of poisoned soils. The only drawback is that the plants accumulate the metals to an extent which makes them unsuitable for grazing. Kitchen vegetables grown on soils containing Lead may also be deleterious.

The highest degree of specialization is probably that found in areas supplied with Sulphur-bearing volcanic waters, where high concentrations of sulphates, sulphuric acid and Sulphur dioxide occur, as well as frequent high temperatures. These soils suffer from an extreme degree of leaching and acidity and it is surprising to find that the vegetation often resembles that of podsols, e.g. in the prevalence of Ericaceae, and *Vaccinium*, *Empetrum* and *Myrica*. Aluminium in soluble forms is washed out of the clay-silicates and is accumulated by many of the plants.

Apart from all these peculiarities stand the salt-soils with a high content of Sodium chloride, which provide a specialized habitat for the plants grouped as *halophytes*. These provide an ecological problem similar to that of the calcicole plants. It is not only maritime soils which are involved, although they have a widespread interest, but saline soils generally, such as occur around salt springs or, more importantly, the alkali soils of arid and semi-arid continental areas, which are climatically conditioned. In these the cardinal salts are sulphates of Sodium, Magnesium and Calcium and Sodium carbonate.

The maritime soils are chiefly intertidal deposits which are fine grained and highly dispersed, with slow drainage and hence of a marshy character. The degree of salinity reached depends, first, on the saline concentration of the sea water, which is lowest in tidal estuaries and highest in semi-enclosed seas, like the Mediterranean, with a hot climate and few large rivers inflowing. Lowest of all is the Baltic Sea, into which the tides scarcely penetrate and which is copiously supplied with river water; it is so low in dissolved salts as to be practically drinkable. Second, on the climatic factor of the amount of evaporation during low-tide periods and, third, on the length of the exposed periods between tides. As this last factor depends on the height of any position above low-tide level, it gives rise to a well-marked and characteristic zonation of plants.

The continental areas of alkaline soil depend, on the other hand, first, on the rainfall and its annual distribution. Second, on the evaporating power of the air, third, on the rock structure of the area and the character of the derived soils, and fourth, on the depth of the water table, which is a resultant of the three primary factors.

The concentration of salts in maritime soils is principally owing to NaCl and $MgCl_2$. The combined amount of the chloride ion in the soil solution does not normally rise above 2·5 per cent in temperate zones and is often much lower, but isolated tidal pools, especially in hot climates, act as evaporating pans and may almost reach saturation. The figures in Table 14, from Adriani, show the percentage distribution of chloride in the soil water of a Dutch salt-marsh, across the principal zones of vegetation, beginning with the lowest. The figures are averages.

TABLE 14

Zone	0–4 cm	4–12 cm	12–20 cm
Salicornia or *Spartina*	1·01	1·22	1·08
Puccinellia maritima	1·34	1·55	1·46
Artemisia maritima with *Limonium*	1·29	1·37	1·28
Artemisia maritima with *Obione*	1·19	1·33	1·43
Armeria maritima	0·14	0·13	0·11

The figures in the table show in the first place a tendency for salinity to increase downwards in the soil to about 12 cm depth, due probably to down-washing of the soluble chlorides, and in the second they show a rise in salinity in passing upwards from the lower zones, followed by a dramatically steep drop where we pass beyond the zone of regular submergence to that which is reached only by the equinoctial high tides. The salt content is thus the resultant of three factors, two of which are variable: periodic renewal by tidal flooding, evaporation from the exposed surface, and leaching of the exposed surface by rain.

Below the lowest exposed zone of *Salicornia*, there is also a sparse phanero-gamic vegetation which is normally submerged or only occasionally exposed. In this situation we find *Zostera* in cool temperate latitudes, *Posidonia* in warm temperate to sub-tropical latitudes and *Enhalus* and *Halophila* in the tropics. With the algae we are not immediately concerned.

The marsh soils are slowly built up by deposits of tide-borne silt. When a level is reached which is only covered by exceptionally high tides the salt is rapidly leached out and mesophytic plants invade the marsh and soon build up the soil above tidal reach. Meadow and freshwater marsh plants then take over and the ground ceases to be halophytic.

On rocky shores the inter-tidal vegetation is almost entirely algal, but above tidal levels there is a spray zone,* with flowering plants which are able to resist an occasional drenching with storm spray. Here we find *Cochlearia*

* The supra-littoral zone of the Stephenson's classification, see p. 3762.

anglica, *Crithmum maritimum*, *Inula crithmoides* and the small *Limonium binervosum*. Where the beach is backed by a zone of sand, this will be found to be practically devoid of salt, except at its seaward edge where the sand-loving halophytes *Salsola kali* and *Cakile maritima* find a habitat which they have almost to themselves.

The flat tidal surfaces which are the normal habitat of temperate salt-marsh plants are, in the tropics, usually occupied by the arboreal mangroves, forming extensive swamps peopled by a few species of small evergreen trees belonging to the genera *Rhizophora*, *Avicennia*, *Bruguiera* and *Sonneratia*, with fringes of the palm *Nipa fruticans* and the giant fern *Acrostichum* (*Chrysodium*) *aureum*. The peculiar characteristics of the mangrove trees, their adventitious stilt roots, their horizontal root systems with the negatively geo-tropic branches called pneumatophores which appear above the soil and the viviparous habit of reproduction in *Rhizophora*, have been so often described in textbooks from the time of Schimper onwards, that we shall not pursue the description further. They are true halophytes, subject to the same conditions as others of the class.

Tropical sandy shores are usually dominated by the ubiquitous *Ipomoea pes-caprae*, a very fast-growing creeper which serves efficiently to bind the sand. Other rapidly creeping plants may serve the same function as the *Ipomoea*, namely, *Canavalia maritima* and *Vigna marina* (Leguminosae) or, in tropical America, *Remirea* (Cyperaceae) and *Clitoria* (Leguminosae). These are all partially halophytic, growing, as they do, on sand below high-tide mark, though salts are much more rapidly leached from sands than from the muddy soils of the salt marsh.

Above high-tide level on tropical beaches a characteristic strand-woodland soon establishes itself. In the East this is called the *Barringtonia* association, from the dominant tree, which is associated with *Pandanus*, *Cocos*, *Hibiscus tiliaceus*, *Scaevola*, *Casuarina*, *Terminalia*, etc. In Brazil the similarly placed woodland is called *restinga*. It has a very similar facies, but *Agave* (*Fourcroya*) replaces *Pandanus* and *Clusia* replaces *Barringtonia*. These beach woodlands, however, cannot be called properly halophytic as their contact with sea water is only occasional.

Two ecological problems present themselves regarding the environmental relationships of true halophytes. First, how do these particular plants resist the effects of salt concentrations which plasmolyse the root cells of normal mesophytes? Second, do they absorb salt in abnormal quantities, and if so what becomes of it?

Regarding the first question there is good evidence that halophytes have unusually high osmotic values in the cell sap. Analysis of twelve North American halophytes of the salt marsh gave an average figure of $31·5$ atm (Steiner) with a maximum of 43 atm in *Salicornia mucronata*. These high values are largely due to the high content of Cl as NaCl. The share of NaCl in producing the osmotic value is between 40 and 90 per cent. A slightly higher average osmotic value was found in trees of the East African mangrove, but in these also the share of NaCl in the total osmotic value was approximately

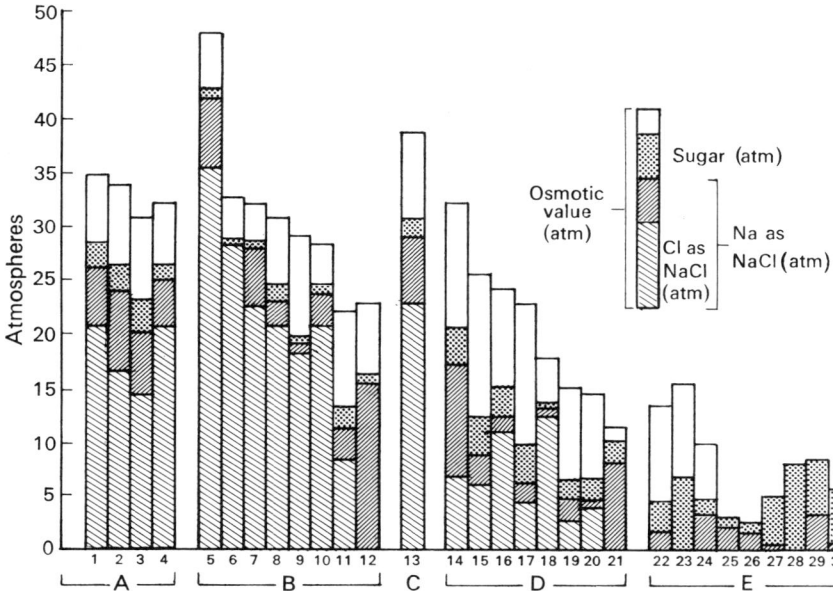

FIG. 4.49. Osmotic values in atmospheres of the cell sap of certain North American plants, with the Na, the Cl and the sugar components. A–C, salt-marsh plants. D, plants of brackish marshes. E, mesophytes.

A. *Salt secreting species.* 1. *Spartina patens.* 2. *S. glabra.* 3. *Distichlis spicata.* 4. *Limonium carolinianum.*

B. *Succulent species.* 5. *Salicornia mucronata.* 6. *S. europaea.* 7. *Plantago decipiens.* 8. *Atriplex patula* var. *hastata.* 9. *Iva ovaria.* 10. *Suaeda linearis.* 11. *Gerardia maritima.* 12. *Aster subulatus.*

C. *Other species.* 13. *Juncus gerardi.*

D. *Plants of brackish marshes.* 14. *Spartina cynosuroides.* 15. *Hierochloe odorata.* 16. *Spartina michauxiana.* 17. *Elymus arenarius* v. *hirsutiglumis.* 18. *Solidago sempervirens.* 19. *Panicum virgatum.* 20. *Solidago graminifolia.* 21. *Baccharis halamifolia.*

E. *Mesophytes.* 22. *Solidago graminifolia.* 23. *Psedera quinquifolia.* 24. *Eupatorium perfoliatum.* 25. *Xanthium canadense.* 26. *Ambrosia artemisifolia.* 27. *Pyrus arbutifolia.* 28. *Sambucus canadensis.* 29. *Rhus toxicodendron.* 30. *Acer rubrum.*

Part of the Na content is united to ions other than Cl, e.g. SO_4 etc. (*After Steiner.*)

the same as in the temperate salt-marsh plants. In a comparable sample of 16 species of mesophytes the Cl share of the osmotic values was under 5 atm in a total average osmotic value of 14 atm, and the percentage attributable to Cl varied between 0 and 40 per cent. There is generally an excess of Na above the Cl equivalent, which shows that Na must enter associated with other ions, in particular SO_4.

These observations help to answer our questions. Typical halophytes develop unusually high osmotic potentials, that is they have abnormally high diffusion pressure deficits and can thus absorb water from solutions as concentrated as sea water. They also affirm that Sodium chloride is absorbed to a considerable extent. The graph by Walter, shown in Fig. 4.50, also demonstrates that the osmotic value of the cell sap and the chloride content rise proportionately to the saline content of the soil externally.

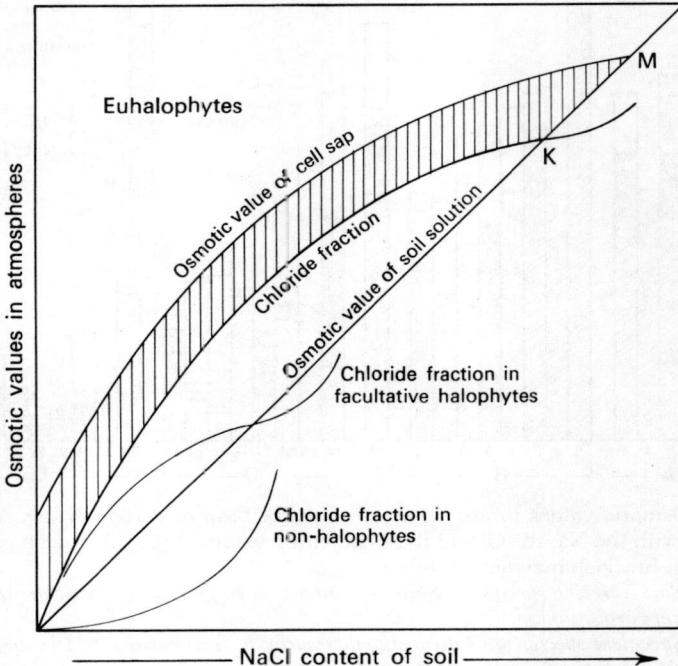

FIG. 4.50. Relationships between the Sodium chloride content of the soil and the osmotic values of the cell-sap in true halophytes, facultative halophytes and non-halophytes. The striped area represents the non-chloride fraction in the cell-sap values. (*After Walter.*) (See also p. 3674.)

With regard to plants growing on any one type of soil we can distinguish two modes of behaviour concerning salt accumulation. There is, first, continued accumulation throughout the growing season, of which *Juncus gerardi* and *Suaeda fruticosa* are examples. The osmotic value and the Cl-content show parallel increases from May to September. The osmotic value rises from 20–40 atm and the Cl-share of the value remains constant. The other type is regulatory; the osmotic value fluctuates irregularly, but shows no overall increase and may drop towards the end of summer. The first type presents no special ecological problem, but the second type inevitably invites the question of how the regulation of the salt content is accomplished. Two methods of regulation are known but they do not cover all the possibilities, for salt regulation among mangrove trees is by some as yet unknown method. The first method is by the external secretion of salt from leaves and young stems. *Armeria, Limonium, Glaux, Acanthus ilicifolius* and *Frankenia pulverulenta* are examples. They possess epidermal salt glands, from which a salt solution is secreted, at the same osmotic value as the mesophyll sap, the concentration of which is not thereby altered. The solution dries in the sun and leaves a superficial deposit of salt crystals, easily removed by rain, but cumulative in some desert halophytes.

The second method is by increasing succulence, in other words by an increase of the water content to keep pace with the rising Cl-content, thus maintaining a fairly steady osmotic value. In halophytes with succulent leaves, the degree of succulence increases throughout the season, the later leaves being more succulent than the earlier. Some facultative halophytes, that is plants not usually halophytic, but moderately salt tolerant, show the same reaction as these halophytes by developing succulence when brought within reach of sea water. *Cochlearia* species, *Atriplex* species and even *Cheiranthus cheiri*, with other plants of the spray zone on rocky shores, show this reaction. Halophytic succulence of the leaves of *Atriplex* in Australia was shown by Black (1958) to be associated with an extended period of ontogeny and reduced general size.

Succulence is not simply a reaction to high concentrations of salts; it appears to be directly dependent on the NaCl concentration. Culture experiments have shown that mixtures of other chlorides or of Sodium sulphate or nitrate with Sodium chloride may induce succulence, but the degree of succulence produced (measured by Delf's quotient: Total water content/Surface area) is independent of the total salt concentration. In halophytes of the Namib desert Walter found that the press juice showed considerable variations in the ratio of Cl to SO_4 and that in the succulent forms the Cl was preponderant while in the nonsucculent forms there was generally an excess of SO_4. Even on desert soils rich in sulphates only exceptional species show a predominance of sulphate in their ash and these are non-succulent. In proportion to their relative concentrations in the soil the chlorides are preferentially absorbed, especially by succulent species.

Since, it is clear that halophytes are able to absorb and accumulate unusually large amounts of NaCl without damage we may ask whether they are specially adapted in this sense. Their high osmotic values ensure, most importantly, their water intake, but these osmotic values depend on the quantity of the salt intake, which thus appears to be the governing factor. This depends mainly on two factors, namely, the high concentration of the external solution, so much greater than in ordinary soils, and the respiration of the root systems which is the activating factor. Here we touch the very controversial ground concerning ionic uptake in general and 'salt respiration' in particular, which we discussed in Vol. 3, p. 2927 *et seq.* Data about the root respiration of halophytes are too sparse to allow of general conclusions about their salt intake, but it is interesting to note than van Eijk found in the succulent halophyte *Aster tripolium* that the factor for salt respiration was abnormally low, about one tenth that in *Triticum*. This suggests that the respiration energy required for salt intake in marsh halophytes is of a low order, which may be connected ecologically with their badly aerated rooting medium.

The experiments of van Eijk with *Salicornia herbacea*, cultivated in an artificial seawater mixture containing varying additions of NaCl, from 0–3·0 per cent, showed that the amounts of Na and of Cl absorbed rose steadily up to the maximum concentration used. At the nil addition level more Cl was absorbed than Na, but even at the lowest addition of NaCl, 0·6 per cent, this was reversed and Na absorption rose with the concentration resulting in an increasing excess of Na over Cl.

These results do not, however, bring out the difference between optimal and maximal concentrations of NaCl, which appears when we look at van Eijk's figures for dry-weight produced at different concentrations, again in *Salicornia*.

NaCl concs. in per cent	0	0·16	1·2	1·8	2·4	3·0
Average dry weight in mg	50	116	128	121	138	68

Van Eijk does not consider that the differences between the three highest figures are significant and concludes that the optimum concentration is about 2·0 per cent. Keller, also working with *S. herbacea*, but in an inland area of South-eastern Russia, found that both Na and K chlorides in small doses enhanced both growth and succulence, but while both these properties were further increased by rising doses of NaCl this was not the case with KCl, which appeared to be toxic at the maximum doses given. He concluded that an NaCl concentration of about 1 per cent was optimal. Seeds germinating on non-salty soil produced weakly plants which were easily suppressed. Dry, hot summers in these salt deserts caused a marked concentration of salt in the soil and the internal osmotic values rose correspondingly to an abnormal extent. Growth was reduced and transpiration became insignificant.

The preferential absorption of Na over Cl by *Salicornia* is not a peculiarity of halophytes, indeed it is somewhat less marked than in some non-halophytes. Overstreet and Handley found that seedlings of *Triticum* absorbed ten times as much Na as Cl from a nutritive solution.

The results of van Eijk and Keller showed that *Salicornia herbacea* definitely benefited from the presence of NaCl in the growth medium, and it should therefore be regarded as *salt-adapted* and classed as s Eu-halophyte or obligate halophyte. This is not true of all species growing in salty soils. Keller showed that *Frankenia pulverulenta*, which is a characteristic plant of the Russian salt deserts, absorbs and secretes externally large amounts of NaCl, which crystallizes all over the plant. Nevertheless, it derives no benefit from such salt absorption and grows equally well on non-salty media. It appears to be only *salt-tolerant* and to belong to a class of Mio-halophytes (Chapman) or facultative halophytes. Investigation may show that this significant difference is reflected in the marked zonation of species on marine salt marshes.

Even with obligate halophytes there is, however, a limit to their absorbing capacity. Reference to Fig. 4.50 will show that the Cl content of the cell sap rises to a point K at which it intersects the rising line of osmotic potential in the soil solution. Any further increase of osmotic value must depend on the non-chloride fraction, chiefly due to Na, and at the point M, where this equals the osmotic potential of the soil solution, the maximum is reached. Beyond this point the osmotic potential of the sap can only be raised by evaporation losses, which implies a loss of hydrature and inhibition of growth.

In connection with all the phenomena of halophytism one must always bear in mind the probability of ecotypic selection among members of a species and the false impression of variability in reaction which such selective differences may create.

All in all, however, halophytes will always have a great interest for ecologists, for there are few accessible types of vegetation in which reactions to habitat changes are more clearly seen or more readily studied than in marine tidal zones.

Air in the soil

The conditions of aeration in soil constitute a factor of leading importance in its influence on vegetation, which one would expect to be maximal in the fine-grained and highly dispersed soils of salt marshes, where reducing conditions often prevail. In tidal marshes colonized by *Spartina alterniflora* examined by Teal and Kanwisher, the lowest topographical levels showed intensely reducing conditions below the surface of the mud and here root formation is exceedingly restricted and the growth of plants under such conditions is only made possible by the well-developed gas transport system in the plants themselves, which allows of diffusion from leaves to roots of more than sufficient O_2 to support respiration. Both O_2 and CO_2 diffuse outwards from roots and rhizomes into the mud. The O_2 is consumed, chemically and biologically as soon as released. The CO_2 is swept away by drainage water during low-tide periods. The outward diffusion of CO_2 is enough to promote mass-flow of air down to the roots in addition to diffusive movement of single gases. These facts illustrate the decisive importance of internal gas transport in the colonization of marsh soils and other badly aerated soils.

In a well-aerated soil the CO_2 content of the soil air remains more or less constant at increasing depths, but in a wet soil the amount may increase downwards to 2 per cent or more, which is enough to have a restrictive effect on deep-rooted trees. *Alnus* and *Salix* appear to be least affected and are often the only trees on marshy soils. *Salix* certainly has an effective gas transport system, like herbaceous marsh plants. (Armstrong, unpublished.) As the CO_2 content rises the O_2 content diminishes, to 16 per cent or lower. The CO_2 production in the soil (soil respiration) reflects the micro-biological activity in attacking organic matter. It is much greater in the F layer than in the mature humus incorporated with the mineral soil. Fungi contribute to it largely. Naturally it is affected by temperature and moisture, and the escape of CO_2 to the air is subject to daily and annual rhythms. Even in desert soils there is some soil respiration, indicating micro-organismal activity, mostly bacterial, in spite of drought.

Two distinct values are associated with measurement of the soil atmosphere. They are, the air-content and the air-capacity. At first sight they might seem to be the same thing, but the latter term is used in a technical sense to mean the irreducible minimum of air which is left in a soil which is water saturated by capillary rise at atmospheric pressure, i.e. when only non-capillary pores are left unfilled. In extreme cases it may fall as low as 1 per cent of the pore-volume of the soil, but in the upper layers of a light soil it may rise to 25 per cent. Heavily saturated or marshy soils may be reduced to their minimal air capacity for long periods, which practically excludes plants without an internal gas transport system to their roots. The comparative fertility of a soil is generally linked to a high air capacity.

The existence of this residual air capacity makes it impractical to estimate the pore-volume of a soil by comparing the weight of a given volume of dry soil with that of the same soil when water saturated. Such estimates will always be too low. A better measure is given by comparison of the volume of a block of fresh soil extracted, for example, by a boring tube of known volume, with the volume of the solid matter of the same sample when dried and powdered. An average figure for the pore-volume of a woodland soil is about 50 per cent, but in a fibrous peat it may rise to over 80 per cent. The values for the pore-volume drop with increased depth in the soil but below 50 cm they remain fairly steady, as below that level the soil is fully compacted.

The actual air content of a soil will only equal the pore-volume in an air-dry soil. In all natural soils air and water share the volume between them and their respective amounts are inversely proportional. If the graphs of the two quantities are constructed for an entire growing season, it will be seen that the falling curve of water content from winter into summer and its reciprocal, the rising curve of air content, intersect at some point or, at least in continuously wet soils, show a point of nearest approach. The earlier this point is reached the drier will be the summer habitat and vice versa.

If the pore volume is known, the air content can be obtained by subtracting the water content, measured by the difference between the fresh and dry weight of the sample or, alternatively, by weighing a fresh soil sample and then completely saturating it with water under an air pump.

The air in soil is by no means stagnant. If it were the Oxygen content would be much lower than it actually is. At depths of over 1 metre there is only a very slow movement of gases and the CO_2 content may rise to well over 10 per cent. As the soil at such levels contains almost no bacteria the high CO_2 content must be due to downward diffusion from more active levels. Diffusion can account for a considerable interchange with the atmosphere, but apart from that the large Oxygen demand of organic substances and micro-organisms in the soil-body creates conditions for mass inflow. Temperature differences between air and soil are also effective in promoting diffusion movement as the rate of diffusion is proportional to the square of the absolute temperature. The part played by changes of barometric pressure has been under-estimated and its influence sometimes denied, but some investigations by R. D. MacLean on air movements in deep wells and borings in North London show a different picture. These were deep wells into the chalk. In one case, where the well was lined with cast-iron to a depth of 26 m, the water level was at first 38 m from the surface. Thus 12 m of chalk was exposed in the well. Subsequently the water rose above the bottom of the iron lining so that no chalk was exposed. In the first case the Oxygen percentages in the air of the well varied between 5·5 and 21 per cent and the changes closely followed the barometric pressure, even small pressure changes causing violent changes in the O_2 content. CO_2 changes were inverse to those of O_2, but on a much more limited scale, the highest concentration recorded being only 0·7 per cent. The striking fact is that when the water rose to cover the chalk walls these fluctuations in both O_2 and CO_2 ceased completely. It is reasonable to conclude that the previously observed fluctuations originated in gas exchange with the exposed chalk, under the influence of barometric changes.

As with soil water, so with soil air, the amount of lateral movement is inconsiderable, so that its composition may vary over quite short distances even in a uniform soil and such differences may be a decisive factor in determining specific distribution or patterning in a community.*

The rate of air diffusion in soil is very sensitive to temperature changes, as we have pointed out above. It is also more or less proportional to the total volume of air-filled pores in the soil. The size of the pores has little effect because the kinetic diffusion path of the gas molecules is minute in comparison with the diameter of the pores themselves. Nonetheless, the apparent diffusion rate in the soil is little more than one-half of that in the free atmosphere, probably because of the tortuous paths that the air is obliged to follow.

While it is possible to calculate theoretical rates of air diffusion when the soil is regarded simply as a porous medium, the fact is that it is not a passive medium and that a steady state of diffusion is rarely, if ever, attained, because localized production and consumption cause constantly varying local concentrations of both CO_2 and O_2. To the ecologist, perhaps the most significant factor is the proportionality between diffusion rate and pore-space, which implies that the rate of air diffusion varies inversely with the water content, the chief factor causing variation of free pore space. Buckingham's early (1904) calculations showed a proportionality between rate of diffusion and the square of the free pore space, implying that if the pore space is reduced to one-half the rate of diffusion will be reduced to one-fourth. The dependence of diffusion upon total pore space rather than upon pore size signifies that at a low level of water content there may be a considerable degree of aeration even in heavy, fine grained soils.

The air in the soil is not all in the gaseous state. The organic colloids occlude considerable amounts, but as the Oxygen in this state is rapidly consumed by oxidation reactions the occluded atmosphere is almost entirely N_2 and CO_2. It is in this anaerobic environment that the anaerobic bacteria of the soil, including the Nitrogen-fixing anaerobes, find a suitable medium.

There is also the air dissolved in the soil water, and this is of particular significance because the contact of the plants' roots with soil gases is chiefly

* Lateral movement both of water and of ions in the soil is greatly influenced by the activity of roots in the rhizosphere. More investigation is needed of the profiles of water content and ionic concentrations in the zones immediately around absorbing roots.

through the water films which surround them. The condition of the gases in solution is very different from that in the gas phase. At a temperature of $15°C$ Carbon dioxide is approximately twenty-five times as soluble as Oxygen. Thus pure water at $15°C$ in equilibrium with atmospheric air will contain 0·6 per cent by volume of O_2, but as much as 0·45 per cent of CO_2, that is, the Oxygen content is approximately 3 per cent of that in air, while the CO_2 present is more than thirteen times as much as in the atmosphere. Moreover, the diffusion rate of O_2 in water is only about one ten-thousandth of the rate in the gaseous atmosphere, which imposes a heavy restriction on the access of the root cells to the soil Oxygen, even allowing for the extreme thinness of the superficial water film on the root. As the solubility of Oxygen decreases with rising temperature and the metabolic demand increases, it can be understood that to produce a given result by soil aeration requires a higher percentage of O_2 in hot climates than in colder regions, and that spring cultivation is more influential in this respect than winter cultivation.

As far as the effects of soil aeration are concerned it would seem that shortage of Oxygen is more likely to be a limiting factor of plant growth than is the amount of CO_2 in the soil air. Amounts under 15 per cent in samples drawn from the soil air must be regarded as suggestive of deprivation.

Sampling is, however, a difficulty. The traditional method is by aspirating a sample from a tube driven into the undisturbed soil to the appropriate depth. This has been criticized as giving O_2 values which are too high, partly because the sample is drawn from all around the end of the tube, from above as well as from below, and partly because of the possibility of a downward leakage along the outside of the tube. The first is a systematic error, the latter can only be minimized by taking great care. Taylor and Abrahams proposed, as an alternative, sinking a porous chamber into the soil and leaving it until the air inside had come to a diffusion equilibrium with the soil air. Using this method in comparison with the other they found O_2 concentrations which were only 45 per cent of those given by the aspiration technique. Probably the most reliable criterion of deficient aeration is by comparing the growth of test plants on a given soil, with and without artificial aeration. For this purpose, however, it is essential to use a test plant which is sensitive to deficient aeration, as the susceptibility of species in this respect varies a great deal. Knight in 1924 showed that *Zea mais* reacted to increased aeration in soil, but not in water culture, while *Cheiranthus* and *Chenopodium album* reacted positively in both media. In 1940 Baver and Farnsworth showed that sugar-beet was very sensitive to increasing pore space (i.e. to increasing aeration) up to about 10 per cent pore space, beyond which it appeared to be unaffected. The improved yield was attributable to aeration alone as the addition of artificial fertilizers brought about no yield increase. Additions of organic manure, on the other hand, reduced the natural loss of pore space which occurs between June and September to nearly one-quarter of the loss in unmanured plots, due probably to the effect of humus in promoting aggregation in the soil.

Aeration in the above experiments was increased by cultivation, by ploughing and in some cases by ridging as well. The benefits to soils from cultivation, especially on heavy soils, are numerous, but increased porosity and hence increased aeration must be ranked highly among them. The extent of the benefit will depend, of course, on the specific sensitivity of the plants concerned to this factor and on the prevailing temperatures. Shallow-rooted grasses are perhaps least affected by low porosity, deep-rooted plants the most, but there are exceptions. Barley and Millet among cereals, Cotton and Potatoes, all give their maximum yields with high porosities, of 30 per cent or over. Apart from direct effects on the higher plants, increased aeration also promotes Nitrogen fixation by *Azotobacter* and increases the rate of mineralization of organic Nitrogen compounds, thus raising the level of available Nitrogen.

At the other end of the porosity scale, especially in submerged or water-logged soils, the Oxygen level may be so lowered that reducing conditions obtain, and Iron and Manganese salts or oxides occur in the reduced or -ous condition, in which state they are more soluble and may become toxic. Ferrous iron is not necessarily associated with low pH, as it is in aerobic soils. Submerged soils with a high pH and calcareous content may be strongly reducing.

Pearsall was the first to show the effect of deprivation of air by soil compaction and its effect in lowering the redox potential. Even surface compaction by heavy rain may prevent the free escape of CO_2 from the soil and the entry of O_2 from the air unless it is broken up.

The microbiology of the soil

Decomposition of the organic increments to the soil from plant and animal debris, is one of the most considerable parts of the working mechanism of every ecosystem. The amount of this increment contributed to the soil varies greatly with the climate and type of vegetation. It is estimated at 2–3 tons per acre in Northern pine woods and between 10–14 tons per acre in tropical rain forests. In a well-established and stable community in a temperate climate, the amount decomposed each year is virtually equal to the annual increment, unless the process is retarded by cold, waterlogging or soil acidity. These factors are all inimical to micro-biological activity which is the agent for almost all decomposition. Under humid tropical conditions in Ghana the rate of decomposition of litter was estimated at an average of 1·3 per cent per day, but the mineralization of humus compounds was much slower, only about 4 per cent per annum. This must be a very low rate for such conditions, as there is little or no accumulation of humus in most tropical forest soils, otherwise the closed cycle of nutrients on which the life of tropical rain-forests depends, would be impossible. All ecological activity is limited to the surface zones of the soil in these forests. The underlying soil may be highly impoverished by podsolization or otherwise, but the further progress of impoverishment is delayed or prevented by the protection of the forest cover. The mineral resources of the subsoil have been largely drawn by root action into the upper, active layers during the long period of forest development and in the mature ecosystem there is little further need for them.

The biological agents of decomposition are very active, ranging from bacteria through fungi (including some yeasts), algae, protozoa and nematodes to higher animals, acarines, centipedes, ants, beetles, grubs, slugs, snails, spiders and woodlice, while outranking all others in size, and nearly all in potential, are the earthworms. Mice, too, though on rather than in the soil, may play no inconsiderable part in the destruction of litter and by the increments of their excreta to the soil.

Soil bacteria form a distinct ecological group, native to the soil, and are not simply a heterogeneous crowd. They are indeed a balanced community about whose ecology far too little is known. Invasion of bacteria from wind-borne dust must be continuous, yet few invaders survive for long and the community retains its distinctive character. Their numbers are, of course, enormous, running into many millions per gram of soil and as their individual requirements are so minute, some bacteria can thrive even in desert soils.

The great majority are heterotrophic, actively concerned, each in its own way, in the breakdown of organic substances of the most varied kinds, from which they derive their Carbon and Nitrogen and their chemical energy. Most of these are aerobic, though some proteoclastic anaerobes are also present, generally spore-formers of the genus *Clostridium*, including the Nitrogen-fixing *C. pastorianum*.

The organisms in the soil can be grouped, on a biological basis, into soil inhabiters and soil invaders. Most of the pathogenic organisms and the mycorrhizal fungi fall into the latter category. The true soil inhabiters were grouped by Winogradsky as *zymogenous* and *autochthonous*. Although he applied his grouping to the soil bacteria it also fits the fungi and protozoa.

The zymogenous organisms are those concerned with the first attacks on fresh organic material. They constitute a wave of activity which is assisted by invertebrate animals while the organic structures are still more or less intact. The animals devour organic structures and replace them by a mass of excreta, which is, in turn, attacked by the zymogens. All this is located in the F layer, immediately below the surface The zymogen attack begins with what are called 'sugar fungi', because they only require simple substances such as sugars, organic acids and some amino-acids. They are mostly Phycomycetes. They are followed by cellulose decomposers, mostly Fungi Imperfecti, and finally by lignin decomposers, mostly Basidiomycetes.

The autochthonous group, the aborigines of the soil, is the basic population, concerned with humification but not with fresh organic matter. They are perhaps more clearly marked as a group in poorer, wild soils than in rich agricultural soil, but it is very difficult to name particular organisms as definitely autochthonous, for too little is known about them. Almost the only fungus assignable to the group is *Zygoorrhyncus vuillemini* which is often found in the lower layers of sandy soils. It will grow in washed sand which is irrigated with a water extract of soil, suggesting that it can live entirely on dissolved substances. Its population in the soil shows little fluctuation, in contrast to that of the zymogenous forms and it may be that this is a distinguishing feature of the autochthones.

A curious point about the soil bacteria is the discrepancy between their numbers as estimated microscopically in dilute soil suspensions and the numbers found by culture of a soil suspension on agar. The former numbers are always much higher than the latter, which indicates that a large proportion of those seen microscopically are either dead or incapable of multiplication on the culture media employed. Dead bacteria however, even before they disintegrate, are capable of acting as colloid bodies in ionic exchange reactions, and when they disintegrate they liberate nutrients to the soil in quite significant amounts.

In comparison with the inorganic soil the *mass* of the bacterial population is small. Jensen has conservatively estimated that 1000 million bacteria per gram of soil only amount to 0·1 per cent of the weight of the soil, but even this small percentage provides a weight of 2·5 metric tons of bacteria per hectare, reckoning the soil, to 20 cm depth, at 2500 metric tons. The bacteria have an average content of about 20 per cent dry weight of body substances, of which 50 per cent is Carbon and 10 per cent Nitrogen. This means that even at this relatively low number of bacteria (a count of 56,000 million per gram has been recorded) some 250 kg/ha of organic Carbon and 50 kg/ha of organic Nitrogen are locked in bacterial bodies and *pro rata* for higher numbers.

There has been little attempt at the laborious and unrewarding identification of species in the bacterial flora of the soil, except for a few outstanding forms with specialized habits, but morphologically they have been grouped by Taylor (1942), from cultures as follows: gram negative rods, 36·1 per cent; gram positive rods, 42·7 per cent; cocci, 3·8 per cent. These figures were based on 625 cultures of British soils. Lochhead and Thexton give somewhat different proportions for Canadian soils: gram negative rods, 13·3–51·1 per cent; gram positive rods (of all types, including some gram variable forms) 21·2–70·6 per cent; cocci, 1·5–7·7 per cent. Both estimates agree in the preponderance of gram positive rods. What value can be attached to these estimates is doubtful, since a large number of pleomorphic forms occur, especially in culture, and particularly among coryneform bacteria and those of the *Arthrobacter* groups. Moreover, Lochhead has shown that the great majority of soil organisms require the addition of accessory growth substances to permit their growth on agar; only about 10 per cent grow satisfactorily without such additions. Finally, these are all aerobic forms, and quantitative information about soil anaerobes is very scanty, even about the forms concerned in Nitrogen fixation. Cultures have yielded numbers of over one million per gram of soil, more than the numbers of *Azotobacter* (see below), but for various reasons these estimates are certainly too low. For one thing many anaerobes require additions of accessory growth substances before they will grow in artificial cultures. Their specific requirements are often different, so that while any one accessory substance, or even a mixture, will greatly increase the numbers developing in culture, we have no means of knowing what the potential total may be. The intrinsic Nitrogen-fixing power of *Clostridium* species is low in relation to the carbohydrate respired and is only occasionally as efficient as that of *Azotobacter*; they tend to flourish in the lower soil layers and *Azotobacter* near the surface. *Azotobacter* is more acid-tolerant than most aerobes, but a strong growth of aerobes can assist anaerobes by de-oxygenating the air in small pockets and producing a suitable redox potential for the anaerobes to grow in. In any case the total N product of the independent N fixers of the soil is small, almost insignificant compared with the product of the symbiotic Nitrogen fixers (see Vol. 3, p. 3043).

Among the aerobes, *Azotobacter* and its tropical correlative *Beijerinckia* are the most powerful and widespread N-fixers, but the power is shared with some Cyanophyceae, and some of the Purple Bacteria and some chemosynthetic forms such as *Methanomonas*, the role of which in the soil is uncertain. *Azotobacter* is widely distributed, except in arctic soils, but is rare below pH 5·8 and is most abundant and active between pH 7·0 and 8·0. Its abundance in the soil varies sporadically, but the highest numbers found are to be reckoned in tens of thousands per gram rather than millions. Its numbers are greatly depressed by nitrogenous manures, but are raised by carbohydrate supplies such as glucose or mannitol, though without increasing the amount of Nitrogen fixed. Even the best recorded yields of fixed Nitrogen would be quite insufficient for the needs of crop plants. The yield is, moreover, expensively produced, the ratio of N fixed to carbohydrate consumed being of the order of 1:40–50. Nitrogen fixation by *Azotobacter* is unlikely as a concomitant of humus breakdown since this process releases nitrate and ammonia, both suppressors of the bacterium.

The symbiotic N-fixing organism *Rhizobium* is primarily a constituent of the bacterial flora of the soil and can pass a considerable part of its life history as a simple saprophyte. It has an apparently complex cycle of stages, including a coccoid stage, a motile stage of small-celled flagellated cells, a non-motile bacillary stage, a filamentous stage of many cells, and, according to Israelskij, a filter-passing stage. It can be isolated from soil samples on ordinary growth media, but prefers a medium with a low N content. There are, however, a large number of biological strains of *Rhizobium*, differing morphologically, as well as in their requirements of accessory growth substances, in the preferred growth medium and in slime formation, and so forth. At one time these were regarded as all variations of one species, *R. leguminosarum*, but some six or seven are now considered tentatively to be specifically distinct. Extensive experiments on inoculation of leguminous host plants have shown wide variation in the infection relationships both of bacteria and of hosts. Some biological strains of the bacterium have a wide range of infectiveness, others are narrowly specific and the same appears to be the case among the hosts with regard to infectability. Even different genetic strains within the same species may differ considerably in their infection relationships. With the reactions of the bacterium within the host plant we are not concerned here (see Vol. 1, p. 359; 2, p. 1663; 3, p. 3043).

The opposite of Nitrogen fixation is de-nitrification, in which nitrates are used as Hydrogen acceptors in bacterial respiration and eventually molecular Nitrogen is released. The process is limited to a group of heterotrophic bacterial species in the genera *Pseudomonas*, *Achromobacter* and *Micrococcus* along with the autotrophic *Thiobacillus denitrificans*. In cultures the process requires a very low Oxygen tension and a neutral reaction and is therefore limited to micro-habitats in most soils, so that it is seldom extensive enough to be a serious factor of loss, but in a water-saturated soil, rich in nitrates and provided with suitable nutrients as sources of energy, the process can rise to high dimensions. Such a medium is, for example, a fresh manure heap, in which the losses of Nitrogen during decomposition can be formidable. For this reason fresh manure should never be added to a soil in cultivation.

The ratio of Carbon to Nitrogen in the soil, the C/N ratio, is often regarded as an index of the microbial activity in organic decomposition and indirectly of the soil's fertility. It is highest in cold, acid, or waterlogged soils, where microbial activity is restricted, and lowest in soils with a high pH, in drier and warmer climates. The loss of nitrates by leaching under heavy rain raises the ratio and there may thus be considerable seasonal fluctuations. Ratios between 12 and 18 may be regarded as normal for the upper layers of brown earths in western Europe; in arid regions dropping to about 6. Tropical soils are often richer in Nitrogen and the ratio may be about 12, but decreasing with depth and in the dry season.

Alongside the bacteria must be reckoned the related groups of Actinomycetes, which are minute hyphal forms, intermediate morphologically between bacteria and true fungi. They are reproduced by conidia which are freely formed in the air spaces of the soil. Owing to difficulties in their microscopic recognition and uncertainties in their development in cultures, their abundance in the soil, estimated at a few million per gram is quite uncertain, but they are biochemically active and apparently play their part in organic decomposition. They are also a rich source of antibiotics and may thus play a role in controlling the microbial population. *Streptomyces*, notably abundant in saline soils, seems to be the most widespread genus and next to it *Nocardia*, which is closely related to *Mycobacterium* and classed with it among Proactinomycetes.

The fungal flora of the soil is very large. Examples of almost every saprophytic group can be isolated from soil, though the number of those which are vegetatively active is much more limited and consist mostly of Fungi Imperfecti, Phycomycetes and lower Ascomycetes including Yeasts. They differ from bacteria in being mostly acid-tolerant and in mor or peaty soils they are the chief agents of decomposition. In numbers they are fewer than bacteria, but in mass they are nearly as great. Their maximum number is about 8 million per gram, but enumeration is suspect, for an unknown percentage is represented only by resting spores, which may or may not become vegetatively active in the soil. They seem to be most numerous and most active in neutral soils which are warm and constantly humid, for they are sensitive to drought when vegetative, though their spores can be very resistant. In addition to fungi identifiable in culture, there are also numerous sterile mycelia, which have been claimed as forming the majority of fungi in woodland soils.

Fungi possess by their mycelial growth the power of extensive penetration of the soil. Their effects are therefore more nearly uniform throughout a soil mass and less affected by local concentrations of nutrients, which are the cause of some discontinuity among the bacterial population. The laborious measurement of the lengths of all mycelial fragments microscopically observed, has yielded estimates as widely different as 25–2640 metres of mycelium per gram of soils in different parts of the world. Probably some percentage of Actinomycetes was included in these measurements.

The usual method of isolating soil fungi has been by plating out drops of a soil suspension on agar, a method which is strongly selective and is biased in favour of quick-growing and sporing species. Many fungi are restricted in regard to the growth conditions they can tolerate or in which their spores will germinate and all workers have found that modification of the growth medium used produces considerable differences in the flora isolated, both quantitative and qualitative. For example, in one test with Coon's agar, lowering the pH from 5·0 to 4·0 doubled the numbers isolated from the same soil. Apparent uniformity of results must therefore be suspected as reflecting uniformity of method, and existing numerical estimates as being too low.

Qualitatively, the species lists may be incomplete, but the constant recurrence of members of certain families and the large numbers of their species can leave little doubt that they are the most important elements among soil fungi. Among the 617 species recorded, as assembled by Gillman, the largest numbers are as follows:

PHYCOMYCETES	Mucorales	97 species
	Saprolegniales	73 species
ASCOMYCETES	Sphaeriales	18 species
FUNGI IMPERFECTI	Moniliales	375 species

These numbers are obviously influenced by the relative size of the respective families and they give no information about the abundance or frequency of individual genera or species.

Specific enumerations give various results. Some workers have found correlations between fungal species and the characters, vegetation and history of the respective soils. Others have found little or no sign of such correlations, but, on the other hand, considerable uniformity between different soils, at least those in the same locality. The conclusion seems to be that considerable diversity exists, but a geography of soil fungi lies in the future.

Certain genera have appeared with such constancy in soils in North America and Europe as to leave little doubt that they, at least, are practically ubiquitous, such are: *Alternaria, Aspergillus, Cephalosporium, Cladosporium, Fusarium, Mucor, Penicillium, Rhizopus, Trichoderma* and *Zygorrhyncus*. Several of these are also very frequent in air samples and are probably wind distributed in dust.

There is some reciprocity of influence between soil organisms and the vegetation of higher plants, so that the soil beneath an old and stable form of vegetation may acquire a special soil flora. For example, *Mucor ramannianus* is said to be characteristic of the soil of northern pine forests, together with several accessory species of *Mucor* (*strictus*, *flavus*, *sylvaticus*) which are associated with it as a 'society'.

A peculiar fact is the almost total absence of Basidiomycetes from the soil flora proper. Considering the abundance of Hymenomycetes, especially Agaricaceae, which grow on the ground there must be a special reason for their absence from soil samples. There is perhaps more than one reason. For one thing, their spores are shed from sub-aerial fruit bodies and are mostly trapped by the litter layer, while it is in the relatively acid conditions of the subjacent fermentation (F) layer that they chiefly vegetate, as can readily be seen by disturbing the litter in almost any woodland. Second, their thick cable-like hyphae are relatively slow growing. The same applies to sclerotia or rhizomorphs formed in the soil, and they are easily overwhelmed by fast spreading moulds in cultures made from the soil. In spite of all this they have their own special importance as agents in the formation of mycorrhiza as we shall see below.

Soil inhabiting algae are relatively few, both in numbers and in species and it cannot be asserted that they play a significant part in the microbial life of the soil. There is more algal activity on the bare surfaces of the soil and some species attain in this way considerable local importance, though it is more correct to treat them as part of the sub-aerial vegetation. Thus, two species, *Zygogonium* (*Zygnema*) *ericetorum* and *Mesotaenium violascens*, form carpets on bare soil in *Calluna* heaths in western Europe, while the red alga *Porphyridium cruentum*, a member of the Bangiales, sometimes forms extensive gelatinous patches on damp soil. Members of the Cyanophyceae are perhaps the algae most frequent in the soil as also on the soil surface. On submerged soils, in particular in flooded rice paddies, filamentous blue-green algae form superficial mats in which Nitrogen fixation is active.

Algae being photosynthetic, they would not be expected to thrive below the soil surface, yet the list of genera of Chlorophyta and Chrysophyta and Diatomaceae which have been found in the body of the soil is quite considerable. In some cases, moreover, the numbers are maximal at distances as much as 10 cm down. The amount of light filtering down to such depths can only be minute at best, though in loose soils some light does penetrate into the soil thus far, but only in scattered spots. The status of these soil algae has not been satisfactorily solved. They may be purely adventitious, due to washing down from the surface, or they may be able to live for a time by absorbing organic substances, semi-saprophytically. In favour of the adventitious view is the fact that most of the sub-surface species are unicellular and of small size and that the few filamentous species found are in the form of fragments. Moreover, Esmarch found that among the blue-green algae found at depths down to 30 cm, the species were 'with rare exceptions' the same as those found on the surface. Against it may be set the findings of Bristol-Roach that some unicellular green algae can live and multiply in the dark if provided with organic

C and N sources, without losing their chlorophyll. A few are also proteolytic and can liquify gelatine.

The Cyanophyceae are in a special position since many species can fix Nitrogen, while some species can undoubtedly live as saprophytes, as, for example, in the tissues of the underground stem of *Gunnera*. They have thus a stronger claim to be treated as truly at home in the soil than have the green forms, and to contribute to the microbiology of the soil. Some of the smaller Cyanophyceae (*Chroostipes*, etc.) can enter into a symbiotic union with bacteria and protozoa, called *syncyanosis*, which presumably enhances the nutrition of both parties.

Turning to the animal population the first group to claim attention is that of the Protozoa. Representatives of three main groups are abundant in soils, namely, the *Rhizopoda* or amoeboid organisms; the *Mastigophora* or Flagellates; and the *Ciliophora* or ciliate protozoa. Nearly all species have the power of encystment. The cysts are dormant and are valuable in enabling the organisms to withstand unfavourable periods. Sometimes, but not always, encystment involves reproduction by the division of the cyst contents into new individuals. Cysts are very resistant and enduring, and they can survive conditions which kill active organisms, a point of some practical importance. Encystment is not necessarily a direct response to unfavourable conditions. It seems to be a regular and periodic feature in the life history of many forms, and at any time the proportion of encysted forms is between one-third and one-half of the population, whether this is large or small.

Populations do vary considerably and can fluctuate very rapidly with changing conditions. This is due to their short lives and rapid reproduction rather than to active migration, which is probably limited to distances of a very few centimetres. The researches of Cutler and Crump on soils at Rothamsted in 1920 showed that populations of amoebae and flagellates could multiply themselves ten times or even more in twenty-four hours and return again to a low level just as rapidly. These fluctuations pose a biological problem not fully resolved. A suggestion that it was due to a mass-germination of cysts is negatived because there is no inverse relation between the number of cysts and the number of active organisms. The fluctuations seem rather to exemplify a phenomenon, often observed in organisms of the most varied kind, when a period of exceptionally rapid multiplication is followed, possibly through exhaustion of food supplies, by a dramatic decline in numbers towards virtual or even total extinction. The only special feature among protozoa is the rapidity of the changes. There is, however, a suspicion that the methods of counting exaggerate these fluctuations.

Some idea of the protozoan soil population is given by the figures quoted by Russell for Rothamsted soils, in numbers per gram of soil.

	Maximum	Minimum
Flagellates	770,000	350,000
Amoebae	280,000	150,000

These numbers are probably far short of actuality, though it is fair to say that much lower numbers have sometimes been recorded. The maximum and minimum averages are apparently seasonal, an autumn peak appearing in late November and a lesser, spring peak about the end of March. The percentage of cysts is highest in the summer months.

The distribution of protozoa in depth is uncertain except that they are most numerous in the top 15–22 cm of soil. Observations of their occurrence in active form have occasionally been made from deeper levels, even down to 2·8 metres, but at the deeper levels almost all the organisms were encysted.

Sandon, who made a world survey of soil protozoa (1927), found that the most abundant species in the Rothamsted soils were also the most cosmopolitan, occurring in soils of all types from all over the world. These species are: *Oikomonas termo, Heteromita globosa, Cercomonas crassicauda* (Flagellates); *Naegleria* (*Dimastigamoeba*) *gruberi, Hartmanella hyalina* (Amoebae); *Colpoda cucullus* and *C. steinii* (Ciliata). Encysted protozoa were found even in desert sands.

This survey also showed that the number of species of protozoa apparently indigenous to the soil is very small. Sandon catalogued about 250 species, but out of these only 8 flagellates and 8 amoebae, with 5 other Rhizopoda, were found only in soil. This is not altogether surprising if we take into account the exceptionally wide tolerance displayed by protozoa, which, in the soil, seem to be indifferent to the composition or reaction of the soil, to moisture content, aeration or even to temperature, since they occur in arctic soils which are frozen for a large part of the year. They are thus well equipped to colonize a wide variety of habitats.

From the point of view of soil ecology perhaps the most significant feature of protozoan life is their relationship to the soil bacteria. In 1909 Russell and Hutchinson propounded the view that as protozoa eat bacteria their numbers in the soil are in inverse proportion. Their proof lay in what they called 'partial sterilization', which involved submitting soils to temperatures (or chemical procedures) which would kill off the protozoa, but leave the bacteria relatively unharmed. Protozoa are very selective with regard to the species of bacteria which they eat or reject, so that their influence on the bacterial population is qualitative as well as quantitative. The thesis of partial sterilization is simply that the bacteria are essential agents of decomposition, humification, nitrification and Nitrogen fixation. Any agent which is inimical to them must also be inimical to these processes. Controlled heating by steam of the soil in bulk (82°C for 10 minutes is now accepted as standard procedure) results in an immediate drop in the number of bacteria, but this is rapidly made up and the final numbers considerably exceed those in the untreated soil. The superiority in numbers is more or less permanent and nitrification is markedly increased, and it is not due to chemical changes in the soil itself, for the introduction of a small portion of untreated soil brings about a fall in bacterial numbers. Exact numerical relationships between protozoa and bacteria have not been established, but qualitatively the enhanced fertility of the partially sterilized soil has been repeatedly demonstrated. That the relationships are not quite so straightforward as they were deemed to be by

Russell and Hutchinson was shown in 1935 by Cutler and Crump, who found that the production of metabolic CO_2 by soil bacteria in the presence of amoebae was almost double that by the bacteria alone, for equivalent numbers. The numbers of the amoebae are not stated. Their figures did, however, also show that the CO_2 production fell off as the numbers of bacteria increased, thus illustrating that a limitation of the bacterial population encouraged more active metabolism. Another peculiarity was brought to light by Nasir in 1923, and later confirmed by Cutler and Bal, that *Azotobacter* was more active in Nitrogen fixation in the presence of protozoa, especially Ciliates, than in pure culture. These results are still enigmatic. It is perhaps material to the question that temporary freezing and temporary drying also cause a rise in the bacterial population and it has been suggested that this may be due to the destruction of fungi and actinomycetes, rather than protozoa.

The invertebrate soil fauna is even more miscellaneous than the flora, but it includes many species which are only found in the soil at one stage of their life history, which may be a stage of dormancy. The proportion of the total which pass their whole lives in the soil is fairly small. Their numbers are not large, only a few thousand collectively of all groups per gram of soil, but they make up in weight what they lack in size and Russell estimates a gross dry weight of 235 kg/ha (210 lb/acre) of which earthworms and myriapods make up about 90 per cent.

Size, however, is not a good guide to activity. Thus Bornebusch in Danish soils found that mites and springtails (*Collembola*) showed a metabolic activity disproportionate to their small size as shown in Table 15.

TABLE 15

	Mull Soil pH 6·1–5·8		Mor Soil pH 5·6–3·8	
	Earthworms	Acarina and Collembola	Earthworms	Acarina and Collembola
Millions per acre	0·72	17·85	0·33	45·5
Percentage of total weight	75·1	0·4	22·4	2·3
Percentage of O_2 consumption	56·2	4·5	12·2	16·1

Thus the ratios of weight to O_2 uptake were, respectively, for earthworms 1:0·75 and 1:0·55 and for acarines and springtails 1:11 and 1:8.

Many insect larvae in the soil consume plant roots and may be regarded as purely destructive, and one of the very numerous and also destructive animal groups is that of the eelworms or Nematodes. They are practically microscopic, most species being less than 1 mm in length. Their numbers are small compared with protozoa, being about 1 million per square metre in woodland and arable soils and from 2–20 millions under grassland. They are, however, very active metabolically. Between twenty and thirty species are common and widespread and seem to be relatively indifferent to soil conditions.

Biologically they form three groups. (1) The most numerous group, prey upon the soil micro-flora, bacteria, algae and fungi. (2) A group which prey upon the soil fauna, including protozoa, other nematodes and small oligo-chaete worms. (3) Those parasitic on the roots of plants, which attract most attention. There is little evidence that any species feed on plant debris or assist the processes of decomposition. They have some effect upon the proto-zoan population, but it is their destructive effects on roots which makes them ecologically interesting.

One very remarkable fact is that some fungi have turned the tables on their predators by developing hyphal traps in which they catch nematodes, invade their bodies and digest the contents. A large number of species have recently been discovered and they are clearly abundant in soils. Some are Phycomycetes and have been formed into the order Zoopagales, others are Fungi Imperfecti.

We can pass briefly over most of the medium-sized or meso-fauna, in-cluding many insects, arthropods and mollusca, because many are only temporary soil inhabitants and there is little evidence that they, or the others which are genuine soil animals, contribute importantly to the economy of the soil except by mechanical burrowing and mixing. This is also true of the ants and termites, although they operate on a larger scale than most of the others, but their burrowing and nest building have little effect on the soil beyond im-proving aeration locally and mixing materials, especially mineral grains, in the upper layers. The same is true of the more extensive burrowing of small mammals, such as Voles, Gophers, Viscachas, etc., or even Rabbits, for in spite of their creation of remarkable tunnel-towns in prairie soils, they are intruders from the surface, not true indigenes. An exception must be made in favour of the insects which live in and on the surface litter. Although not strictly soil organisms, they contribute influentially to soil development by eating the litter and reducing it largely to a mass of excreta which is easily de-composed and incorporated in the soil.

The Earthworms, however, are in a different category. Truly indigenous to the soil they dominate the whole soil fauna where they flourish. Ever since the pioneer experiments of Charles Darwin (p. 3631), they have been accorded a leading place of interest among soil animals. Their importance lies in their active consumption of plant debris and their habit of dragging it down into their burrows, both actions which promote humification and the mixing of humified material with the mineral soil. They are sensitive to pH and the critical level is at pH 4·5, below which they are scarce or altogether absent. Their absence allows the development of mor soils in which there is a sharp separation between the acid humus layer at the surface and the mineral soil below. It might be said, indeed, that peat only forms in the absence of earth-worms.

Although the group of earthworms is worldwide in distribution the genera and species are very varied geographically. In northern Europe the species all belong to the family Lumbricidae, of which the large *Lumbricus terrestris* is the most important. In Britain there are some eight genera, with twenty-five species, but the commonly widespread species, along with *L. terrestris*, are *Allolobophora longa, nocturna, calignosa* and *chlorotica*, the last two usually in arable soils, the former under grass.* In New Zealand, where the introduced European species have practically ousted their native competitors, it has been noted that the total weight of earthworms per acre of soil is close to the total weight of sheep which the land will carry!

Apart from their function in mixing organic debris with the soil, earthworms also increase aeration and drainage by their extensive burrowing, mostly in the top thirty centimetres, but often much deeper, which in the case of *Allolobophora* goes on continuously. These burrows all connect to the surface and the number of such openings per acre can be reckoned in millions, each one a channel for penetration by air and water.

In 1881 Darwin published the results of many years observation of the habits of earthworms. He observed the average annual weight of wormcasts on the soil to be from 7·5–18·0 tons per acre. The actual weight produced varies with the nature of the soil and the average rainfall. It may be as low as 3 tons or as high as 20 tons. All this represents soil substance passed through the body of worms and excreted at the surface. Not all earthworms produce surface casts. *Allolobophora* species generally do, *Lumbricus* occasionally. Other species leave their excreta in their burrows, where it forms a smooth coating. In feeding earthworms ingest much soil along with plant debris. The amount discharged underground varies inversely with the weight of casts formed, according to the Rothamsted investigations, and the combined figure for soil consumed annually is about 20 tons per acre. The worms require a high proportion of Calcium in the soil, which accounts for their preference for a high pH. This is excreted as Calcium carbonate in the digestive tract and mixed with the excreta. The casts are very much richer in exchangeable bases than the original soil; i.e. Calcium, Magnesium, Potassium and Phosphorus together with nitrate Nitrogen; they also a higher pH, and an increased C/N ratio and percentage base-saturation. Their structure is also more resistant to weathering than that of the surrounding soil, and they probably contribute largely to the formation of the clay-humus adsorptive complex in the soil.

Darwin also observed many examples of the rise of soil levels due to the accretion of mould from worm casts. Where old surfaces could be recognized and dated the accretions showed rises of from 1·9–2·2 inches in 10 years, from which he deduced that the burial of ancient objects below the surface was chiefly attributable to worm action and was independent of their weight.

It is an important fact that there is no general correlation between microbiological activity in the soil, as measured by the size of the population, and soil fertility. The fertility depends on a plexus of factors many of which affect both the crop and the soil population. A correlation of plant yield and bacterial

---

* The worm which multiplies most readily in old manure heaps is usually *Eisenia foetida*.

numbers has exceptionally been found, but there is usually no simple cause-effect linkage between them.

That plant roots had a special and intimate relationship to the soil population has long been realized, and Hiltner in 1904 coined the term 'rhizosphere' to denote the zone of soil occupied by and influenced by the root systems. There are difficulties in defining such a region, for it varies with the soil. In poor soils the effects of an isolated root on the soil-flora may extend around it for 10 cm or more, so that, in effect, the entire root zone is included in the rhizosphere, while in rich soils the effects may be confined to a millimetre or so, that is to the immediate neighbourhood of the root surface. This has given rise to a distinction between 'inner' and 'outer' rhizospheres.

The rhizosphere effects are shown both in the increased numbers of micro-organisms as compared with soil outside the root zone and also qualitatively in the selective nature of the increase, which does not affect all the soil organisms indiscriminately.

Starkey, who laid the foundations of this study, used an R/S ratio to evaluate the root effect, that is, the ratio between numbers of bacteria and fungi, within the rhizosphere and in the general bulk of the soil. This ratio can vary greatly. It is influenced not only by the character of the soil, but also by the age and identity of the plants. The lowest values were found for cereals and the highest for legumes. For bacteria alone Starkey found ratios as low as 2 for cereals and as high as 50 for clover. The abnormally high ratio of 1400 was found by Webley for *Atriplex* in dune sand, but ratios over 100 are not uncommon. Nearly all groups of soil organisms are affected, the highest ratios being shown by bacterial numbers, except for Nitrogen-fixing and cellulose-attacking species, for which the ratios may be actually negative around young plants though becoming positive as the plant ages. The ratio tends to increase with the age of the plant, especially among annual or biennial species, but it is also positively correlated with the vegetative vigour of the plant. This picture is generally true for uncultivated soils, the addition of nitrogenous manure increases the ratios in almost all groups except for *Azotobacter*, where the ratio is scarcely affected.

As moribund and dead roots do not show a rhizosphere effect it would seem that materials provided by the living roots are responsible. There is good evidence for the excretion of amino acids from some roots and, in some species, of accessory growth factors such as aneurin and biotin. Lochhead, however, showed that the majority of the rhizosphere bacteria grew well on ordinary media without special additions, though some required additions of amino acids, and Lochhead and Thexton suggested that the latter group probably depended on amino acids synthesized by other bacteria rather than on the root supply. On the other hand Rovira showed that not only amino acids but also sugars were present in root exudates from oats and peas, and that the exudate from pea roots was sixteen times more concentrated than that from oats, which agrees with their respective rhizosphere ratios (see above).

As the plant grows older an increasing proportion of the diffusing material is derived from the decomposition of old cells sloughed off the root surface.

There is also an absolute increase in the amount, which may be as much as 100 per cent. That this is not the whole attraction is shown by the preference of some organisms, especially Mucoraceae, for the root-tip region, where only excreted materials are available. The association of soil fungi with roots varies from simple surface growth to root parasitism, e.g. *Pythium*, *Fusarium*, *Verticillium*, etc., and to symbiotic union in mycorrhiza.

The term 'mycorrhiza' does not properly apply to either the fungus partner or to the root as such, but to the associative union of the two, the infecting fungus and the (often morphologically specialized) roots on or in which it lives, which together form the 'fungus-roots', that the Greek name implies.

Mycorrhizal fungi are generally divided into two groups: First, *ectotrophic*, which form a mycelial mantle on the root surface, but do not invade the living tissue. This is generally the case in conifer mycorrhiza, and it may be regarded as an extreme case of the rhizosphere effect. Second, *endotrophic*, in which the fungus invades and lives in the living tissues of the host, linked only by hyphae with the soil outside; the orchid type. There is, however, a third form, *ectendotrophic*, which combines the growth of an external mantle with cellular invasion by hyphae, which later fragment and degenerate. This type occurs in some conifers and *Betula* and in Pyrolaceae.

Ectotrophic mycorrhizal fungi can penetrate among the cells of periderm, but there is no intra-cellular penetration. The mycorrhiza formed are usually short lateral rootlets which are much branched, usually dichotomously. This type of growth can be called out by dosage with auxin and so is presumably due to excretion of auxin by the fungus. Long roots can also be infected.

More is needed for mycorrhizal formation than simple contact. Strains of fungi vary in their infectiveness and conditions may affect the resistance of the root in one way or the other. The entry of the fungus is strictly an infection of the root, but it is a limited and controlled infection, in which the resistance of the root is the dominant factor.

Mycorrhiza are extraordinarily widespread. Apart from certain families, such as the Orchidaceae, the Ericaceae, the Pyrolaceae and the Coniferae, in which the habit is well nigh universal, it has been found that between 80 and 90 per cent of all flowering plants, in all climates and in all kinds of community, except pioneers and plants of submerged soils, possess mycorrhiza. There can be no doubt therefore of the importance and the advantageous nature of the association.

A mycorrhizal association also occurs in all saprophytic flowering plants, such as *Neottia* and *Monotropa*, Burmanniaceae, etc., and in a variety of Cryptogams, to wit in the saprophytic prothalli of many Pteridophytes and among Bryophyta in *Aneura* and its saprophytic relative *Cryptothallus*, also in many of the Jungermanniaceae, including *Marchantia* and *Pellia*.

The ectotrophic and ectendotrophic mycorrhiza of conifers and some other tree species have proved to be almost all formed by Hymenomycetes. More than fifty species have been shown to form mycorrhiza, involving some eleven genera, among which *Amanita*, *Boletus* and *Tricholoma* contain a majority of the species. Two Gasteromycetes are also involved, *Rhizopogon* and *Sclero-derma*. Thus nearly all the common toadstools of woodland are involved in mycorrhiza, though some which live exclusively in litter, as *Mycena* and *Collybia*, appear to be independent. To what extent the Hymenomycetes are dependent on their mycorrhizal connections is not certain. Even the vicious parasite *Armillaria mellea* is symbiotic with a Japanese orchid, *Gastrodia elata*, which only flowers when its saprophytic tubers are invaded by the rhizo-morphs of this fungus.

The truly endotrophic mycorrhizal fungi do not form so clearly defined a group as the ectotrophic forms. They include both septate mycelia, character-istic of Basidiomycetes, Ascomycetes and Fungi Imperfecti, and the non-septate mycelia of Phycomycetes. Some of the septate forms have been identified as Basidiomycetes, in particular *Corticium* from some Orchids and possibly from *Aneura*. The majority are only known as sterile mycelia, which are grouped into the form-genus *Rhizoctonia*. Among the non-septate forms the only systematic attribution has been to *Endogone*, which forms mycorrhiza with *Fragaria*. The sterile forms have been lumped together as *Rhizophagus*, by analogy with *Rhizoctonia*. Such intracellular symbionts are also known in fossil plants from the Devonian upwards.

As in the case of the nodule-forming bacteria (p. 3682) the specificity of fungal infection is variable. Some species of *Boletus* are specialized to certain genera of conifers, others are limited to conifers as against angiospermic trees. *Amanita muscaria*, on the contrary, is common to a wide variety of trees. Among endotrophic fungi it is rarely possible to make systematic identifica-tions, but undoubtedly biological strains exist, many of which, especially among orchid mycorrhiza, are highly specialized. So much so that it was formerly the practice of orchid growers to sow seeds only in soil in which the parent plant had grown, to ensure the infection of the seedlings. Today it is customary to isolate the fungus, whenever possible, from roots of the parent plant and to sow seeds into a culture of what may be safely assumed to be the 'right' fungus.

It is natural to ask whether the whole business of mycorrhiza is just an accident of nature or has it any biological value to either of the partners? The answer, so far as the host plant is concerned, is certainly, 'Yes, it has', and less certainly, though very probably, also 'Yes', for the fungal partner. Among orchids the dependence of the plant on its associated fungus seems to be well-nigh complete. It is true that under conditions of artificial culture, on media enriched with sugar and yeast extract or extracts of green tissues, it is possible to raise uninfected (asymbiotic) seedlings of many orchids, though some, especially the terrestrial orchids, will not respond. Such experiments demon-strate that the orchid seedling has requirements of sugar and accessory growth substances over and above its own resources and that these can be artificially supplied. They do not prove that the fungus is unnecessary under natural conditions, for it is true that the fungus partner does in some way satisfy the requirements to tide the seedling over its initial shortages. Once raised as far as leaf production a sterile seedling may become independent of external supplies of organic substances and some have been raised to flowering in a sterile condition.

There is much evidence to show that solutes can be translocated to the seedling from the external mycelium and that they can be transferred from the internal hyphae to the orchid cells, just as substances can pass from one parenchyma cell to another. The digestion of the fungal hyphae by the cells of the higher plant has been regarded as a means of assimilating its materials. This, no doubt, happens, but it is not so effective a means to this end as the continuous transfer by diffusion from living hyphae. Digestion of the internal fungus has also been regarded as a defence or immunity reaction on the part of the host and, as some at least of the fungal partners are potential parasites, such a defence reaction seems definitely needed.

The ectotrophic mycorrhiza of conifers, and of some broad-leaved trees such as *Fagus*, seem to be essential to their healthy development. This has been shown repeatedly in nursery beds of conifer seedlings grown in soils known to be free from fungi of tree mycorrhiza. Seedlings intentionally infected develop strongly while the sterile controls remain weakly and stunted.

The analysis of mycorrhizal and uninfected seedlings of *Pinus virginiana* by McComb showed that the former are notably richer in Nitrogen, Potassium and Phosphorus, especially the last of these. Melin also showed by the analysis of young *Pinus* growing in a prairie soil that plants with mycorrhiza absorbed 86 per cent more N, 75 per cent more K and 234 per cent more P than plants of equal age without mycorrhiza. This has been repeatedly confirmed and it is clear that phosphate absorption is a critical factor in the fungal-host relationship. The absorbing surface of mycorrhiza is greater than that of uninfected roots and hyphae spread from them to a varying degree in the surrounding soil and it has been proved that these hyphae act like root-hairs in absorbing and translocating solutes. This gives the mycorrhiza a competitive advantage in absorption over the uninfected roots. There is also some evidence that the fungus may increase the solubility of soil phosphate. However, it has been shown by using P^{32} that 90 per cent of the phosphate absorbed by the mycelium in the soil is retained by the fungal coating of the mycorrhiza and only 10 per cent passes into the root tissues and is further translocated. This great disproportion seems to be peculiar to phosphate, the alkali metals are not similarly retained by the fungal sheath. A similar restrictive action on the absorption of phosphate has been attributed to rhizospheric bacteria coating a root surface, when the phosphate is only available in small amounts. When the phosphate concentration is raised sufficiently to saturate the bacterial coating the restriction disappears. In view of these observations it would seem that the increased phosphate content in mycorrhiza as compared with non-mycorrhizal roots represents chiefly the accumulation of phosphate in the ectotrophic fungal sheath and that this sheath has a restrictive action on root absorption. Further, when the sheath is stripped off the rate of absorption of phosphate by the root itself is about doubled. The restriction becomes ineffective at high phosphate levels and the retention of the absorbed phosphate by the fungus is not necessarily permanent. If the mycorrhiza are in a phosphate-free medium, i.e. when no absorption is going on, phosphate is then passed on to the host from the fungal sheath by a metabolic process which is temperature and Oxygen sensitive. This phosphate appears to be more readily translocated to the plant shoots than that from uninfected roots. The overall action of the fungal sheath therefore appears to be that of a reservoir, controlling phosphate supply to the host.

So far as the fungi themselves are concerned the evidence of Melin strongly supports the belief that the mycorrhizal fungi are in a special biological category and that most of them have only a restricted life apart from their hosts, in other words that their symbiosis is obligate. What it is that brings about the close association of the fungus with a susceptible host is not certainly known. It is related to a high carbohydrate status in the host tissues, but this is quite unspecific and the action of specific secretions from the roots which affect only selected fungi is little more than an attractive idea. Once association is established, however, Melin and Nilsson have proved that considerable amounts of organic photosynthates pass from the host to the fungus, if the host is an intact green plant. Melin has shown that materials exuding from roots have a stimulating effect on fungal mycelia growing in plate cultures but it is not clear that this effect is confined to mycorrhizal fungi, nor is it only shown by the roots of mycorrhizal plants. The nature of Melin's hypothetical 'M-factor' or attractive agent, therefore remains obscure. Melin also showed that there were anti-fungal exudates. During life these only tended to reduce the effectiveness of the stimulating substances, but after the death of the root they persisted and showed a clear effect in fungal cultures. The specificity of attraction between root and fungus may therefore be a compound effect of balanced attraction and repulsion.

Ecologically the most important distinction is between *primary root fungi* and *primary soil fungi*. In the first category are the obligate parasites and obligate mycorrhizal fungi, which have little if any capacity for independent growth. In the second category are those whose life is normally saprophytic. These may be divided into those which are facultative parasites, that is to say weak parasites, which sometimes attack living roots, gaining their introduction by exploiting the dead or moribund tissue associated with root lesions, and 'true soil fungi' which are exclusively saprophytic and which Burgess divides into sugar fungi and humus fungi. Some degree of synergism must exist between these two groups for Doryland has shown that as little as 0·2 per cent of glucose reduces ammonification practically to zero.

The virulence of parasites is well known to vary greatly between strains of the same species and it can be changed in the same strain by appropriate treatment. Prolonged saprophytism weakens virulence, or what Garrett calls 'inoculation potential', passage through a living body increases it. In this way a cleavage may arise between the habitual root-invading fungi, whether parasitic or mycorrhizal, and the actively saprophytic species, among which an inoculation potential has been lost or has never arisen, while in the former group the potential is renewed or reinforced in each generation.

When we look at the soil population as a whole, from the point of view of its distribution, we will be struck by two apparently contradictory facts. On the one hand is the irregularity of its local distribution both in space and time and on the other hand is its cosmopolitanism. Local irregularities are related to the heterogeneous character of the sites available in the soil for exploitation by micro-organisms. Particles of decaying organic matter of the most varied composition are scattered through the soil and every one, however small, can be a microhabitat for organisms. Direct observation microscopically has shown that soil bacteria tend to occur in clumps of one species, the average size and number of which is characteristic of the soil. These clumps are associated with exploitable organic particles and as these can be moved about in the soil by drainage water, small animals or root growth, so the distribution of bacteria and fungi changes. Motile bacteria and fungi can also migrate spontaneously from one 'nidus' to another.

Every substrate particle passes through three phases; initial colonization, exploitation and exhaustion, at the close of which the micro-organisms concerned either form resting cells or migrate. During the exploitation period chemical and physical changes in the substratum will induce a succession of different species of organisms, involving fluctuations of the population in time. The history of each species of micro-organism follows a wave pattern in time, increasing to dominance at certain spots and certain moments and then falling away into a trough of the wave as it is succeeded as dominant by another species. Succession does not here involve the total suppression of earlier organisms but only their overshadowing. Dominance is transient and cyclic. If we remember that a cubic centimetre of soil may contain hundreds if not thousands of micro-habitats, each with its developmental history, it is easy to grasp the immense complexity of the distributional pattern of micro-organisms and how gross is the information obtained by examining the flora of bulk samples. 'Pattern' is the key word, but it is a microscopic and ever-changing pattern and only micromethods can hope to analyse it, even to a limited extent. This kind of information is still lamentably scarce.

Over the micro-pattern there is superimposed a macro-pattern due to environmental influences. Differences of water content, texture, aeration and reaction bring about variations of the soil population as a whole, in different areas or different zones of the soil, which are easily traced in bulk samples. From the point of view of the ecology of the surface vegetation these are more immediately significant than the variations of the micro-pattern, though the latter may have the greater ultimate importance in conditioning the availability of soil nutrients.

Large-scale movements of bacteria and fungal spores can occur by down-wash of water. Burgess showed that among fungi the spores with wettable coats moved fairly readily downwards, reaching depths of over 40 cm from the surface, but that non-wettable spores tended to remain near the surface, though a very few may penetrate below 20 cm. Bacteria may be similarly affected but there are no data.

Distribution in depth is influenced by the Oxygen tension and by the amount of CO_2 in the soil air The zone of greatest O_2 depletion is often between 10 and 28 cm, which is correlated with, and probably caused by, the large bacterial population at this level. It is also the level at which the greatest number of anaerobic bacteria are found. So far as fungi are concerned penetration in depth is controlled not so much by the Oxygen deficiency, which is rarely below 10 per cent, as by the CO_2 content. Carbon dioxide tolerance varies between species, but in the soil the percentage of tolerant species increases downwards as might be expected, though it does not, as might equally be expected, reach 100 per cent, but is only about 50 per cent of the fungal flora in the C horizon. *Penicillium* is particularly sensitive to increased CO_2 percentages, and it is interesting to correlate this with the fact that its spores are non-wettable and cling to the soil surface.

A basic consideration in soil ecology is, as Russell pointed out, the total energy available in the organic substrates which serve bacterial and fungal metabolism, since this is a fixed datum which has to be shared by the whole soil population either directly, or secondarily by the protozoa and other predators. The addition of farmyard manure to a soil very greatly increases the amount of energy dissipated in a given time (about 1:15), but does not cause a corresponding increase in the bacterial or fungal population (less than 1:2). The conclusion seems to be that the organisms after manuring are living more actively, but activity is not easily estimated. The CO_2 production has been generally used for this purpose, but there is not a complete correlation between this value and the energy output. On the other hand the numbers of protozoa give a useful hint, for their activity is observable and they do in fact increase greatly in numbers after manuring (1:4 or 5).

Every energy-yielding process of organic break-down is linked to an increase of the microbial population, either in numbers or in metabolic activity or both. Some percentage of the products of dissimilation is therefore used metabolically by the micro-organisms and is thus withdrawn from mineralization. This percentage has been expressed as the 'economic coefficient' of the organisms. We have already met similar coefficients in connection with the growth of higher plants. Among micro-organisms it varies from a maximum of 50 in aerobic moulds and yeasts to a minimum of 1 in anaerobic butyric acid bacilli. Average values lie between 10 and 20, signifying the percentage of the organic material undergoing decomposition which is embodied in bacterial substance.

The cosmopolitanism of soil organisms is very striking. Identical or closely similar forms can be recovered from soils in the most varied parts of the world. There is some degree of diversity between climatic zones, but it is more quantitative than qualitative: the number of genera known to be limited to the tropics, for example, is very small. This is attributable in part to the great facility of dispersal of small spores and partly to the fact that soils are well buffered against climatic changes and are therefore more uniform as habitats than the surface of the ground. Another factor is the stability of soil populations once established. Permanent changes in the population can only be brought about by procedures which cause fundamental changes in the constitution of the soil. Introductions of non-indigenous species into a soil have generally been found to produce a temporary stimulation of certain indigenous species which act as antagonists and destroy the invaders. This is true even when the introduced species are pathogens of the roots of higher plants. Only in cultivated soils or after the long cultivation of one crop, do the pathogens succeed in building themselves up to epidemic proportions.

As many soil fungi and actinomycetes are known to produce antibiotic substances, some of which have become well known in medicine, it is natural to suppose that they might in this way control the soil population. Experiments *in vitro* and in cultures have amply shown the activity of antibiotics in antagonizing the growth, usually of a restricted number, of other species but such tests are of limited use as indicators of what goes on in the soil. *Penicillium* and *Trichoderma* are powerful antibiotic producers and are abundant in soils, but their products can be neutralized or decomposed in neutral soils and there is only doubtful evidence that they have any effect upon root-pathogens living in the soil. In acid soils the antibiotic substances are more stable and their influence may therefore be proportionately greater. In spite of the number of species in the soil known to produce antibiotics it seems probable that their influence on other species is limited and local.

CHAPTER XLIX

THE FRESH-WATER ENVIRONMENT

As plants of an aquatic environment we shall consider those which are rooted in, or fixed to a submerged substratum, under standing or running waters, whether they emerge above the water surface or not, also those which grow floating freely in water or on the surface of water. Included among these are the majority of algae, some bryophytes and a few ferns, together with a great variety of angiosperms of very varied habits of growth. The margins of bodies of water very often pass by indefinite gradations into marsh or fen, with no clear boundary line, and we prefer to designate, as above, the kinds of plants which we shall consider as aquatics, rather than to attempt a physical definition of their environment, a definition which is bound to be complex and probably insecure.

The obvious primary division of the topic is between 'fresh' waters and 'sea' water and the separation seems valid in spite of the many gradations between the two categories which exist in coastal areas. Conditions in the sea, as such, differ fundamentally from those in any body of inland waters, even the largest (see p. 3742).

The subject is a vast one and has, in fact, been divided into two distinct sciences, *limnology* dealing with inland water and *oceanography* dealing with the sea. There are numerous textbooks and a very extensive literature in both branches and we can do no more than give an outline of the principal phenomena, as a guide to those who are interested in more detailed study.

Lakes, ponds and rivers

The plants of fresh waters avoid the perpetual problem of water conservation which dominates the physiology of land plants, but in doing so they submit themselves to another problem, Oxygen supply, which can be serious, quite apart from other concomitant problems such as Carbon dioxide supply, illumination, the supply of solutes, water pressure including currents and waves, and the temperature range including ice formation.

We might therefore expect that aquatics would show considerable differences in habit and life-form from related land plants. A biological classification in this respect starts with the separation of those plants which are attached to the substratum, the *benthon* (*benthos*) and those which float freely in the water and are carried about by water movements, the *plankton* and *pleuston*. A fourth category, of self-motile swimmers, the *nekton*, applies only to animals. The term plankton is generally used for the swarms of microscopic life, but logically it includes free-floating higher plants which are similarly situated, though these are generally separated under the name of *pleuston*.

Benthon

Among the higher plants there are five life-forms, all having attachments or roots which penetrate the bottom deposits. (1) Totally submerged. Among familiar British plants there are not many of these, since sub-aqueous pollination of flowers is difficult. One or two species of *Callitriche* (*C. autumnalis*); *Najas* and *Hydrilla*; the non-flowering *Chara* and *Nitella* and the moss *Fontinalis* are in this group. (2) Vegetative organs submerged, but flowers emergent above water. This is a larger group, including several species of *Potamogeton* (*P. praelongus*), *Ranunculus* (*R. circinatus*); *Lobelia* (*L. dortmanna*); *Myriophyllum*; and *Elodea*. (3) Plants submerged but with floating leaves. An abundant group. It includes many species of *Potamogeton* (*P. natans*); *Nymphaea*, *Nuphar*, *Nymphoides* and *Ranunculus* (*R. peltatus*). All have emergent flowers. (4) Plants mostly submerged, but the upper parts emergent. Not infrequently the submerged leaves and the emergent leaves differ in form (this may also apply in group 3). *Sagittaria*; *Sparganium*; *Hippuris*; *Polygonum* (*P. amphibium*); *Nasturtium* (*N. officinale*) and many others. (5) Plants of shallow water with only roots and rhizomes submerged and all the foliage emergent. *Alisma* (*A. plantago-aquatica*); *Typha*; *Scirpus* (many species); *Carex* (many species); *Menyanthes*; *Phragmites*; *Butomus*. This last group shades off into marsh plants.

Among benthic plants must also be included the many diatoms and other microscopic algae, with a few lichens, which grow attached to submerged stones or form a coating on the surface of submerged sediments.

Plankton and Pleuston

A comparatively small number of flowering plants are floaters and here again there are totally submerged and emergent types. The distinction may seem trifling, but the difference of habit makes a profound difference in the life of a species. Submerged floaters are represented in Britain by *Ceratophyllum* and *Utricularia* (flowers emergent), which are not only submerged but rootless. Among emergent or surface-floating plants we have *Hydrocharis*, *Stratiotes* and the ubiquitous *Lemna* as well as the water-fern *Azolla*. In warmer countries this group is often more abundant and includes such widespread plants as *Pistia*, *Eichhornia* and *Salvinia*. The bulk of the plankton is made up of the multitudes of microscopic and filamentous algae and microfungi, which often flourish in swarms among the submerged parts of the higher plants, depending on the Oxygen the latter release.

As the depth of the water generally determines the distribution of the different life-forms in a body of water, we find that *zonation* is a marked feature of aquatic life. The pleuston plants are generally in the deepest water where they have the surface to themselves, though they often stray into shallower places, where some of the more aggressive species like *Eichhornia*, the Water Hyacinth, are well able to compete with rooted plants. The benthon zones generally follow the order 1–5 of the groups we have described, in decreasing depth towards the shore line, though some with floating leaves, like *Nymphaea*, can adjust themselves, by elongation of their petioles, to quite deep water. Their large floating leaves, however, stand in need of shelter both from wave action and from wind, and they are therefore often found growing among reeds rather than in exposed positions. Hydrophytes as a class do not exercise the degree of stomatal control over transpiration that is characteristic of land plants. Although the stomata are often highly light-sensitive, they react very slowly to water-loss, if at all. Porometer readings on *Sagittaria* and *Nymphoides* in air, change only slowly and regularly with time and show no change of rate attributable to stomatal closure. Similarly, the loss of weight by drying in detached leaves is completely linear in *Alisma* and *Nymphoides*, while *Potamogeton natans* and *Eichhornia crassipes* only show a check to water loss when their water content is reduced to between 25 and 30 per cent. Not only is there little control, but the extent of the transpiration is quite surprising. *Potamogeton natans* transpired 6480 mg/dm^2/h as against 1280 mg from *Syringa vulgaris*. There is considerable variation between hydrophytes in this respect. *Nymphaea* transpired only 1760 mg and *Nelumbium*, 1750, while *Sagittaria*, at 1380 mg. lost only slightly more water than *Syringa*. The comparatively low rate of water loss in *Nymphaea* is associated with a very small area of conductive xylem in the long petiole, the ratio of xylem area to the cross-sectional area of the petiole being only about 0·020, less than one-tenth of the ratio in a normal land plant like *Syringa*. This may imply either an unusually high rate of water conduction in the petiole or, more probably, that the leaf supplies itself with water by direct absorption through the lower surface, without calling for much water from the petiole. This idea is supported by the presence on the lower, submerged, side of many floating leaves, of very numerous minute glands, called *hydropotes*, consisting of small nests of cells, with thin, perforate cuticle, which are active in the absorption of water. As the porometer readings on the leaf of *Nymphaea* increase about ten times between the point of attachment and the margin of the leaf, without a corresponding difference in the number or size of the stomata, it would seem that under the suctional influence of the porometer the petiole acts as a better source of air than of water.

Zonation in depth indicates a succession not only in place but in time. As lakes, ponds and river-beds are depressions in the land surface, they receive drainage from round about and thus tend to accumulate silt, which means that they all have a natural tendency to fill up. The extent and the rate of silting depend on the nature of the surrounding rocks, whether hard or soft. In a lake among igneous rocks the silting process may be so slow as to be almost negligible and the bottom remains rocky indefinitely. The same is true of rivers where the current is strong enough to carry silt downwards towards the sea. But in most quiet waters the process of silt accumulation advances measurably. To this process the plants contribute greatly, by the building up and subsequent decay of organic matter which is added to the mineral silt on the bottom. The process is quite analogous to that of soil formation on land except that it is here concentrated in a limited area and is consequently more influential locally.

As silting progresses in the deepest area among submerged and floating plants, their assemblage will be invaded by rooted plants, which become increasingly emergent as bottom deposits accumulate and the water gets shallower. The earlier species disappear and the marginal reed-swamp of *Equisetum*, *Juncus*, *Scirpus*, *Typha* and *Phragmites* gradually takes over the whole area, which soon ceases to be aquatic and becomes a swamp. There has been thus a succession or *sere* of species, the *hydrosere*, one of the most clearly marked and observable examples of ecological succession. Nor is it unduly prolonged. To anyone who knows the Norfolk Broads the rate at which reed swamp can invade and take over any areas of water sufficiently shallow is almost frightening. The process is self-promoting for the growth of the reeds increases the rate at which silt is captured and stabilized.

The succession in a hydrosere does not always proceed smoothly. It depends on a sufficient supply of mineral silt to mix with the organic matter, and it is greatly aided by some degree of water mixing, either by influent water or by periodic flooding. If water is entirely stagnant an excess of organic matter may accumulate, depriving the water of Oxygen and generating H_2S, resulting in practical sterility. Such waters shade off into *foul waters* which are contaminated with organic influents or by cattle, and in which the O_2 content and sometimes the penetration of light are too low to admit the growth of higher plants.

Other factors, too, may influence the hydrosere, for example, the ionic concentration in the water; likewise the reaction, which in practice usually means the Calcium bicarbonate supply; the prevailing temperature, especially the extent of freezing; the illumination and the Carbon dioxide supply, but in spite of all local variations and vicissitudes a hydrosere will almost always prevail. Finally this leads up to either *marsh*, with a mineral soil or *fen* with a peaty soil. No clear line can be drawn between them, they are all part of one long succession and the closing stages of the aquatic succession are the opening stages of another.

As Tansley and Pearsall have both emphasized, the silting factor is a dominant influence in aquatic vegetation. The amount and the character of the silt arriving determines the nature of the substratum and the rate of shallowing, it also determines the rate of, and the completeness of humification. It determines also the supply of nutrients for the rooted plants and in these and other ways it determines the all-important 'ageing' of the water.

Physical conditions

Having sketched above some of the prevailing ecological features of fresh-water vegetation, we may now turn to a more detailed consideration of the state of the water itself, physically and chemically.

First of the physical factors we will take *temperature*, since plants vary greatly in their tolerance range for this factor. Whatever their optimum may be, whether high or low, plant species are either *eurythermic*, with a wide tolerance, or *stenothermic*, with a restricted tolerance range. Variations of temperatures are thus of prime importance.

The distribution of temperatures in a body of water can only remain exceptionally and in very shallow water, uniform throughout, or *homoiothermal*. Surface evaporation causes cooling and any change of temperature causes changes of density in the water, which at once introduce convection currents, while wind on the surface generates turbulent motion.

If we imagine a lake to start (say, in early spring) with a zero temperature, then any heating of the surface by radiation will *increase* the density of the surface water until it reaches 4°C. At this, its maximum density, the surface water will sink and be replaced by colder water from below and the process will be repeated until the whole lake becomes homoiothermal at 4°C. Further heating of the surface water will, however, *decrease* its density and a stratification results, with dense water below and less dense above. Any mixing that now takes place will demand work. In very shallow water turbulence induced by wind may be enough to effect mixing throughout the water, so that it remains homoiothermal even though its temperature is rising. The deeper the water the more work is required to effect mixing and as wind-generated turbulence is of limited depth, deep water may remain cold throughout the season, its temperature corresponding approximately to the annual mean temperature of the surrounding area. In temperate zones this means about 4–6°C, but in tropical lakes it may rise as high as 24°C.

This process of surface warming by radiation results, wherever the water is deep enough, in a temperature stratification. The warmed and less dense surface layer is called the *epilimnion*, which rests upon the deep, denser water of the *hypolimnion*. The epilimnion is a zone of free circulation maintained by wind-generated turbulence and by convection currents, the result of radiation and of surface evaporation and nocturnal cooling. In shallow water the whole mass belongs to the epilimnion. In this zone water temperatures are therefore more or less uniform and show little, if any, variation with depth.

The epilimnion is separated from the hypolimnion by a narrow zone in which the temperature drops steeply. This zone, the *thermocline* is defined as the zone of maximum rate of decrease of temperature. At the top of the thermocline there is often a very sharp turn in the temperature curve, the point at which temperature begins its rapid fall. This has been called the *knee* of the thermocline, which generally lies between 1 and 10 metres deep. The whole thermocline zone has also been called the *metalimnion*. At the bottom of the thermocline we enter the comparatively stable hypolimnion, in the upper part of which there is still a decrease, though much gentler, of temperature with depth. This has been distinguished as the *clinolimnion*. Further down the temperature becomes uniform irrespective of depth, i.e. the temperature curve becomes vertical

Density differences due to temperature increase progressively with rise of

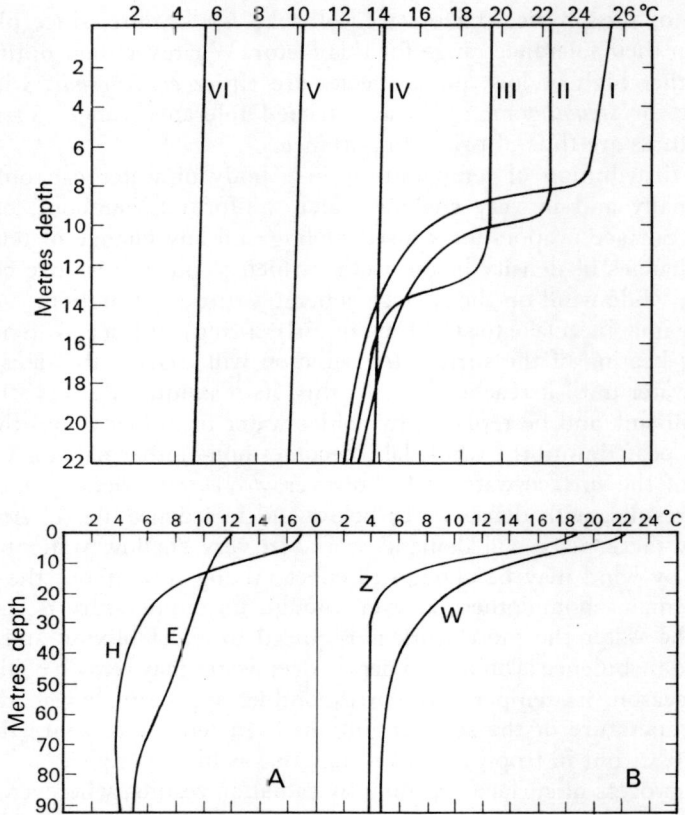

FIG. 4.51. *Upper figure.* Cooling in Lake Mendota (Wisconsin). I. July; II. August; III. September; IV. October; V. November 2; VI. November 24. Free circulation in the isothermal lake from October onwards. (*After Briggs and Juday.*)

    Lower figure. Summer temperature curves. A. The effect of oceanic and continental conditions. H. Hölsfjord (continental) and E. Evangervatn (oceanic). B. The effect of water volume. Z. Lake of Zürich, W. Walensee. The latter has a much greater mean depth than the former (103:44) and has an abnormally gentle curve in consequence of its greater volume development. (*After Halbfuss, from Hutchinson, 'Limnology', Wiley.*)

temperature. The density difference between 15 and 16°C is seven times that between 4 and 5°C and between 25 and 26°C it is thirty-one times. This has also a marked effect on volume. A rise from 5°C to 15°C causes a volume increase of nearly 0·1 per cent, i.e. 1 m^3 of water would increase by nearly 1000 cm^3 and 1000 m^3 in a pond would increase by nearly 1 m^3.

    Here we have induced stratification dependent on temperature differences, the stability of which increases as the temperature range involved rises, so

that in warm lakes even a small temperature range can induce a stable stratification. In temperate lakes the thermocline becomes established at the end of
June or the beginning of July. During the summer its depth varies between
3 and 10 m, but it tends to sink as the surface water is warmed up and by
September it may have dropped to 12 to 14 m. By the middle of October the
water will usually have become isothermal and the thermocline disappears.
In warm seas it lies exceptionally deep, about 27 to 30 m. The depth of the
thermocline can be affected by steady wind pressure from one direction,
which entrains the surface water and piles it up to leeward and it can be
temporarily dispersed by wind-force turbulence. Although it depends on a
temperature difference it must not be overlooked that the separation of two
distinct bodies of water involves all the other qualities of the water which are
important to life so that epilimnion and hypolimnion are in effect two different
habitats.

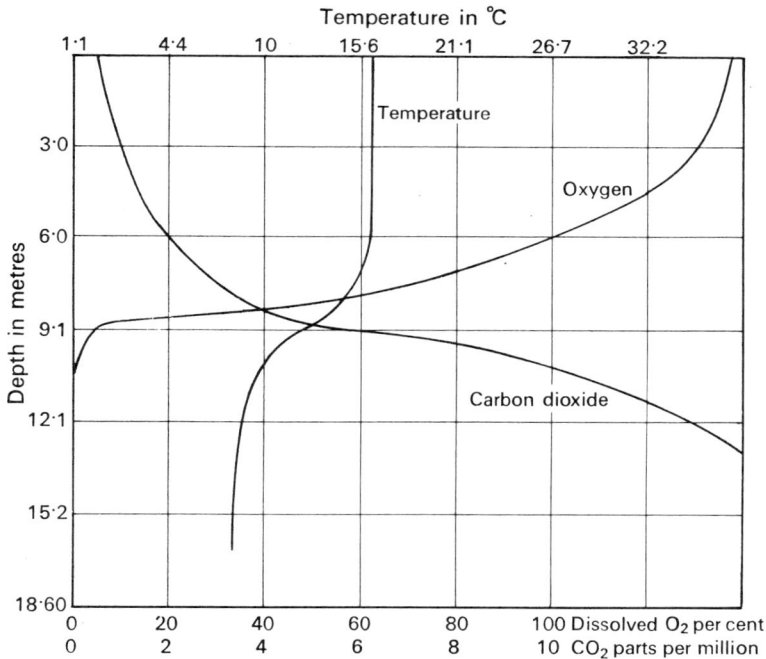

FIG. 4.52. The relationships of temperature, Carbon dioxide content
and the content of dissolved Oxygen to depth in a deep pond at
Cambridge, Mass. in October. (*After Kendall.*)

In a cold climate the process of stratification begins with the warming up of
surface water to 4°C (39·2°F), the temperature of maximum density. The dense
water sinks and is replaced from below. This involves the more or less
complete mixing of the water, the *spring turnover*, which brings to the surface
deep water often rich in solutes and dissolved CO_2. In consequence many

lakes (and shallow seas) show a *vernal outburst* of plankton and higher life-forms. Stratification begins when the surface is warmed above 4°C and continues through the summer. In the tropics stratification may be permanent and there is no spring turnover.

The existence of stratification depends on the bulk of the water being free from lateral movement. The water must therefore be (a) stagnant or (b) so large that influent and effluent streams have only a local effect on the stability of the water. It is impossible in rivers and in those lakes which are merely enlargements of a river bed.

What happens at the end of summer in a cold climate? As the prevailing temperature of the surface-water drops the difference between the strata will diminish and the thermocline eventually disappears. If the surface water eventually reaches 4°C, there will be a second or *autumn turnover*, the water becoming again homoiothermal. Further cooling of the surface below 4°C, will put a stop to the turnover and there will be a short period, in the range between 4°C and 0°C when a reversed thermocline may develop, the surface water now being colder, but lighter, than the deep water.

A similar reversed gradient on a small scale exists at the water/air interface. Cooling of the surface occurs, particularly in the autumn, where the air is colder than the water. This results in differences of as much as 1°C between water at 3 mm and water at 30 mm depth. The sinking of the cold water may carry dissolved gases through the surface layer.

The order of events will obviously depend on the local climate and several attempts have been made to classify lakes into thermal categories One useful grouping is that of Yoshimura (1926) which uses geographical terms.

1. *Tropical lakes.* Surface temperatures high, with little variation and a slight temperature gradient with depth. If any circulation occurs it is only once annually, at the coldest period.
2. *Sub-tropical lakes.* Surface temperatures high, never falling below 4°C, with considerable variation and a marked gradient in depth. Only one turnover annually, in winter.
3. *Temperate lakes.* Conditions as we have described above with two annual turnovers.
4. *Sub-polar lakes.* Surface temperatures only rising above 4°C for a short time in summer. Thermal gradient with depth small and stratification poor. Two turnovers in early summer and early autumn.
5. *Polar lakes.* Surface temperatures always below 4°C. Circulation only during the short ice-free period.

These groups are physically satisfactory, but the geographical terms may be misleading unless they are accepted as simply descriptive, since lakes of the types described are not limited to single geographical zones, but may be determined by all sorts of local conditions, e.g. sub-polar lakes among the high Andes or lakes of tropical type in western Scotland. A simpler classification was proposed by Hutchinson, namely, *monomictic* for lakes with only one annual turnover, divided into *warm* and *cold* monomictic; and *dimictic* where there are two annual turnovers.

A non-thermal factor which may interdict the thermal programme is that of deep water the density of which is raised above that of water at 4°C by its content of solutes. Such cases are exceptional, but when they occur they impose a permanent stratification, the hypolimnion remaining stable and the thermal turnover only descending so far as the density gradient permits. Such lakes

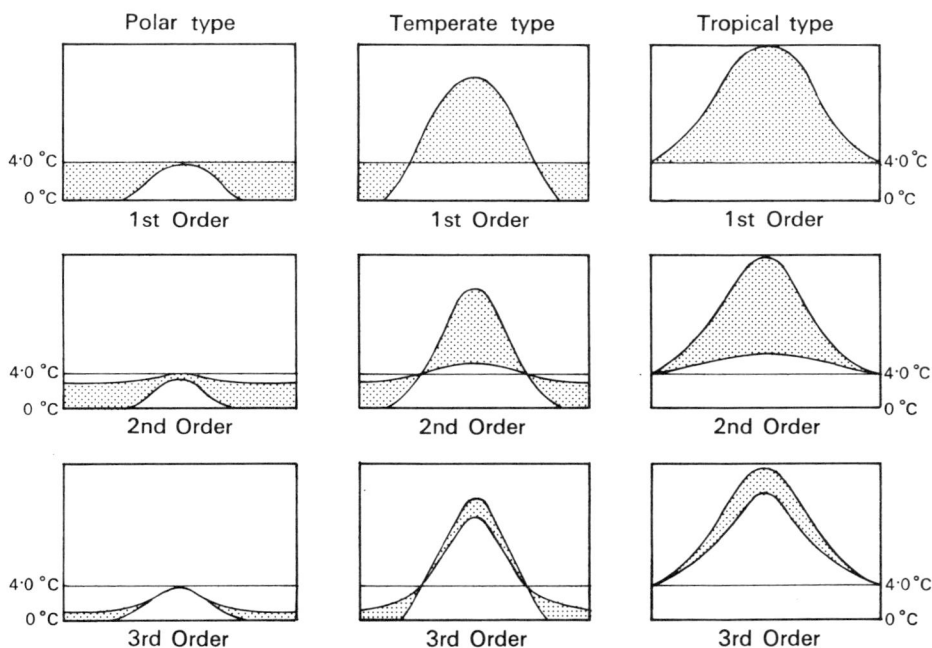

FIG. 4.53. Diagrams after Whipple, illustrating the annual sequence of thermal events in lakes according to the Forel–Whipple–Welch classification. Striped areas show the epilimnion. For details see text.

are called *meromictic* (i.e. partially mixed) in contrast to those where the water turnover is complete, the *holomictic* lakes.

Another factor promoting stability is only appreciable in very deep lakes, namely that the temperature of maximum density in water is lowered by increased pressure. Thus bottom temperatures less than 4°C can exist without causing instability. The fall in the point of maximum density is linear, with increasing depth from 3·98°C at the surface to 3·84°C at 100 m, 3·73°C at 200 m and 3·40°C at 500 m depth.

Nothing has been said so far about the depth to which the epilimnion may extend and, indeed, it is so variable that average figures would be meaningless. The complexity of the determining conditions have stimulated several mathematical attempts to analyse them so as to test the weight of the various factors concerned, with varying success. Experiment is the obvious resource in any given case and if our expectation lies between 0·5 and 10 m it will probably cover the majority of cases.

The most extreme type of polar lake is that which is permanently frozen and hence *amictic*. One such lake, in Antarctica, was investigated by Murray in 1909. During the summer it showed only superficial thawing. Boring showed 4·5 m of ice, which could never wholly melt under present conditions, yet from the muddy substratum Murray recovered living rotifers and other microscopic organisms. These survived drying on filter paper and when

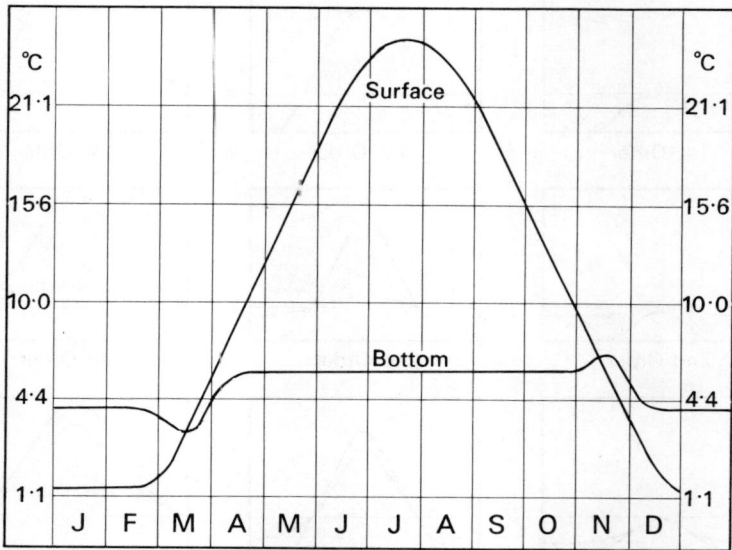

FIG. 4.54. Annual temperature chart of a dimictic lake 18 m deep.
(*After Whipple.*)

brought back to this country and soaked out showed unimpaired vitality, when seen by the present author, a marvellous demonstration of indefinite tenacity of life.

Another, opposite, extreme is shown by equatorial lakes. In these surface temperatures may be as high as 30°C and bottom temperatures from 20–27°C, which decrease linearly with altitude and presumably represent the minimum climatic temperature recently experienced. Small and deep lakes become more or less permanently stratified even with temperature differences of only a degree or two. Only some violent or unusual climatic disturbance will disturb the hypolimnion which otherwise remains stagnant and devoid of Oxygen. Such lakes have been called *oligomictic*. Large shallow lakes, on the other hand, heat quickly in the daytime and cool quickly at night, when mixing occurs. These are known as *polymictic*.

In sub-tropical areas certain lakes have been called 'solar lakes'. These are lakes with a sharp increase of salinity and density with depth. Consequently a very stable layer develops at the bottom which traps solar heat and retains it. This phenomenon has been suggested as a source of thermal energy in areas where fuel is scarce.

The *annual heat budget* of a lake is defined as the total amount of heat entering a lake between the time of its lowest and its highest heat content. This may be calculated by plotting against depth (Z) the product $A_z(\theta_{sz} - \theta_{wz})$, where A_z is the area at depth Z and θ_{sz} and θ_{wz} are the summer and winter temperatures respectively at that depth. This is calculated for a series of depths and the results plotted. This is integrated by measuring the area of the curve and dividing by A_0 the surface area. Hutchinson* has tabulated the data for all the lakes for which they are known. The highest total is for Lake Baikal, 65,500 cal/cm^2, the second being Lake Michigan, 52,400 cal/cm^2 (no data for the other Great Lakes). In Britain our largest lake, Loch Ness, has also the highest total, 37,200 cal/cm^2. It is difficult to draw comparisons because latitude, altitude, area and depth have all to be taken into account. There is no consistent relationship to either latitude or area, but polar lakes and lakes at high altitudes show generally low budgets. In both these cases, however, the latent heat of melting of ice is a very important part of the budget and if this is allowed for the budget figure rises to something comparable with temperate lakes, i.e. over 20,000 cal/m^2. Equatorial lakes also have small budgets owing to the high 'winter' temperatures, unless they lie at high altitudes like Lake Atitlan in Central America (22,110).

As nearly all the heat increment comes from radiation it is possible to compare the above figures, in some cases, with the radiation values measured at land stations in the same area, when it becomes evident that a 'first class' lake, with a budget between 20,000 and 40,000 cal, retains 40–60 per cent of the received radiation, rising in some to about 70 per cent.

The heat increment affects not only the water, but, in shallower waters, the substratum as well, especially in polymictic lakes, the heat thus absorbed being returned to the water in cold periods. The amplitude of the variation falls off rapidly with depth in the sediment, just as in a terrestrial soil, but depending also on the depth of the water. With increasing depth the total variation of temperatures in the sediments is reduced and there is a progressive delay in reaching the maximum, which may not be reached till October, while below 2–3 m from the surface of the sediment the seasonal variation may be practically nil.

The high specific heat of water as compared on a volume basis with air inevitably brings about marked differences in the thermal conditions of the two bodies. In spring the air temperatures rise much more rapidly than the water temperatures, which may not reach their maximum until the late summer. In autumn the contrary happens, the fall of water temperatures lags well behind the fall in air temperatures. The winter minimum of the water temperatures may not be reached until March. The amount of these effects

---

* HUTCHINSON, G. E. (1957) *A Treatise on Limnology*, Vol. 1, Wiley, NY.

is naturally proportional to the volume of water concerned. In small ponds they may be almost negligible, but they are most pronounced in large and deep lakes and most of all in the sea.

Bodies of water, either very extensive or very numerous, have thus an effect in modifying the climate of their neighbourhoods. By taking up or giving out heat they make the climate more equable thermally and prevent extremes. This is best exemplified on a large scale in the difference between an 'oceanic' and a 'continental' climate, but even isolated lakes have local effects which may be reflected in the local distribution of plant species. The difference of specific heats implies that a drop of 1°C in the water temperature in autumn or winter releases enough heat to raise by 10°C a volume of air many hundred if not thousand times greater. This may prolong the growing season or prevent severe frost. The vicinity of a lake has always a milder and more equable climate than obtains at a distance from it. Such thermal effects are apart from other climatic effects on humidity, rainfall or cloudiness which may also be of importance to plants.

Illumination. As we have seen the thermal properties of lakes depend on the radiant energy they receive. This is not all absorbed at the surface, but penetrates more or less deeply into the body of the water, so that its absorption is a complex differential process depending on its degree of penetration to different levels. This penetration is also indispensable for all photosynthetic organisms below the surface and has some importance for most of the others.

We have already discussed the terrestrial supply of radiant energy from the sun (p. 3539) and we need only recapitulate one or two points of special importance limnologically. The atmosphere cuts off most of the ultra-violet component of sunlight, so that the amount of energy received from wavelengths below 400 nm is very small at sea level. This amount rises considerably with height above sea level as the density of the atmosphere falls off almost logarithmically with altitude in the lower range of heights, so that at 8900 m one-half of the mass of the atmosphere is already below one. The angle of the sun is also effective in this connection, since with the sun at 30° above the horizon the path of its rays through the atmosphere is twice as long as when the sun is at 90°.

The total radiation energy received by a horizontal surface consists of direct radiation and diffuse radiation from the sky. The proportion of the latter is naturally very variable, depending on the state of the atmosphere, for instance clear or cloudy, and on the sun's angle and on the altitude of the site. It is greatest when the sun is low, reaching a value of about 40 per cent at sea level with the sun at 10° and about 20 per cent with the sun at 45° which may be taken as an average value for midday on a clear summer day in temperate latitudes. The values fall off considerably with altitude and at 3000 m are only about half the above.

The total radiation energy flux in calories per square centimetre per day at different latitudes and altitudes can be estimated by integration from the solar constant, the solar angle, the length of the day, and, for altitude, the transmissivity of the atmosphere. Perl (1935) has calculated the figures for various latitudes, which show that the *annual* flux is greatest at the equator, owing to the absence of seasonal differences, but that the highest *daily* totals increase with latitude, owing to greater day-length, the maximum of 667 cal being for a June day at the Pole, with a length of twenty-four hours.

Steinhauser has produced calculations for variation with altitude at lat. 47°N which show an increase over the sea level value of 12 per cent in the first 1000 m, but only 13·6 per cent at 2000 m for the summer season April–July, the total ranging from about 60,000 cal (extrapolated to sea level) to 86,699 cal at 3000 m.

These are theoretical figures for direct radiation and take no account of atmospheric losses. Actual measurements taken at sea level and at 3000 m gave values of 57,060 in both cases and it is safe to say that 60,000 cal/cm²/year is about the maximum that a surface in temperate latitudes is likely to receive. Some part of this radiation income is lost by surface reflection and by re-radiation of absorbed energy. After allowing for these losses we can estimate the *net radiation surplus*, which represents the actual gain of energy.

Reflection affects both direct sunlight and diffuse sky light. Reflection is a function of the angle of incidence and the angle of refraction according to Fresnel's law. It is greatest at low angles, but as comparatively little energy is received at low solar angles an approximate value of 10 per cent may be assumed for midday radiation. The values for diffuse light can only really be treated empirically, and they appear to be rather lower than for the direct radiation, so that an integral figure for the total loss by reflection and scattering is about 6 per cent in summer and perhaps 10 per cent in winter. Moderate surface disturbance does not seem to affect these values to any great extent.

Reradiation may be taken as closely approximate to black-body radiation, that is, following the usual rule, it is proportional to the fourth power of the absolute temperature, but its amount is complicated by back radiation from clouds and non-radiant heat losses by convection and evaporation.

The integrated result of the various processes involved is the net-surplus radiation, the actual income of the water in energy. This is not quite the same as the heat budget (p. 3709), although it is involved in the heat budget, since it is possible for water to be receiving a positive increment of radiant energy at times when the water is cooling, not heating, and this is regularly the case in autumn, when a lake may have a radiation surplus of more than 200 cal/cm²/day, but is in fact losing more heat by non-radiant conduction and convection than it is gaining.

The greatest surplus is in July, but it has fallen relatively little by September and under a clear sky may continue at small positive values throughout the year. Under a cover of 10/10 cloud the peak values in July are about the same as the clear sky values in December, though they extend over a much longer day. The net losses in winter (January) are, however, much less under a cloud cover than under a clear sky, which we have already seen applies to land as well as water (p. 3542). For practical purposes a peak surplus of 400 cal/cm²/day is a fair approximation for a clear summer day in temperate latitudes, or 500 cal in lower latitudes, with values falling away normally on both sides of the peak.

The penetration of light into water is of the greatest importance to living organisms and it has been the subject of a great deal of work. In the first place some absorption is due to the water itself, in other words is inherent in pure water. This is a matter for laboratory study, some of the best values having been established by James and Birge (1938) for a water-length of one metre. Absorption depends on the wavelength of the light. Beyond 700 mµ it rises very rapidly, reaching 91·4 per cent at 760 mµ. The far-red and infrared thus have very little power of penetration. In the visible spectrum, between 400 mµ and 700 mµ the amount of absorption at the short-wave end is very low, the minimum being in the blue at 470 mµ, where it is less than 1 per cent. It remains low until the yellow-orange is reached (600 mµ) where it is 20 per cent. Above this it increases very rapidly and at 700 mµ is 45 per cent. The obvious consequence is that the long wavelengths have a very low power of penetration compared with the shorter waves. Of the photosynthetically potent light (at 660 mµ) 69 per cent of the incident value survives passage through one metre of water. Integrating the values for the whole range of wavelengths shows a general absorption of 53 per cent in the first metre.

The above remarks apply only to pure water and therefore express maximal quantities. In natural water four factors may diminish them: content of solutes, suspended colloidal matter, suspended plankton, colouring matters. Their combined effects are summed up as *transparency*. The effect on transparency of suspended matter is generally non-selective, it diminishes illumination as a whole. The effect of solutes is chiefly on the refractive index and hence on the reflectivity of the water surface. Plankton may have selective effects, depending on its spectral absorption, while colour is generally absorptive for the short-wavelengths. Transmission cannot, of course, be treated as if all light were direct sunlight normal to the surface. Diffuse light comes at all angles and low angles of the sun lengthen the path of its rays through the water. A further complication is the fact that there is considerable horizontal variation in transparency. In a freely circulating epilimnion the transparency is practically uniform, but a layer of minimum transparency may occur just below the thermocline and in the deep hypolimnion there can be a complex pattern of transparency due to drifts of suspended matter.

It cannot therefore be assumed, except perhaps in some very clear lakes, that the percentage absorption of light remains constant with depth, so that the amount of light is directly proportional to the depth. Some coloured lakes show a *decrease* of the percentage of absorption with increasing depth. This is apparently due to the selective absorption in the top metre, in which the colour screens out all but the wavelengths to which the water happens to be transparent. The percentage of these wavelengths lost in subsequent increments of depth is thus less than in the top metre. Counter to these conditions there are others in which the hypolimnion becomes less transparent with depth, due to increased suspended matter, or from some other cause, and therefore the percentage absorption increases with depth and the illumination is less than would be expected.

The very clearest lakes approximate to pure water and in them, as in open sea, it is the blue light which penetrates deepest. In most clear lakes, however, without measurable colour, the green is the most penetrating. Slight coloration tends to exclude the shortest wavelengths and the deep water light is yellow. Increasing colour excludes more and more of the short-waves and in highly coloured water very little light and that all of the long-waves, gets below the metre.

The colouring matters in natural water are derived partly from the soil and partly from plankton, but chiefly from the former. They are naturally not of uniform constitution. Inorganic solutes may affect colour indirectly by changing the state of aggregation of coloured colloids, but not otherwise unless they are themselves coloured, which very rarely happens. Colour here means transmitted colour as seen through a column of water, but it can also refer to the reflected colour as seen by the eye. In the latter case the lakes of maximum transparency are blue, like the blue of tropical seas, signifying in both cases the almost complete absence of plankton life. From blue the range is through blue-green and green to yellowish or brownish, the yellowish being, as in the Arctic seas, the most highly productive.

Intrinsic, transmitted colour is measurable against a standard which can be a mixture of Potassium-Platinum chloride (2·492 g) and Cobalt chloride (2 g) dissolved in hydrochloric acid and made up to 1 litre. This is called '1000 unit strength' and is compared colorimetrically with the water, the colour of which is expressed in 'Platinum units'. A simpler alternative is to dissolve 3·6 mg of methyl orange in 1 litre of a dilute alkali. One unit (Ohle) is the equivalent of 0·01 mg methyl orange per litre or 2·8 pt units. Colour, like transparency, is not uniformly distributed throughout the water and there may be strata of varying colour in the hypolimnion. It also tends to increase near the bottom due to materials diffusing from the substratum.

Light penetration can be measured by means of photoelectric cells, but a simple and practical measure of vertical illumination is the Secchi disc, which has already been mentioned (p. 3553) and is widely used. This is a white disc of 20–25 cm diameter (accuracy is not important) which is suspended horizontally and lowered into the water until it just disappears. The depth is noted on the cord used for lowering. The disc is then lowered further and brought up until it just re-appears, the mean of the two depths being recorded. The surface of the water should be calm and shaded for observation.* It should be remembered that the level of visual extinction of light will be at twice the depth measured by the disc, since the light reflected from the disc to the eye has traversed the given depth of water twice over.

For shallow ponds and small lakes, or if the water is very turbid, another measure of tranparency is more suitable. A piece of platinuum wire 1 mm in diameter and a few centimetres or so in length is inserted into the side near

---

* Greater accuracy may be obtained by observation through a 'water glass' which is simply a box with a glass bottom, the use of which eliminates surface reflection.

one end of a rod about 1·5 m long, so that it projects horizontally. At the other, upper end of the rod a small wire ring is fixed, directly above the platinum wire. The rod is immersed and the platinum wire observed through the upper ring, the depth at which it disappears being recorded in millimetres.

The turbidity is expressed as the equivalent of parts per million of a standard suspension of precipitated silica. These values increase logarithmically as the depth of disappearance diminishes. At 1 m depth a difference of 100 mm only implies one integer difference in turbidity, but at 10 mm depth a difference of 2 mm raises the turbidity by 1000 units. Depths of disappearance of less than 100 mm are not likely to be met, but if they are they should be treated as very approximate.

The most transparent lakes are those formed in ancient volcanic craters, the so-called 'caldera' lakes, where there are no influent or effluent streams and the water level is maintained only by rain and by underground seepage. The clearest of all is Masyuko Lake in Japan with a Secchi disc figure of 41·6 m. Next to it is the famous Crater Lake in Oregon with figures up to 40 m. In most of the larger lakes of Europe and North American the values generally lie between 10 and 20 m, and some at least of the African lakes attain the higher figure. Many observers have found higher figures in winter than in summer, probably due to the influx of snow-water and the decrease of plankton. Small lakes are generally rather more turbid, with Secchi values of only 1–3 m. (See also p. 3554.)

The depth of penetration of light is obviously of vital importance to the photosynthetic life in the water and hence indirectly to all. The photosynthetic zone extends from the surface downwards to a level where the light factor corresponds to the compensation point, at which photosynthetic gains are balanced by respiration losses. This level depends not only on the transparency of the water, but on all the vagaries of climate and on the nature of the organisms themselves, which vary considerably in respect of their relationship to light intensity, so that a horizontal layering of species may result from their reactions. Most submerged plants are 'shade plants' and high intensities of light are as unfavourable to them as those too low. Bottom-living plants (benthic) have a downward limit of 8–10 m in most lakes and the phytoplankton range is similar. As at 10 m depth the total light intensity in the clearest water is only about 10 per cent of the surface illumination and may be as low as 0·01 per cent in coloured water, such a depth distribution for the plants implies low compensation points, as in other shade plants.

Photosynthetic algae have sometimes been dredged up from surprising depths, e.g. below 400 m, where they could not possibly live by photosynthesis. Either they are living saprophytically as algae in the soil are suspected of doing, or they have been accidentally and temporarily carried downwards by water movements. Convection currents are set up by temperature changes, especially between night and day, and their range can be considerable. There is a general tendency of plankton to move upwards at night and downwards in daylight, but the recorded range of some organisms between day and night is far beyond their own capabilities of movement and can only be attributed to convection currents, which conceivably may affect a buoyant organism more than others, thus producing a selective effect.

An additional factor affecting the sub-aqueous illumination is the virtual shortening of the day by the increased amount of light reflected from the surface of the water at low angles of the sun. When the angle is above 30° with the horizon variations of angle make little difference to the amount of light reflected, but below this angle the percentage reflection rises very rapidly and at an angle of 10° reaches 36 per cent for water, and as the amount of radiation at these low angles is, in any case, small, this means virtual darkness under water for an hour after sunrise and an hour before sunset, a considerable shortening of the day.

The *relatively* deeper penetration of long wavelengths in coloured waters may have an effect in increasing the photosynthetic zone, if they are sufficiently powerful, but it may be surmised that this is rarely the case. Actual tests in America with suspensions of algae, which were immersed at different depths, showed that in clear, colourless water the compensation point might be as deep as 17 m, but in brown coloured lakes it did not usually exceed 1 m.

Hydrostatic pressure as such is of little direct ecological consequence as it only becomes an effective factor in the deepest lakes. It can, however, have effects on the physical properties of water which may condition the equilibrium of the water. Pressure also often increases the solubility of substances, which may be important in the water layer next to the bottom. Increased pressure lowers the temperature of greatest density in water below 4°C (3·98°C). Thus, in the deepest water temperatures may be well below 4°C (p. 3707) and yet the water be of maximum density and not dislodged by water at 4°C sinking from the top. The stored solutes in this deep water are not involved in any overturn of the water and only reach the zones of life in the water by slow diffusion or by convection currents. The thermal differences which set these currents in motion have only a minimal effect at low temperatures. The difference in the density of water for a change of 1°C from 4°C to either 3° or 5° is the same and is very small, only 0·000008 of a gram per cubic centimetre. The decrease of density per 1°C gets rapidly greater as the temperature rises, and over 20°C it becomes more than twenty-five times as much as at 4°C. As a consequence small thermal differences in tropical lakes are much more effective in promoting water movements than in colder climates (see also p. 3708).

A peculiar phenomenon affecting the stability of lakes is that of the *seiches*. The word, supposed to be derived from the Latin *siccus*, 'dry', is applied to slow, oscillating surges of water level, of which the amplitude, that is the rise or fall below the equilibrium level may vary from a few centimetres to a metre or so. Seiches have been noticed for a long time, as they are of widespread occurrence, and raised much speculation, but the investigations on Lake Geneva begun by Forel in 1869 brought them into the realm of mechanical analysis and a proper understanding of their characteristics, which make them important to hydraulic engineers. They are in fact waves, of great length and small amplitude, like tides. A simple wave which travels the length of a lake and is reflected back is called *uninodal*, but like other wave movements they tend to set up harmonics so that seiches with 2, 3, 4 or 5 nodes occur and as many as 15 nodes, with progressively shorter periods, have been observed in Loch Earn, superimposed on the uninodal wave. In very wide lakes transverse seiches may also occur.

Seiches have been observed in lakes of all sizes with periods varying from several hours in the Caspian Sea and the Aral Sea, to seconds in small ponds, where their small amplitude makes them difficult to detect.

The short-period oscillations are parasitic on the energy of the uninodal wave which thus tends to die out until renewed by a fresh input of energy. Two such sources of energy have been found mechanically adequate. One is from a change of barometric pressure, the other is from the piling up of water to leeward by strong winds, called 'denivellation'. The relative importance of each cause depends on the geography of the lake. Another, less common, cause has been traced to earthquake waves which can start lake water oscillating at great distances from the epicentre.

In some very large lakes with a long east–west axis, true lunar tides have been detected, but with amplitudes of only a few millimetres.

Of some interest ecologically are the internal seiches which occur in waters with well-marked stratification, which may occasion large-scale oscillations of temperature and of stratified plankton. The ranges and periods are far greater than in the surface seiches and it is probable that such oscillations are very widespread and are a largely unrecognized feature in sub-aqueous ecology. The range of such seiches, that is their total movement from lowest point to highest point are truly astonishing. Ranges of up to 10 m are not uncommon, while in Loch Ness a range of 60 m and in Lake Baikal of up to 150 m has been recorded. The water movements are not purely oscillatory, but are accompanied by complex turbulent, spiral and horizontal water movements, which can cause a periodic tilting of the thermocline in two directions and can set up currents in the deep hypolimnion, altering completely the lake isotherms and bringing up deep water laden with solutes. The internal seiches are not a separate phenomenon, but are part of the general seich system and traceable to the same causes as the surface oscillations.

Wave action though not normally of any great force in lakes can, of course, be greatly magnified in exceptional storms. In small lakes and ponds it is not the strength but the continuity of wave action which is effective and the shore to leeward of the prevailing wind will often present a bare appearance, devoid of any marginal zones of plants, in contrast to the sheltered shore where silt lies undisturbed and marginal plants flourish.

Chemical conditions

Natural waters have two origins. First, and predominantly, they come from precipitation of water evaporated from the sea or the land surface and recondensed. Much of this water makes a passage through rocks and soils of the earth's surface before it emerges to become surface water. As this process is continuously repeated such water may be classed as superficial or external. The other category is deep water, released by deep-seated chemical processes, especially in volcanic regions, and classed as juvenile water. It is very rich in Carbon dioxide and often charged with minerals and mineral acids. Of such a type are many 'mineral springs'.

The quantity of inorganic solutes in most superficial water is very low, of the order of 0·1–0·5 g/litres, and depends on the nature of the surrounding rocks, on the CO_2 content of the water and on whether the water has drained through mature soils or not. The silty substratum of fresh waters is usually much richer in nutrient solutes than the water itself and complex ionic exchange relationships exist between them. This implies a great ecological difference between floating plants and those rooted in the silt, for while the former draw their nutrients from a meager supply in the water alone, the rooted plants are much better supplied and are much more dependent on the substratum than on the ambient water they absorb.

It is commonly said that superficial water starts its cycle as distilled water in the process of evaporating, but this is probably never quite true, for in rising the water vapour entrains with it an aerosol of solute particles derived from many sources, and at its initial condensation in clouds the water already contains measurable amounts of solutes, notably Cl and SO_4, and it is upon these ionic nuclei that condensation takes place. During their downward journey raindrops pick up more and by the time it reaches the earth the rain is already a solution of potential nutrients. We have previously spoken (p. 3655) of the chemical constitution of rain water and need only add here a few particulars.

Chloride is almost universally present and may all be traced to the sea. Coastal rain during storms is practically dilute sea water, but there is a rapid drop inland for about 10 km and then a much slower drop to a minimum of about 0·3 mg/litre in central continental areas, where some Cl may come from salt deserts. The alkali metal ions show a similar distribution. Magnesium, however, does not travel far from the coast. Its salts are very hygroscopic and easily washed down.

Sulphate also seems to be universal, but the ratio of SO_4 to Cl is much greater than in sea water. It is about 1·5:1 in coastal areas but much higher (4·5:1) in continental rain. This suggests that chlorides may be more quickly washed out of the air or that excess sulphate is perhaps generated by the oxidation of H_2S. It is noteworthy that a similar excess of sulphate is also found in softwater lakes.

Phosphate has sometimes been found, but only in trace values of about 0·01 mg/litre. Bicarbonate is also scarce, less than 1 mg/litre which accords with the relatively low pH of most rain (4·0–5·0).

Nitrogen occurs in both oxidized form and as ammonia, but it is interesting from the point of view of the bacterial flora of the air that one-quarter of the nitrogen is in organic compounds. The quantities of nitrate and of ammonia found show close correlation, and it has been suggested that the nitrate is chiefly formed by the photo-oxidation of ammonia: the total amounts are, however, small, less than 1 mg/litre.

Among metals, Na, K, Mg and Ca occur, the latter alone in notably higher concentration than in sea water. This and the fact that the amounts found increase with distance from the sea, suggests that there is a large contribution from calcareous dust. Potassium is in slightly higher concentration than in the sea, but perhaps not significantly; there are hardly enough analyses available to settle the matter.

Analyses of lake and river water show that in both the predominant element is Calcium and that in both there is a considerable increase over the amount present in rain water. The kationic order in river systems is usually $Ca > Mg > Na > K$ and the anionic order, $HCO_3 > SO_4 > Cl > CO_3$.

The total 'salinity' of fresh water is highly variable but a maximum of 300 mg/litre is acceptable for uncontaminated waters. The lowest figures are those of waters from bare igneous rocks and anything less than 50 mg/litre may be taken to indicate drainage from such rock types. Nevertheless solution from the rock base is only a first step; thereafter the composition of the water is modified by base-exchange reactions with clay colloids, as in the soil, and there seems to be a general tendency towards approximation (see also p. 3649). Indeed, soil solutions and natural free waters show striking similarities.

The enrichment of rain water with CO_2 as it penetrates soil leads to increased solution of rock carbonates as bicarbonates of Calcium and, to a lesser extent, Magnesium. This determines the slightly alkaline reaction (circa pH 7·5) of most free waters. Equilibrium depends on the CO_2 content and if vegetation is abundant in the water the withdrawal of CO_2 by photosynthesis leads (a) to a rise of pH towards 9·0 or more and (b) to the dissociation of bicarbonates and the precipitation of insoluble carbonates. Water with an acid reaction occurs in deep waters, where an excess of CO_2 may be present, or in peaty pools where free humic acids are present, which may depress the pH to below 4·0. Extremely pure waters, with hardly any buffering, may also approach pH 5·0 due to CO_2 in solution. There is no strict relationship between pH values and the salinity, but waters with the highest solute contents show a tendency to rise to pH 8·0 or over (sea water averages pH 8·2 with a solute content ten times as great as the richest fresh waters).

As might be expected, it is the free-floating plants which are most affected by the solute content of the water. An interesting analysis by Lohammar of twenty-five Swedish lakes, classed according to their solute content, showed an overall increase in species between the poorest and the richest lakes in the proportion of 100:178. The species of pleuston floaters increased in the proportion of 100:290, while the species of rooted, *Nymphaea*-like plants remained constant, and the species of *Isoetes*-like plants (*Isoetes*, *Lobelia*, *Subularia*, etc.) dropped in numbers. As the second group can grow healthily in richer water their apparent preference for the poorer habitat may be accounted for by their inability to compete with more luxuriant growths. Also they are bottom-living plants and the richer waters tend to be less transparent.

Numbers of species do not present a complete picture of the situation since we must also reckon with their mass development, which is undoubtedly greater in the richer waters though there are few figures to measure it by.

The differences in total salinity are mirrored in the conventional ecological distinction between *oligotrophic* and *eutrophic* waters, which merely represent two extremes of a series. Nevertheless, the differences are not only seen in the comparatively meagre vegetation of the former group, which is obvious to inspection, but also in its taxonomic composition. Not only are the species of oligotrophic waters fewer and less luxuriant than those of eutrophic waters, but they are largely different, so that two clearly recognizable plant communities are represented.

The electrolytic conductivity of the water offers an excellent guide to the chemical composition. Not only is there a linear relation between conductivity and the total salinity, but a closely linear relation to the concentration of each of the leading ions, the correlation coefficient between the values being of the order of 0·9. Only for SO_4 is there any marked divergence, the coefficient dropping to 0·73. This implies that from the conductivity it is possible to make a close prediction of the amounts of each ion present, indicating an astonishingly constant proportional constitution of the solute content, which Rohde calls the 'standard composition'. Rohde attributes this to an equilibrium in ionic exchange between the water and the colloids of the soil and the lake sediments.

Clarke in his book *The Data of Geochemistry* gives an analysis of lake water which represents the average of all lakes in the world hitherto investigated, in other words the standard composition. Rohde has tabulated Clarke's figures, as shown in Table 16, in comparison with those found for twenty-five Swedish lakes by Lohammar. The correspondence is so close that they strongly suggest that the 'standard composition' has objective reality.

TABLE 16. IONIC COMPOSITION IN MILLEQUIVALENTS PER CENT

	Averages for Lakes of the World (Clarke)	Averages for 25 Swedish Lakes (Lohammar)
Ca	63·5	67·3
Mg	17·4	16·9
Na	15·7	13·6
K	3·4	2·2
HCO₃	73·9	74·3
SO₄	16·0	16·2
Cl	10·1	10·1

The figures shown in Table 16 are not to be understood as meaning that the ionic concentration of lake water is constant. We have seen above that it is not, but they do imply that in spite of great variations in total salinity the relative proportions of the ions remain substantially the same.

The outstanding feature of the chemistry of fresh waters is that they are almost universally dominated by Calcium bicarbonate, so much so that they can be designated as the Calcium bicarbonate water-type.

It is interesting to notice that, in contrast to the land vegetation, we do not find among water plants a marked division between calcifuge and calcicole plants. The reason for this is obvious when we realize that the bicarbonate in the soil is just another anion, but that in the water it is the chief source of Carbon dioxide for assimilation and practically indispensable. Waters poor in Calcium bicarbonate are generally poor in all nutrients. Plants which grow in oligotrophic waters do not do so because they are calcifuge, for they can grow in richer waters if they get the chance, but because in the poorer water they find a field for extensive development which is denied to them by the competition in richer water. Rooted and emergent plants of the littoral are in a different category as they assimilate CO_2 from the air and are not true aquatics, but plants of waterlogged soils. A calcifuge submerged aquatic would be a natural paradox. The Desmidiaceae, it is true, show a preference for water with an acid pH, but not necessarily Calcium-free.

It is worth while to notice here the opposition between the two great classes of natural waters, fresh and sea. The one is dominated by the Calcium bicarbonate complex, the other by the Sodium chloride complex. The diagrams in Fig. 4.55 illustrate graphically this opposition.

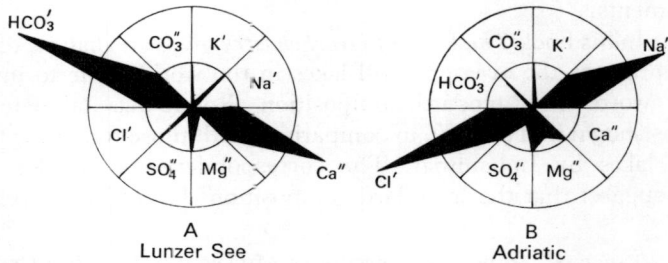

A
Lunzer See

B
Adriatic

FIG. 4.55. Diagrams illustrating the contrast in the ionic composition of a freshwater lake (Lunzer See) and sea water (Adriatic). (*After Maucha.*)

The evolutionary movement from the sea to the land which is often regarded as involving merely accommodation to lower osmotic potentials, would be, in reality, a much more difficult transition to an entirely different ionic environment. We know, of course, that some animals habitually make this transition, but these are physiologically closed systems which do not use the water for nutrition. For the open systems of plants such a transmigration is a much greater problem.

Owing to the predominance of Calcium bicarbonate in the total salinity there is a correlation between this latter value and the pH. Sjörs (1950) has given an extensive comparative analysis of the water of Swedish 'mires'* in

---

* The Swedish *myr* is usually translated as 'mire' or 'fen'. The former is ambiguous in English as it is commonly applied to mud. The latter term is better, but as the water-table is often above the ground surface they approach Tansley's category of *swamp*.

this respect and of Finnish mire water from data of Kivinen (1935). Total salinity is represented by the conductivity at 20°C, less the conductivity of H ion alone $[K_{20} - (K_H)_{20} \times 10^6]$. It must be granted, however, that pH measurements of the water may not represent accurately the conditions of many rooted plants if their roots are deep enough to penetrate to the reducing zones of the substratum (p. 3730).

For conductivities below $K = 50$ Sjörs' figures show a rapid rise of pH from 4·0 to about 6·2; the curve is there inflected and the subsequent rise is much more gradual, but practically linear, up to $K = 400$ and pH 8·0. The rising curve for Finnish water is more gradual with the point of inflexion about pH 5·3 and the maximum at pH 7·2 in an exceptionally rich water ($K = 651$). Sjörs also correlates the different classes of fen with the pH (Fig. 4.56) and concludes that there is more correlation between vegetation of fen type, (within wide limits) and the pH than with the electrical conductivity (salt content), which, except at the extremes, is not always closely related to the vegetative type.

Iversen, in Denmark noticed the connection between the pH of fresh water and the photosynthesis of submerged vegetation. In a thick bed of *Elodea* he found that the highest pH (8·8) was in the upper 10 cm, where assimilation was most active and that it dropped downwards to about pH 7·0 as the light intensity diminished and less CO_2 was being abstracted from the water. The surface pH was well above the normal for fresh waters so that it had been raised by active photosynthesis.

The same effect may be seen in tidal pools on the foreshore where algal vegetation is vigorous. If these are left isolated by the tide during daylight, the pH of the water rises steadily from pH 8·2, that of the open sea, to 9·6, which represents the complete dissociation of all the soluble bicarbonates. Pools beyond the reach of neap tides retain this high pH for many days, which is a powerful factor in limiting their population. Low tides at night have the opposite effect. The release of respiratory CO_2 causes a drop in pH below 8·2. (See also p. 3735.)

A feature which may be noticed in ionic analyses of lake waters and especially in acid humic waters, is an apparent imbalance between the total equivalents of anions and of kations, with often an excess of anions. It has been suggested by Gorham that an appreciable part of the anion content, where there is an excess, is in the form of organic acid anions. The opposite condition may be due to the adsorptive concentration of kations.

Variations in the amounts of individual ions present are usually so dependent upon geological and topographical conditions that they can only be discussed on a local basis, but there are one or two climatic influences which can be widespread. Among these are the direction and strength of prevailing winds and the dust supply. How significant the latter may be is shown by the observations of Eriksson that in southern Sweden the rain content of Mg and Cl only accounts for about one-third of the total amounts brought by atmospheric transport.

Above all, however, is the effect of precipitation. Just as there are ombrotrophic bogs so there are ombrotrophic lakes. Forham and MacKereth have both shown that in the English Lakes much of the ionic material in solution, except for Calcium bicarbonate and to a lesser extent that of Magnesium, appears to be derived from rain. Fluctuations of composition are more marked

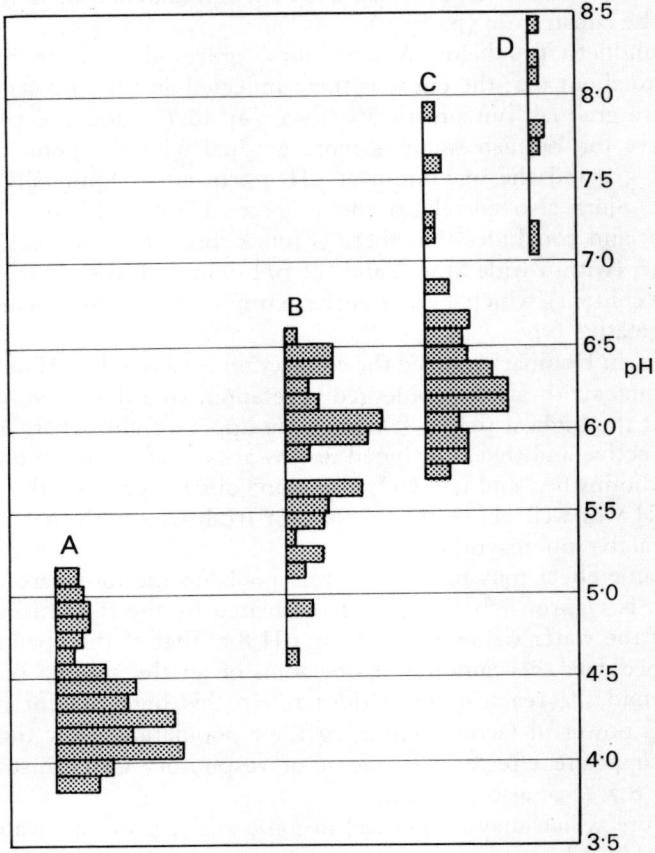

FIG. 4.56. The relation between vegetation and pH in Swedish fen waters. A. Extremely poor fen. B. Intermediate fen. C. Transitional rich fen. D. Extremely rich fen. Although there is an overlap in the pH ranges the general differentiation is clear. The size of the rectangles is proportional to the number of observations. (*After Sjörs.*)

in small pools than in larger lakes. In the former the composition is considerably affected by the weather, while the latter usually have several sources of telluric water and represent a synthesis of influent supplies. Gorham investigated pools in peat in northern England and found that in wet weather the ratios of Sodium and Magnesium to chloride were close to those in sea water, while in dry weather they rose much higher. As he says, 'A sea spray origin seems inescapable'. In Swedish inland water also a west–east gradient has been found by several investigators, especially in Na, Cl and SO$_4$, which are attributed to atmospheric transport from the sea. Malmer pointed out that the effect on waters probably extended also to the surrounding soils and Troedsson showed that drainage water from soaked podsolic soils corresponded in its major ionic constitution with that of the lake waters. Some phytogeographic significance has been attributed to this ionic gradient in determining the frequency of oceanic (atlantic) species. Gorham, however, says that a marine origin cannot be generalized for other major ionic materials, since similar west-east gradients exist for K, NH$_3$ and NO$_3$, but in concentrations which make an origin from the sea highly unlikely. What influence if any, the differences in the ions of marine origin in rain have upon the distribution of species between west and east remains an open question. There are other, e.g. climatic, factors which might equally be responsible for the distribution of species. For example, Kotilainen's index of oceanicity gives higher values over the whole of Scotland than those which prevail even on the west coast of Norway or Sweden.

Much interest has centred on the chloride ion as one of distinctively marine character. Inland waters not related to salty sediments have ordinarily very little, but they do consistently have some. All the Wisconsin lakes studied by Juday, Birge and Maloche contained Cl, usually less than 1 mg/litre but rising to 4·5 mg. This lake region is 900 miles in a direct line from the nearest sea and one may conclude that in such a situation the chloride is of telluric origin.

All rain water affects primarily the epilimnion and during the summer stagnation that layer only. Vertically there is often a decrease of chloride in the hypolimnion, but as the bottom is approached there may be a secondary maximum of chloride content, derived by diffusion from the substratum. Where the depth and pressure are sufficient this water may remain undisturbed by the annual overturn and chloride may accumulate up to 40 mg/litre.

There is a wide difference between closed lakes, with no effluent, and lakes which are little more than local expansions of a river, in which total circulation is continuous. Most of the considerations we have already dealt with refer to lakes which are intermediate between these two extremes, with influents and effluents which bear only a small proportion to the volume of the lake and in which therefore the period of storage of water in the lake is long enough for local conditions to develop.

Closed lakes are those in which water loss is almost entirely by evaporation. In arid climates they may disappear entirely leaving a salt bed. In more humid regions influent water or rain prevents desiccation, but continuous evaporation may concentrate the water to the point of saturation with some solutes. The solubilities of most salts of Na and Mg is so high that saturation is seldom reached for them, but it may occur for carbonates of Ca and Mg and for Calcium sulphate. In 83 per cent of the closed lakes examined, total ionic concentrations lie between 0·1 and 100 g/litre. The relatively low solubilities of Calcium carbonate and Calcium sulphate indicate that they will be among the first salts to be precipitated during evaporation, but the chemical interactions which take place as concentrations increase make it very difficult to foresee what will result in particular cases. There are three main types, but with many intermediates. These types are based upon the principal ion surviving in solution and they are respectively, chloride, sulphate and carbonate waters. Chloride waters owe their character to influents enriched with Cl from salt beds. Thus the Dead Sea has salinity of 226 g/litre, the highest known, of which 66 per cent is Cl. Its influent, the Jordan, has a salinity of only 7·7 g/litre, but of this the extremely high percentage of 41·5 is chloride. Sulphate waters are mostly in prairie regions and seem to owe their peculiarity to sulphate from the prairie soils. The carbonate waters arise from the substitution of Sodium for Calcium, if the latter is all precipitated as Calcium carbonate and sulphate, with the result that the water is finally dominated by Sodium carbonate and bicarbonate. Among kations, solubility determines the final predominance of Sodium and Magnesium.

It may have been noticed that in the foregoing discussion of the mineral constitution of natural waters there has been little or no mention of certain ions which are of vital importance to plants. Such are phosphate, nitrate and ammonium and, to a lesser extent, silica. The reason for this is that they are usually present in such small and fluctuating amounts that it is necessary to treat them as if they were trace elements, though there are, of course, many other ions present in even smaller amounts, some of which e.g. Manganese, are veritable trace elements, as in soils.

Phosphate. The scarcity of phosphate was well illustrated in the analyses by Juday, Birge and others of the Wisconsin lakes (494 altogether). In 47 there was no detectable phosphate while in the remainder 56 showed only a trace and measurable amounts in the others ranged from 2–7 µg/litre. Similar results were reported from Indonesia, Japan and Europe. In the latter, improved methods were unable to detect even 0·1 µg/litre of soluble phosphate in many lakes. This poverty is characteristic of the surface waters and is attributed to absorption by the phytoplankton in the illuminated region, since the hypolimnion may show considerable enrichment with phosphate, up to 0·4 mg/litre, especially in autumn, attributable partly to its release from decaying planktonts and partly to the release of adsorbed phosphate from the bottom sediments.

The spring overturn of water in many lakes and the restriction of plankton growth during the winter often lead to a modest increase of phosphate in surface water, which encourages the revival of plankton growth for as long as it lasts but its exhaustion brings on a summer minimum of plankton. There can be no doubt that in this respect it acts as *a* limiting factor, though not necessarily the only one, or *the* limiting factor in Blackman's sense.

Many experiments have been made involving the phosphate manuring of waters, with rather unexpected results which show that the relationship of phosphate concentration to plankton and other growth is by no means simple. Quantities of superphosphate added to small lakes which are poor in Phosphorus lead, it is true, to temporary outbursts of plankton life. Phosphate was incorporated into algae and sank with them into the hypolimnion. Released by organic decay it either remained sunk, was drained away or passed into the sediment, the upshot being that twelve months after manuring the P content of the water was back to its original level. Einsele, after an experiment of this kind in which the expected plankton increase did not appear suggested that there was some self-regulating system in the water, which automatically maintained a P level suited to the vital economy of the lake. Other experiments by Nelson and Edmondson, at Bare Lake in Alaska, did in fact cause a peak of plankton production of some duration, which was accentuated by treatment in a second year.

The difference in the two results may have been due to geographical differences between the lakes or it may have been due to the fact that Einsele's lake was thermally stratified while Bare Lake was not. A negative result may also be due to over-manuring, since Rohde has pointed out that phosphate can be a maximum factor as well as a minimum factor. He divides the plankton algae into three groups according to their P requirements. For group 1 the upper limit of P is less than 20 μg/litre while groups 2 and 3 have an upper limit above this quantity.

Phosphate storage in the algal cells has also been shown to be important. This accumulation in the cells may quickly deplete the water of P, without necessarily implying an increased number of cells and this storage may enable the plankton to continue in apparent vigour through periods when the water contained no detectable phosphate. These facts help to make clear that although phosphate may act as a limiting factor there is no simple or direct relationship between P level and plankton growth.

Gardiner (1941), working on a reservoir with an initially high P level, found that the greatest multiplication of the diatom *Asterionella formosa* coincided with a fall of the phosphate level to its minimum. Mackereth in Windermere, which has a very low P level (below 1 μg/litre in summer) found an *Asterionella* maximum in April and May without any change in the phosphate concentration. This diatom appears to be one capable of phosphate storage, since in Windermere the cells contained about four times as much P as the water.

Three other factors have tended to confuse the relationship between plankton algae and Phosphorus. The first is that amounts which are chemically undetectable (0·01 μg/litre) may be enough to support a fairly large population of some species. For example *Asterionella* requires only 0·06 micrograms per million cells. Second, organic compounds of P may be utilized, e.g. phytin and adenylic acid, which are not included in the usual estimates of soluble P, though they may greatly exceed it in amount. Hutchinson has calculated from the data of Juday and Birge on the Wisconsin lakes, that the ratio of soluble to organic phosphate may vary from 1:1 to 1:89.

Third, there is the curious point that *Asterionella*, and possibly other species, are unable to assimilate P from a solution in distilled water though it is readily absorbed from a solution of the same strength in lake water. It is not known what factor is at work in the lake water.

Phosphorus is not only abstracted from the water by plankton, but also by higher plants growing in it. This has been followed by using radioactive P^{32}, as it was found that phosphate was abstracted (adsorbed) by the 'solids' (*seston*) in contact with the water, including both plants, animals and the bottom sediment. The diminution of P^{32} in the water was at first rapid, but the curve of loss flattened out and remained steady at a concentration about one-tenth of the original. The conclusion was that this represents an equilibrium level between adsorbed and dissolved phosphate and that any further fall in the latter amount is compensated by liberation of adsorbed P from the solids, including the bottom sediment.

The situation of knowledge regarding the relationship of phosphate to aquatic life in general is plainly still imperfect and obscure, but, as we have seen, it is certainly neither simple or direct, and that simple chemical analysis of the water is an insufficient criterion for judgement.

Inorganic Nitrogen. Nitrate is of chief importance in the upper layers of water and ammonium Nitrogen in deep waters and bottom sediments. Nitrate estimation is not as accurate as P estimation and the reactions are interfered with by some other ions, notably ferric, so that older estimations are probably too high. It has lately been found more satisfactory to reduce the nitrate to nitrite, the estimation of which is more accurate. Quantities found are greater than those of P, but in uncontaminated waters are still low. In the 472 Wisconsin lakes analysed the average was 18 µg/litre and in European lakes the average is about ten times that of phosphate.

Pearsall, working on the English lakes from 1921 onwards, founded a classification of lake waters on the *basic ratio* $(Na + K):(Ca + Mg)$. Lakes with a high ratio have usually a low pH and a low nutrient status, and are mostly of the rocky, sub-alpine type with little sediment. They would generally be called oligotrophic. This type of lake is favoured by Chlorophyceae, especially desmids. Lakes with a low ratio are the 'calcareous' waters, with higher pH values and a higher nutrient status, coincident with abundant silt, i.e. they are eutrophic. In these waters phosphate and silica are generally low, but nitrate high. They are favoured by diatoms and, among Chlorophyceae, by the motile Volvocales. The third great group of lake algae, the Myxophyceae, favour organic waters, particularly among the eutrophic waters of the second group.

That nitrate absorption is affected by the kationic environment is shown by some figures of Gessner's based on *Potamogeton perfoliatus*. These showed that the absorption of KNO_3 was not affected by the age of the leaves but that absorption of $Ca(NO_3)_2$ fell off by nine-tenths with increasing age. One would expect therefore that nitrate absorption would be more active in Pearsall's 'alkali' lakes, where $Na + K$ predominate, even with a lower nitrate content, than in the 'calcareous' lakes.

The nitrate in lake waters comes chiefly from the influent waters, which bring with them the drainage of soils and rocks. In oligotrophic lakes the effluent contains the same amount as the influent or only very slightly less, whereas in the more productive lakes the amount in the effluent water is only about one-half that in influent water. Much of that retained in the lake is utilized by plants but their consumption does not account for the whole of the nitrate retained. Some is lost to the bottom sediments, but the greater part seems to be reduced by de-nitrifying bacteria and remains dissolved in the water as molecular Nitrogen. A subsidiary but by no means unimportant source is from Nitrogen fixation by many bacteria and some Myxophyceae both in the upper waters and also at and near the surface of the substratum. Some nitrate is also released by decaying vegetation.

Nitrite is occasionally present in small amounts, probably only as a stage in de-nitrification of nitrate, and it is not important as a N-source. Ammonium ion in the epilimnion is only found in very small amounts, varying from zero to about 0·03 mg/litre and compared with nitrate is of little importance as a source of nitrogen for organisms. In the hypolimnion, by contrast, substantial amounts may accumulate from the bacterial decomposition of dead organisms. A minor maximum may be reached in early spring, followed by a rapid drop and a subsequent rise to an autumn maximum which may be over 3 mg/litre. Both NH_4 ion and undissociated NH_4OH are present, and both can be used as N sources, but there is conflicting evidence about their utilization and about the preferences for NH_4 or NO_3 which are shown by different organisms. Some algae prefer one and some the other, but the pH of the water has an effect, and above pH 7·0 there seems to be an advantage with NH_4, especially with the hydroxide, which applies to higher plants as well as algae. In view of the existence of such specific preferences it may well be that the ratio of NH_4 to NO_3 in the water may have an influence on the taxonomic character of the aquatic populations.

The increase of ammonium in the hypolimnion seems to accord with the summer disappearance of Oxygen from the deep water. Mortimer has shown that when anaerobic conditions are established large quantities of ammonia are liberated from the bottom sediment, due, it is suggested, to the disappearance of the oxidized surface layer of the sediment, which had hitherto formed an adsorptive barrier preventing the diffusion of ammonia into the water. (See also ferrous Iron, p. 3729.) The autumnal turnover may distribute the ammonia throughout the lake.

Silica. Another nutrient which is often in minimal supply in water is silica. While commonly referred to by this name, it is more correct to say that 'soluble silica' mostly consists of undissociated, crystalloidal, silicic acid or silicates in true solution while colloidal silica and suspended silica particles are also present, often to a much greater extent than the soluble silica. Moreover, in neutral or acid waters the silicate ions tend to aggregate into a colloidal sol or they may form colloidal complexes with ferric or Aluminium hydroxides. The overall situation is therefore complex and liable to change.

The range of ionic concentrations occurring in surface waters, even in one region, is very great so that average figures lack significance. In milligrams per litre the range in north temperate lakes is from zero to about 3·5, but subtropical and tropical lakes have often much greater amounts and a greater range, from 1 to over 70. Volcanic regions with voluminous ash deposits yield the higher figures, but the intensive leaching of silica from tropical soils (p. 3607) is also responsible.

Silica is extensively derived from bottom sediments, and it is believed that the state of oxidation of the mud-surface is a controlling factor as it is with phosphate and ammonium. Mortimer found in Lake Windermere that in two experimental tanks placed over sediments, one aerobic and the other anaerobic, three or four times more silica diffused into the anaerobic water than into the aerobic. Yoshimura also showed that in a Japanese lake a zone of oxygen-free water existed over the bottom during the summer stagnation and that it was in this zone that the highest concentrations of silica (16 mg/litre) were reached. Rapid cooling of the surface water in August–September (30–20°C in one month) brought some of this silica upwards into the photosynthetic zone. A similar cycle has been reported for other temperate lakes, namely, a general increase at the time of the spring overturn, followed by a drop to a minimum in May or early June and from then onwards a rise continuing into late September.

It is the Diatromaceae above all which are intimately linked to the silica cycle. As their silica demand is great it can act for them as a minimal factor, even though there may be much more silica available than phosphate, for the silica requirement of diatoms is much higher than that for phosphate. Culture experiments by Jörgenson showed that growth did not start below a concentration of 1 mg/litre and that it proceeded until a limiting or threshold concentration of 30–40 μg/litre was reached. The silica demand is not, however, constant and at times of rapid cell multiplication the Si content of *Asterionella formosa* may be only one-half the initial value. Mackereth has shown a corresponding drop in the P content of multiplying *Asterionella* cells, down to a limiting value of only 0·06 micrograms per million cells.

Silica only becomes a limiting factor in the true sense when cell production is very high and this, as Lund points out, depends on other factors, phosphate supply and perhaps the supply of trace elements, since heavy metals such as Titanium and Zinc are found in the cells in much greater amount than in the lake water. A good phosphate supply can, as Mackereth found, push up the diatom population to a point at which silica does become limiting and the de-silification of surface waters following on a spring outburst of diatom growth is a commonplace. As the phosphate demand is so very small it does not need much phosphate to produce this situation. The limiting value of 30–40 μg/litre below which, as Jorgensen found, Si appear to be practically unassimilable, is rarely found in natural waters, but even concentrations of 100–200 μg/litre are sufficiently low to reduce diatom growth to an uncompetitive level. The apparent inability of diatoms to take up silica from the high dilutions from which phosphate can be absorbed may perhaps be connected with the observations of Lewin that Si absorption is linked to aerobic respiration, in other words it is an active process requiring an energy supply.

The relatively high concentrations of Si in the deep water, reach a maximum near the bottom level, and in the months of summer stagnation when the Si-content of the epilimnion is lowest. This has been commonly attributed to the re-solution in deep water of the shells of dead diatoms falling from the epilimnion (the 'plankton rain'). Gessner has pointed out that the abundance of diatom shells in the lake sediments negatives the idea that there is any large scale re-solution. Mortimer attributes the source of the silica to the break up of ferri-silicate-humus complexes. If these enter the reducing zone of the sediment (see below) the Fe^{+++} is reduced to Fe^{++} and passes into solution, thus breaking up the complex and releasing the silicate from adsorption so that it is free to diffuse into the water, where it is dispersed by eddy currents. These events are linked to the depletion of Oxygen in the hypolimnion, beginning in late spring and reaching its lowest level in July, at which point ferrous Iron begins to be liberated, the oxidation of which still further reduces the O_2 tension in the water so that by the end of July free ferrous Iron begins to circulate. These conditions last till the beginning of October, when the autumn overturn recirculates O_2 in the hypolimnion and reverses the processes of the spring period. All available analyses show that the silica content of the water increases rapidly as the bottom is approached, which strongly suggests its origin from the sediment, but the effect is most marked in deep hollows, which also sugggests that some at least of the highly charged water has accumulated there by creeping flow.

With regard to silica in the epilimnion water it is generally agreed that it begins to drop in March with the minimum in May accompanied by a decline of diatom growth, but after this there may be in some lakes progressive increase in silica content, accompanied by renewed diatom growth, which continues until the autumn overturn although the diatom population has by then again declined. The source of this increase may be traced to influent water, but Hutchinson considers that a much more general cause is the release of silicates from shallow water sediments, i.e. shoreline sediments, within the epilimnion zone. This release may be started by rising temperatures, but whether it is chemically a parallel to what we have described above for deep water sediments is not known.

Oxygen and the Iron-cycle. The oxidation-reduction conditions in lake sediments have been the subject of extensive studies, which in Britain have centred chiefly on Lake Windermere and Esthwaite Water in the Lake District. They are closely bound up with the Oxygen content of the water and the Iron cycle. In summer, in a stratified lake, the O_2 concentration declines with depth from surface values of about 10–12 mg/litre (saturation = 13 mg) to zero. At the spring and autumn overturns the concentrations of O_2 are high and are effectively the same at all depths. The deficiency in deep water increases as time passes after the spring overturn, when stratification begins, and the anoxic condition in the bottom 3–4 m of water is established by July and persists till October. This implies complete stagnation. Any downward flow from influent streams or from shallow marginal water would tend to maintain a deep water O_2 concentration of at least 1 mg/litre, or probably more, and eddy currents may have a similar effect. The normal decline of O_2 content in the water is attributable to the oxidation of organic matter in the water, not to respiring organisms, since it only becomes appreciable well below the plankton zone. It is naturally most noticeable in waters rich in sediment. In rocky lakes where any sediment that forms is almost entirely mineral, the decline with depth may not occur at all, as there is little or no material to oxidize.

The surface of a lake sediment which is in contact with oxygenated water containing not less than 5 mg/litre of O_2 maintains an oxidation–reduction (redox) potential of $E_7 = 0.5$–0.52 volts,* which is the value usually found in well-oxygenated water. This oxidizing surface layer is coloured with ferric hydroxide (see p. 3648). Below this narrow zone the redox potential drops rapidly and reaches zero at about 2 cm. The sediment at this level is coloured grey to black and the Iron is all in the reduced ferrous state. So long as an oxidizing potential exists at the sediment surface the ferrous Iron cannot diffuse out, as on contact with the water it is oxidized to insoluble ferric Iron. The decline of O_2 concentration in the water and the consequent fall in its redox potential and at the surface of the sediment, releases ferrous Iron from the sediment to the water and it first begins to appear in the water when the $E_7 = 0.18$ volts. Ferrous Iron replaces ferric at the sediment surface at about $E_7 = 0.25$ V and the brown layer disappears. The E_7 then falls rapidly to the value where the ferrous Iron begins to diffuse, though it does not rise in the water higher than the level where redox potential is 0.25–0.3 V, as above that level it is oxidized. When the autumn overturn occurs dissolved Oxygen again reaches the bottom water and by late autumn a high O_2 concentration and the full potential of $E_7 = 0.5$ V is restored throughout. There is a rapid oxidation and precipitation of the ferrous Iron which is deposited on the surface of the sediment, restoring the brown surface layer. These conditions are maintained throughout the winter, even when the surface is frozen. In June the Oxygen content and the redox potential in the deep water begin to fall and this continues until the summer conditions we have described are re-established. As the surface water may become super-saturated with Oxygen as a result of photosynthetic activity of the plankton, especially in the spring months, the redox potential at the surface rises to 0.52 V.

It should be noted that although the O_2 concentration is related to the redox potential it is not the only controlling factor. The variation in the two values with depth is often, as in Esthwaite Water, in general accord, but in a few American lakes, a vertical drop in Oxygen content is not reflected in any variation of redox potential with depth. For some chemical reason reducing conditions in these lake sediments are weak, and they retain an oxidized surface even under conditions of very low Oxygen concentration. In some lakes, as in Windermere, turbulent water movements may maintain a sufficient O_2 concentration to preserve the oxidized surface of the mud, even though reducing conditions prevail below. The opposite type of lake, where there is a marked drop of E_7 with depth (sometimes called 'clinograde'), which does not parallel the course of the oxygen concentration, is due to the operation of other reductants in the water, either the ferrous iron we have

---

* E_7 is the potential corrected to a pH value of 7·0, from the observational potential, the correction being -58 millivolts for each rise of one integer of pH.

mentioned, or organic substances, or sometimes H_2S, which can all cause a drop in redox potential apart from the oxygen-hydroxyl system.

Below the surface of lake sediment, provided it contains Iron and organic matter, the redox potential drops rapidly and zero potential ($E_7 = 0.0$ V) is reached about 2 cm down. Below this level negative potentials prevail. As summer stagnation sets in, the equipotential levels ('isovolts') in the sediment rise and when the 0.2 V level reaches the surface ferrous Iron begins, as we have seen above, to diffuse out and the brown ferric surface disappears. By July the zero potential level reaches the surface of the sediment and may include the bottom centimetre of water. By September even negative potentials may reach the sediment surface. The autumn overturn of water reverses the whole process.

Although the Fe ion is predominant in these reactions it is not the only one affected. The divalent Manganese ion may also be similarly prevented from diffusing out of the sediments so long as oxidizing conditions prevail and phosphate, combined as ferrous phosphate in the mud, will be precipitated as insoluble ferric phosphate in the water. Alkalis and silicate will not be so affected.

It is too easily assumed that an Oxygen deficit in lake water is a simple function of depth, but this is not so, as Thienemann (1926) pointed out. A consideration which tends to be overlooked is that of the volume of water involved. If we imagine two lakes with an identical surface area, but one deep the other shallower and of smaller volume, in the photosynthetic (trophogenic) zones of the two lakes the production of O_2 and of plankton cells may be more or less equal and also the descending plankton rain of dead cells. These fall into the lower (tropholytic) zone of the hypolimnion and are there decomposed and their materials oxidized. In the deep lake these materials are dispersed in a large volume of water and their inroads upon the available reserve of Oxygen may consequently be slight or inappreciable but in the smaller tropholytic zone of the shallower lake the extent of organic oxidation may seriously deplete or exhaust the supply of dissolved Oxygen. In the former case the graph of the O_2 concentration, from top to bottom of a section of the lake, will be almost straight, i.e. *orthograde*, while in the latter it will slope towards lower values as the bottom is approached, i.e. it is *clinograde*.

The conditions we have described hitherto are those reported from lakes of the north temperate zone, where seasonal differences of light and temperature are considerable. Quite different conditions obtain in *tropical lakes* so far as our information extends. Moreover, there is a great difference between lakes at high altitudes where cold nights are customary and those at low levels, where temperature variation is slight. The high-level lakes are subject to nightly surface cooling, which promotes regular circulation. This may affect the whole volume if the lake is shallow and consequently overturn of the water is more or less continuous (polymictic), but, if they are deep, the overturn of water may only affect the upper zone and the deep water may be stagnant, except that in large lakes disturbance by storms and the massive seiches (p. 3715) they can produce may cause considerable turbulence in the hypolimnion. The division of epilimnion and hypolimnion which we have formerly used is not strictly applicable to these lakes. In some there is no thermocline and where one occurs it is slight, not sudden, and, owing to the greater intensity of surface heating, lies much deeper than is usual in temperate lakes.

The stratification in the low-level lakes is normally very persistent. The temperature of the bottom water represents the lowest surface temperature which has been recently experienced, the cooled, denser water having sunk to the bottom and stayed there. Thus any overturn experienced is a comparatively rare event. While temperature differences are only slight there is usually a sufficient gradient of temperature, even if it is only a few degrees, to ensure stability (p. 3708). On the other hand there may be a level with a very sharp drop in Oxygen content, reaching zero at varying levels. In one Indonesian lake the anaerobic level was reached at 14 m. In Lake Tanganyika it was found at 20 m in the hot season, but in the cool season there was enough thermal overturn to drive the anaerobic level down to 110 m. The deepest water in all these types of tropical lake is semi-permanently anaerobic and may contain H_2S.

Carbon dioxide in water is chiefly important in two ways. First in its effect upon the pH and, second, and more importantly, as the raw material of photosynthesis.

The free CO_2 in solution in natural waters comes primarily from the atmosphere and the amount held in solution in the surface layers of water is in equilibrium with the atmospheric content. As the absorption coefficient at 15°C and 760 mm pressure is practically unity* the CO_2 tension in the water will correspond with that in the air, 0·03 per cent, which means 0·6 mg CO_2 per litre. The CO_2 molecule is, however, highly hydrophilic and its potential solubility is much greater than the above figure implies. Water percolating through rocks and soil or in contact with decomposing vegetation, will therefore take up much larger amounts of CO_2. Water in contact with soils in which microbiological respiration is active may contain more than 50 mg/litre, even though it is necessarily losing CO_2 to the air with which it is also in contact.

The affinity of CO_2 for water leads to chemical combination in the form of carbonic acid, H_2CO_3. This is subject to a double dissociation, with the successive loss of the two hydrogens, yielding a bicarbonate and a carbonate ion respectively. The hydrogen ion, when split off is supposed to be transferred to a water molecule, forming a hydronium ion, H_3O. The two stages are therefore as follows:

1. $H_2CO_3 + H_2O \longrightarrow H_3O^+ + HCO_3^-$.
2. $HCO_3^- + H_2O \longrightarrow H_3O^+ + CO_3^=$.

In most waters these ions exist as anions of Calcium salts and, in a lesser degree, of Magnesium. The carbonates of these metals in the rocks are only

* For pure water, 1·002; at salinity 35, 0·844.

very slightly soluble in pure water. Their solution depends on the carbonic acid in the percolating water which dissolves them as bicarbonates. In underground waters the amount of solution depends on the amount of carbonate and on the amount of carbonic acid present, but in water exposed to the air a theoretical limit is imposed by the necessity of maintaining a CO_2 equilibrium with the atmosphere. This requires the presence in the water of a surplus of free CO_2 over and above that in bicarbonate combination. Any imbalance in this respect will be automatically corrected by gas exchange with the air. The amount of free CO_2 is called the CO_2-equivalent. We have therefore a bilateral equilibrium in the water, between a relatively large amount of bicarbonate in solution, with a relatively small amount of free CO_2 on the one hand and with a relatively small amount of carbonate on the other. Any deficiency of free CO_2 will be made up by dissociation of bicarbonate, thus increasing the amount of carbonate. Any surplus of free CO_2 will be adjusted by the formation of more bicarbonate at the expense of the carbonate.

We have seen above that the equilibrium amount of free CO_2 at $15°C$ and 760 mm is 0·6 mg/litre. The corresponding concentration of Calcium bicarbonate is only 1·0 me/litre. In springs issuing from calcareous rocks the amount of bicarbonate in solution may be much greater than this, from 2·0–7·0 me, though siliceous, acid rocks yield much less, from 0·3–0·7 me. These springs represent underground waters and there is nothing surprising in their high concentrations of bicarbonate. What is surprising is that amounts much greater than 1·0 me are to be found also in lake waters in all parts of the world, values of 2·0 or more being not uncommon. One reason advanced to account for this is the very slow rate at which adjustment of the equilibrium proceeds, estimated rather vaguely as 'several days', which allows a condition of imbalance to persist for a long time after ground water has reached the surface.

In some 'hard' waters a condition of super-saturation with Ca ion and with bicarbonate may persist indefinitely. One suggestion is that an excess of Ca salts is present in a stable colloidal form. Another is the supply of CO_2 to the water, not from the atmosphere, but from the bottom sediments. Much of the hypolimnion is not effectively in contact with the air at all and is therefore more like underground water. The solubility of CO_2 in water increases with diminishing temperature and with increasing pressure and as both these conditions apply to the deep water of lakes, the CO_2 content of the water can there rise considerably above that of the surface water (which is in equilibrium with the air) and can carry a larger amount of Calcium bicarbonate dissolved from the sediment. Even in shallow water a rich supply of CO_2 from below can maintain a condition of bicarbonate imbalance, even allowing for the CO_2 lost to the air.*

The loss of CO_2 to the air, involves, as we have seen above, an increase

* Variations in the CO_2 content of the air would make a difference to the balance. There has been a rise during the last half century and a modern figure of 0·330 per cent. is now widely accepted. This is equivalent to 0·65 mg CO_2 per litre in water, which would raise the corresponding millequivalents of Calcium bicarbonate from 1·0 to 3·0.

in the amount of carbonate present, which leads to super-saturation and consequently precipitation of the excess carbonate, but this takes place very slowly. Quite otherwise is the extensive precipitation which occurs through the dissociation of bicarbonate ions in the water by photosynthesis. This leads to many submerged plants being encrusted with Calcium carbonate. That all are not so encrusted is apparently due to the superficial secretion of mucilage which sloughs off the precipitated carbonate from plants of most species. Precipitation can also occur on a geological scale where volcanic springs emerge, the carbonate deposited from the super-saturated water building up into craters or terraces of hard rock called travertine. Old lake or pond sediments sometimes contain enough precipitated carbonate to form a soft calcareous material known as 'tufa'.

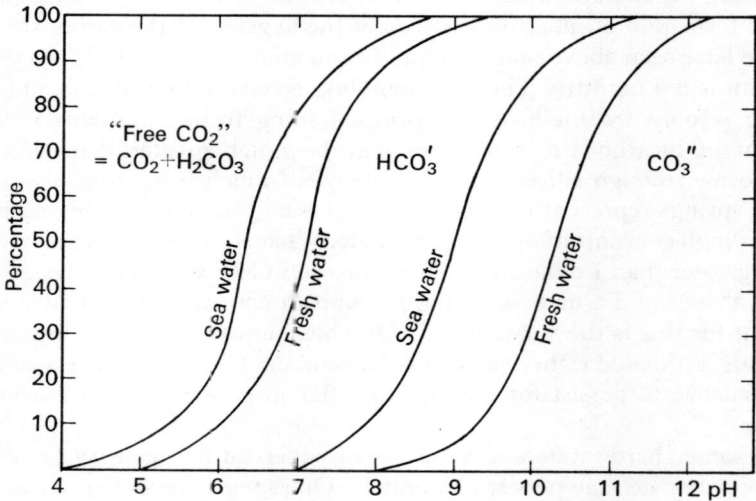

FIG. 4.57. Diagram showing the percentage of free CO_2, of HCO_3' ions and CO_3'' ions in fresh water and in sea water of salinity 34 per thousand at various Hydrogen ion concentrations. (*After Kalle.*)

The same equilibrium between free CO_2, bicarbonate and carbonate which occurs in fresh water operates on an immense scale in the sea. The maintenance of a 1:1 equivalence between the atmospheric CO_2 and the free CO_2 in solution over the vast extent of ocean surface, facilitated by the continuous agitation of the water surface, and regulated by either emission of CO_2 from water to air or solution from air to water, is certainly a major factor in preserving the minute but astonishingly constant percentage of that gas in the atmosphere, on which our planetary life depends. This is safeguarded by the reserves of available CO_2 in the bicarbonates, which are estimated to be of the order of twenty-seven times the amount needed to replenish the entire atmospheric content.

We have seen (p. 3722) reason to consider the prevailing pH of fresh waters as a factor of importance in determining their populations. The prevalence of a high pH is in general attributable to the supply of base-rich silt and basic material in solution in the influent waters. Water on acidic rocks, where such a base supply is wanting, till tend to have a low pH and fall into the category of oligotrophic water. There is, however, another factor which brings about cyclical fluctuations of pH and this is the CO_2 content of the water. A solution of CO_2 in distilled water has a pH of about 5·0. Thus distilled water left in contact with air, the so-called 'equilibrium water', is distinctly acid in reaction. Buffering materials in solution will, however, limit the extent to which carbonic acid can change the pH, and in most natural waters there are two categories of substances chiefly active as buffers. In eutrophic waters phosphates are usually the most influential in this respect, in oligotrophic waters buffering is considerably less and is due to carbonates, so far as they are present.

There is a marked difference between epilimnion and hypolimnion with regard to the part played by CO_2. The epilimnion is the productive layer, where photosynthesis is active, and as free CO_2 is assimilated by all photosynthetic organisms, there will be periods when the free CO_2 in this layer is well below its equilibrium level and may be entirely exhausted. In standing water this imbalance is only slowly regulated and the pH may rise temporarily to 9·0. Running water, especially fast running water, is not affected in the same way, since equilibrium is rapidly restored. Many, if not most, submerged plants can also assimilate the bicarbonate ion, which they dissociate internally with the liberation of CO_2 and the excretion of OH^- ions (from the upper sides of the leaves only, as Arens has shown). These react with $CaCO_3$ to form Calcium hydroxide which may lead to further rise of pH to the extreme of pH 11·0. As the amount of carbonate in solution, in comparison with bicarbonate, is usually very small, such a great rise of pH does not often happen and the dissociation of the bicarbonate usually results, as we have seen, in the precipitation of supersaturated carbonate.

At night, with the cessation of photosynthesis and the excretion of respiratory CO_2 by the plants, the situation can be rapidly reversed. The pH drops and some at least of the carbonate is again brought into solution as bicarbonate. There is not a complete reversal, however, for at the lower night temperatures, the elimination of CO_2 is not as active as its assimilation during daylight and a surplus of undissolved carbonate may remain. Moreover, the respiratory quotient is not always unity. In the complex metabolism of a cell the quotient may be as low as 0·69 (Meyer) and an average value is about 0·85.

Some observations in rock pools on a tidal shore throw light on the conditions. These pools lay on a limestone foundation and were rich in bicarbonate. When the low tide interval left the pools isolated in the daytime the pH rose steadily. The pH of the open sea water was 8·2 and the rise shown by the pools was proportional to their position on the shore slope and the length of time they remained isolated. Those highest up the slope, isolated for 6 hours, rose to a pH 9·6, which was immediately restored to 8·2 when the rising tide again reached them. If the low-tide interval was at night there was no rise of pH, but, on the contrary, a drop to a minimum of pH 6·5.

It has been shown that submerged mosses are unable to utilize the bicarbonate ion and require free CO_2, for which reason they can only grow in rapidly running water or in bog pools where they are supplied with CO_2 from below.

Other bicarbonates besides that of Calcium may be present especially in permanently alkaline water. Magnesium bicarbonate is usually present only in small amounts, but, if present, photosynthesis may lead to an increase of its concentration relative to Calcium, since its carbonate is more soluble than that of Ca and it is not precipitated. Sodium or Potassium bicarbonate may also occur. As their carbonates are completely soluble they lead to no precipitation, but they also cause a rise of pH to above 9·0 when dissociated. Sodium bicarbonate is, however, somewhat toxic to plants.

In the hypolimnion, where, as we have already mentioned, (p. 3733) the CO_2 is increased by micro-biological respiration, little if any remains free. It maintains equivalence ('attached' CO_2) with the bicarbonate it forms from the carbonate precipitated from above, If, however, there is very little calcareous material available or if respiration is very active in the sediments, an excess of free CO_2 may occur, the so-called 'aggressive' CO_2 which is available to attack any source of calcareous material and increase the bicarbonate equivalence. In general the distribution of attached CO_2 with depth is inverse to the distribution of Oxygen. Where an excess of free CO_2 occurs in the deepest water it can be attributed to anaerobic respiration in or above the sediment surface.

At pH < 6·0 the presence of acids additional to carbonic acid is probable. The highest acidities are generally found in lakes of volcanic areas, where mineral acids (e.g. sulphuric) are present and pH values of 2·0–4·0 are found. Bog pools, in which acidity is generally attributed to organic acids may sometimes show a pH well below 4·0, and in these extreme cases the presence of sulphuric acid is probably accountable. This acid can be produced from the sulphate in rain water by exchange of kations for H ion in the peat, or by the oxidation and hydrolysis of ferrous sulphide, forming ferric sulphate and sulphuric acid. The former may hydrolyse in the acid solution to precipitate ferric hydroxide (see lake sediments, p. 3730).

PHYSIOLOGICAL FACTORS IN THE AQUATIC ENVIRONMENT

Among such factors one of the first things to attract attention is the relationship of the plants to the water itself. The paucity of xylem in the stems of many aquatics, especially submerged aquatics, led to the belief that water was absorbed indifferently all over the surface of the plants, that there was no conduction of water and that the roots served merely for anchorage.

This view was negatived by Wielder's demonstration in 1892 of root pressure in such plants as *Elodea* and the further demonstration by Pond (1905) that even in free-floating plants (except *Lemna*) the roots were also active in mineral absorption, which would tend to produce root pressure.

It had long been known that the leaves of most submerged plants have apical or superficial openings, formed by the collapse of cells, which were called *hydropotes* because they were supposed to be the sites of water absorption as indeed they may be. Riede (1921) showed, however, that they were also portals of escape for exuded water, comparable to the hydathodes of land plants and like them capable of guttation. By using two potometers Riede was able to equate the amount of water absorbed by the roots with the amount exuded from the leaves and concluded that there was a normal but very small upward movement of water. Other forces besides root pressure have been suggested as taking part in this water movement, such as: the withdrawal of water from the xylem by growing tissues; the adjustment of water balances between different tissues; electro-osmosis, and temperature differences between the plant and the external water due to the absorption of radiant energy by the leaves.

When we turn to consider plants with exposed or floating leaves we return once more to the realm of normal transpiration losses but with this difference, that there exists a 'surface climate' over water, a microclimate like that near the soil which Geiger explored (see p. 3530). It has higher humidities, even at midday, than those in the general climate and there is little resistance to wind velocity, except in reed swamps which are marginal, in both senses, to the aquatic habitat. Among reeds the humidity is in fact higher than over the open water surface, especially at midday.

The ratio of transpiration to the evaporation from an equal area of open water (T/E) is in all cases but one below unity. The highest ratio measured is for *Pistia stratiotes*, 0·66, the lowest is for *Caltha palustris*, 0·08. The sole exception is *Lemna*, where the ratio is 1·0, but *Lemna* shows a marked resistance to the effect of wind. Its transpiration rate is so little controlled by wind that at a wind velocity of 3·7 m/sec its T/E ratio has dropped to 0·33. Nevertheless, the effect of wind on the transpiration rate of hydrophytes, is in general, greater than its effect in raising the rate in mesophytes or xerophytes. The relatively high rates of transpiration from the leaves of hydrophytes may be attributed to the simplification of their stomata and the prevailing lack of a closure mechanism and also to the extent of trans-cuticular water loss, the cuticular covering being in most cases very slight or absent.

Rooted, emergent plants of the reed type can transpire very vigorously and their upright habit provides a much greater area for transpiration than the horizontal area of water which they occupy, which, moreover, they protect. Thus *Phragmites* may, according to the weather, transpire from 3–16 kg of water per m² of foliage per day, with T/E ratios up to 7·0. There are many regions where reed swamps cover enormous areas, the Danube delta, the lower part of the Shatt el Arab and the *sudd* area of the Nile, for example, and it is evident that these vigorous rates of transpiration must have a considerable effect on the whole water situation of the region, for, unlike land plants, they transpire several times the amount of the local rainfall.

We discuss the water losses of seaweeds undergoing desiccation on p. 3764 and it is interesting to note in comparison that leaves of fresh water plants in drying generally show a continuous and regular loss of weight, but that *Potamogeton natans* behaves like a seaweed, losing water rapidly to the extent of nearly 70 per cent of its fresh weight, at which point the curve bends sharply to the horizontal and no more water is lost, so that it has finally a higher water reserve than some much more substantial plants such as *Eichhornia crassipes*, the water hyacinth.

Another big physiological factor in aquatic life is, of course, the peculiar conditions governing Carbon assimilation in an aqueous environment. This should not affect the emergent plants whose foliage is effectively aerial, especially as they suffer no interruption of their CO_2 supply through water shortage, as do many land plants. When therefore we find that their production of dry matter per unit area is less than that of corresponding types of land vegetation, it would seem that a shortage of mineral nutrients may be responsible.

We have already spoken (p. 3735) of the importance of the dissolved bicarbonates as a source of CO_2 for submerged plants, but this may also be a factor for plants with floating leaves, which can assimilate bicarbonates from the water through the under surfaces of their leaves. Otherwise these plants have access to atmospheric CO_2 through the stomata on the upper surfaces of their leaves and are thereby brought more or less into line with land plants.

Plants like *Nymphaea*, with floating leaves, have one experimental advantage, that they enable comparisons to be made of leaves below water with those above water, which throws some light upon the relative disadvantages of submergence. The respiration of exposed leaves is only slightly greater than that of submerged leaves, but the net assimilation rate of the adult exposed leaves is between four and five times as great as that of similar but submerged leaves. Apparently a net positive yield by photosynthesis only appears when the leaves reach the surface and this seems to be linked to the extraordinary habit of these plants of adjusting the length of the petiole to the depth of the water, not exactly but with an excess growth of 10–20 cm, which enables the petiole to take up a slanting direction and thus avoid the overcrowding of leaves by distributing them in a rosette. That this adjustment is really connected with the free access of CO_2 has been shown experimentally by covering young plants with glass cylinders, one of which was deprived of CO_2 and the other was enriched with the gas. Where the CO_2 was available the leaves stopped growing at the water surface, but where it was not, they continued to elongate as if they were still submerged.

Although the absorption coefficient of water for CO_2 is practically unity and the concentration of the gas in water is equal to that in the air, i.e. about 0·6 mg/litre, there is often a surplus of CO_2 in surface waters due to the dissociation of bicarbonates, liberating CO_2 in excess of that dissolved from the air. This does not necessarily confer any advantage on submerged plants, because the diffusion rate of CO_2 in water is only $1/1000$ of that in air. The plants are thus greatly dependent on water movements to present them with Carbon dioxide in sufficient amounts. Even if this condition is assured there remains a thin film of water held electrostatically to the surface of the plants, which is unaffected by water movements and through which molecular diffusion is the only path of entry; this film is, however, very thin and mass water movements do have a considerable effect in raising the level of photosynthesis, by from 15 to 30 per cent.

In these circumstances there is a definite advantage in increasing the area of contact with the water and at the same time reducing the diffusion path internally, by subdivision of the assimilating tissues. Accordingly we find that a great many submerged plants have finely divided, almost feathery, leaves and that emergent plants often have leaves of this kind below water while producing undivided leaves in their aerial portions. Goebel has justly compared the internal adsorptive surface in an aerial leaf to a lung and the extended, external absorptive surface in underwater plants to gills, since the same basic physical conditions operate both on animals and plants.

The advantage of a maximal ratio of surface to volume is illustrated by Hammann's findings that, in proportion to chlorophyll content, the assimilation of unicellular algae, such as *Chlorella*, is of the order of ten times as great as that of submerged phanerogams. Gessner has also shown that the feathery divided leaves are more responsive to increased illumination than undivided leaves from plants of the same species. The advantage gained by increased external surface appears not only in leaves but sometimes in submerged rhizomes also, as in *Neptunia* and *Jussieua*, which develop thick external layers of loose, spongy tissue, called by Schenck *aerenchyma*, sometimes as a modification of the primary cortex and more usually developed secondarily like cork, from a phellogen layer. This facilitates gas exchange with the water, but in this case it is absorption of Oxygen which is important.

Many floating plants prevent submergence by the development of buoyant organs such as inflated petioles or leaf-sheaths, thus making the best of both worlds by keeping their roots in water and their leaves in air. This is especially advantageous in the frequently muddy water of tropical rivers, which may also be, thanks to the high temperatures, poor in dissolved Oxygen.

Pollination presents certain difficulties to aquatic plants. Most floating plants, such as *Nymphaea* or *Potamogeton*, produce their flowers above water and rely on insect pollination. Even submerged plants such as *Elodea* and *Vallisneria* elongate their flower pedicles to reach the surface. The latter has long been celebrated for its curious mechanism. The female flowers are raised to the surface by a long, flexible pedicel. The male flowers are detached and float about until they make contact with the stigmas of the female. Once pollinated the female flowers are withdrawn downwards by the coiling of the pedicel and complete their maturation below water.

Totally submerged aquatics may depend simply on water currents to convey their pollen (hydrophily) and in some cases the efficiency of the method is increased by the development of filamentous pollen grains, which are more readily caught by the stigmas.

A further concomitant of aquatic life is the use made of water currents for dispersal by means of buoyant fruits and seeds, many of the latter with water-hard coats which prevent quick germination. We need only to mention this feature for it has been fully dealt with in other books.*

LAKE FORM AND VEGETATION

Lakes are of many physiographical types, Hutchinson has classified them under 76 headings, dependent on all the various geological processes or accidents which have produced them. They are of all ages, from very new to very old and during its lifetime a given lake passes through the phases of a development which depends on so many factors, physiographical, hydro-logical, chemical and biological that it can be quite idiosyncratic. One feature, however, all lakes have in common; they accumulate silt, sometimes rapidly sometimes very slowly, depending on the nature of the surrounding area and of the lake bottom. If the lake is isolated, without drainage, there is still a slow deposit of atmospheric dust and of organic matter bred by the lake itself.

Silting is biologically important; it is the biggest single factor determining the establishment and the distribution of the plant population in a lake. Its extent increases with the age of a lake and so does the variety of its plant inhabitants. The latter phenomenon has usually been ascribed to the opera-tion of chance governing the, probably infrequent, arrival of new recruits to the population. No doubt this is operative, but it is not the only factor, since some at least of the later comers could not have established themselves earlier if the requisite amount and quality of silt was lacking.

In the deepest lakes much of the silt lies below the biological level, and its effects on the physiography of the lake may be negligible, though as a source of soluble bases it may have great chemical importance.

The ideal lake form is that of an inverted cone. If $\bar{Z}=$ the mean depth and $Z_m=$ the maximum depth; then ideally $\bar{Z}:Z_m=0.33$. The majority of lake basins have a higher ratio, those with very steep sides may have ratios above 0.5. In such lakes the plant populations may be limited to a very narrow girdle.

In shallower lakes, on the other hand, silting may in time diminish the volume of water to such an extent that vegetation can spread over nearly the whole area and, once established, vegetation increases the rate of silting by checking water movements and thus favouring the deposition of sediment. The process is thus self-accelerating, the lake becomes dead in the hydrographic sense and passes through swamp to marsh and eventually to 'dry' land.

* See especially, H. N. RIDLEY (1930) *The Dispersal of plants throughout the World.* L. Reeve.

This transformation is accompanied by a series of successional changes in the vegetation which have long been known under the name of the 'hydrosere'. The later stages can be surprisingly rapid, as has been distressingly seen in the shallow waters of the Norfolk Broads.

We have already (p. 3700) spoken of the different life-forms which make up the vegetation of fresh waters. Each life-form has its preferred depth of water, hence the concentric zonation which is often conspicuous. The greatest depth which can be colonized depends chiefly on the transparency of the water. Pearsall showed that in the English lakes the depth limit was set by an illumination which was 2 per cent of the incident light on a bright day, but that the vegetation did not, in fact, generally go down so far, especially where the light limit lay deep, since other factors, temperature or Oxygen supply, may become limiting. Thus in the clear Wastwater the light limit lies at 10 m, but the vegetation does not extend below 7·7 m, whereas in the peat-coloured Esthwaite Water the light limit lies at 4·1 m and the vegetation extends to 4 m.

The submerged plants are often spoken of as the pioneers of the hydrosere but this is not strictly true, since whenever a lake exists there must be a gradation of depth around the shores and in the shallow edges emergent plants will establish themselves from the beginning. Submerged plants are only the deep water pioneers, where they persist until the accumulation of inorganic and organic silt raises the bottom to a level which makes possible its invasion by emergent plants of the swamp type, which replace the submerged plants. In large lakes the succession may only occur locally, in sheltered bays and inlets or where an inflowing river, its current checked by entering the lake, builds up banks of sediment. Sometimes these may fill in the whole end of a lake and a level strath is formed. This is all part of the dynamic process of accommodation between water and land which goes on all through the history of a lake.

What are sometimes called 'primitive' lakes are those lying on very hard rocks where silting is at a minimum and dissolved salts are scarce. Their rocky or pebbly shores offer little foothold for plants. Under such conditions the lake sediment is chiefly organic and acid, a sub-aqueous peat, on which, in Britain, the submerged *Lobelia*, *Myriophyllum* and *Potamogeton natans* (in decreasing order of depth) build up to a marginal vegetation of *Equisetum limosum* and *Carex inflata*.

The type of lake where inorganic silt is abundant and soluble bases plentiful, shows a different sequence of *Littorella*, *Potamogeton* (several species) both submerged, then *Nymphaea* and *Nuphar* with surface-floating leaves and finally *Phragmites*, *Scirpus* and *Typha*. These are, however, only samples of the various successions which may occur, influenced by differences of silting or exposure to wave action on lee shores. Pearsall has shown that it is the finest fractions of silt which are the chief source of potash and phosphate and as these fractions tend to settle in deeper water he has maintained that many features of aquatic zonation are related to differences in the distribution of the silt.

CHAPTER L

THE MARINE ENVIRONMENT

WHILE sea water has many properties in common with fresh water, it also has many striking differences. We cannot here open out the whole field of Oceanography, which is a science in its own right, with its own literature and distinctive textbooks, but we can consider some aspects of inshore life, which is literally marginal to our landward preoccupations.

The most outstanding character of sea water is naturally its salinity, with which are bound up its osmotic potential, its pH and its content of dissolved gases. The standard analysis of the principal salts in sea water is that made by Dittmar on the famous 'Challenger' expedition of 1872. It gives the proportions of the salts as percentages of the total solids.

Sodium chloride	77·758	Potassium sulphate	2·465
Magnesium chloride	10·878	Calcium carbonate	0·345
Magnesium sulphate	4·737	Magnesium bromide	0·217
Calcium sulphate	3·600		

Later analyses have shown the presence of minute amounts of nearly every element, including Radium, many of which are important as trace elements in plant metabolism, but the order of predominance of anions is clearly, chlorides–sulphates–carbonates. There is a constant ratio of chlorides to the total salinity and the latter is generally estimated by the empirical relationship of Knudsen and Sorensen:

$$\text{Salinity} = 0·03 + 1·805 \text{ Cl, in parts per thousand.}$$

Total salinity varies geographically, 35 parts per thousand being an average figure. The diagram (Fig. 4.58) shows the variation from N–S in the Atlantic and Pacific Oceans. Rivers pour immense amounts of fresh water into the sea, estimated at 30–40,000 km^3/year, and consequently coastal waters are generally less saline than the open sea. The whole of the Baltic is 'coast water' in this sense, while the Mediterranean has little or no coast water and has high prevalent salinities, owing to its hot climate which promotes evaporation and the scarcity of big influent rivers; with the Red Sea it shares the highest general salinity.

At a salinity of 35 parts per thousand (corresponding to a chloride concentration of 19·375 parts per thousand) the following are the amounts of kations in solution, in grams per kilogram (litre) of water.

Sodium	10·752
Magnesium	1·295
Calcium	0·416
Potassium	0·39

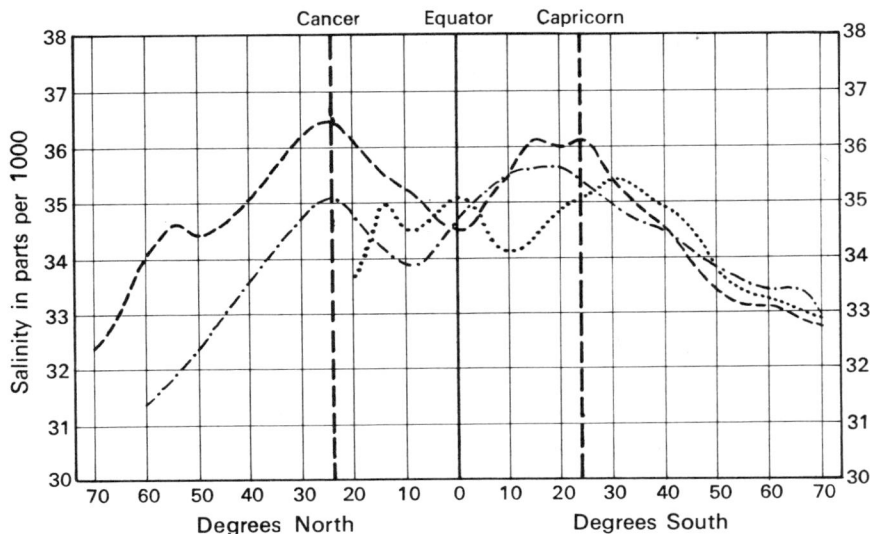

FIG. 4.58. Distribution of salinity with latitude in the surface water of three oceans:
Atlantic — — — —, Pacific —·—·—, and Indian ·········. The tropics are
marked by broken vertical lines. The average value for the Atlantic Ocean is 35·37,
for the Pacific 34·91 and for the Indian Ocean 34·81. Salinity is highest at the
tropics. (*After Krümmel.*)

The striking contrast with inland fresh waters is best brought out by a
comparison of the relative proportions of the anions as shown in Table 17
(Helland–Hansen).

TABLE 17

Anion	Sea Water	River Water
Chlorides	88·64	5·2
Sulphates	10·80	9·9
Carbonates	0·34	60·1
Various	0·22	24·8
	100·00	100·00

To an organism the sea is a chloride environment, while fresh water is a
carbonate environment. In considering movement from one medium to the
other this great chemical difference is not sufficiently appreciated. Migration
involves not simply adjustment to a different osmotic value, it involves a
complete change of ionic relationships and a consequent change of metabolism.

The high osmotic potential of sea water is nevertheless an influential factor
in its biology. It increases with the salinity and with the temperature, though
the latter has less effect than the salinity. In the Baltic Sea the salinity is so

low (it varies between 5 and 15 parts per thousand) that the osmotic potential is little above that of fresh water, 5–10 atm. The OP of open Atlantic water is 23–25 atm. and that of the Mediterranean is higher still, 25–27 atm. It is normally higher in the surface water than at depth, e.g. in the Sargasso Sea it is 25 atm. above and 23 atm below, but in the Arctic and Antarctic Oceans both temperatures and salinities are lower at the surface than in the depths and hence the OP is lower also, probably due to the flotation of fresh water from melting ice. High temperatures with intense radiation increase surface salinity and ice-formation produces an analogous effect by lowering the amount of free water present. Both factors therefore increase osmotic potential. As shown by Fig. 4.58 the highest salinities occur in the sub-tropics where radiation is greatest.

Allowing for some variation it still remains true that the OP of sea water is at least twice that of the cell-sap in most mesophytic plants, hence the destructive effect on such plants of any chance encounter with sea water. We have already spoken about this when discussing halophytes which are habituated to sea water (see p. 3672).

TABLE 15. OSMOTIC POTENTIALS IN ATMOSPHERES OF SEA-WATER AT DIFFERENT SALINITIES AND TEMPERATURES (Helland–Hansen)

Salinity p.p. thous.	Freezing Point in °C	0°C	10°C	20°C
30	−1·627	19·67	20·39	21·11
32	−1·740	20·95	21·72	22·49
35	−1·910	23·12	23·97	24·82
37	−2·024	24·50	25·40	26·30

The figures shown in Table 17 of the salts in sea water make no mention of two substances which are of vital necessity to organisms, namely Nitrogen and Phosphorus, to which we may add silica. This omission is due to the fact that these three are, in fact, the problem substances of the sea and deserve separate treatment. All are present in surface waters only in minimal amounts and may even at times disappear altogether. Approximate average amounts are, in milligrams per litre: Nitrogen, as NO_3, NO_2 and NH_4, 1·0; Silicon, 1·0; Phosphorus, 0·6. The amounts found are often less than these.

All three substances share the same cycle; they tend to disappear from the surface when plankton life is most active; they are carried downwards as dead plankton organisms sink; they are liberated in deep water as a result of decomposition; they are eventually restored to the surface, at least in part. The last process can be very slow and is a severe limiting factor on organic productivity. (See also p. 3724.)

Nitrates are all soluble, so there is no question of their deposition as sediment, but the amount held in solution in deep water is enormous compared with that available in the photosynthetic zone of the sea, which we may estimate as the upper 50–100 metres. The following analysis by Gran of Atlantic water near the Azores shows this:

Depth (metres)	Nitrate (mg/m^3)
0	12
10	7
50	6
100	55
1000	265
3000	265

In shallow waters the vertical circulation is practically continuous and the avoidance of any delay in the restoration of nitrate to the surface leads to much greater fertility in such areas as compared with deep seas.

Nitrate-forming bacteria are active in deep water and transform into nitrate the ammonia liberated by protein decomposition. The cycle may be shown thus:

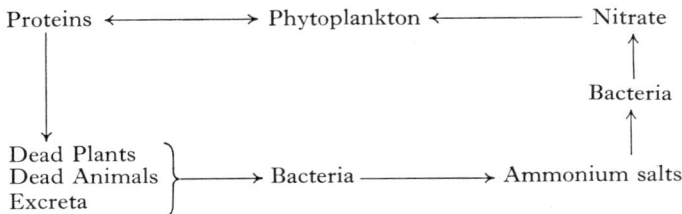

Proteins ←————————→ Phytoplankton ←———————— Nitrate

Bacteria

Dead Plants
Dead Animals }————→ Bacteria ————————→ Ammonium salts
Excreta

Ammonia may also be contributed to the cycle by direct N fixation.

Winter cooling of the surface water in cold seas reduces or reverses the density gradient from surface to depth and so promotes restoration. Conversely, intense surface heating in some areas of tropical seas may produce a flow-off current and so cause up-welling of deep water.

In temperate seas the consumption of nitrate by plankton is restricted during autumn and winter by the limited amount of sunshine, with the result that regeneration overtakes consumption. The worse the winter the greater the amount of regeneration and storage of materials at depth. These reserves are available for return to surface water as the cold surface water sinks and displaces the deep water in December–January. The replenishment of the surface water paves the way for the big spring outburst of plankton life in April–May, when the increasing sunshine permits consumption by plankton to overtake release of materials by decomposition.

What has been said about nitrates also applies in great measure to phosphates and silica. The phosphate supply, especially, is a limiting factor on productivity. During the summer in temperate seas it may be completely removed from the surface water. In March, in the English Channel there was 0·036 mg/litre (as P_2O_5), but by July the amount was zero, the period of time during its disappearance coinciding with maximal development of plankton, especially diatoms.

Sea water is almost always slightly alkaline, that is to say that the bases present are slightly in excess of the available acids in solution. This *excess-base* or alkalinity represents the amount of bases available for combination with CO_2, or the alkaline reserve of the sea. The carbonates are the most important materials involved, though ammonia may sometimes be a minor contributor. The excess base bears an approximately constant relation to total salinity and also to the CO_2 content of the water, which are the controlling factors in determining pH, so that there is an alkalinity (or salinity)–pH–CO_2 equilibrium enabling any one to be calculated from the values of the other two (Fig. 4.59).

The relationship of alkalinity to total salinity in the north Atlantic was expressed empirically by Schulz as:

$$A = 0 \cdot 06788S$$

where S = the salinity in parts per thousand. Other estimates lie very close to this, e.g. Brennecke: $A = 0 \cdot 6779S$. For an average salinity of 35 this corres-

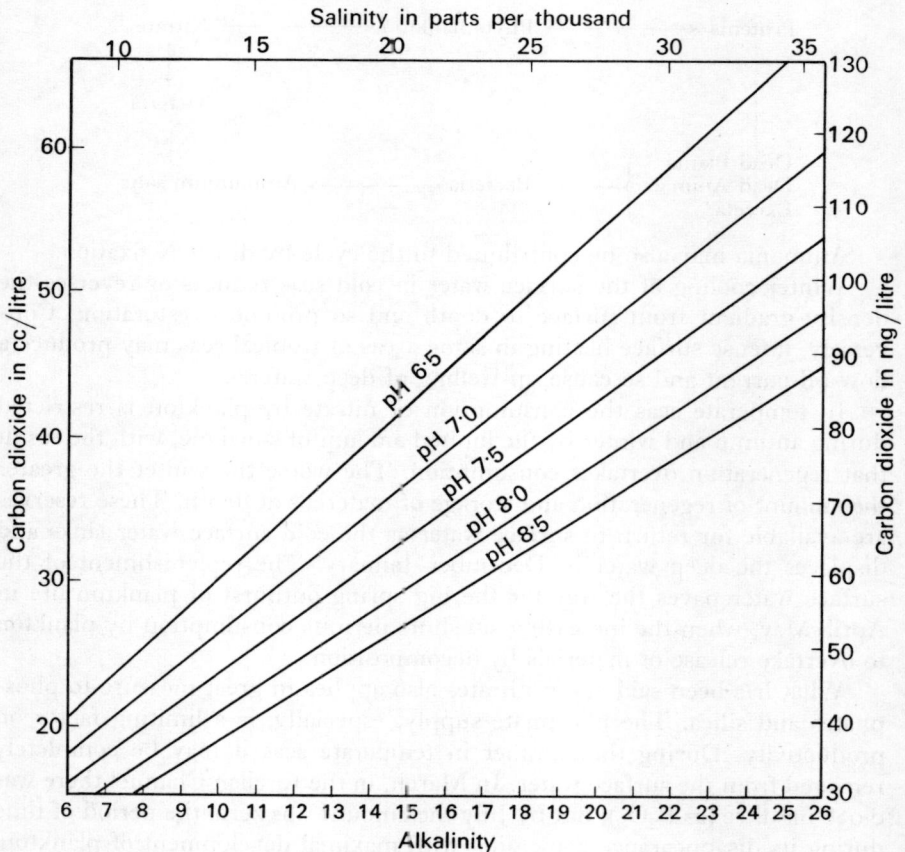

FIG. 4.59. The interrelationships between alkalinity, salinity, pH and CO_2 content in sea water. (*After Kreps.*)

ponds to an excess base value of nearly 24 of titratable alkalinity. The usual method is to titrate 100 ml of sea water with $N/100$ HCl, using brom-cresol purple as indicator. After the first titration the water is boiled to drive off the CO_2 liberated by the acid, then the titration and boiling are repeated until a stable acid (yellow) colour is formed. The number of ml of acid equals the alkalinity in terms of $N/10,000$ calcium carbonates and bicarbonates present.

Open seas show little geographical variation. In the Pacific Ocean off the California coast the value is 23·5. At Woods Hole on the Atlantic coast it is 24. Off Florida it is 23–25. In the Baltic it is as low as 5, while in fresh river water it may be only 1–2. For this reason sea water can hold much larger reserves of CO_2 than fresh water. The Mediterranean shows increasing salinity and alkalinity eastwards, and in the Adriatic,which receives extensive limestone drainage water, the alkalinity reaches 55. In the Black Sea peculiar conditions obtain. The surface water is highly alkaline but there is a very steep downward gradient (negatively clinograde) with depth. There is a stagnant mass of water below 90 fathoms, which is the depth of the rock-sill in the Bosphorus. This stagnant water contains free H_2S and its lower reaches are anaerobic and support only sulphur-oxidizing bacteria. Similar conditions are found in some Norwegian Fjords which have a rock-sill at their seaward end and deepen further inland. Nevertheless, these waters are still slightly alkaline. Truly acid sea water has only been found temporarily amidst melting ice (pH about 6·5).

Regions with a markedly seasonal climate show variations of alkalinity (and pH) with the season. Atkins showed that in the English Channel the alkalinity increased progressively till August in good summers, but that in wet summers the maximum was in May, followed by a steady decrease. The rise is attributed to abstraction of CO_2 by phytoplankton, the subsequent fall to the upwelling of CO_2-rich water from below.

In deep water the relationship of alkalinity and CO_2 content characteristic of surface water does not hold. The amount of dissolved CO_2 increases with increasing pressure and lower temperature and at 1000 fathoms the amount is about three times that at the surface. At the same time alkalinity is increased by solution of sediments. The picture is therefore generally one of alkalinity and dissolved CO_2 *increasing* with depth but pH *decreasing*.

An alkalinity of 26 and pH 8·1 corresponds to a CO_2 content of about 48 ml/litre of water at NTP of which only 0·5 ml is in true solution and 47·5 ml is combined with excess base. A difference of one unit in alkalinity corresponds to a difference of about 2 ml/litre in the content of CO_2.

The sea water is far from saturated with bicarbonate and carbonate and will dissolve considerable amounts when shaken with precipitated chalk (e.g. 64·9 mg/litre/hr). For this reason carbonates are not easily precipitated even when the CO_2 content is diminished, except for drastic reduction by dense photosynthetic algae. As we have pointed out above, the establishment of equilibrium with the atmosphere is very slow, so that sub-normal contents of free CO_2 in the surface water may persist for long periods. For example, in the English Channel the free CO_2 in winter was found to be 0·022 per cent and in winter only 0·017 per cent with an air tension of 0·033 per cent. The surface is

TABLE 19. DISTRIBUTION OF CO_2 IN SEA WATER AT DIFFERENT SALINITIES AND
TEMPERATURES (Hamberg/Krogh)

Salinity	A	$t°C$	a	b	c	B
17·78	13·47	0	1·38	24·18	0·42	25·98
		10	2·31	22·32	0·29	24·92
		20	2·94	21·06	0·23	24·23
26·58	20·26	0	3·52	33·48	0·40	37·40
		10	4·49	31·54	0·27	36·30
		20	5·51	29·50	0·22	35·23
35·13	26·96	0	5·07	43·78	0·38	49·23
		10	7·06	39·80	0·26	47·12
		20	0·64	34·64	0·21	44·49

A = Alkalinity. B = total CO_2. a = amount in carbonates, b = in bicarbonates, c = free
in solution. The latter amounts decrease with rising temperatures but, owing to
chemical combination a and b rise with increasing salinity, unlike oxygen.

thus always ready to absorb CO_2 from the air and this condition is very widespread in temperate latitudes, whereas in the tropics (CO_2 being less soluble
at higher temperatures) the surface water is often super-saturated with CO_2
and emission to the air is the rule. There is thus a general cosmic circulation
of CO_2 from the equator towards the poles.

We must distinguish between CO_2 content, measurable by weight or
volume, in the sea and the partial pressure of the gas. The latter is dependent
on temperature because the coefficient of absorption decreases with rising
temperature. Thus, if water at 30°C has a partial pressure of CO_2 of 0·045, i.e.
more than the atmospheric level, the same water, cooled to 0°C would only
have a partial pressure of 0·015; much below the atmospheric level, though the
actual content is unchanged. The warm water would, however, be apt to emit
CO_2 to the air, the cold water to absorb it. This is what happens to the water
of the Gulf Stream as it flows northwards and gives up heat.

Like Carbon dioxide the sea contains large reserves of Oxygen, which are
naturally of paramount interest to sub-aqueous life. Oxygen is more soluble

TABLE 20. DISSOLVED OXYGEN IN ML/LITRE OF WATER AT
0°C AND 760 MM PRESSURE (Krümmel from analyses by
Jacobsen)

Temperature	Salinity: 30	35 per thousand
0°C	8·4	8·0
5	7·4	7·1
10	6·6	6·4
15	6·0	5·8
20	5·5	5·4
25	5·1	4·9
30	4·7	4·5

in water than Nitrogen, so that while the proportion of O_2/N_2 in air is $1:5$, in water it is more nearly $1:2$. Thus water at salinity 35 and 15°C contains 5·8 ml of O_2 per litre at NTP and 11·25 ml of N_2. There is, however, considerable variation, as the solubility varies inversely with the salinity and with the temperature.

It is evident from these figures that the amount of dissolved Oxygen is lowest in tropical seas and increases with latitude. Arctic and Antarctic seas are the richest in Oxygen and also the richest in plankton and in the animals which live on plankton.

Large bodies of the sea are deficient in Oxygen. In the tropics especially, between 100 and 400 fathoms, there may be less than 2 ml/litre which is below the minimum requirement for fishes (about 2 ml/litre). Below this level there is a slow increase and at 1000 fathoms the content is about 5 ml/litre in most parts of the Atlantic. This is partly due to the increased solubility of O_2 at high pressures and partly due to the fact that much of the bottom water in the Atlantic is dense, cold, polar water which has crept southwards under the lighter, warmer water, bringing its Oxygen with it.

Although there may be considerable amounts of O_2 present in deep water (or in any water) there may be nevertheless an Oxygen deficit. If $O_n =$ the saturation value and O the observed value, then $O_n - O$ is the Oxygen deficit and the expression $100(O/O_n)$ is the percentage saturation value or degree of saturation, analogous to the relative humidity values in the atmosphere, which expresses the physiological state of the water for aerobic organisms.

The prevalent degree of saturation varies not only geographically and with depth, but also seasonally according to the intensity of plankton growth. Indeed the surface water in productive regions in spring and summer is often super-saturated and emitting O_2 to the atmosphere. Putter calculated that in Kiel harbour in summer there is an algal production of 0·4 litres of O_2 at NTP per square metre of surface per hour (p. 3758). At the opposite end of the scale there are bodies of stagnant, 'old' water in deep basins (such as the Black Sea and the 'threshold fjords' already mentioned) which are totally deficient in oxygen.

It is interesting to note that the total content of Oxygen in the sea is never high enough to support warm-blooded animals. Were all the Oxygen completely respired, the total rise of temperature calculated as simple oxidation, would only be of the order of 0·04°C. Hence gill-breathers can never be warm-blooded.

This is perhaps just as well considering the relatively low average *temperature* of the sea. The highest average, surface temperature at the equator is 27°C. Downwards from a surface at this temperature we would find at 200 fathoms 10°C and at 1,000 fathoms only 2°C. In the great depths (3000–5000 fathoms) there are extensive masses of water below 0°C. The average temperature for the whole mass of the sea is only 3·8°C, and three-quarters of the mass is under 3°C.

Heating therefore comes by radiation on the surface and there is little or no heating from below. The limit in depth of the direct effect of radiation is about 50 fathoms, but wave action and currents cause radiant heat to be carried down to 100 fathoms. Somewhere about this level we find the same discontinuity of temperature that we find in fresh waters—a thermocline layer marked by a rapid drop of temperature and dividing the water into surface water, the *epithalassa* and the deep water, the *hypothalassa*. Temperatures in the hypothalassa are very constant and very small differences may be important because of their effect on density. Taking into account the huge

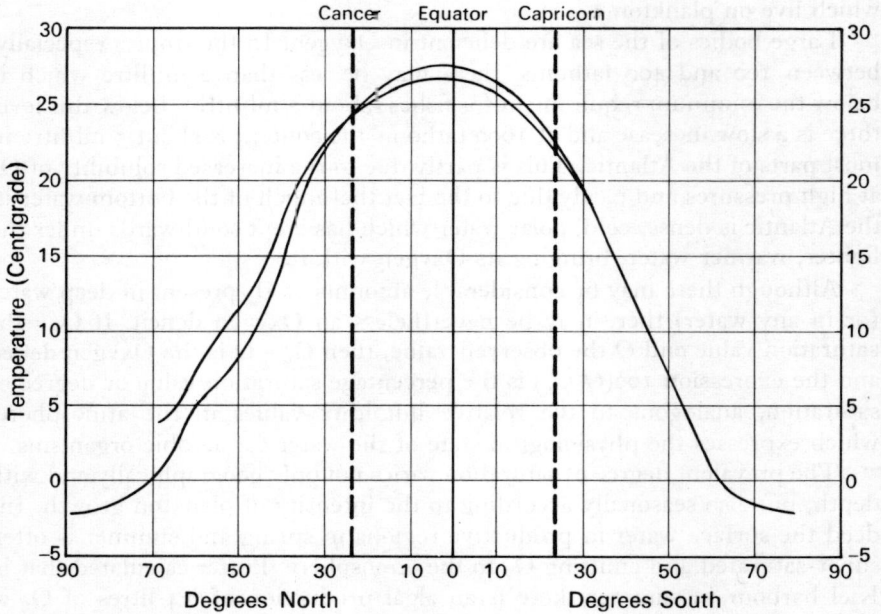

FIG. 4.60. Yearly averages of surface temperatures in the sea according to latitude. The longer curve represents the average for all seas, the shorter curve shows where the Atlantic Ocean differs from the average. The Indian Ocean reaches a peak about 1°C above the average at 5° North. (*After Krümmel.*)

volumes involved and the infinite time available, even the smallest difference in density may initiate great movements which can have widespread effects. It is the heating of the shallow water in the Gulf of Mexico which so reduces its density that it is able to overflow the colder Atlantic water and preserve its identity for so long that it has momentous effects on the climate of western Europe. That, of course, is a surface phenomenon, but similar great movements take place in the hypothalassa from slighter causes. It has been said that the temperature of deep water in the Atlantic is the surface temperature of the area where it was formed, generally supposed to be north of Jan Mayen Land, where it sinks and creeps southwards into the Atlantic basin, displacing upwards all warmer and lighter water and thus promoting some degree of interchange with water in the upper zones.

Daily variations of temperature in the open ocean are usually limited to fractions of a degree, nor is the seasonal variation great, the average being only about 5°C. Very different is the state of shallow coastal waters, where there may be a maximal seasonal change of 30–40°C.

This is very destructive of life, but the winter killing paves the way for the spring regeneration and total productivity is actually greater in such waters than in the deep sea. It is over submarine plains, such as a continental shelf, where the depth is under 100 fathoms that productivity is greatest, for in such waters an early spring or late winter turnover occurs, like that in inland lakes, ensuring the return to the surface of regenerated food materials from below.

The really 'deep' water of the oceans is remarkably stable, and the small seasonal variations of temperature (and density) limit the range in depth of any seasonal turnover. Return to the surface of deep water reserves of food materials is exceptional, as for example, where a deep current is deflected upwards by a submarine ridge, but where this occurs, as at the Newfoundland Grand Banks, productivity can be immense. Otherwise the deep water is almost undisturbed and being saline and under great pressure may have a constant temperature as low as − 2·0°C. It is over such depths that the thermocline is best developed, with a temperature change of 10°C in a few metres of depth. Shallow coastal waters only develop a thermocline at about 10 fathoms in very calm weather, and it is easily broken up by storms, which unify temperatures down to about 50 fathoms.

The relative stability of ocean water so reduces productivity that large areas, especially in warm seas are relatively poor in plankton, which Schmidt, as already mentioned, has expressed in his dictum that 'Blue is the desert colour of the sea'.

The specific heat of sea water at a salinity of 35 is 0·93, that of air only 0·20 (by units of weight). For this reason sea water is much slower than air in responding to temperature changes. In spring it warms up slowly, in winter it cools slowly so that it is always behind the atmosphere in changing. The coastal waters only reach their coldest point in March and they are warmer than the air during the winter months. This is the reason for the ameliorating effect of the sea on coastal climates. Calculated on a volume basis the difference between water and air is much greater and the sea can act as a vast reservoir of heat, much of which is given up to the air as it cools in winter. Thus a drop of 1°C involves such a loss of heat from 1 m³ of water to the air that it would raise 3000 m³ of air by 1°C.

When it is considered that the Gulf Stream and the Gulf Stream Drift together move about 150,000 km³ of water annually into the Norwegian seas, it is not surprising that Nansen and Helland–Hansen found close correlations between winter sea temperatures off Norway and climatic conditions in north-western Europe in such features as the date of the disappearance of snow in Sweden or the growth of pines and crop yields in the ensuing summer.

We have already spoken of the importance of changes in the *density* of sea water and the movements which may ensue, both vertically and horizontally. Light coastal water may overflow the denser ocean water for great distances. In the Great Belt opposite Copenhagen there is an upper low-salinity current running out of the Baltic and a lower high-salinity current running towards the Baltic, with only a distance of a metre or two separating them. The cold arctic water sinks and spreads over the bottom of the north Atlantic. The great ocean currents, which are often distinguishable by temperature or salinity, flow in the epithalassa over the stable deep water.

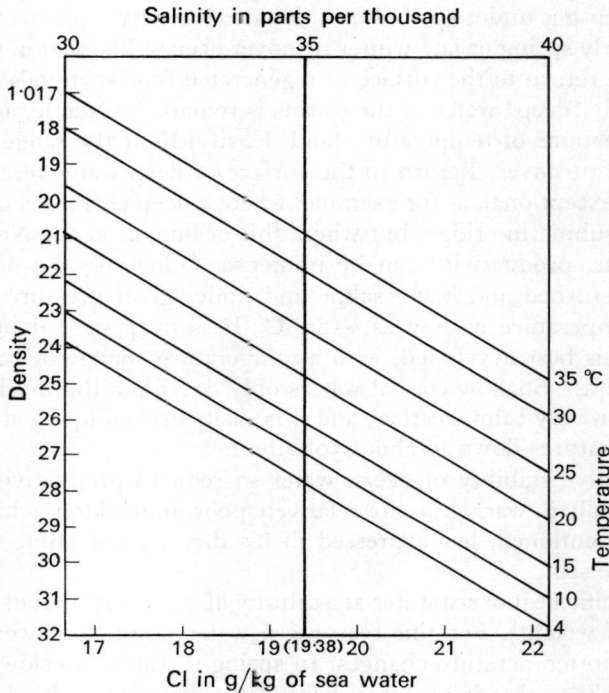

FIG. 4.61. The interrelationships between density, salinity, Chloride content and temperature in sea water. (*After* McClendon.)

Crossed influences are at work on the density for it varies directly with the salinity and inversely with the temperature. At a salinity of 35 and temperature 15°C, the density is 1·0255.* The upper limit is about 1·030 (which at 15°C would imply a salinity of nearly 41) found in the eastern Mediterranean. The highest figure in the open ocean is 1·028 (at 10°C = salinity 37 or at 4°C = salinity 36). A change of 1°C implies a change of 3 units in the fourth place of decimals in the density (e.g. 1·0270–1·0273) while a change of one unit of salinity causes a change of one unit in the third place of the density value, but in the reverse sense. Salinity is therefore rather more influential than temperature in its effect on density, but on the other hand is much less changeable.

So delicate is the balance of sea water that even minute changes of salinity may be of great importance. For charting 'isohalines' or lines of equal salinity,

* Distilled water at 4°C being taken as unity.

the unit employed is usually 0·1 parts per thousand. A change of only one-tenth of this, 0·01 per thousand, produces a change of 8 units in the sixth decimal place of density measurement and even this may be enough to cause big water movements. Hence the need for great accuracy in salinity measurements. Temperature becomes the controlling factor on density in areas of open ocean where salinity is uniform.

A feature of the seas which is normally absent or slight in inland waters is the great pressure developed at the greatest depths. Water is only very slightly compressible but the effect is enough to cause increased densities with increasing depth and the density may rise to 1·06. The effect is increased by the decrease of temperature downwards. Such high densities exceed anything that can result from greater salinity, and in these circumstances the saltest water may lie over less saline deep water, the opposite of what obtains in shallow coastal waters.

The problem of *illumination* exists in the sea as in fresh waters, but with a larger depth-scale. In most coastal waters the burden of suspended silt in the water imposes a considerable restriction on the penetration of light, but in the clearest of ocean water algae grow at much greater depths than in fresh waters. In the hard, highly saline water of the Adriatic the transparency is such that algae can grow at 250 m depth and yellow-green cells, presumed to be photosynthetic, have been dredged in the same sea from depths of over 1000 m. A very feeble illumination does reach great depths, for a photographic plate is reported as darkening at 1500 m under a vertical sun in the tropics. The effective limit for algae is set, however, by their compensation point rather than by the absolute amount of light, and in this respect they vary greatly (see p. 3754). It is difficult to set an extreme limit to the possibility of photosynthesis in the sea, but for practical purposes the limit is generally set at 70 m in temperate zones.

The loss by reflection from the surface is very difficult to estimate exactly as it varies with the solar angle and with the proportion of general sky light in the total radiation. At a solar angle of 40° the loss is probably between 5 and 10 per cent, depending on the angle of refraction in the water, a function of density, and on the disturbance of the surface, which diminishes reflection to some extent when it is considerable. Foam also is markedly effective in increasing losses by reflection.

As in fresh water there is marked selectivity of the different wavelengths absorbed by the water. Suspended matter, other than plankton, is non-selective and simply causes a general diminution of the light. This determines the transparency of the water, which can be measured, as in lakes (p. 3713) with a Secchi disc. Blue water is generally the most transparent, green less so and yellowish water the least transparent. In blue water the Secchi disc may be visible to 35 m and in the exceptionally transparent water of the Sargasso Sea, to 66 m. Under a vertical sun the clearest water may look black, as there is no internal reflection and the light vanishes into the depths.

Differences of colour in the open sea, where there is little or no silt in suspension are chiefly due to the relative abundance of plankton. As we have already said, blue seas are the least productive and the yellowish seas of the high latitudes are the richest. This is partly due to the greater solubility of Oxygen and Carbon dioxide at low temperatures and partly to the long periods of illumination in summer. There is indeed considerable productivity even under the ice-fields, which is particularly marked towards the end of summer, when the ice floes are thin enough to allow a considerable amount of light to penetrate to the water beneath.

The remark sometimes printed on Admiralty charts 'Discoloured water seen', usually refers to floating fields or 'rivers' of surface plankton (Peridineae?), often of a striking orange colour.

Atkins and Poole measured the absorption of *radiant energy* as distinct from visual illumination. They found that, even with a high solar angle, 70 per cent of the incident energy was absorbed in the first quarter of a metre, owing to the almost total absorption of the red and infrared rays, which have the highest energy content. Some of this absorption must, however, be credited to the chlorophyll absorption of the plant plankton, so that it is not all loss. The wavelengths of the orange-red band absorbed by chlorophyll are effectively removed by the combined effects of the phytoplankton and the water in the first 25 m, but wavelengths of the blue absorption band of chlorophyll, at 450 nm, penetrate down to the photosynthetic limit.

At 10 m depth only 1 per cent of the original radiant energy remained. This loss is greater than the visual measurements would suggest, since at this depth the greater part of the light consists of the less energy-rich short wavelengths. These rays retain a visual intensity of 8–10 per cent at 10 m depth in coastal waters. In the open Atlantic they may still retain the 10 per cent level at 30 m and in the clearest water (Sargasso Sea) down to 60 m. They can be utilized by the accessory pigments of red algae, which at such depths have the field to themselves and often reach their greatest luxuriance. They are for the most part shade-plants and on the tidal beach are usually sheltered by large brown algae or on the vertical sides of rock pools. Only a few can colonize exposed rock surfaces, like the tough *Laurencia* or the cushions of *Rhodocorton*, which are largely protected by the entanglement of drifted sand.

The lower limits of photosynthesis are determined practically by the *compensation point*, that is the value of illumination at which photosynthesis just exceeds respiration loss. This does not mean that photosynthesis may not occur at still lower levels of light, but it will not result in any net gain to the plant and cannot therefore form a permanent basis for life. The depth at which the compensation point is reached depends on a multiplicity of factors affecting the illumination and also on the specific characters of the organisms themselves, which vary greatly. Not only so, but in cells affected by nutrient deficiencies photosynthesis declines more rapidly than respiration, which naturally raises their compensation point, so that plankton cells in oligotrophic seas will only be able to photosynthesize at a productive rate during the height of summer and near the surface. Direct estimates of photosynthetic rates at different depths have often been made with algae enclosed in flasks, lowered to the appropriate depths, but in the absence of any similar direct method of estimating respiration the values of real assimilation deduced rest largely on assumptions which are disputable. For these various reasons the depth at which the compensation point is reached is a matter for individual experiment and can only be stated here in the most general terms. Ketchum has collected a number of such estimates of which the following are examples:

Location	Depth (m)
Sargasso Sea	Over 100
English Channel, summer	45
Loch Striven (Clyde) summer	20.30
Loch Striven, winter	5
Puget Sound, summer	10.18
Oslo, Fjord, spring	10
Woods Hole Harbour, summer	7

The comparative summer and winter values for Loch Striven show that below the 5 m mark algae must either be annuals or assume a winter resting form, if they are not to suffer fatal losses of weight during the dark months.

CHAPTER LI

THE PRODUCTIVITY OF THE SEA

THOSE who contemplate with dread the unrestricted growth of the world's population often turn their thoughts to the possibilities offered by the sea as a source of food. It is not the *productivity** of the sea that is in question, but rather the feasibility of harvesting the product, the great mass of which consists of microscopic plankton. This name, first given by Hensen in 1887, applies to all passively floating organisms, but is now generally restricted to microscopic organisms. It is becoming increasingly familiar.

It is a very miscellaneous population. Temporary plankton is mostly animal and includes the larval stages of many crustacea and mollusca as well as embryonic stages of many worms and some fish. The permanent plankton includes protozoa and many other minute animals and a great variety of microscopic algae, the phytoplankton, which are the primary producers for the whole population. Many plankton organisms are in fact self-motile, but their range of movement is very limited.

The three most important groups of algal plankton are: (1) Diatomaceae (Bacillariaceae). (2) Peredineae (Dinoflagellata). (3) Algal Flagellata. Chlorophyceae and Myxophyceae are relatively minor constituents. It is by the diatoms chiefly that the rise and fall of populations can be followed, as they are very numerous and practically ubiquitous.

There is no direct relationship between total salinity and productivity, since the major elements are always abundantly available and, indeed, coastal waters with sub-normal salinity are often the most productive. The controlling nutritive factors are, as we have already mentioned (p. 3724) the Nitrogen and phosphate supplies, which may be run down by plankton consumption to very low levels, even to zero level in the case of phosphates.

Considering the immense stores of nitrate in the deep waters of the oceans the amounts available within the photosynthetic zone are very small, especially in the tropics where the uniformity of the climate greatly limits the possibilities of vertical mixing. Brandt also attributes the relative poverty in

---

* It has long been proverbial. In *The Faerie Queen*, Book IV. Canto XII, Edmund Spenser delivered himself thus:

> O what an endless work have I on hand,
> To count the sea's abundant progeny
> Whose fruitful seed farre passeth those on land,
> And also those which wonne in th'azure sky!
> For much more eath to telle the starres on hy,
> Albe they endlesse seem in estimation,
> Than to recount the sea's posterity:
> So fertile be the floods in generation,
> So huge their numbers and so numberlesse their nation.

plankton of many tropical seas to the greater activity of de-nitrifying bacteria at higher temperatures. Hensen, on the other hand, considers that a shortage of atmospheric precipitation bringing additional nitrates, is a contributing factor. Quantitative plankton estimations are made by filtering measured amounts of water through a fine net (No. 20 bolting silk netting with 1/100 inch mesh). The residue is taken up in neutral formalin and the suspension allowed to settle in a graduated cylinder. The volume of sediment is the fresh volume (*rohvolum*). Krämar found, in the central Pacific, near the Tonga Island, as little as 0·15 cm^3 of fresh-volume plankton per cubic metre of water and further south, in Australasian waters, only 1·84 cm^3. These are the smallest measures recorded. The Atlantic is somewhat richer, the minimum amount, in the Sargasso Sea, being 3·5 cm^3 while the maximum, off the Greenland coast, was 116·9 cm^3 per cubic metre of water (German National Deep Sea Expedition, 1889).

Atkins, working in the English Channel, found a phosphate content (reckoned as P_2O_5) in March of 0·036 mg/litre in the surface water; by July there was none left. Silica was also lost by July. The March amount looks very small, but Atkins reckoned that it would suffice for 26·8 million diatoms per litre, i.e. that number of diatoms would exhaust the supply. Atkins used the phosphate consumption as a measure of productivity. He found in cultures that phosphate represented 0·15 per cent of the fresh weight of the algae. Down to 70 m the annual consumption of phosphate was 30 mg/cm^3. This means a production of 1·4 kg of algae per square metre of the surface down to 70 m. Transcribed into English measures this means a fresh weight of 5·6 tons per acre* (approximately 14·2 metric tons per hectare).

Alternatively, Atkins used CO_2 consumption as a basis for calculation. He found an average carbohydrate content in fresh algae of 15 per cent, reckoned as hexose sugar. The annual abstraction of CO_2 from Channel water was enough to form 200 g of hexose per square metre of surface, or 1·34 kg of algae; this equals 5·4 tons per acre, a striking, if perhaps fortuitous agreement between the two estimates. It is a fair conclusion that, in terms of organic production, the sea can well compare with agricultural land. The plankton algae are the true measure of productivity, since the whole animal population simply represents transmuted phytoplankton. Fungi are quantitatively negligible, except for yeasts, which are fairly abundant and certainly permanent planktonts, while bacteria are, of course, ubiquitous.

Mention of bacteria is a reminder that filtration is a very imperfect means of evaluating the total plankton population, for there is a very large number of organisms too small to be thus caught. The netted species form the *macroplankton*, the smaller organisms are grouped as the *nannoplankton*. Though more numerous than the macroplankton their total volume is only a small fraction of the latter. The nannoplankton can be recovered by centrifuging, but for a complete enumeration cultures are necessary. For example, netting in April in the Irish Sea yields an average of 1 organism per cm^3.

* These figures do not take into account respiration losses, which might amount to 2·5 per cent. The fresh weight of a Maize crop is given by Transeau as 6·6 tons per acre.

Centrifuging raised this to 15 organisms per cm^3, but cultures yielded 464 organisms per cm^3 exclusive of the macroplankton. Bacteria, of course, account for a large part of the highest number.

Even among macroplanktors netting is inefficient, and its only defence is convenience as a standard of comparison in different seas. The following figures quoted by Johnstone (per cubic metre) show this clearly:

Producers	Netted	Centrifuged
Diatoms	108,000	902,000
Peredineae	2,315	439,385
Consumers		
Rhizopoda	2,650	3,335
Flagellata	20	264,000

As can be seen the diatoms far outstrip the other groups in numbers. Huge catches are sometimes netted. In Kiel Bay, where the surface water is super-saturated with oxygen (p. 3749), an annual production of 6000 million per cubic metre has been recorded; Hansen estimated in the Baltic an average of 500 million per cubic metre. The relatively small numbers of the zooplankton, in the above table, is at first surprising but it has been observed that dense growths of phytoplankton tend to exclude the zooplankton by some un-identified chemical action. This has a curious consequence. Phytoplankton produces a phosphate shortage, the more phytoplankton the lower the phos-phate and the fewer the zooplankton. Whales feed on the zooplankton. The phosphate content of the water is thus a guide to the likelihood of finding whales.

Periodicity in plankton production is well marked, especially in northern waters, and a regular calendar of events exists.

1. *November to March.* Poverty months with minimal sunshine. Diatoms and Peredineae present. Copepods scarce. Fish ova begin to appear. Nitrate, phosphate and silica accumulate in epithalassa.
2. *March to June.* The vernal outburst, when at least three hours daily sunlight enables assimilation to overtake dissimilation. Diatoms dominant.
3. *July to August.* Phosphate and silica exhausted. Diatoms rapidly de-crease. Zooplankton abundant.
4. *September to October.* Secondary maximum of diatoms and of copepods, numbers gradually decreasing as sunshine diminishes. Nitrate and phosphate being liberated in hypothalassa.

The figures given in Table 21 illustrate these phases, and are the averages of netted catches at Port Erin in the Isle of Man, for the years 1907–1920 (each figure is for eight cubic metres of water). The figures for the diatoms are the combined figures for the three chief genera, *Chaetoceras*, *Rhizosolenia* and *Coscinodiscuns* and take no account of the smaller forms. Geographical varia-tions are much less known than the seasonal ones, but Cleve was able to map regions in the North Atlantic which were distinguished by their species populations, especially by the dominant species of each area. How stable these may be is not known and without doubt ocean currents can move plankton organisms about on a large scale. Information from other oceans is too scanty for similar generalization. (But see p. 3403 on South China Sea.)

That the biochemistry of algae and of unicellular planktonts should be dominated by the ionic environment of the sea is scarcely surprising, but that

TABLE 21

Catch	January	April	June	August	October
Diatoms	22,614	5,197,084	5,513,409	63,363	445,934
Ceratium tripos	3,037	2,968	42,255	21,380	6,429
Noctiluca miliaris	259	29	173	332	1,773
Balanus larvae	404	11,021	13	1	0

it should also be reflected in ourselves was unexpected when Macallum in 1908 drew attention to the similarities both qualitative and quantitative between our blood salts and the salts of the sea. Palitsch has shown that the blood salts in marine crustaceae vary with the salinity of the surrounding sea water, and Macallum suggested that our present equipment of blood salts represents that of the Palaeozoic ocean in which our ancestors first acquired a closed circulatory system. It certainly seems indicative of a marine origin of life, rather than a terrestrial origin, that Aluminium and Silicon, two of the most abundant elements in the earth's crust, play, at the most, a very minor part in physiological chemistry. Al is never an essential element and Si is so only for the diatoms and Radiolaria whereas the low atomic weight elements which predominate in the sea are everywhere essential to living cells.

There are striking analogies between the conditions in the sea and those in living cells. First, there is the matter of chemical composition we have touched upon. Second, there is temperature regulation by surface evaporation. Third, there is the production-storage cycle. Fourth, the association of Iodine with reproductive activity in many algae. Last, and most importantly, there is the regulation of CO_2 content and pH in both sea water and blood by the carbonate-bicarbonate equilibrium, associated in the blood with haemoglobin and in the sea with chlorophyll, the two chief organic pigments playing analogous parts; iron porphyrin for animals, magnesium porphyrin for plants.

CHAPTER LII

SEA BEACHES

THERE is one important aspect of marine life that we have so far left untouched, that is the semi-aquatic life of the tidal beach, the aspect indeed most accessible to land dwellers. The existence of a zonal distribution of seaweeds was first made clear by Börgesen in 1904, on the shores of the Faröe Islands, but the first investigation of the causes of zonation was by Sarah Baker in 1909 in the Isle of Wight. She showed experimentally that the survival of different species was related to the length of exposure during low tide, which is a function of their vertical height above low-water mark.

That the matter is very complicated is evident when we consider the variety of conditions involved. First, there is the nature of the shore material, whether solid rock (and in this case whether hard igneous rock or soft sedimentary rock), or isolated boulders or shingle, sand or mud. Second, there is the variation of tide levels between neap and spring tides, and the wide variation of tidal range in different areas. Third, there is the varying effects of exposure to wave action, which, on very stormy coasts, has not only a powerful mechanical effect but carries the influence of the sea much higher up the beach. Last, there is geographical variation, which brings in entirely different algal populations in different areas.

Exposure during low tide is undoubtedly the dominant factor on the beach. Variations of salinity or pH in the water have comparatively small effects. This is evidenced by the population of rock pools, in which there is no exposure. Even pools lying high up on the beach are not populated by the species characteristic of the neighbouring exposed rocks, but by species which otherwise are only found near the low tide level.

An exception to the comparative indifference of shore plants to changes of salinity is that shown in river estuaries, where a progressive disappearance of marine species occurs in advancing up to the estuary from fully saline to brackish water and eventually to fresh water. A study of the estuary of the river Exe in south Devon was made by Gillham. The Phaeophyceae disappeared in the following order from the mouth of the estuary inwards: *Fucus serratus*, *F. spiralis*, *Ascophyllum nodosum*, *F. vesiculosus* and finally *F. ceranoides* which is an estuarine species. Among Rhodophyceae *Catanella repens*, *Rhodocorton rothii* and *Bostrychia scorpioides* penetrate farthest, but the green alga *Enteromorpha*, which possesses astonishing adaptability, goes furthest of all, into practically fresh water. As the algae grow mostly on the lower part of the shore line the effects of intertidal exposure in this locality are diminished, so that not only lower average salinity, but the great and rapid variations of salinity with tidal movements must have a maximal effect. In the brackish water zone variations between 5 and 50 per cent of sea water strength occur and even 1 mile from the mouth the salinity drops from 33 per cent at high tide to 20 at low tide. Such conditions are utterly different from those of the open seashore. Algae which can tolerate wide variation of salinity are called *euryhaline*, those with only a narrow tolerance, *stenohaline*. Similar terms, *eurythermal* and *stenothermal*, denote different degrees of tolerance for temperature changes.

The difference between an exposed, turbulent coast and the quiet waters of protected inlets has many effects. Turbulence prevents sedimentation and maintains temperature and salinity at steady levels as well as ensuring Oxygen saturation. These factors and the mixed aerial and submerged life suit some species and there are 'surf-loving' plants which prefer such conditions, for example, the robust brown alga *Postelsia palmaeformis* on the California coast. Quiet waters are more liable to surface heating, which excludes some stenothermal species, and are often burdened with suspended sediment, which forms a muddy substratum unfavourable to colonization by seaweeds, and may build up into a distinctive salt marsh. We have discussed the Angiosperms of the salt marshes (halophytes) before (p. 3669) but some specialized algae also colonize these marshes, and like the flowering plants show some vertical zonation, though the differences of level involved may be very small.

On the exposed flats, below the phanerogamic levels, there is a general covering of *Enteromorpha* and *Cladophora* species, which prefer the more sandy types of sediment, or of sheets of dark green *Vaucheria thuretii* on fine-grained mud. The latter also colonizes the muddy edges of creeks in the marsh, often accompanied by sheets of filamentous Myxophyceae, especially *Lyngbya*. The free-living, i.e. unattached, fucoids are, however, the most striking and peculiar group of salt-marsh algae. At the lower fringe occurs *Fucus vesiculosus* forma *volubilis* which is spirally twisted, but the main mass is in the *Salicornia* society where thousands of tiny plants of *Pelvetia canaliculata* forma *libera* occur packed together among the *Salicornia annua* plants. Among them is also the red seaweed *Bostrychia scorpioides*. Other free-living forms are *Ascophyllum nodosum*, forma *scorpioides* and forma *mackaii* and *Fucus vesiculosus* forma *muscoides*, which are all of local occurrence in salt marshes.

For the period between high tides, when algae are left in the air, the term *emergence* is better than *exposure*, since the latter is also used for the incidence of wave action at certain places.

The effects of emergence are threefold. First comes desiccation, which can be very considerable. *Pelvetia*, for example, may lose nearly 70 per cent of its fresh weight during one emergence period and become quite horny and brittle. This is an extreme case, but losses such as no phanerogam could tolerate are experienced by many other seaweeds. Second, there is intense insolation and concomitant heating, both highly inimical to most seaweeds. Third, there is exposure to a high Oxygen tension which has disturbing effects on respiration, though little enough is known about them. What happens to respiration and photosynthesis, if any, in a dried, warmed, radiated and Oxygen bathed seaweed we simply do not know. Nevertheless, these conditions appear to have some beneficial effects on the intertidal species of the seashore, since for many of them a period of emergence is essential and if permanently submerged they die. It is not therefore in their case a question of exceptionally wide tolerances enabling them to avoid the competition of more restricted species, as is so often the case in phanerogamic vegetation, which gives rise to the marked zonation of species with reference to tidal levels, but of a positive preference, or rather preferences, for they differ among themselves.

These zones on the beach, or to give them the term now generally preferred, *belts*, have been the subject of interest since the beginning of the century and workers in different regions have not always agreed as to terminology, which has resulted in a certain amount of confusion. We must confine ourselves to rocky shores since it is on them that most attention has been fixed, where the features are most visible and where the general principles involved are clearest. We think it will be generally agreed that these shores show topographical *zones*, related to tidal levels, within which, or corresponding with which, there are biological *belts*, definable by the plants and animals which constitute them. Any scheme which seeks to unify the treatment of shore vegetation in different parts of the world must be cast on very general lines, since tidal levels are only one factor out of many which are influential. The prevailing climate, sunshine, rain, temperatures and photoperiods are all additional causal factors which have a wide, as well as a local incidence and determine both the presence of species and their relation to emergence. Furthermore, any scheme must be such as will allow for local modification by special local conditions, shade, wave action, nature of the rock, etc.

It will be seen that we have, though in a relatively small area, a variety of ecological factors as complex as any we have encountered in land vegetation, if not more so, and with problems of the same order but which so far have barely been formulated, let alone investigated.

In 1949 T. A. and Anne Stephenson put forward a draft of a generalized scheme of shore vegetation which they suggested might be applicable anywhere. This was based on exceptionally wide knowledge of beaches in many parts of the world, and although criticized in some quarters it has been strongly supported by others of wide experience.

Its basic feature is the *littoral zone*, extending from extreme low-water of spring tides (ELWS) to extreme high water of spring tides (EHWS). This is the main shore-zone, in which all organisms are liable to periodic submergence, even if only once a fortnight. Above this lies the *supra-littoral zone* which is never wholly submerged but is under the influence, to some extent, of waves and salt spray. This zone extends upwards indefinitely, merging gradually into land vegetation. Below lies the *sub-littoral zone*, which is never wholly emergent, extending indefinitely downwards as far as benthic growth can go.

These are simple physiographical concepts, easily applicable in theory, but subject to fairly wide local modifications which influence the distribution of the various belts of organisms, and tend to blur the precise limits of the zones.

For this reason it has become necessary to recognize the existence of two 'fringes', the *sub-littoral fringe* and the *supra-littoral fringe*.

The sub-littoral fringe lies just above extreme low-water mark and may extend some little way below it. It is characterized by being emergent at some tidal periods even if only for a very short time in each full tide cycle. Even if not fully emergent the lower part of the fringe is only covered by very shallow water at low tide and its conditions of illumination and temperature (and wave action) approach those of the Littoral Zone. Experience shows that on any given shore there is a group of species, both plant and animal, which is characteristic of these fringe conditions. The group of kelps, i.e. the large daminariacae, are the most conspicuous occupants of the sub-littoral fringe both in northern and southern hemispheres, though absent from the tropics. They extend both above and below the lowest tide mark and are so large that they may be emergent even where the ground is not. The beginning of the supra-littoral fringe is marked by the upper limit of barnacles (of whatever taxonomic category) which usually lies well short of the upper limit of the Littoral Zone. Above the barnacles extends an area of the shore-living mollusc *Littorina*, which reaches into the Supra-Littoral Zone, where it meets the lower limit of phanerogams.*

This must not be taken as a blue print of shore vegetation everywhere. No such claim has been made for it and variations in the dispersal of species through the zones are endless, but as a skeleton scheme it has much to recommend it.

Starting from the upper beach levels on British coasts, a feature which can be very striking is that of the lichen belts in the Supra-littoral Zone, or Maritime Zone of Lewis. These belts form a conspicuous colour feature running for long distances along the cliffs. Where these are hard, acidic rocks a grey belt of the shrubby lichen *Ramalina fastigiata* stands just below the phanerogamic level. On softer, calcareous rocks this may be replaced by a golden belt of *Xanthoria parietina*. Below this there is often a conspicuous black belt of *Verrucaria maura* with *Lichina pygmaea*. Whereas the first two are almost certainly Supra-littoral the black zone probably belongs to the supra-littoral fringe and extends downwards to overlap the top of the belt of *Pelvetia canaliculata*, which is often associated with *Fucus spiralis* in its lower part, though the latter is also found much lower down. These are all above the barnacle limit and therefore in the fringe area. At very exposed positions *Pelvetia* and even *Verrucaria* may be absent and the fringe is then largely bare.

The Stephensons consider the Littoral Zone proper (or their mid-littoral) to begin at the upper barnacle limit, where the shells *Littorina neritoides* and *L. rudis* of the upper shore replace barnacles, but the relation of plants to the barnacle limit is very variable. This zone is the region of the large brown seaweeds, which show distinct though overlapping belts, in descending order, of *Ascophyllum nodosum*, *Fucus vesiculosus* and *F. serratus*. Below this we enter the sub-littoral fringe, where barnacles may be replaced by *Lithothamnion* as

---

* This scheme of zonation seems to follow closely that of Kjellman (1875–83), who distinguished supralittoral, littoral, infralittoral and elittoral zones, in descending order.

the principal rock cover, and *Laminarias* make their appearance. Here also grow, exposed to emergence, many of the typical rock-pool Rhodophyceae of the Littoral Zone, *Polysiphonia* and *Ceramium* species among them, extending down into the Sub-littoral Zone where they may be joined by two Phaeophyceae, *Halidrys siliquosa* and *Cystoseira ericoides*, at least in the south. Where laminarians are scarce or absent a rich fringe flora of smaller species takes their place, including such Rhodophyceae as *Laurencia*, *Chondrus* and especially *Corallina*, which may be regarded as typical of fringe conditions, and do not penetrate deeply into the Sub-littoral Zone; while true sub-littoral species, such as *Haliseris polypodioides* or *Delesseria sanguinea*, scarcely emerge at all into the fringe and become larger and finer the deeper they go below tide level. As the Stephensons have emphasized the Zones and Belts can only be related to tide levels in the most general way and their position on the beach is greatly affected by local conditions whether of exposure or shelter. Their essential character is biological not topographical.

Zonation such as we have described obviously reflects differences in the constitutions of the species concerned but, in spite of massive research on the physiology of the algae, we have really very little idea of what the important differences are, for only a few of the experimental and biochemical investigations have been ecologically slanted. Perhaps also ecologists have failed to analyse the available information from their own standpoint.

The most obvious consideration is that of the physiological effects of emergence. There is evidence that water is lost more slowly by algae of higher belts than by those of lower belts, but even so, algae like *Pelvetia* and *Fucus spiralis* have to survive not just a single intertidal period, but the whole period between high spring tides. which may mean eleven or twelve days continuous emergence, during which their water content inevitably drops to a very low level. Partial desiccation suppresses photosynthesis and recovery after submergence may be slow. *Pelvetia*, after eleven days emergence recovered 70–80 per cent of its normal assimilation rate in nine hours. *Fucus vesiculosus* recovered 72 per cent of its assimilation rate in four hours after five hours emergence, but after three days it only recovered 20 per cent in nine hours. *Fucus serratus* recovered slowly after five hours emergence, but was killed by three days, while *Laminaria digitata* could not tolerate even two hours emergence (values from Chapman). These are plainly significant facts with regard to zonation, but they throw no light on the physiological mechanisms involved. There is a great tolerance for freezing as well as drying. Kauwisher showed that 80 per cent of the water in *F. vesiculosus* could be frozen without any ill effects. The presence of geloid carbohydrates in the tissues, which have great water-holding powers, is probably associated with slow drying. Among these substances in the Phaeophyceae are the dextran laminarin, which is widespread, and fucoidin, a mixed salt of fucoidic acid in which there are sulphuric ether linkages; also algin, a Calcium-Magnesium salt of alginic acid, which is a condensation product of the sugar acid, mannuronic acid. Among the Rhodophyceae there are carrageen (from *Chondrus*) and agar from various algae, both Calcium salts with sulphuric ether linkages.

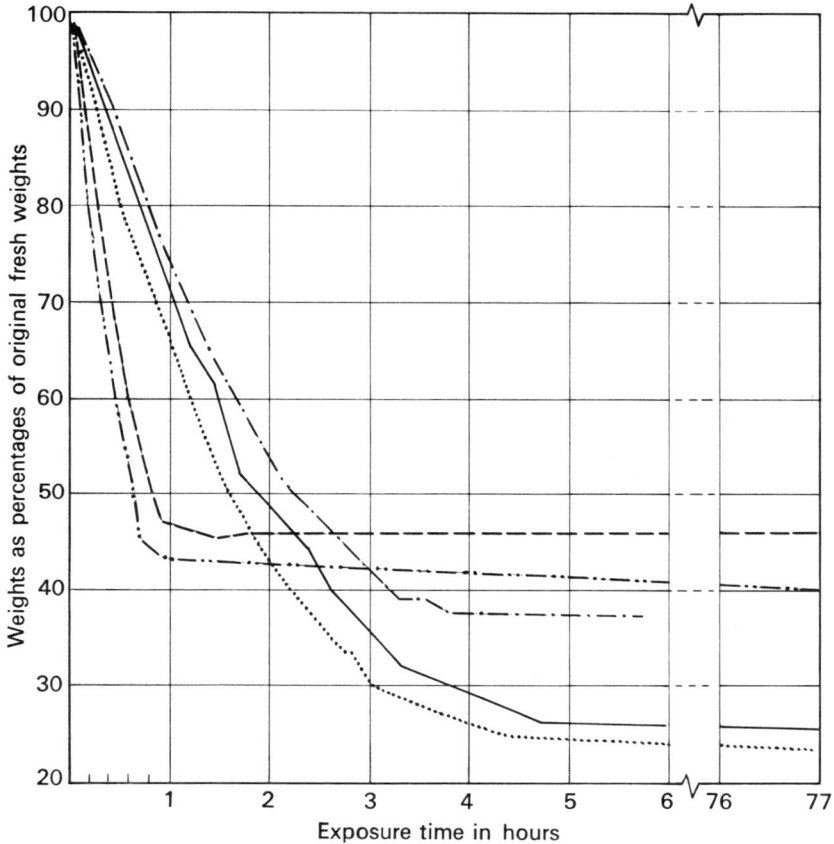

FIG. 4.62. Loss of weight, expressed as percentages of the original fresh weight, of some seaweeds on prolonged exposure on Heligoland at 25°C and RH 74.

Ulva lactuca — — —
Porphyra laciniata — · · —
Cladophora gracilis — · —
Polysiphonia nigrescens var. = P. regularis ————
Plocamium coccineum · · · · · ·
(After Biebl.)

The nature of the cell walls should be important, but has been little studied, though cellulose or a close analogue occurs in brown algae. Haas and Hill have also shown that the ether extract, including true fats, decreases in the species of successive belts as we go downwards. This could be important as a means of resistance to the effects of emergence, but we do not know. Inclusions of fatty oils are a regular feature in the cells of other plants, such as mosses, which are highly resistant to drying.

Reference to Fig. 4.62 shows that there are two types of water loss. First there are the thin thalli of *Enteromorpha* or *Porphyra*, with a relatively high water content in relation to dry weight, in which drying proceeds very rapidly for two to three hours (dependent on the atmospheric RH) and then sharply ceases, the curves becoming practically horizontal. The other type, of leathery seaweeds with a lower initial relative water content, lose water more gradually

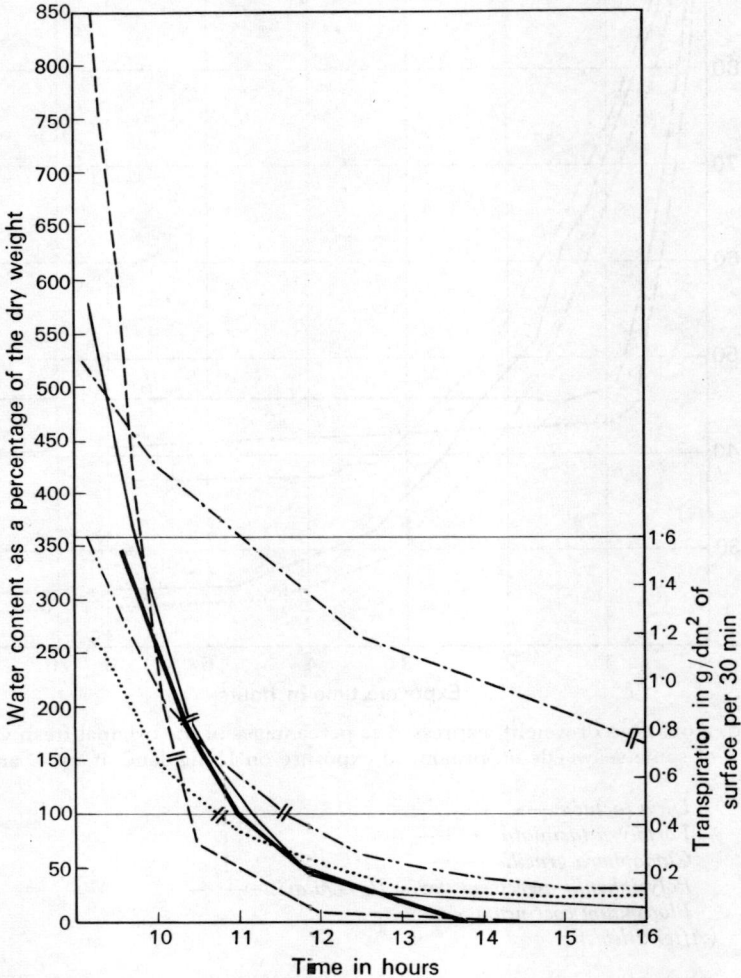

FIG. 4.63. Drying curves of certain common seaweeds on exposure on a sunny day in July in Heligoland. (Temp. 18°C, RH 78–83 per cent.) The heavy line shows the transpiration of *Fucus vesiculosus* per unit area of surface. The double cross lines mark the compensation point of each species.

     Fucus platycarpus —·—
     Fucus serratus —··—
     Enteromorpha linza — — —
     Porphyra atropurpurea ———
     Fucus vesiculosus ······
(*After Stocker and Holdheide.*)

and retain a somewhat higher water reserve. When measured for unit surface area the water loss is seen to fall at a constant rate for about an hour and thereafter diminishes more slowly towards a limiting value.

These are values referred to dry weight as standard. On a fresh weight basis *Porphyra* and *Ulva* still show the same type of drying curve, but the limiting value is now seen to be at 40–50 per cent of the original fresh weight, in other words, even these thin, cellular thalli retain a certain water reserve. The initial rates of water loss per unit area are closely similar for both types, e.g. per square decimeter, in the first 15 minutes, *Enteromorpha* lost 0·51 g and *Fucus vesiculosus* 0·52 g, *F. platycarpus* is exceptional, it lost 0·90 g/dm^2 in 15 minutes, but retained a much higher water reserve than other species (Stocker and Holdheide 1937). Drying affects the water absorbing power of the membranes so that there is a marked hystereris in reswelling on submergence, which may account for the slow recovery of photosynthesis in some species (see above).

The beginning of emergence causes a short rise in both photosynthesis and respiration, but the former falls off with time more rapidly and more drastically than the latter and at a certain degree of desiccation the two curves meet, that is to say they arrive at a compensation point. On submergence both processes increase their rates, photosynthesis more rapidly than respiration, to a transient peak value followed by a decline. An alga too far up the beach and subjected to an injurious degree of exposure may, however, retain a high rate of respiration after submergence, with no decline of rate, and the regular repetition of such an event is likely to prove fatal. At low beach levels photosynthesis may also be depressed at high tide due to the depth of water, so that the curve of photosynthesis rate shows a rise on submergence followed by a trough of depression at high tide and this again by a transient rise as the tide falls, with a subsequent rapid fall after emergence.

Something should also be said here about the theory of complementary coloration in seaweeds associated with the name of Engelmann (1883). He suggested that the yellow-brown pigments in the Phaeophyceae and the red pigments in the Rhodophyceae were accessory to the chlorophyll, to which energy was transferred from the short-wave light which prevails below the sea and which these pigments were able to absorb, thus enabling the plants to live in deeper water than purely green algae and giving them a biological advantage. Engelmann himself showed that these accessory pigments were photosynthetically active and, indeed, if they were not so they would be a positive disadvantage by screening the chlorophyll from the available blue light which it can and does absorb almost completely. Subsequent work has not refuted Engelmann's views, and it is a fact that these 'coloured' algae can grow at depths which are beyond the reach of Chlorophyceae. For example, *Macrocystis* (Phaeophyceae), on clear water, maintains the same rate of photosynthesis as at the surface to a depth of 12 m and does not reach the compensation point until at a depth of 40 m (North).

Finally, the question arises whether the methods and concepts of plant sociology, as recognized among land plants can be applied to the seashore algae. It is apparent from what we have said that certain 'dominant' species characterize certain areas of the beach, and investigation shows that a number of minor species are associated with them, but whether these groupings constitute distinct 'associations' or whether they are aspects of a continuum is still an open question. The relationships between algal species are very different from those between phanerogamic species. Moreover, the number and variety of micro-habitats on a beach is so great that sociological uniformity over an area of any large size is highly improbable. A beginning has been made with the application of the quadrat method and with statistical analysis, but far wider knowledge of the sociology of algae is required before we can see whether the existing ideas of plant sociology are applicable or whether we need a new set of concepts.

The fucoids of the shore have been used since time immemorial for the organic manuring of soils, but recent times have seen the large-scale exploitation of the 'kelps' for Iodine; for the versatile alginic acid or algin, used in many ways; for agar and for carrageen or Irish Moss. This interest is concerned chiefly with the Sub-littoral Zone, where the biggest beds are, and extensive submarine surveys have been made to delimit suitable fields, for which special methods have had to be used, such as aerial photography, echo sounding, trawling, etc., which serve well for commercial ends, but for scientific examination of the sea floor in detail nothing can replace personal examination, with the aid of aqua-lung techniques.

IN completing this study of the realm of Plant Ecology one is left with two main impressions. One is that within its wide sweep it embraces knowledge which is crucial to the understanding of plant life. The other impression is one of regret that so much of that knowledge is still superficial and in some respects old-fashioned. Schimper made a great step forward when he adopted a physiological basis for his treatment. Perhaps someone in the next generation will make a further advance and present us with an account of Plant Ecology on a molecular basis.

PART TWO

PRINCIPLES OF PLANT GEOGRAPHY

CHAPTER LIII

A RETROSPECT

A practical concern with Plant Geography, or Geobotany as it is sometimes called, must be nearly as old as mankind, or at least as old as the Neolithic Revolution which transformed men from primitive hunters into settled cultivators. The discovery of the staple crop plants, the search for others and the colonization of new areas all drew attention to the fact that certain plants were only to be found in certain areas and that they could not be successfully grown in places where conditions were radically different from those in their native countries. The realization of a linkage between plants and climate must have dawned very early on men's minds.

The array of food plants primarily concerned was very limited, and a mere handful quickly became distinguished as of universal importance, supplemented here and there by local discoveries of regional value but which were for the most part non-exportable. Interest in plant exploring was, however, greatly extended by the demand for plant materials: timbers, fibres and condiments, for example, and for medicinal and 'magical' plants whose real or supposed qualities were physically or socially beneficial to their possessor.

The fact that plants of interest were not to be found everywhere yet were widely in demand made them the objects of adventure journeys or of inter-tribal barter, even in prehistoric times. We have evidence of the importation of flax to the north of the Alps at the period of the Swiss lake-dwellings and there are many legends of plant hunters, such as the fabled introduction of wheat into Egypt 'from the north' by the goddess Isis, or the voyage of Hercules to the Canary Islands (or possibly Libya?) to collect oranges (v. *The Golden Apples of the Hesperides*).

In antiquity some of the centres of developing civilization, especially Phoenicia and Egypt, ventured on distant voyages which made them aware, not only of the different climates of other lands, but also that these climates were associated with different forms of vegetation from which they were able to draw new and precious plant products, 'cedar wood and sandal wood and sweet white wine', together with spices and sugar or 'Indian Salt', which became matters of continuing international trade.

Although this overseas merchant trade dwindled almost to nothing in Christian Europe during the Dark Ages, it was kept alive by the Arabs, who inherited it from Egypt, and its earliest European revival was in Venice, Genoa, Pisa and Naples, whose trade, after the period of the Crusades,*

---

* The Crusades were part of an expansionist movement in Europe, and in their intermissions trade with Africa and the Levant flourished and was encouraged by many European governments. The Treaty of Tunis (1270), for example, opened trade with the Moors of brought them into close contact with the Arabs around the Mediterranean. The fabulous wealth acquired by these cities, especially by Venice through their Levantine trade, became widely known, because the Venetians extended their marketing all around the northern seaboard, and it excited rivalry which, in the reviving activity of the fifteenth century, started the enthusiasm for exploratory voyaging that continued until the nineteenth century.

In earlier times explorers accepted, for the most part, unreflectingly the differences of vegetation which they observed, but they brought home collections of plants and in the period of 'Enlightenment' in the late eighteenth century the new 'scientific mind', pondering over these collections,* begin to wonder on what principles plants were distributed as they seemed to be.

This is more than a purely academic question, since the social organization of peoples depends markedly on their chief food materials and on how these are obtained. We have had one striking and fairly recent example of this in Britain in the far-reaching social and political changes which resulted from the large-scale importation of American and Canadian wheat, which began in the latter half of the last century and dislocated the old economy of rural life.

Any consideration of the geography of plant distribution necessarily depends on knowledge of the plants themselves, their identities and their relationships. To this knowledge Linnaeus contributed immensely with his Sexual System, which, although artificial, better expressed the relationships of the higher plants than any previous system. It was almost universally used for nearly a hundred years and during a period of worldwide extension of botanical information it enabled new discoveries to be indexed or, in current phraseology, 'slotted in', to the system almost mechanically. The later Natural Systems, beginning with de Jussieu, appear not so much as a revolution but as upgrowths from a Linnaean root.

It was perhaps inevitable that geographical considerations should first be applied to the best-known floras in the best explored countries and here again Linnaeus was a pioneer. Although written before his system was formulated, his *Flora Lapponica* (1737) and *F. Suecica* (1745) are not confined to descriptions of species, but embody reflections on the vegetation as a whole, its origins and geographical relationships. A similar outlook inspired Gmelin's *F. Sibirica* (1757).

'God Almighty first planted a garden', said Bacon, echoing the Genesis story, and the fundamental belief of early speculators was that the dispersal of

---

North Africa. The Venetians, who provided transport during Crusade wars, acquired several ports in the eastern Mediterranean which served them profitably in later time. Trade with the Arabs, besides bringing gold to Europe, which established gold currencies for the first time since the seventh century, also introduced the knowledge of Arab science and of Arabic translations of lost Greek works which led to an early revival of learning in the twelfth century. Many of these Arabic manuscripts were tranlated into Latin by Constantine Africanus at the monastery of Monte Cassino, which thus became the door through which science and philosophy re-entered Europe.

* The collecting of foreign plants began in the sixteenth century, partly in the form of 'herbaria' and partly as living seeds and plants. The first Botanic Garden was established at Padua in 1545.

plants, like that of mankind, had started from this centre. Why then was their distribution now so diverse?

The first to write upon the relation between the distribution of plants and climate and situation was Willdenow in 1792, who deserves to share with Humboldt the honours of founding Plant Geography. Alexander von Humboldt, whose *Essai sur la Géographie des Plantes* appeared in 1805, was the first to base his opinions on worldwide travels, which gave them great authority and indeed celebrity, for he became one of the leaders of European science and founder of the University of Berlin.

His views, as we have explained more fully on p. 3317 of this volume, laid chief stress on physiognomy as a vegetational character and he showed that this overrode floristic distinctions, in as much as similar climates in different parts of the world produced vegetation with similar physiognomies, although made up of quite different species.

Humboldt took, however, a very broad view of his subject, and he confronted the problems of physiography which are raised by the facts of plant distribution: e.g. the former connection of land masses now widely separated (which Willdenow had also suggested), the separation of islands from continents and the evidence (including fossils) for large-scale changes of climate. In other words he clearly distinguished between what we would now call the ecological and the historical influences on plant distribution. He stigmatizes as mythical the idea of the whole world of plants having originated from a single centre. He thus clearly established the outlines of botanical geography as a branch of science and the nature of its subject matter.

Acceptance of the idea of plant migration naturally drew attention to their available means of migration. This also Willdenow had sketched out in his short work, and it was elaborated in the work of Augustin de Candolle, from 1820 onwards, who laid down the principle that the slow but universal spreading of plants combined with the selective effect of climate on their differing constitutions was the effective cause behind the facts of distribution. He considered that not only broad climatic distinctions, but also that the 'habitat' limitations of plants were explicable in the same way (see Beyerinck, p. 3784).

The works of Schouw (1822–33) and of Meyen (1836) were both concerned with the ecology of plant distribution, the influences of climates and soils, which Meyen considered in relation to the Humboldtian 'physiognomy' of vegetation. They did not regard the history of floras as part of their subject matter.

When Alphonse de Candolle published in 1855 his *Géographie botanique raisonée*, the subject may be said to have come of age. He envisaged two fundamental questions: 'Why is this plant here?' and the converse, 'Why is that plant *not* here?' The answer to the first he found chiefly in climatic considerations, but the second question was more puzzling, for he found that it required a knowledge of the past history of plants; not only their recent history but even far back in geological time. The historical aspect had also an inportant bearing on the first question, since the species which make up the flora of any locality form only a sample of the very large number which could thrive in that particular climate and the reasons for their presence, as well as the absence of others equally suitable, must be historical.

De Candolle developed these theses in detail, which was made possible by the new understanding of geological changes which Lyell had introduced in 1832 and by the idea of the Ice Age which Edward Forbes had postulated in the 1840s. It was a two-way benefit, for both Lyell and Forbes emphasized the important evidence which the distribution of plants and animals offered in their support.

De Candolle's book was pre-Darwinian, but, like all students of Plant Geography, he could not but be impressed by the evidence of change which made the orthodox dogma of the eternal fixity of species impossible to accept. We find evidence of this state of mind as early as Willdenow in 1792 and it becomes explicit with Meyen in 1836. He quotes Link's assertion: 'The species is that which is constant in nature,' which carries the implicit corollary that that which is not constant is not a species. Without venturing to refute this he says:

> ... probably some individuals have changed so much by reason of the influences of a different climate and of peculiarities of soil that they now, as permanent varieties, must appear to us distinct species. It would certainly be easy to explain many things if this latter opinion could be supported by observations; many naturalists would also be very much inclined to admit it, as the great influence which variety of climate and local conditions exercise on the form of plants is universally known, and thus a great number of forms have arisen which it has been attempted to make into species, ...

I have quoted this at length because it illustrates two things; that the facts of Plant Geography had early led to a belief in transformism and the intellectual quandary to which this gave rise.

It was in fact his own observations on plant and animal distribution, during the voyage of HMS *Beagle* in the years when Meyen was writing, which determined Charles Darwin's views on evolution, views which he did not invent, but for which he provided what had hitherto been lacking, an acceptable, logical mechanism, in place of the vague Lamarckism which had haunted and disturbed the minds of Mayen and many others. Darwin himself regarded the facts of the geography of living creatures as practically the key to many of the problems of evolution, just as they had previously been the stumbling block to the acceptance of the dogma of the fixity of species.

After the Darwinian period works on Plant Geography multiply, but for the most part they tend to follow the systematic (floristic) line or else the ecological line, seldom both. De Candolle's comprehensive survey remained a landmark and retains its evidential value even today.

Grisebach (*Vegetation der Erde*, 1872) based himself on an elaboration of Humboldt's twelve physiognomical classes, which he extended eventually to sixty, but his outlook was floristic and his chief contribution was the concept of the floristic provinces, areas delimited by natural barriers, within which the vegetation was more or less homogeneous, floristically and physiognomically. Engler, in his *Entwicklungsgeschichte der Pflanzenwelt* (1879–82) analysed Grisebach's provinces floristically, illustrating their homogeneity and their differences from one another. He attempted to relate these differences to the history of the plant world since the Tertiary Period, a task which was somewhat premature. Grisebach's concept of floristic provinces remains valid, though increased knowledge has led Good (1964) to extend the original number of twenty-four to thirty-seven.

Drude, a disciple of Grisebach, published a number of works on Plant Geography from 1883 onwards, of which the *Handbuch der Pflanzengeographie* (1890) is the most substantial. He also started with Engler a series of specialist monographs under the collective title of *Die Vegetation der Erde*, of which only a few volumes were completed.

Drude follows Grisebach in dividing his subject matter under three headings which may be summarized thus: climatological geobotany (ecological), geological geobotany (historical), and topographical geobotany (floristic and also partly historical). To these he added a fourth division: human influences (historical). He devoted considerable space to climatic factors but said very little about geological history. From the floristic angle he recognizes fifteen provinces, based on community of floras, but his chief claim to originality lies in the recognition of formations, synthetic units with an ecological basis but with physiognomical and floristic characters. These were intended to be expressions of a social analysis of the vegetation in which the floristic, ecological and physiognomical characters of the constituent species all play a part .This is a distinctively naturalistic and modern-looking picture and it reduces 'physiognomy', under the newer guise of 'growth-form', to a subordinate rather than a primary role. Species are to be evaluated for their importance as determinants of a formation on five grounds: (1) Frequency. (2) Growth-form. (3) Their climatic boundaries. (4) Their habitat relationships. (5) Special biological relationships, especially with other plants or with animals. Frequency includes the consideration of sociability (*Geselligkeit*), also arranged in five grades (see p. 3437). In all this we can recognize a modern idiom in which we can share. Naturally, so detailed a programme of vegetation analysis has only been partially fulfilled for the world as a whole, but such progress as has been made has been largely along the lines that Drude foresaw.

One of the most comprehensive and certainly one of the most influential books of Plant Geography is that of A. F. W. Schimper, *Plant Geography on a Physiological Basis* (1898. Eng. Trans. 1903), which was revised and republished, after the first author's death, by F. C. von Faber (Third edn, 1935). Schimper's previous botanical work had chiefly dealt with plants of special situations, desert plants, mangroves and shore plants, myrmecophilous plants and epiphytes and this had impressed him deeply with the importance of adaptation, which, in his time, still meant the direct response of the plant to its environment. It was natural therefore that he would make adaptation to environment the keynote of his treatment of plant distribution. He was not concerned with the history of floras and very little with Plant Sociology. He recognized two types of social vegetation, that determined by climate and that determined by edaphic factors, and the formations he distinguished under these headings were largely physiognomical. These views have already been discussed on p. 3325. He simply took the plants as he found them and endeavoured to show that their characteristics were related to the conditions under which they grew. This was by no means a new thesis, but the immense information which he imparted, its attractive presentation and its copious illustration, made possible by the introduction of the half-tone process, was unprecedented and is still unique. Despite outdated viewpoints it remains an indispensable work of reference.

The influence of Schimper diverted attention from Plant Geography, in the older sense, towards Ecology, which grew into a dominating concernment as the twentieth century proceeded. Ecology having thus developed and separated itself from Plant Geography, the older subject was left to its original resources and a revival of interest in historical, floristic studies became manifest. Recent literature therefore contains two categories of books; a considerable number of which are geographically oriented and are mainly descriptive accounts of the vegetation of different climatic regions; a smaller number are of a more analytical and theoretical character, and are concerned with the principles of the developmental history which underlies the present distribution of plants. Four of these may be selected for mention: Wulff, E. V. (1943) *Historical Plant Geography* translated from the Russian; Cain, S. A. (1944) *Foundations of Plant Geography*; Good, R. (1964) *The Geography of the Flowering Plants*, third edn.; Croizat, Leon (1952) *Manual of Phytogeography*. Mention should also be made of the controversial but highly informative works of J. C. Willis (1922 and 1949) *Age and Area* and *The Birth and Spread of Plants*.

The extension of our knowledge of Quaternary vegetation now gives a secure basis for judging the chronological sequence in the floristic changes of that era, at least in certain areas. The migrations of Tertiary plants are less assured, but the uncertainties of identification which for long entangled their investigation are being slowly cleared up by intensive methods of examination. The material available is abundant and we should some day have a reliable chart of the floristic changes in this period. In some areas, notably in California, it has been possible to correlate, not only species from Pliocene beds, but whole plant associations with those still living (see Vol. 3, p. 2565 and for details of Tertiary and Quaternary floras).

The theory of Continental Drift which originated with Wegener in 1924 gave a great stimulus to inquiry into plant distribution, the facts elucidated being held by some to support the theory, though others denied this and controversy flourished. On the whole biologists found the theory satisfactory, but as most geologists would have none of it the biologists were left somewhat in the air. Now, however, as it has become geologically respectable, indeed orthodox, at least in modified forms, biologists feel themselves entitled to invoke its aid where it seems to be necessary (see further p. 3795).

In addition to the books mentioned, there is an extensive periodical literature on Plant Geography. Reference may be made to the bibliography in the third edition of Good's *The Geography of Flowering Plants* mentioned above.

CHAPTER LIV

DISCUSSION OF THE AIMS, METHODS AND PRINCIPLES OF PLANT GEOGRAPHY

PLANT GEOGRAPHY is a term which has been in use since early in the last century. It has a clear significance and is preferable to the alternative 'Geobotany', which has borne more than one meaning. The latter term has been applied to a study of plant distribution as related to soils in a specific region, and has also been applied to the distribution of the various plant associations, world vegetation, in fact, as well as having been used by Grisebach to cover the whole subject of Plant Geography.

Plant Geography is itself really a compound subject. There are three distinct elements in it, any of which may be separately pursued. First is the floristic element which is concerned with the geographical distribution and the areas occupied by systematic units; species, genera or families. This is at bottom a factual study, but it can hardly be practised without some speculation arising as to the why and wherefore of the facts recorded. To support this it must draw upon materials from the other two elements, namely, the ecological element and the historical element. The nature of the former of these is self-evident, as it is simply a part of general Ecology. The latter, historical Plant Geography, is not so clear in meaning. It may be, and sometimes is, the history of the changes wrought by human action: the clearance of forests, the draining of swamps, the diversion of rivers, or large-scale irrigation, industry and the growth of cities, not to speak of the intentional or unintentional carriage of plants and their parasites about the world and the suppression of species by the plough or by grazing flocks. All these afford ample material for study and yield important practical lessons. Nevertheless, the subject goes deeper; it is also concerned with happenings before the advent of man, with the climatic and topographical revolutions of the Quarternary Era and even before that, with the geological changes of the Tertiary Era which determined the broad outlines of distribution while the angiospermic families were comparatively young. This type of geological history of recent plants was called by Diels 'Genetic Plant Geography' but this was in 1908 before Genetics had developed into its present significance and the term is now inappropriate since its implications are more *phylogenetic* than genetic in the modern sense, i.e. it is concerned with the ancestry of living plants. Genetics, in the modern sense, does also touch upon problems of distribution in studies of populations and the genetic changes taking place in them, as we shall see later (p. 3864).

Here then are three elements which are all needful for a full understanding of the plant populations of the world, each being to some extent dependent on the information contained in the other two. Floristic studies are indeed basic, for a knowledge of what plants grow where is essential before one can begin to consider why they grow where they do and not elsewhere. Floristic Plant Geography is therefore an expansion of classical taxonomy with its morphological criteria. Indeed, geographical data have become an essential part of taxonomy. The distribution of a group of related species often gives a picture of their probable relationships and of their relative ages, which is a valuable supplement to cytological and other evidence.

As we have remarked before, the discrimination of taxa from one another must be critical, since ecological, including climatic, preferences are often distinctive where morphological differences may be very small. The ecotype concept (p. 3359) has fruitful applications not only in ecological Plant Geography, but also in historical studies, where the survival or disappearance of a species under the adverse effects of climatic or topographic changes may often depend on the existence of a suitable ecotype, viable under changed conditions, which may be the only survivor of the species and thus give rise to the appearance of an alteration in the habitat preferences of the species, which in the past was often mistaken for direct adaptation.

Alongside of the knowledge of the composition of a flora comes the determination of the areas occupied by its constituent species, information which is still very scanty for large portions of the globe, where the distributional areas of only a few of the most conspicuous species are even approximately known. In only a few relatively small countries has the whole flora been analysed for distribution. The results have been very interesting for not only do the patterns of distribution reveal clearly historically different 'elements' in the flora, but they also reveal curious lacunae in the areas inhabited by species, which demand either an ecological or an historical explanation. Were similar details available for areas such as Brazil they would undoubtedly shed a flood of light on the history and affinities of the flora of a whole continent.

Distributional areas of species, genera and even of families are therefore one of the primary objectives of floristic Plant Geography. Such areas are not, however, always easily defined. The tendency of every plant is to spread, and estimates of its reproductive capacity and evaluation of its means of dispersal are important indicators of the reasons for areas being of very different sizes and of why some species are 'common' and some are 'rare'.

Quite generally we regard the boundaries of the area of a species as representing the limits of its climatic tolerances, but there are many exceptions. First there are peculiar soil preferences, e.g. calcifuge plants or serpentine plants (p. 3662), dependence on certain pollinating insects, parasitism, epiphytism, or other specialized habits as well as many other causes may limit the spread of a plant quite apart from meteorological data. Where these latter are operative, however, the real limits may sometimes be directly related to such factors as decreasing rainfall or minimum winter temperatures.

Second, areas may be limited by physical barriers, such as high mountain ranges, wide seas, densely forested areas, deserts and other barriers which inhibit the passage of germinable seeds for large numbers of species. The number of species which can properly be called 'worldwide' is very small indeed, and some at least of these are now weeds of human cultivation, whatever they may have been originally.

A third factor which makes for instability of areas is the ability of many plants to overstep their climatic limits, at least temporarily. We know very well that many plants can thrive when cultivated in places well outside their natural areas, and this sometimes happens also in nature, but such outlying individuals do not often reproduce themselves, so that their growth is limited to one generation. Even if they succeed in reproduction, the fact that they are growing near the limit of their ecological tolerance makes them very vulnerable to the competition of native species, and only rarely do they succeed in settling permanently in their new area.

It happens, however, that there are, in different parts of the world, climates which are essentially similar, carrying vegetation which is very much the same physiognomically, though composed of different species. For the most part these areas are too widely separated to permit migrations between them by natural means of transport, and the causes of their differing floras are purely historical. Man can, and does however, bridge these gaps and species thus introduced may soon become an integral part of the vegetation in their new habitat. Such are, for example, the Mexican *Agave* and *Opuntia* in the Mediterranean region, where they are so much at home that it is only botanists who know that they are not original inhabitants. Sometimes, unfortunately, they may prove more aggressive in their new home than in their country of origin and, like the Californian *Pinus radiata* in New Zealand, the Australian *Acacias* in South Africa, or the European blackberry (*Rubus*) in the southern hemisphere generally, they prove themselves either a blessing or a curse.

The ecological treatment of Plant Geography comes into play when a sufficient knowledge of the floristic composition of the vegetation is available, even if this is not complete, for much can be done with only knowledge about the most sociologically important species. If these govern the character of the vegetation then ecological information about their environmental relationships can shed light on those of the vegetation as a whole. Indeed, in little-known territory, valuable information can accrue with very little floristic knowledge, simply on the physiognomy of the different forms of vegetation observed, which gives at least a foundation for future investigation.

The methods adopted in this connection are those of general Ecology and may involve a climatological assessment of the vegetation of a large area, sociological analysis, or the soil and water relationships of a selected type of vegetation or of a single species. These are lines which can only be followed with a good knowledge of Ecology, and here the preceding chapter (Chapter LIII) may be helpful. All this is, in effect, an aspect of Ecology itself into which the special consideration of geographical distribution has been introduced.

Historical plant geography, on the other hand, has to deal with very different considerations. On the one hand there are studies in palaeogeography and palaeoclimatology. The records of the first of these, of the various submergences and emergences, of the tectonic upheavals of land surfaces and the movements of land masses, are purely geological. Palaeoclimatology depends much more on biological evidence. Even the history of the most recent revolution, the Pleistocene Ice Age, rests to a considerable extent on the changes of vegetation found in successive beds. Indeed the accepted stratigraphical level of its commencement rests largely on botanical evidence (see Vol. 3, p. 2579). How trustworthy plant remains are as indices of climate depends on their age. Remains in younger beds, which may be of species still living or closely related to living species, may be relied upon, but the further back we go the less we know of the ecological preferences of the plants concerned, and we can only make deductions from their structural features. Fortunately it is the plants in the later Tertiary and Quaternary beds which are most significant from the point of view of Plant Geography.

Considering the immense multiplication of species which appears to have occurred among Angiosperms during Tertiary time, their distribution must be considered with regard to the changing dispositions of land and water and the known changes of climate at different epochs. It is obvious that such changes must have impressed a pattern of distribution both on new, developing species as well as upon older survivors from the past, often on so large a scale that the effects have persisted to the present day. One has only to mention the continuity of the circumpolar arctic flora in the northern hemisphere, due to the contiguity, age and stability of the land masses in that region or, on the contrary, the wholesale disappearance from Europe of the warm-temperate plants of the Pliocene and Miocene with the progress of the frozen Pleistocene. Their elimination from the European flora was assisted by the barriers of east-west mountain ranges and by the width of the Mediterranean Sea. Most of these plants have never come back but now appear as if they were characteristically American or East Asiatic species.

Historical studies in Plant Geography must begin with a knowledge of the geological history of the area under investigation and particularly of the fossil plants from its recent past, since these are the most direct evidences of former distribution. Such data often give positive evidence of changes which have taken place; they also form a basis for judging the possibility of changes which may be otherwise inferred but are not directly demonstrable.*

More immediate knowledge of plant history can often be derived from living plants themselves, but the conclusions drawn are subject to checks either from Palaeontology or from Palaeogeography. In the absence of such checks the conclusions can only have varying degrees of probability. One method, which has been successfully employed in Britain is to analyse a flora with regard to the geographical distribution of the species in other areas or in the world as a whole. This may lead to the recognition of certain groups of species with similar geographical affinities and those 'floristic elements' give strong indications of the derivation of the species in the given area. Taken in

---

* See especially with regard to Gondwanaland on p. 3807. The evidence for its existence based on the distribution of fossil plants was considered so irrefutable that botanists refused to concede the geological objections. Now the situation is changed.

connection with palaeogeographical data they can sometimes also be assigned the chronological order of their migrations.

Another method which has given useful information is the intensive study of the taxonomic affinities of species in connection with their geographical distribution. Just as geography can assist taxonomy so can the relationship be reversed. Taking into consideration all the relevant information, morphological, cytological, genetic and biochemical on the relationships of the species and comparing these with their distributions, a pattern may become evident giving strong indications of their relative ages and origins.

Such investigations are only legitimately applied to closely allied groups of species, either a small genus or a group within a larger genus, whose origins and separation are presumably recent. It is unsuitable for application to whole families, whose separation and history are so ancient that present day appearances may be quite misleading. For example, the Sapotaceae are now limited to the Tropics and it has been argued that their centre of origin lay in Malaysia, from which they spread east and west, but when we consider the number of genera of that family and their abundance in the European Miocene beds one cannot but feel that their presumed history is doubtful. The Fagaceae would be considered by most people, familiar with *Fagus* and *Quercus*, to be a typically north temperate family, but if concentration of species be regarded as indicating a centre of distribution the fact is that the number of species of *Quercus* and *Pasania* in Malaysia probably outnumbered all those in the rest of the world. The genus *Carex* is universal in north temperate regions, but *Unicinia*, which alone possesses the ancestral floral structure of the Caricoideae is of far-southern distribution. It was at one time believed that *Nothofagus*, on the ground of its antarctic distribution, was a southern derivative of the northern *Fagus*, regarded as modified from the better-known *Fagus* during its long journey south. We know now, however, that *Nothofagus* was probably a member of the European Tertiary flora and apparently of equal antiquity with *Fagus* itself.

These facts show that large-scale deductions from present distributions have their pitfalls and need palaeontological backing before they can be unreservedly accepted.

There is insufficient palaeobotanical evidence from Africa to verify conclusions about its floristic history, but the permanence of the continent as a unitary land mass, since a period before the present climatic zones were established, seems to be indicated by large-scale floristic links. For instance, the concentration of *Erica** and *Pelargonium* both around the Mediterranean and in South Africa, implies a north–south continuity, though the tropical belt now interposes a climatic barrier. Continuity in the east–west direction is also predicated by the close affinity between *Dracaena draco* of the Canary Islands and *D. ombel* in Ethiopia and the endemic *D. cinnabari* on the island of Socotra.

Even where a continuous land mass is geologically assured, its topography

* Some 15 species of *Erica* (out of a total of 700) are found on mountains in tropical E Africa, including the widespread *E. arborea*, which makes a link between the southern centre and the Mediterranean group.

may facilitate or, conversely, oppose, distributional movements and may thus have widespread effects upon present day distributions, so that they must be taken into account in any attempt at historical reconstruction. The post-glacial history of vegetation in north Europe and in North America re-spectively shows this clearly. In Europe, as we have previously mentioned, the prevalence of east–west mountain ranges, not only opposed the retreat of warm temperate species before the southward advance of the Pleistocene ice, thus contributing to their extinction, but they also hindered their re-migration northwards from southern and eastern refuges. The plant population of northern Europe was thus permanently impoverished. In North America, on the other hand, the principal mountain chains run north and south and no such impoverishment occurred, for, as the ice retreated, forests quickly be-came re-established and warm-temperate genera like *Liriodendron, Menisper-mum* and *Calycanthus* were able to migrate northwards without barriers.

A system of Plant Geography complete in itself has been laid down by Leon Croizat. He isolates Ecology entirely from Plant Geography, assigning to the former the study of the way in which given plants fit in to their environ-ment, while the latter is specifically the study of how those plants came to be where they are. Taking the purely descriptive data of distributions as material, the task of the Plant Geographer is to interpret the data as evidences of an historical process which only began with the appearance of Angiosperms in the evolutionary sequence.

Towards this end Croizat adopted as his leading principle the concept of the *genorheitron* which originated with Lam in 1938. This term is used to signi-fy the evolutionary track of a taxonomic group of plants through time and space.* The time sequence can only occasionally be derived from direct palaeobotanical evidence but, by Croizat's technique, is deduced from the distribution of species along divergent lines of migration. The author accumu-lated an immense body of facts on which his views are based and, without binding himself to complete accuracy in details, he maintains that the total consensus of evidence all points the same way, southwards. He rejects con-tinental drift but, at the same time, is insistent that the migrationary tracks have little or no relation to the present-day geography of the continents and should be viewed as being determined by the geographical changes which have occurred since Upper Jurassic times, especially to what he calls the 'crum-bling' of former land masses.

Croizat's main argument, slightly paraphrased, is as follows. An ancient 'antarctic' shore connected every land in the deep south of our maps. The lands, once extant but now vanished, between the approaches of the Kerguelen Islands and Fuegia, were the ultimate hub of angiospermous dispersal. It was from this centre that all the angiospermous 'gates' of dispersal were fed with the genorheitra of the seed plants.

In short, his is a Holantarctic as opposed to a Holarctic theory of the genesis of the Angiosperms. On a Jurassic antarctic continent much larger

---

* Croizat's own definition is 'a stream of plant life in motion, laden with evolution potential'.

than at present, he contends, the Angiosperms arose and their rapid emigration and entry into the world's flora was conditioned by the geological revolutions of late Cretaceous times. Most of the families and many large genera are supposed to have been already separated before migration on a large scale began, so that their wanderings are traceable from the time of their entry into the continents onwards. The 'gates' he refers to are the regions where the southern tip of continental masses approach most nearly to the assumed shoreline of ancient Antarctica. These are, Fuegia, South Africa and Australasia (New Zealand, SE Australia and the area of New Caledonia and New Guinea which is now a complex of islands). One or more of these may indeed have been formerly joined to the Proto-antarctic block. Some genorheitra led through only one of these gates others used two or even all three, with very different subsequent histories. The gates are in fact compound areas and are analysed into sections (p. 3816).

Croizat presents plant migrations as 'orderly, precise and repetitious', involving vast stretches of time and being essentially mass phenomena, in relation to which varying means of dispersal have little significance.

These sweeping theories are obviously controversial, but overlooking their polemic and dogmatic presentation, they merit serious consideration by plant geographers since they do present, for the first time, a logically coherent story of angiospermic evolution and dispersal as a unitary process. Their author uses them to account for some examples of wide discontinuity in distribution for which no alternative explanation has been offered, a strong point in their favour. Apart from everything else, the work shows good reasons for abandoning a number of misconceptions which have tended to cloud phytogeographical thinking. People concerned with presenting a case are apt to overstate it, but this does not necessarily mean that the case is a bad one.

Taxonomy is theoretically a static entity. It represents an immediate cross-section of a genorheitron of which the continuing dimension is time. It follows therefore that it is relative to its time-period and that at different time-periods there have been or will be different taxonomies. As Adolf Meyer has pointed out, the history of organisms is, at the same time, the history of their dispersal and that without this information biogeography is incomprehensible. We come back therefore to the conclusion that taxonomy, history, dispersal and evolution are all involved in one complex which may be called Phylogeny and that Plant Geography is its present and immediate expression.

The opposite, strictly ecological, view of Plant Geography is summed up in the apothegm of Beyerinck: 'Everything is everywhere, but the environment selects.'* It is difficult to see how this can be interpreted literally, if it refers to organisms rather than to genetic potentialities, in view of the notorious facts of migration, invasion and acclimatization, which are not historical deductions, but matters of direct observation. In a limited sense, if we replace 'is' by 'arrives', it comes nearer to truth. Far-straying individuals, seeds or spores may appear almost anywhere. Flamingoes, coconuts and *Entada* seeds may reach the coasts of Britain, but 'the environment selects' and they disappear. This is a very different thing from the ordered mass advance of

* Originally postulated for micro-organisms, but elevated into a law by Baas-Becking.

genorheitra in space, which we have no reason to believe have all as yet finished their course, still less that they have all become cosmopolitan. The space-time course of a genorheitron depends on its genetic potentialities. The greater its reservoir of genes the better able it will be to generate entities which fit into the multifarious environments it encounters in its migration or, in other words, to spawn species. Its advance is thus not so much environmentally *selected* as environmentally *directed*. It will advance in that direction, or those directions, in which the environment offers niches for its genetic potentialities. It will not advance into regions where this is not the case.

In the contrary sense, genetic impoverishment limits the evolutive capacity of a species and hence limits its spread. This may come about either by the action of highly unfavourable conditions (as, for example, a glacial period, see Vol. 3, p. 2591) or the long isolation of a small population. There is probably a critical limit to the degree of impoverishment, below which survival is impossible and a level anywhere near this limit will probably confine a species to one narrow habitat (see under 'areas', p. 3835). As Lam points out, 'all grades of areas may be imagined as there are all kinds of taxonomic units, rich and poor, large and small, long-living and soon-disappearing.' If a distributional area shows a decrease this may be due to one of two causes; either, loss of genetic potentialities leading eventually to 'relic' distribution in a very limited area and finally to extinction, or, it may be due to rejuvenation, when the original taxon is giving rise to new taxa and is disappearing because its original taxonomic definition no longer fits.

H. B. Guppy in 1918 writing, as he says, from the 'standpoint of an idealist' argued that the evolution of Angiosperms has proceeded from the family level by differentiation downwards into genera and species, not by the building up of genera and families from related species. The family archetypes are supposed to have originated in a very stable geological period with comparatively little surface relief and highly uniform climate; the middle Jurassic is suggested. Under such conditions, he agreed, life was like a ship drifting in a calm. Mutation was unrestricted, any viable mutation might survive for there was practically no environmental selection. If the polar axis of the earth happened to stand at right-angles to the orbital plane at that period there would be no annual seasons and the climate in any given zone of latitude would be highly uniform. On the other hand zonal differences would be accentuated. Under such conditions the main family types were established. If this be granted, then the pattern of family distribution is the oldest. Guppy examined the distribution of families, genera and species respectively with regard to their occurrence on the two great land-masses, the Old World and the New, which diverge from the north southwards. Of the 272 families in Engler's system 70·5 per cent are common to both land-masses. A small majority of these families are northern in distribution, where the continents are closest. Of families with a northern, extratropical distribution 77 per cent are common to the two Worlds, while the percentage for tropical families is 69. Guppy does not consider the difference to be significant as the latter figure is so close to the percentage (70·5) for the Angiosperm families as a whole.

There is a residue of 30 per cent, or 80 families, which are not common to the two Worlds. They are presumably of later origin. They might be expected to occur in proportion to the relative sizes of the Old and the New Worlds (35:15 or 2·3:1) but actually they are proportionately more abundant in the New World (45 Old to 35 New or 1·3:1) as if the New World had been more prolific in producing new families.

While the families largely ignore the land cleavage, the distinction is more marked among smaller units. The proportions common to the two hemi-spheres are less than 20 per cent for genera and about 1 per cent for species. This is in accord with expectation if we consider the primary distribution to have concerned family types, which reached very wide or worldwide dispersal in olden times, and differentiated successively into genera and species in accord with the diversification of habitats associated with geological revolutions. The family characteristics are, the author contends, very ancient, very stable and but little affected by ecological factors such as climate, though there has been a broad selection into tropical and non-tropical families, a distinction which is by no means complete and is frequently overstepped by genera. Species, on the other hand, are markedly sensitive to environment in their distribution and experiments at the Earhart laboratories in California have shown that the distribution of ecotypes is extraordinarily sensitive to climatic conditions affecting, first, germination and, second, flowering. If we regard ecotypes as being potentially embryonic species then their differentiation is practically an affair of today.

There is little to distinguish the families of Monocotyledons and Dicoty-ledons with regard to intercontinental dispersal. The percentages of those found in both hemispheres are 76·8 (out of 43 families) for Monocotyledons and 69·4 (out of 229) for Dicotyledons. The proportion of tropical families is almost the same in both groups, about 60 per cent, but the remainder of the Dicotyledons is almost entirely temperate, while of the remainder of the Monocotyledons only one-third is exclusively temperate. Assuming, as Guppy apparently does, that the tropical environment is the primary one, this would mean that the Dicotyledons are more advanced than the Monocoty-ledons in developing temperate families. This may be a misreading of the facts, however, since, with the climatic selection of ecotypes in mind, it may only mean that the Dicotyledons had more families which were inherently capable of living in temperate conditions.

We should remember that the above was written before the Wegener theory of continental drift was made public (1912). It does not figure in the views of either Guppy or Croizat, yet the latter insists on the necessity of envisaging large-scale rearrangements of land and sea in former times and such rearrangements are implicit in Guppy's views. He points out that the distribu-tion of families between eastern and western hemispheres would not be sensi-bly different if the two land-masses were united. Moreover, he points out that the extent to which families are shared between the two continental masses is not a function of the distance now separating them. Among purely tropical families 61 per cent are shared, while of purely northern temperate families, 64 per cent are shared, not a highly significant difference.

To what extent do the views of Croizat and Guppy harmonize? Both look back to an early period of Angiosperm evolution in which the family characters became established. Both consider their subsequent subdivision into smaller entities, genera and species, to have been bound up with their experiences during migration and with progressive climatic differentiation from the early Cretaceous period onwards. Both consider that the early, pre-migration phase of evolution took place in a highly favourable climate, but while Guppy accepts the tropics as they are today, Croizat chooses a former Antarctica as this Garden of Eden.

It is difficult to see how Antartica as we know it could ever have had such a climate, but we know that Greenland and possibly the rest of the Arctic, most probably did have such a climate in late Jurassic–early Cretaceous times. Under present cosmographical arrangements this could only have occurred by the rare coincidence of an extreme inclination of the earth's axis coinciding with the perihelion of the orbit during the southern (or northern) summer. Even this, though it might give a hot summer would also give a long and very cold and dark winter, more likely to produce a desert than a tropical forest. Yet we have palaeobotanical evidence (*Antarcticoxylon* and coal seams) that vegetation once flourished in Antarctica. How could it happen? It looks as if we must accept Wegener's concept of the continents shifting their positions. Croizat's objection that if America was joined to the Old World it leaves the Pacific gap twice as wide as now does not really matter if Antarctica was available as a land bridge and a source of plants.

Plant Geography has inevitably a close link with taxonomy. From the taxonomists' side the distribution and habitat preferences of a species are an important part of its description. The Plant Geographer, on the other hand, requires to know about the nature and relationships of the species which are, in a sense, his raw material and what connection, if any, there is between distribution and the formation of species, the process called 'speciation'. The species defined by practical taxonomists are, for the most part, morphological entities but they are by no means the monololithic entities that the formal 'diagnosis' of a species suggests.

Biologically every species includes a considerable variety of biotypes, differing genetically within limits, and many of these biotypes may also be ecotypes, that is biotypes with specialized ecological preferences or tolerances, each of which will therefore have its own pattern of distribution, sometimes only on a local scale, but sometimes also on a large climatic scale, so that two different biotypes may be segregated at opposite ends of the area of distribution of the species as a whole. Some biotypes or ecotypes show recognizable morphological variations but many do not, their differences being purely functional or biochemical. It is important to the Plant Geographer to grasp the qualities of the species he is concerned with and the range of its potentialities, information which must derive largely from observations in the field. It is illegitimate to deduce the area of distribution of a species from the known preferences of some of its biotypes only and to conclude, for example, that a species generally calcicole cannot occur in areas where the soil is non-calcareous. The relationships of plants to their environment are much too complex to allow such simple generalizations. One might add that general 'laws' of distribution should be accepted with caution for all have their exceptions.

The Plant Geographer must accept the taxonomist's species as a basis, including taxonomic sub-species and varieties. His analysis of the extent to which variation within the species affects distribution must be his own study, and it is one to which too little attention has been paid. Both ecologists and physiologists have been too ready to accept the taxonomist's species as units for purposes of research, at least as far as wild plants are concerned, though with cultivated plants the importance of identifying the exact strain, or biotype, is generally recognized.

Biologists accept as fact that new species have constantly been arising in the course of evolution, the evidence is too strong for denial. Many people, however, think of evolution as something that *has happened*, not as something that *is happening*. There have been periods in the past when evolution, the production of new species, was exceptionally rapid and multiple advances were made. It may also be accepted that such periods were generally associated with periods of extensive geological change, but evolution is not confined to such periods and there is no reason to believe that it has ceased. Genetic discontinuities in a population can still arise and new, true breeding entities be segregated, which can only be regarded as new species. Indeed, to detect the origin of a new species could be one of the most exciting adventures in genetic Plant Geography.

Speciation involves two distinct phases. First is the origin of a new genotype, second is its segregation and survival. Genotypical variation among bisexual organisms is well nigh universal, as it shows itself in individual and biotypic variations, but for a new species to arise there must be more than this. There must be either gene mutation or interspecific hybridization. The latter sounds like a contradiction in terms since the essence of a species is supposed to be its genetic isolation. But we must remember the difference between taxonomists' species and natural species and many taxonomically valid species nevertheless may belong to what Danser has called a 'commiscium',* a group capable of at least a limited degree of interbreeding. Even generic limits are not too great to be included in such a group and bigeneric as well as bispecific hybrids do occur. Some degree of gene exchange is possible across some taxonomic boundaries, even when these boundaries delimit morphologically valid species.

Isolation is one of the most important influences on the chances that a new genotype has for survival. It may be either *geographical, ecological* or *genetic*. Geographical isolation is rarely possible at first when a new genotype arises amid a population of the parent species; it is usually achieved by continental species at a later date, if at all. Only where a new genotype arises at the boundary of a species area is it possible for it, provided it has suitable tolerances, to achieve such isolation rapidly. Among islands the case is altered. On remote islands, or on islands of an archipelago between which opportunities of transit

* Also called a 'syngameon'.

are rare, a new genotype may have an island field to itself from the start and may there establish itself as a distinctive element of the local flora.

Ecological isolation is more frequent and arises from the possession by the new genotype of a new range of tolerances which enable it to colonize habitats from which the parent form is debarred. In the case of a genotype which is potentially a new species this is only an amplification of the sorting out by habitats of the ecotypes within an existing species. The constitutions of both ecotypes and of new genotypes, whether arising by mutation or hybridization, are genetic in basis, the difference between the two cases being quantitative rather than qualitative.

Genetic isolation is the most secure of the three, though it is not possible to say whether it is the commonest. The change in the genotype has rendered it incompatible with the parent type, or at most with only limited possibilities of gene exchange. Thus isolation is immediate, but it may be followed by isolation of either of the other kinds as well.

Mutation is not uncommon, but the great majority of mutations are either useless or harmful. Mutations which are valueless from the point of view of survival have been probably retained in great numbers, but one which endows the changed organism with outstanding success is indeed a rarity.

Whether the frequency of mutation is directly encouraged by changed conditions of the environment is an open question and there may be no general answer, but it is at least probable that the chances of a mutation surviving (and so becoming observable) are increased by an environmental change, especially if the change is detrimental to the parent type.

From what we have said it will be seen that the origination of new species is often reflected in the areas they occupy in relation to those occupied by closely related species, so that it has a geographical importance which we shall come back to when discussing the facts about areas (see p. 3835).

We have already emphasized that the underlying basis of plant distribution is historical, or in other words, that the provision of the plant material which constitutes a flora depends upon the history of the species involved. This cuts two ways; it implies that some species which could inhabit a given area have not arrived there in the course of their historical migrations, and it also implies that other species may arrive in areas where, because of the prevailing conditions, they cannot survive. Given that a certain array of species is historically available in any area it is the environment that sorts them out and determines the eventual composition of the flora and the distribution of its species.

The primary controlling factors in this process are undoubtedly those of climate, while within any climatic area edaphic influences are the main secondary factor in controlling local distribution. A third influence, is that of biotic factors, arising either from animal activities or from inter-competition between plants, including parasites. This is a variable influence, being sometimes of decisive importance and sometimes only of minor effect. We have already discussed these influences, from the ecological point of view, in the previous section of this Volume.

The influence of environment on the plants is usually called *action*, while the plants' responses are referred to as *reaction*. Clements added a third term, *co-action* to describe the relationships of the plant and animals in a biome.

Exceptional extreme values of a climatic factor, for example, temperature or rainfall may kill individuals or even, though rarely, eliminate a whole species from an area, but they do not permanently change the vegetation as a whole. However destructive temporarily they may be, once they have passed the natural vegetation of the area will re-establish itself. Thus, paradoxically, while extremes may determine exclusion it is the climatic means which determine inclusion.

Variable as each factor of the environment is, every plant must possess a certain range of *tolerance* for such variations, if it is to survive. It would be wrong, however, to regard these tolerances as absolute, inherent qualities of an individual or a biotype. If, indeed, as physiologists, we isolate a single variable, a plant may show a definite range of tolerance between a minimum and a maximum, but this is not an ecological situation. In nature the plant is multi-conditioned, all the factors influence not only the plant, but each other (see Fig. 4.17, p. 3494), a state of affairs called *holocoenotic*. In these circumstances tolerances, too, are variable, they are *conditioned* by the consensus of factors and their range contracts as their extreme limits are approached. While we can envisage the situation, its analysis in quantitative terms is practically impossible.

There is, moreover, a further complexity. Tolerances for particular factors are often different at different periods of the life history, or they may be different for vegetative and for reproductive organs. This is specially well marked with regard to the position of the *optimum*, that is the value of any factor at which its action is most favourable. This is naturally some intermediate point on the scale of tolerance. Again, it is a physiological concept relevant to one factor and one process or one event, considered in isolation. For example, careful analysis of the relationships between flower development and temperature in bulbs grown in Holland showed that each phase of development had a different temperature optimum for growth rate. Further, the temperature optima for germination, for seedling growth, for leaf development, for fertilization and for fruit growth may all be different.

Obviously there can be no absolute optimum of all conditions for a plant in nature and the survival and success of a plant in any habitat is inevitably based on compromise. The 'best' habitat for a plant is therefore that in which the seasonal relationships of the various phases of its life-cycle accord most closely with the seasonal conditions of the environment.* An example of this can be found in photoperiodic relations. Plants of the tropics are necessarily those which flower with short days or else are neutral to photoperiod. Moving north or south from the tropics the proportion of long-day species

---

* Went has recorded a number of examples of this. For example, *Bellis perennis* has optima of 14°C day temperature, night 8°C and a 10 hour photoperiod. This corresponds to the climate of Pasadena, California in February and March, and it is in those months that the species flowers at Pasadena.

increases, while short-day species in temperate latitudes are those which flower in early spring or in autumn, when days are short.

Only in the case of old and well-established species is there a reasonable presumption that the area which it occupies corresponds to the limits of its climatic tolerances and even for many such old species edaphic preferences may prevail over climatic tolerances to limit or distort the area. New species or new migrants to a locality will tend to increase their areas progressively, though sometimes very slowly. On the other hand, there have been many instances of new migrants spreading over a country almost explosively. Old species, too, may extend their known areas if a barrier is removed, such as the destruction of forest or drainage or irrigation by man. This kind of alteration may break down the ecological isolation of two related species and lead to new hybridization, a proceeding which has been called 'hybridization of the habitat'. Spreading an area may have dangers as well as gains, it may bring a species within the range of a new parasite or predator, as happened in the downward spread by cultivation of Andean *Solanum* species, which brought them into contact with the Colorado beetle.

Many Plant Geographers have emphasized the view that efficient mechanisms of seed dispersal are important in determining the extent of the geographical distribution of species. The idea was regarded as so self-evident as to need no demonstration, which unfortunately it cannot get. Bentham in a comprehensive survey of the distribution of the Compositae found that the relative efficiency of the means of dispersal could be ignored as a determining cause of distributional areas. Dispersal mechanisms, of whatever kind, are species characters which seem to have appeared late in the history of evolution, after the main lines of distribution of the larger taxonomic groups had been settled by events in far-off geological periods. While an efficient dispersal mechanism may have an effect on the relative abundance of a species within its existing area, the long-range effect of such mechanisms in extending the area or in the colonization of entirely new area has been subject to some imaginative exaggeration. Events like these may indeed happen, but they are unusual if not rare and do not provide a basic explanation of distributional areas. Indeed, when concrete cases of the rapid spread of a species over new areas of occupation are considered it will be found that they are rarely, if ever, associated with highly developed means of seed dispersal. In general such rapid spread is passive, due to external agencies combined with a wide tolerance and a high degree of vegetative vitality. Normally dispersal is slow, even where dispersal mechanisms exist. (See further p. 3858.)

One ecological factor which does not seem to have been given sufficient weight in Plant Geography is the effect of vegetational succession on the distribution of species. Pioneer species disappear locally as succession proceeds and their place is taken by species characteristic of later successional phases until the population reaches a metastable climax. This may be dismissed as being of purely local significance, which is often true, but not always. It may lead to the widespread disappearance of certain species, which are no longer able to find suitable habitats. This is all part of the secular changes which constitute the historical side of Plant Geography, but it is not all historical, it is a continuing process and has its effects in the present, effects directly observable in the retreat or advance of species. It may be difficult to assess its general importance, but the consideration of what is actually happening may serve to remind us that history goes on for ever and to keep in check a tendency to think of present areas of distribution as static entities, arrived at by historical plus ecological processes but now magically stabilized. Succession both has modified and is modifying distribution.

Plant Geography has one simple though laborious task, which requires no theoretical framework, namely, the collection of accurate and detailed information about plant distribution at the present time, which still offers an immense field of investigation and combines very suitably with taxonomic studies. It is not enough to know in which countries a species occurs, plant geography requires to know exactly where each species is found, and there are only a few countries which have anything like a complete knowledge of the flora in this sense. Yet it is only on such a basis that the geography of a species can be founded. From this starting point two lines of inquiry diverge. One is ecological and the other is historical, the latter being the true interpretative role of Plant Geography. It must be apparent that the two are quite distinct. The relationships of a species to the factors of the environment in which it finds itself is ecology pure and simple, whereas how the species comes to be there in the first place is a problem for the plant geographer. It might be said that Ecology seeks to know *how* the plant *grows* here and plant geography to know *why* the plant *is* here.

It would be extreme and hypercritical to assert, as some writers do, that Plant Geography and Ecology are two completely distinct subjects. True, their approaches to the subject matter are distinct. The Plant Geographer is primarily a taxonomist, while the Ecologist is basically a physiologist, but this must not obscure the fact that the territory they are approaching is common ground, namely, the distribution of plants in nature. Both approaches are valid, both are necessary and each possesses a body of data which is essential material for the other. The taxonomist analyses the plant population and decides the identity of its constituent elements. When we say 'decides' we mean that this is in fact what he must do, since for all practical purposes the 'species' he uses must be morphologically determined units which are the common currency of biology, and which a taxonomist is trained to discriminate.

The existing flora of any region may be looked upon as the survivors of an army of migrants which, in the course of evolutionary history, either drifted into the region from outside or were born in it from a migrant parentage. Their survival and their distribution in the region are, for the most part, ecologically determined. This environmental determination is so general, so obvious and so overwhelmingly important that the Plant Geographer cannot ignore it. He must absorb the ecological data into his theoretical scheme, taking them from ecologists or else observing them for himself. The ecologist, on the other hand, must take his floristic data, that is the identification of his subject material, from the taxonomist. So immediately relevant are the ecological factors in distribution that they tend to obscure the importance of the underlying historical factors which have, in a sense, prejudged the composition of the flora.

We have previously referred to the influence of vegetational succession on distribution and another consideration for the Plant Geographer is the distribution of plant communities. Species are influenced not only by climatic and edaphic factors, but also by biotic factors and a species cannot exist in an area, however otherwise suitable, if an inappropriate type of plant community already occupies the ground. This is really an aspect of the question of succession for it means that species are, in general, unable to colonize an area already occupied by a more developmentally advanced community. To take a simple example, dry-land species and a great many herbaceous species will be excluded from areas under a stable forest community by the biotic conditions there obtaining, not by climate or soil, as may often be seen by their rapid colonization of the area following forest clearance.

Turrill in 1939 has summarized clearly the aims and scope of Plant Geography so much in accord with the views we have expressed that we cannot do better than to record them, with approbation.

Aims

1. The determination of the geographical distribution of taxonomic units, of which the species is the most generally important.
2. The determination of the distribution of plant communities.
3. The relating of such distributions to the environmental factors, past and present, and to the evolutionary history of the taxonomic unit or plant community.
4. The correlation, one with another, of all the facts thus obtained and their expression in the form of a description of the flora and vegetation of the earth on a geographical basis.

Methods (Also from Turrill with slight amendment)

1. Listing all the species known for the area with their exact distribution within the area and with the following particulars concerning them: their duration (i.e. life-period), life-form, flowering period, habitat range, altitudinal range and distribution outside the area, if any. To the above we might add that if a species or a group of species is the object of study, rather than an area, then it is their world range that must be considered. This modification also applies to the following.
2. An investigation of the distribution and floristic composition of the prevalent plant communities and of their developmental history.
3. An examination of the physical features of the area and of their possible influence on plant distribution.
4. A study of the soils and of the geological history of the area.
5. A consideration of the meteorological data for the area, including what is known of its climatic history.

6. A study of the human history of the area and its effect on the flora and the vegetation.
7. A review of the cultivated crops of the area and methods of agriculture, horticulture and forestry. Whence come crops thence come weeds.

This is indeed a wide programme, but a synthesis of the data acquired may show the probable lines of migration, the relative importance of environmental factors in the area and the whole character of its plant life. If, in addition, the data reveal characteristic differences which enable phytogeographical regions to be recognized, it will have very valuable practical applications for the planning of developments within the area.

CHAPTER LV

THE GEOLOGICAL BACKGROUND

FROM the point of view of Plant Geography there are two main fields of geological inquiry which profoundly affect the ideas prevailing in the subject. One of these is the Quaternary Era, the immediate geological past, for which our information about earth movements and climatic changes is naturally fullest and the data about the living inhabitants is so copious that biology and geology go hand in hand. Here we must look for our knowledge of the recent history of the comings and goings of species and populations which have resulted in the present pattern of distribution.

The second field of interest is much more remote in time and is concerned with the history of the land masses of the globe and the changes which they have experienced, especially since the Jurassic Period when the Angiosperms first became recognizable as a new component of the world flora. This is a field of much greater uncertainty and controversy than the first mentioned, but it is rapidly becoming less speculative and more factual and we must consider it, with reservations, as basic to our understanding of the dispersal of the great families of plants about the world.

THE QUATERNARY ERA

In Volume 3 we have already discussed the botanical history of this period, covering the best part of the last one million years, at some length, so we shall confine ourselves here to a brief recapitulation of the main points. The Era is generally divided into two periods, an early period, the *Pleistocene* and a later period the *Holocene*. The Pleistocene is usually dated by the beginning of glacial conditions in the northern hemisphere and it is often equated with the Ice Age. Although the extensive glaciation of important parts of that hemisphere was the outstanding feature of the period it must be remembered that, from the point of view of the world as a whole, it was a local phenomenon, with only limited secondary effects on the rest of the globe. At the height of the glaciation ice sheets covered all northern Europe and a part of Siberia and in North America down to about latitude 40°N, though not so far on the west coast as on the east. These ice sheets were continuous with the polar ice cap.

In the southern hemisphere there is also evidence of less extensive glaciation in the Pleistocene, which may have been just as intensive but was restricted in area by the smaller extent of land. There was glaciation in New Zealand, in Tasmania and the extreme south of South America. It is uncertain whether these were contemporaneous with the northern glaciation, but it is probable that they were. Apart from these was the formation, or at least the extension, of the ice sheet over the whole of Antarctica, which, with the ice sheet over Greenland, is the only part which has remained to the present day. These events had the most catastrophic effect upon the flora in the lands affected. They changed permanently the plant populations of immense areas and in the case of Antarctica extinguished the flora completely. The present aspect of Antarctica must not blind us to the fact that it was not always the ice desert it now is, but that in the Tertiary and earlier periods it was a temperate land with an abundant flora which may well have been a reservoir from which migrant species reached the continents to the north of it. It cannot be left out of account in considering the distribution and migration of plants. The cause of this change in the climate of Antarctica cannot yet be stated with certainty, but it seems very probable that it did not always occupy its present polar position and that in the Cretaceous and earlier periods it was closely approximated to Australia on one side and South Africa on the other.

The Pleistocene Ice Age was by no means continuous. Over most of the glaciated region there is evidence of four separate periods of major advances of the ice sheet with long inter-glacial periods in between. In some areas the evidence is confused and there may have been fewer than four maxima. During the interglacial periods the ice retreated or even disappeared. Especially was this the case in the middle or Great Inter-glacial, which lasted for some quarter of a million years, for a large part of which a temperate or warm climate prevailed. As this period was at least ten times as long as the interval separating us from the 'end' of the glacial period, we cannot feel confident that we are not ourselves living in another inter-glacial period. As well as inter-glacial periods there were also minor fluctuations during the major periods of advance. These are called inter-stadial periods, during which there was sufficient amelioration of climate to allow hardier plants to establish a covering of vegetation, which seems to have been extinguished when the ice again advanced.

Altogether the Pleistocene seems to have lasted for some 600 to 700 thousand years and the end or Fini-glacial epoch is dated from the break up of the main Scandinavian ice sheet somewhere about 8000 BC.

The prevalence of such widespread glacial conditions had secondary effects which were felt to some extent all over the world. In the first place the atmosphere over the ice sheet is permanently anticyclonic. Cold winds pour out from all around the ice creating a wide zone of sub-arctic conditions, the periglacial zone, extending in some regions for hundreds of miles. This anti-cyclone occupied the latitudes in which cyclones normally circulate, the cyclonic belt, which was displaced southwards. Thus in lands beyond the peri-glacial zone there would be a great increase in rainfall, and occasionally snowfall, which would be especially marked when glaciation to the northwards was most intense. Southerly lands would therefore experience pluvial periods corresponding to the northern glacial periods, and there is evidence that such pluvial periods occurred even far to the south, perhaps as far as the present tropics.

Similar phenomena affected the southern hemisphere during the Pleistocene, but we do not know so much about them. They were of smaller extension and from what is known they seem, at least in some areas, to have synchronized with the northern periods.

Pollen analysis of peat deposits in tropical Africa have shown that even in such latitudes there was a substantial drop in temperature at periods which seem to synchronize with the glacial advances in Europe. This is shown by the variations in level of the montane zones of vegetation, which came down to lower levels during the cold periods and retreated upwards as the climate improved. Among other effects the cold periods led to a great expansion of areas occupied by Ericaceae.

A more widespread effect of glaciation was that of changing sea levels, which affected the whole world. Such a vast bulk of water was locked up in the ice sheets that periods of increasing glaciation were marked by falling sea levels everywhere, leaving new land surfaces exposed. Inter-glacial melting of ice produced the opposite effect, large transgressions of the sea over what had formerly been dry land. As the differences sometimes amounted to hundreds of feet the changes in land area were quite formidable. They were, however, very slow, a cycle from one inundation to the next covering thousands of years, as long as the time separating us from the great flood of ancient mythology. Even today the melting of the polar ice caps would suffice to submerge the majority of the world's great cities.

Land exposed by the retreat of the sea would be covered by flourishing vegetation for long periods and its remains, covered and preserved by marine deposits when eventually the sea crept back, provide some most valuable evidence of the nature of the vegetation and sometimes a record of the cycle of the climate. As regression was accompanied by active advance of the ice, conditions for colonization were less favourable than during inter-glacial periods when the sea advanced inland and it is from freshwater beds formed inter-glacially that the most complete knowledge of the flora is derived.

The worldwide changes of sea level during the Pleistocene were accentuated in the glaciated lands by the depression of the whole land level under the weight of the superincumbent ice and the rising of the land when the ice retreated. This isostatic change of level under pressure is significant evidence about the nature of the earth's crust, as we shall see later (p. 3802).

During the Tertiary Era there was a progressive cooling of the climate in northern Europe, but even at the end of the period the climate in many areas was at least as favourable to plant growth as at present, perhaps even more favourable. The pre-glacial flora of Europe was much richer in species than now and much of it possessed a very wide geographical extension, indicative of a high degree of climatic similarity over large parts of the northern hemisphere. The glacial period banished many of the Tertiary species permanently from the European flora and their gradual disappearance can be traced in successive deposits of plant remains. Some of the tenderer forms disappeared almost at once, other hardier species reappear more than once in interglacial deposits and survived until well on into the Pleistocene.

Some genera which vanished from Europe have survived in North America, others in Asia, giving rise to the denotation of 'American' and 'Asiatic' elements in the Tertiary flora of Europe, though these geographical labels imply no more than that they survive in those regions, not that they originated there. It has been claimed that they were unable to escape the effects of glaciation in Europe by southward migration beyond the periglacial zone, because of the transverse barriers of the Alps and the Mediterranean Sea. A very few, like the 'Asiatic' *Forsythia*, managed to survive in the extreme south-east of the continent, but most eventually perished. Their place was taken by an 'arctic' element of species driven southwards by the increasing cold, which brought a new component to the flora of north and central Europe. At the close of the glacial period this arctic flora was the first to colonize the newly exposed land surfaces. The species had suitable tolerances and they were more or less on the spot so that they preceded the invasion of more temperate species which advanced northwards from more sheltered refuges on the continent of Europe.

As time went on, however, the first phase of tundra vegetation had to give way before the establishment of woodland trees, first of *Betula* and then *Pinus*. The steady improvement of the post-glacial climate encouraged the advance of broadleaved trees which gradually replaced *Pinus* all over the southern part of the formerly glaciated area and led, at the period of climatic optimum, to the predominance of a rich mixed forest over a large part of Europe, while *Pinus* retained its hold only over the northern lands. Indeed, at this time the mixed forest extended much further north than broad-leaved forest does now, for a subsequent deterioration of climate has caused its retreat and *Pinus* regained some of the area it had lost.

Broadly speaking there were thus three main phases of postglacial climate. First a period of slowly increasing warmth, then a peak period called the 'climatic optimum' and lastly a deterioration towards a colder and wetter climate, with corresponding shifts in the character of the vegetation and the distribution of its constituent species.

All these post-glacial migrations only concerned a limited flora of species which had managed to survive somewhere in northern Europe. The impoverishment of the European flora has been permanent, much of the rich Tertiary flora had gone for good. We may ask why these genera and species have not returned, seeing that many of them survived in Asia? The question cannot be answered with certainty, and any answer would be complex and would involve many factors about which we know very little, but two things cannot be overlooked. One is the widespread influence of man in the agricultural development of Europe, involving the destruction of many habitats which might have served migrating plants. Another is the slowness of migration and the great distance to be travelled. We shall return later to this matter (p. 3859), but we must remember that the few thousand years which have elapsed since the return of reasonably temperate conditions to central and northern Europe are only a moment of geological time. A species can be rapidly extinguished, but its re-establishment by its own efforts might stretch over a hundred thousand years.

North America presents a different story to that of glaciated Europe. Here there were no effective barriers to southward migration, since the main mountain systems run north and south, not transversely. There was also a much greater area of unglaciated land to the south which provided a refuge for warm-temperate species. Consequently there was no large-scale impoverishment of the flora as in Europe and the re-advance northwards, when it came, was the mass advance of a whole flora on a wide front.

Another feature in North America was that of the drainage structure of the country; the retreating ice left behind it enormous flooded areas, of which the Great Lakes are only a shrunken remnant. By the time these floods had subsided enough to allow large-scale colonization by vegetation the climate had already considerably ameliorated and there was no opportunity for an extensive tundra phase. In some places no more than fifty miles separated the ice margin from established forest. Of course, over so large an expanse, climatic and topographical differences had their effects in differentiating the type of vegetation which established itself, so that the north-eastern states in particular still present a mosaic of forest types about the characters and boundaries of which there is some difference of opinion.

The western states escaped glaciation almost entirely, as the limit of maximum advance of the ice came little south of the Canadian border. There was some local glaciation in the mountains and naturally a deterioration of climate, but no ice-sheet. There was, however, the same southward invasion of northern species which we noted in Europe and the sub-tropical rain forests finally retreated into southern Mexico, leaving large areas permanently occupied by what had been in Tertiary times an Alaskan flora. This was a shift of population which had begun in the Pliocene or even before that and in southern California, about 1500 miles from the ice sheet, one might almost say that there was no Pleistocene, for assemblages of Pliocene fossil plants can be matched with associations of living plants still existing in the area. There seems to have been a prevalent continuity of conditions, though some large areas have been affected by post-glacial desiccation and the incoming of a semi-desert flora from the south.

THE HISTORY OF THE CONTINENTS

Turning now to the second field of geological inquiry which, as mentioned on p. 3795, deeply affects our ideas of plant distribution, we move much further back in time, to the epoch when the race of Angiosperms was beginning the dispersal which was eventually to cover the world. That the Angiosperms came from some common centre in which they had evolved has long been an article of faith among botanists, though no one has been able to show precisely where their Garden of Eden lay or even, within wide limits, when they first appeared. Their ancestry is dubious and how far back in time it reached and what the earliest Angiosperms may have been like are still unanswerable questions. They appear in the geological record, like Minerva from the head of Jupiter, full grown, and if any record of their evolution exists it must lie in some land which is as yet geologically unexplored. There are a few relics of what may have been Angiosperms in the Upper Jurassic, some bits and pieces which have an angiospermic look, as we understand that at the present day, but are quite uncertain. In the lower Cretaceous, however, they are recognizably present and by the end of the Cretaceous period they were already abundant and, what is more, their remains can be assigned to existing families or even genera. In a fraction of one geological period, the Cretaceous, in the few million years between the Aptian and the Cenomanian, the Angiosperms had advanced far towards dominance in the world's flora and had assumed a structural and taxonomic character which has not fundamentally changed throughout the 50–100 million years of subsequent time. From the evolutionary point of view this is a hard doctrine to accept. It is much more probable that when they appeared in the Cretaceous they had a long but hidden history behind them and that they had already advanced far towards their present character.

Hidden history certainly, but hidden where? The question we must ask is whether there are any indications traceable in present distributions and in the history of migrations which can provide clues to the direction in which we should look for a solution to the mystery. Theories have not been lacking and the Arctic, the Antarctic and equatorial Africa have all been advocated as possible nurseries of the Angiosperms. The first two suggestions obviously predicate great climatic changes having occurred since Mesozoic times and for the truth of this there is fossil evidence, abundant from Greenland though still scanty from Antarctica.

It is not, however, the evolution of the Angiosperms which is our present concern, rather it is the discovery of their true birthplace, for that must necessarily be the foundation for whatever views are to be held about the dispersal of these plants over the world. Does the history of the present areas of distribution of families and genera or even of species point northwards or southwards or towards some central area? Some, perhaps many, taxonomic units are of too recent an origin to help us in our search, but if we consider their taxonomic affinities may we not get some indication of *where* they have come from while investigating from *what* they have come?

We have been presented with the concept of the evolutionary track through space and time of a genealogy of plants, migrating geographically, differentiating and evolving as they went and throwing off or leaving behind them specialized offshoots which mark the line of their passage to the understanding eye. Such is the *genorheitron* of Lam and Croizat, which may prove in practice to be a most useful gateway to enlightenment.

Obviously such considerations involve decisions about taxonomic relationships which may be questioned and about which different opinions may be held. Unanimity on all details is not to be expected, and it can only be by a massive consensus of evidence that the conviction of truth can be attained.

We are also involved here with a general view of the progress of evolution among plants which, though acceptable to many, is a matter of argument. It has been called *differentiation* and has been expounded by Guppy. In their homeland the earliest Angiosperms are supposed to have enjoyed a long period of geologically tranquil and widely uniform conditions in which the plants lived under little environmental pressure and mutation had an open field. Some such concept is needed if we are to account for their apparently rapid evolution. The first result of these early mutations was to create the proto-types of the chief families. Their mutations in turn segregated the prototypes of new genera, which at length differentiated into constituent species, the last comers in the evolutionary story.

Uniform environmental conditions do not imply uniformity in vegetation. For example, the insectivores of swampy, peaty ground; *Drosera, Utricularia, Aldrovanda, Pinguicula, Dionaea, Cephalotus* are all quite different in their mode of adaptation. Rapid differentiation among Angiosperms during the early phase was probably encouraged by reduced competition between them because they were still relatively thin on the ground.

This must have been a very prolonged process and for long after migration began families and genera may have been represented by their prototypes alone.

The evidence is that successful mutations are rare and many have won-dered whether there has really been time and opportunity enough for the evolution of the multitudinous forms of the Angiosperms. A simple calcula-tion shows that this doubt is baseless.

If we accept a period of 140 million years from the beginning of Cretaceous time, we divide this by an average generation-time. A generation, from seed to seed, varies from 1 year in annuals to about 30 years in large forest trees. Hutchinson divides the families of Lignosae and Herbaceae in the proportion of 54 to 57, so if we take an average generation-time as 15 years this will probably be a high if not an extreme figure. At the rate of one 'good' mutation per hundred generations, that is one every 1500 years, there would have been time for 94,000 mutations in any genetic line. For annuals alone the corre-sponding figure would be nearly 1·5 million.

Steenis has emphasized the teratological nature of the characters distin-guishing many isolated taxa of tropical forests. Given a low level of selection pressure these could become the doors to supra-specific variations. 'Hopeful monsters'.

Subsequent differentiation into smaller units is supposed to have been connected with the variety of new conditions encountered in the course of migration, which provided niches uniquely suitable to smaller variants, which thus survived through ecological isolation. Were it not for such ecological con-siderations, were conditions everywhere uniform, it would be easy to draw the logical conclusion that the number of subdivisions of any taxonomic group and the area which it occupies would both be simple functions of its age. This thesis has indeed been advocated as a general principle of distribution (see p. 3841), but however attractive as a generalization its practical application is attended by great difficulty.

Whether the process of progressive differentiation is peculiar to the conditions under which the Angiosperms began their history or whether it is applicable generally to yet older groups of plants it is impossible to say. Generic and specific differences are recognized among older fossil plants and names given accordingly, but we cannot be sure that these distinctions correspond to those which we apply to living plants.

In summary, we may say that two sets of conditions conducive to rapid differentiation have been operative at different periods in the history of the Angiosperms.

1. Early phase. Widespread uniformity of environment and minimal competition, so that selection pressure is low and the possibilities of isolation are high.
2. Later phase. Increased competition, but a highly diversified topography with many relatively small, separated and often specialized habitats which also lend themselves to spatial isolation.

If we may assume that recognizable Angiosperms were in existence at the close of the Jurassic Period then the history of their early development and their early migration lies within the Cretaceous Period, since the earliest Tertiary beds show them to be already established as dominant in the vegetation and with many genera which are still living.

Before it becomes possible to think of lines of migration and dispersal we must try to envisage what the world was like into which the Angiosperms were born, some 160 million years ago. That there were many differences in the distribution of land and water is plainly shown by the vast marine deposits of Cretaceous origin which are now elevated to dry land and similar subsidences and elevations have altered land outlines many times since then. But were such changes only local incidents, not radically affecting the main continental land masses, or were there more fundamental differences? It is becoming continually more evident that there were.

The outer crust of the earth is very thin. The earth has been compared to a soft-boiled egg, the shell representing the solid crust, but it would really be more correct to regard the crust as represented by the membranous skin of the egg, relative to the bulk of the planet. The crust has been called the *sial*. It is supposed to represent an accumulation of the lighter elements, among which Silicon and Aluminium are conspicuous in amount. Below this lies the mantle or *sima* of a thickness about half the radius of the planet. The mantle is formed of denser materials, with Silicon and Magnesium prominent. It is solid but plastic and very hot. There is evidence that a thin layer of it, below the crust may be semi-fluid. Within the mantle is the core or *nife* (Nickel and Iron), white-hot and fluid though there may be a central part of the core which is solid. All this information is deduced from the behaviour of earthquake waves in their passage through the earth.

Below the continents the crust thickness is greatest, from 30 to 70 km, but under the great oceans its thickness may be no more than 5 km a basaltic ground-layer which also underlies the continents. On the sea-bottom it is only slightly covered by sediment.

The exposed land-surface of each continental mass rises from a broad 'continental shelf' which slopes gently outwards from each coast to a maximum depth of about 500 fathoms. Beyond this the slope suddenly drops to great depths, the oceanic abyss of 2000 fathoms or more. The continental shelf thus constitutes the true continental mass, the exposed coast line being, geologically speaking, only temporary and mutable.

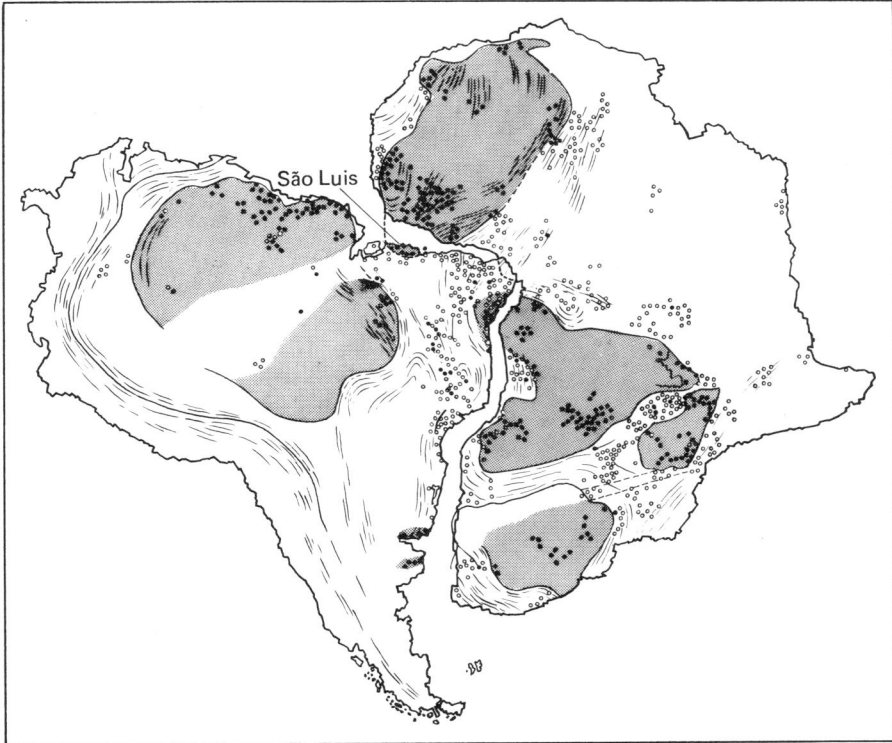

FIG. 4.64. Tentative matching of the geological structure of Africa and South America when in contact some 200 million years ago. The dark areas with black dots are the ancient continental nuclei or 'cratons'. The white dots are areas of younger Palaeozoic rocks. Portions of African cratons appear as if stranded on the coast of Brazil. (*After Patrick M. Hurley, from* 'The Scientific American', *by permission.*)

Included in each continental mass there is a very ancient nucleus of rocks dating back between 2000 and 3000 million years (a thousand million years is conveniently abbreviated to one aeon). Sometimes, as in Africa, there is more than one nuclear mass. These nuclei, or *cratons*, are of igneous origin and are chiefly granitic, that is acidic rocks rich in quartz. The underlying platform is basaltic, a basic rock. This may extend downwards to the plane of separation between crust and mantle, but at a depth of 100–200 km the study of seismic waves has shown the existence of a layer, the *asthenosphere*, which is semi-fluid and on which the crust may be said to float. In this layer there appear to be large-scale convection currents caused by loss of heat upwards, which form large, flat convection cells with rising currents, lateral movements and compensating lines of sinkage. This layer is the source of the fluid magna from which volcanic rocks are formed.

It was a dogma of classical geology that continents and oceans were permanent, that no deep ocean bottom had ever been dry land and no continent had ever been deep ocean. This might be compared to the dogma of the fixity of species in biology. Both concepts have had to be modified by the recognition of the action of developmental or evolutionary processes over immense periods of time.

The primaeval cratons of the continents were subject to rupture, folding and erosion during their early existence and they show the structure of ancient mountain belts. Around them and partly over them in their old troughs have accumulated the sedimentary rocks, the oldest being only about 600 million years old. There is thus a wide gap in age between the two types of rock and this can be measured by various physico-chemical tests. The immense thickness of sedimentary rocks which has accumulated on the continents has, however, no parallel in the oceans. During the whole of geological history only insignificant sediments have been formed there. In the deepest depressions there may be only a few inches of the 'red clay' over the basalt foundation. This certainly seems inconsistent with the belief in their geological permanence. Either the ocean floor is really young, or it has been swept clean in the last few million years.

The idea that the continents may have shifted their positions is an old one and had occurred to many people who noted the striking way in which the coasts of America and of Europe/Africa seem to fit together. A coherent theory of continental drift was first put forward by Alfred Wegener in 1912. It was welcomed by many biologists who saw in it a possible solution of many difficulties in the inter-continental distribution of plants and animals. Geologists, however, were profoundly sceptical, one of their clinching objections being that there was no known force which could have shifted continents. In spite of this some geologists felt that coincidences in the distribution of rocks and structural features on both sides of the South Atlantic demanded explanation. Controversy flourished.

Wegener had postulated a single primaeval continental mass, Pangaea, partially covered by shallow seas which were areas of sedimentation, Pangaea had been ruptured and split apart, the segments gradually separating. This was combined with changes of the earth's axis of rotation so that the poles had 'wandered' in an irregular fashion, thus accounting for great apparent changes of climate from one period to another. Specially significant in his argument was the coincidence of a Permian glacial period in South Africa, Australia and India, all of which were, according to his theory, closely grouped in a position near the South Pole of that epoch. When the polar axis was very oblique to its present position the equator would also have been obliquely inclined, carrying the possibility of a tropical climate far to the north and south of the present tropical limits. This also coincided to some extent with fossil evidence.

Later, in 1937, a South African Geologist du Toit, modified Wegener's original hypothesis of Pangaea by the suggestion that there were two primaeval land masses, a northern and a southern, the former he called Laurasia and the latter Gondwanaland. The latter was a name that preceded Wegener. The coincidence of geological deposits and plant fossils in the Permian and Trias of South Africa, India and Australia had long before suggested the existence of a vanished continent in the Indian Ocean to which the name had been given. Between the two chief masses lay the Tethys Sea, that ancient ocean to which we have often referred in the Palaeobotany section of Volume 3.

Two great discoveries have altered the whole position regarding the theory of continental movement and brought it well into the realm of practical science. One of these is the discovery of residual magnetism in rocks of all ages, the other is the magnetic analysis of the ocean floor.

Before discussing the results of these investigations we should say that recent analysis of the fit between the two sides of the Atlantic Ocean, taking the edge of the continental shelves, not the existing coastline, as the datum and using computer methods, has shown that the fit is much closer than had been supposed, hardly a degree out at any point, which practically rules out coincidence. At least in this area some union there must have been.

Palaeomagnetism depends on the discovery that the Iron or magnetite components of ancient rocks acquired magnetic polarity from the earth's magnetic fields at the time the rocks were formed and that they have retained this polarity ever since. Comparisons of a succession of rocks thus enable the investigator to plot the position of the magnetic poles at different periods. Such comparisons reveal remarkable apparent shifts of the poles, but the significant thing is that these shifts and the apparent wandering of the poles are in almost all cases different for different continents. As there cannot have been a multiplicity of poles the only possible conclusion is that it is not the poles which shifted but the continents in their relation to the poles.

The curves of polar 'wandering' for South America and Africa coincide for a period of 100 million years from the Upper Palaeozoic onwards, strongly indicating that over that period the two continents were either united or closely associated together. Another line of evidence in this respect comes from the off-shore deposits on the continental shelf of West Africa. Bores drilled through these show that they are all geologically young and that there is none earlier than the middle of the Mesozoic. It looks therefore as if the final separation of the two continents had occurred not earlier than the Cretaceous Period, though it may have begun in North Africa as early as the Trias. The separation of East Africa from adjacent lands began earlier still, in the Permian.

Localization of the south magnetic pole deduced from magnetic indications in Early Tertiary rocks in Australia, India and Antarctica are fairly concordant. Those from Late Tertiary rocks, on the other hand, are concordant from Europe and North America, but the indications from India and Australia are widely divergent. To bring them into concord with the other indications

India would have to have been in 34° south latitude and rotated 25° clockwise. Australia would have to have been close to Antarctica. The change to their present positions would have had to occur in *Middle Tertiary* times. Du Toit had previously suggested the same movement and the same timing on purely geological grounds. Continental movements are no affair of the remote past, but affect the period of the full tide of Angiosperm migration.

We are here in the realm of factual evidence, not theory, but it is evidence which supports the theory of continental movement. It is further worth noting that the evidence shows that much of this movement took place in comparatively recent times, within the lifespan of the Angiosperms, and must therefore have had a great effect on their routes of dispersal.

The second great discovery which concerns continental movement came from the survey, especially the magnetic survey, of the ocean bed. First, the sediments on the sea floor at the present rate of deposition would not extend the process of sedimentation further back than Cretaceous times. The sea floor seems to be relatively young, there is no indication of a history going back for aeons.

A further discovery is the existence of a mid-oceanic mountain ridge which extends the whole length of the Atlantic from north to south, mid-way between the continents. This ridge appears to be continuous with a southern ridge encircling Africa and running up the middle of the Indian Ocean. A branch from this runs south of Australia and across the Pacific, then turns northwards towards the west coast of North America. The mid-ocean ridge is crossed by many lines of fracture and shows considerable lateral movement associated with these fractures. In 1962 Hess and Dietz in the United States suggested that the oceanic ridge was being torn apart rather than compressed, as is the case with land mountains, and that this was due to the upwelling of fluid magma along the line of the ridge, the upward flow of a convection current in the asthenosphere forming a ridge of volcanic rock which spread laterally outwards from the ridge, pushing earlier outpourings before it, cooling and solidifying as it went. Thus the youth and presumed expansion of the ocean floor could be accounted for. Proof of this came from magnetic surveys. It had been shown from the residual magnetism in land deposits that the earth's magnetic field had undergone periodic reversals of polarity and from careful assessment of the ages of successive deposits a period of 3·6 million years between reversals was deduced.

Later the work of several American investigators, Mason, Raff, Vine and Wilson, brought to light the extraordinary fact that the sea floor on each side of a mid-ocean ridge was magnetized in a striped pattern showing successive reversals and that the pattern was symmetrical on each side of the ridge. As the period of reversal was already known, these strips could be dated and the rate of spread measured. The picture is that of hot fluid material rising along the line of the ridge, becoming magnetized and then spreading outwards in both directions. We need not suppose a quiet and continuous outflow; it is more likely to have burst up in periodic and catastrophic outbursts associated with the fracture lines across the ridge. The Atlantic, Pacific and Indian

Ocean ridges all show similar patterns. The dated history of these magnetic reversals in the sea floor goes back to the Cretaceous Period, a matter of 80 million years. Once again the Cretaceous appears as a significant epoch in the history of the continents.

The story as it now stands is something like this. The convection currents in the asthenosphere are seen as the motive force. If a rising zone is established under a continental mass, it will split apart and the separated portions will be gradually forced apart as the sea floor expands. A sinking zone, on the contrary draws the surface together towards itself. A land mass will therefore tend to move away from a rising zone and towards a sinking zone. A sinking zone under the ocean forms a deep trench, with chains of volcanic islands, as in the

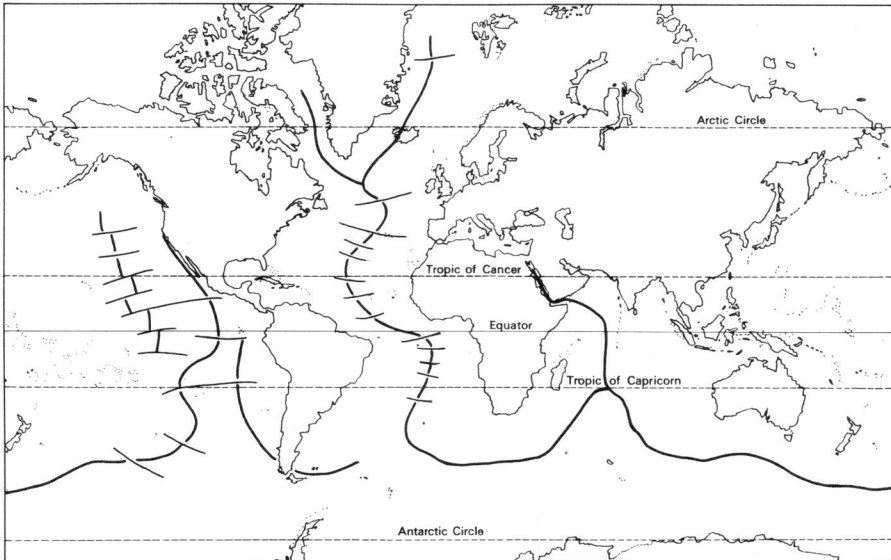

FIG. 4.65. The line of the mid-oceanic ridges, showing the transverse faulting in the Atlantic and Pacific Oceans. (*After S. K. Ransom.*)

western Pacific, off the east coast of Asia. When an ocean floor expands towards a continent which is already advancing, it descends, passes under the continental edge providing a vast heating and remelting of rocks and causing the upheaval of a mountain chain with volcanoes. The classical and best example is the west coast of the Americas. The Andes, indeed, are of such recent origin that the 'collision', so to speak, of the two forces does not date much further back than the Pliocene.

Turning now to the southern hemisphere, the evidence seems to point to a Palaezoic Gondwanaland in which South America with South Africa, Australia, Antarctica, Madagascar and the southern part of the Indian peninsula formed a close huddle. The western part of Antarctica and eastern Australia are marked by two large and very ancient synclines or troughs due to pre-Cambrian folding and filled by sediments dating from late pre-Cambrian up to Permian times and a later succession of sediments common to the two continents. Although the outlines of the continental shelves of these masses can be fitted closely together, the grouping is still hypothetical though studies of correlations in the boundaries of deposits and comparison of fossil plants between Antarctica and the adjacent lands may soon give definite evidence.

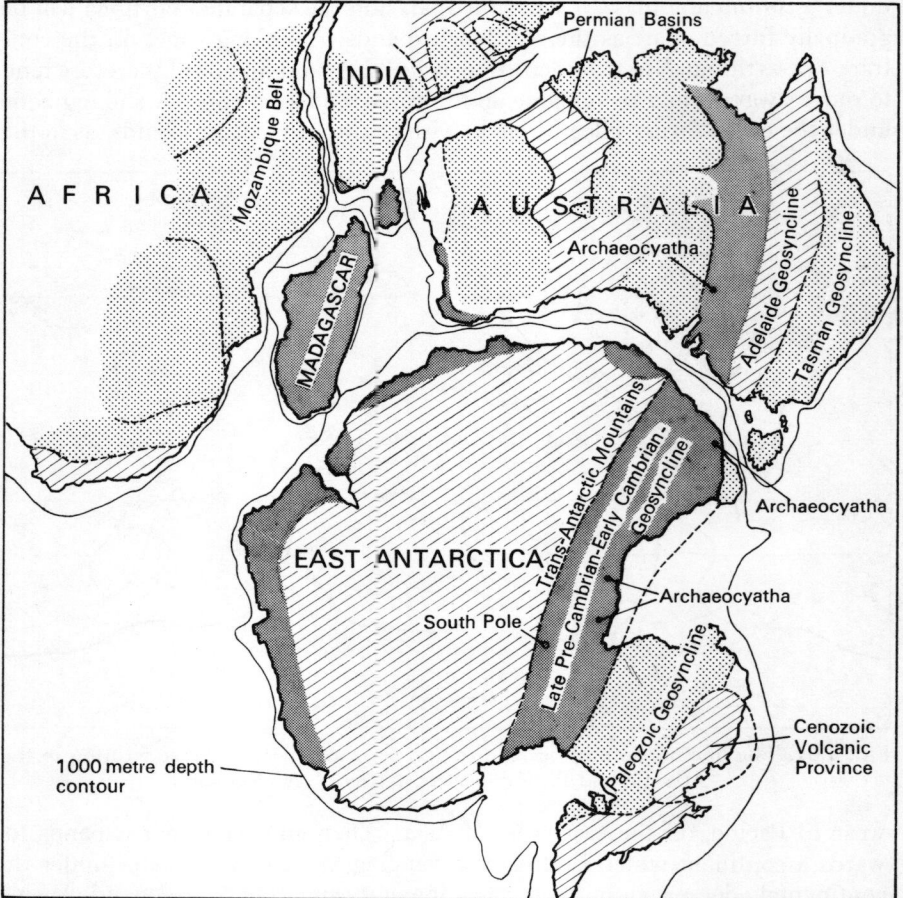

FIG. 4.66. Tentative reconstruction of Mesozoic Gondwanaland. The close fit is shown by the lines of the 1000 m depth contours. The Pre-Cambrian and Palaeozoic beds in eastern Australia match those in Antarctica and the Permian of north western Australia is matched by detached basins in India. (*After P. M. Hurley, from 'The Scientific American', by permission.*)

There is enough to show, however, that Gondwanaland was no hypothetical vanished continent but an actual grouping of existing land masses in what is now the basin of the Indian Ocean, which began to expand as a sea floor somewhere after the end of the Permian Period, that is to say that Gondwanaland began to break up after or at the end of the Palaeozoic Era.

Australia was moved eastwards, Antarctica southwards and India northwards, where it cut across the ancient Tethys Sea and, meeting the Asiatic coast ploughed under it and caused the great folding movement which raised the Himalayas.

Although this is a very simplified account, omitting a great deal which happened and without discussing the objections which can be raised, there is enough to show us that in early Mesozoic times Antarctica was in temperate latitudes and closely associated with other great land masses. From the point of view of plant dispersal this is the most important point, as we shall see when we come to discuss migration and dispersal (p. 3857).

With so much converging evidence in favour of continental movements, a mere denial will no longer serve to oppose it. For example, the claim made by Beloussov that the continents are so deeply rooted that they could not have been moved, is negatived by the evidence that the glaciated areas of the Pleistocene sank beneath the weight of the super-incumbent ice and rose again when it was removed. This is entirely inconsistent with a view of continental bases 1000 km deep, but is consistent with the idea of a light-weight crust 'floating' on a semi-fluid asthenosphere.

If Pangaea ever existed it was a very long time ago, in the Pre-Cambrian Era. Studies by Creer at Newcastle by manoeuvring models of continents to bring their lines of palaeomagnetic drift into coincidence, show that the primaeval mass must have already divided into Laurasia and Gondwanaland some 450 million years ago, in the early Palaeozoic and that the beginning of the main drift was the separation of North America from Europe about 250 million years ago, i.e. in the Coal Measure Period.

But was Pangaea the beginning? It seems unlikely, for there is an enormous stretch of time lying beyond the age of even the oldest known rocks. Furthermore, the oldest mountain chains, for example, the Urals and the Appalachians are interior to the continents and it has been suggested that they date back to an earlier coming together of continents to form the Pangaea super-continent. If we look back far enough we glimpse a vision of earlier continents and earlier oceans performing even stranger gyrations than those of later times. The continental shift we have been speaking about may be only the latest of a series. Australia is an ancient continent and Runcorn's observations of palaeomagnetism in Australia indicate not simply one shift since Pre-Cambrian times but an extraordinary sequence of positions, carrying the continent right round the world and back to near its original position and varying in latitude from the equator to the South Pole. This is a staggering story, and how it fits into the general scheme of continental drift we simply do not know.

The present continental surface occupies only about 30 per cent of the surface of the earth. Creer has made the suggestion that the present land surface is the remains of the original sialic skin which formed the first crust, but he points out that for the present continents to cover the whole surface of the globe postulates a globe only about half (0·55) the present diameter. Subsequent expansion would crack the crust and separate the pieces. When water condensed on the earth it would fill up the interspaces. Expansion thus becomes the motive force of continental movement during the earliest times. Whether such expansion is a physical possibility cannot be definitely asserted though Creer suggests various possibilities involving cosmological theories. What can be said categorically is that no part of the existing land-surface is anything like old enough to have been part of the earth's first crust. If any such still exists it is buried very deep.

A recent view has modified Creer's hypothesis by regarding the earth's primaeval surface as covered, not by the existing continental areas joined together, but by a mosaic of very large 'shields' of which the continents were only parts. Each shield is a unit, which has been moving, rupturing and colliding from the beginning. Continental drift, as now understood, would thus be seen only as a phase of a continuing process.

Where two shields meet, as on the western coast of the United States the edges may be forced downwards and create lines of subterranean tension which can be traced in the earthquake zones of the present day. Direct measurements on the coast of California show that the two shields there in contact are moving relatively to one another at the rate of 6 cm a year or nearly 30 cm in four years.

The postulate of shields rather than continents as the geomorphological units also harmonizes with the evidence that some continents, notably South America and Africa, are not monolithic masses but are compounded of pieces which came together at a remote date, a view which goes back as far as von Ihering in 1893.

It may be objected that however interesting continental movement may be it is much too ancient history to have affected present day distribution of plants, though it may be relevant to the distribution of ancient fossil floras. This is not so, however. To begin with, we do not know how old the Angiosperms as a class really are. Some workers would push back their origin into the Palaeozoic. In any case their earliest progenitors would form such an insignificant percentage of the contemporary flora that their discovery is subject to long odds. In this they resemble the position of mammals in the Jurassic. Nor is it at all certain that when found we would recognize them for what they were. For example, the *Williamsonia* 'flower' has an apparent resemblance to an angiospermic flower, but who can say whether it had any relationship to the Angiosperms. Probably not but we do not know. Among the Pteridosperms also there are approaches to floral organization, but could this be simply parallelism or were they progenitors of the higher seed plants? Far more extensive information is needed before we can say positively when or where the Angiosperms arose.

What can be said with more assurance is that it was during the Cretaceous period that the Angiosperms performed their great migrations and assumed a position of universal importance in the world's flora. The separation of continents seems to have begun before this, probably in Permian times, but the movement was episodic rather than continuous and was far from complete in the Cretaceous, if indeed it is yet complete. Most importantly, South America and South Africa were still in close contact or proximity at that time and the

India–Australia complex had not yet broken up. Even if separation had taken place the intervening gaps must still have been narrow so that inter-continental migration was relatively easy. When we come to consider routes of migration it is the ancient geography, not that of the present day, which we have to bear in mind.

We have seen how doubtful is the 'when' of angiospermic origin. Is there any better indication of the 'where'? It must be confessed that this is still a subject of theory, but it is important to form some opinion on the subject if we are to try to interpret plant migration.

Three views have been promulgated which seem to include all the possibilities. All three postulate climatic changes on a great scale. The first is Holarctic, deriving from fossil evidence of a warm temperate climate in Greenland in late Jurassic and early Cretaceous time. This implies a general southward migration of Angiosperms. The second is inter-tropical, but envisages a vastly extended tropical belt, reaching from 45° north to 45° south latitude.* This view would include the region of south-east Asia favoured by Takhtajian on the ground of the concentration of 'primitive' genera in that area.

The third theory claims that the general direction of migration was northwards and that the Angiosperms were evolved in the southern hemisphere and probably in a temperate Antarctica.

None of these theories really throws any light on the antecedents of the Angiosperms or solves the puzzle of their astonishingly rapid rise to prominence in apparently fully developed taxonomic forms. Each points a finger at a relatively unexplored part of the world and bids us look there for what we want to find. Either the rise of the Angiosperms was in fact sudden or else their early phases were passed through on some unexplored territory or on some mountain terrain where there was little or no chance of fossilization. Both views have sympathizers. In the absence of evidence one can put it no higher than that.

One advocate of the 'sudden rise' theory is Whitehouse at Cambridge, who has suggested that the appearances of multiple-allele incompatibility between pollen and the diploid tissues surrounding the ovule could at one step put a stop to gymnospermic pollination and would practically inaugurate angiospermy by strongly favouring mutations which excluded self-pollen but allowed pollination by cross-pollen. There is genetic evidence for the existence of this incompatibility in early times, i.e. before the separation of Scrophulariaceae and Solanaceae, and Whitehouse suggests that it originated, once and for all, in a single species of Gymnosperm, which already (like *Gnetum*) had angiospermic vegetative characters, in which case its discovery or recognition as a fossil is most unlikely, though descendants of the prototype, still primitively angiospermic, might still be living.

The Arctic theory rests chiefly on the fossil evidence of a temperate to warm climate having prevailed in the north at the period of the early advance

---

* Such a wide swing of the sun's apparent path would involve an almost impossibly large change in the inclination of the polar axis.

of the Angiosperms and that the circle of land masses surrounding the Arctic (at the present day) offers excellent facilities for inter-continental dispersal.

The tropical theory is chiefly associated with Axelrod, who bases his views on the percentage of Angiosperm fossils (in many cases carefully and personally revised) in deposits of different ages arranged according to latitude. The results are expressed in a series of curves drawn for different geographical regions. See Fig. 4.67. The period covered is the early Cretaceous, Neocomian, Aptian and Albian stages, in which a considerable number of plant deposits are available for comparison, mostly reliably dated and distributed over three regions in the northern hemisphere and two in the southern. Looking at the figures given for the percentage of Angiosperms present in each assemblage of fossils, it will be seen that very few occur in the Neocomian and those only at relatively low latitudes, though none actually tropical. The percentages increase rapidly in beds from low latitudes throughout the Aptian and Albian stages, reaching 85 per cent in some places by the end of the latter stage. The sudden spreading of the Angiosperms is borne out by such examples as a change from 1 per cent at Lakota to 45 per cent at Fernson, in North America, both within the Aptian stage.

Localities further north, including some in the Arctic or sub-arctic, were apparently slower to become populated with Angiosperms. At latitudes above 60°N there are no Angiosperm floras at all until the beginning of the Albian stage, though by the end of that stage percentages of from 40–65 have been recorded from 67°N, equalling in this respect beds 30° further south. In the critical Aptian stage the figures are consistent in showing angiospermic floras south of 50°N and none north of that latitude.

These figures seem to negative an Arctic origin for the Angiosperms, but while they make a general northward migration very probable, there are local inconsistencies which are perhaps only to be expected, and it is fair to say that the results fall short of a proof of *tropical* origin. The corresponding figures for the southern hemisphere are concordant in the Australia–New Zealand region, but they are few and in South America far too few to support a generalization. They leave the question of general southward migration open.

Takhtajian, who also supports the theory of a tropical origin, has preferred South-east Asia as his choice of a region of origin, largely on account of the concentration there of a large number of species of the presumptive 'primitive' families of Angiosperms. As we shall see, however, when we discuss distributional areas, a local concentration of species is not always a sure indication of a centre of distribution.

Axelrod offers us a possible explanation of the sudden appearance of Angiosperms in the fossil record. His theory is that the early stages of their evolution took place in upland or mountain regions where there was small chance of their being preserved by fossilization. It is indeed notorious that, among earlier fossil floras, plants which grew in upland areas are poorly represented or, if at all, only by drifted fragments carried down by floods to the lowlands or out to sea.

The 'sudden appearance' Axelrod attributes to the descent of the Angio-

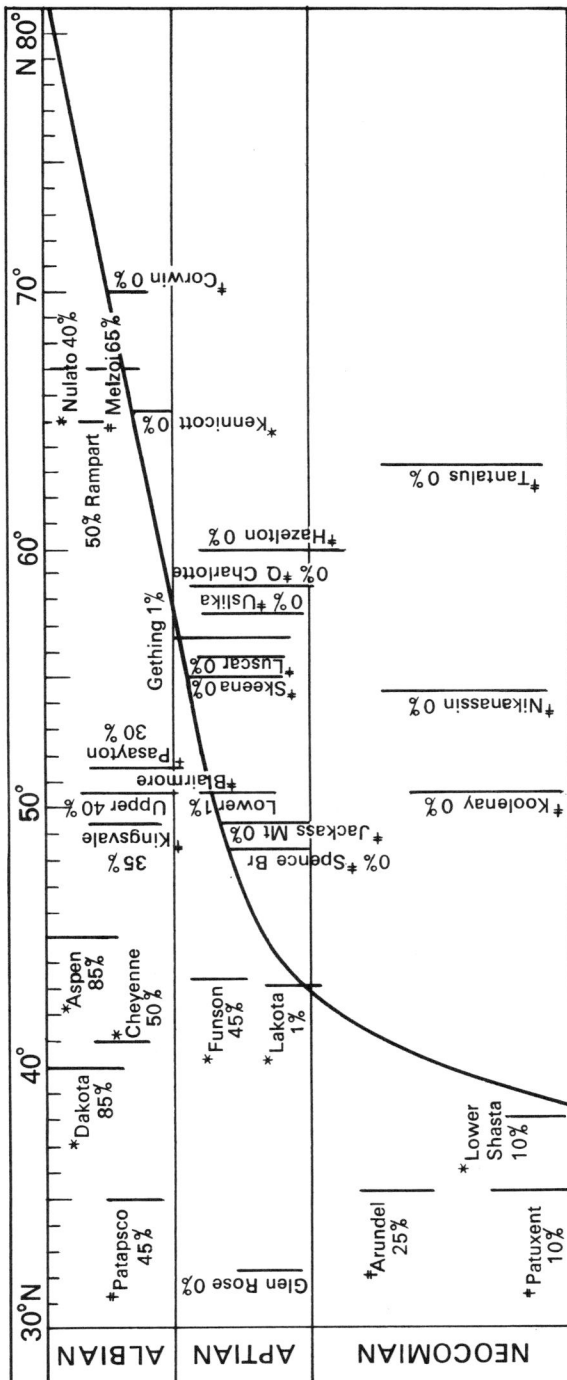

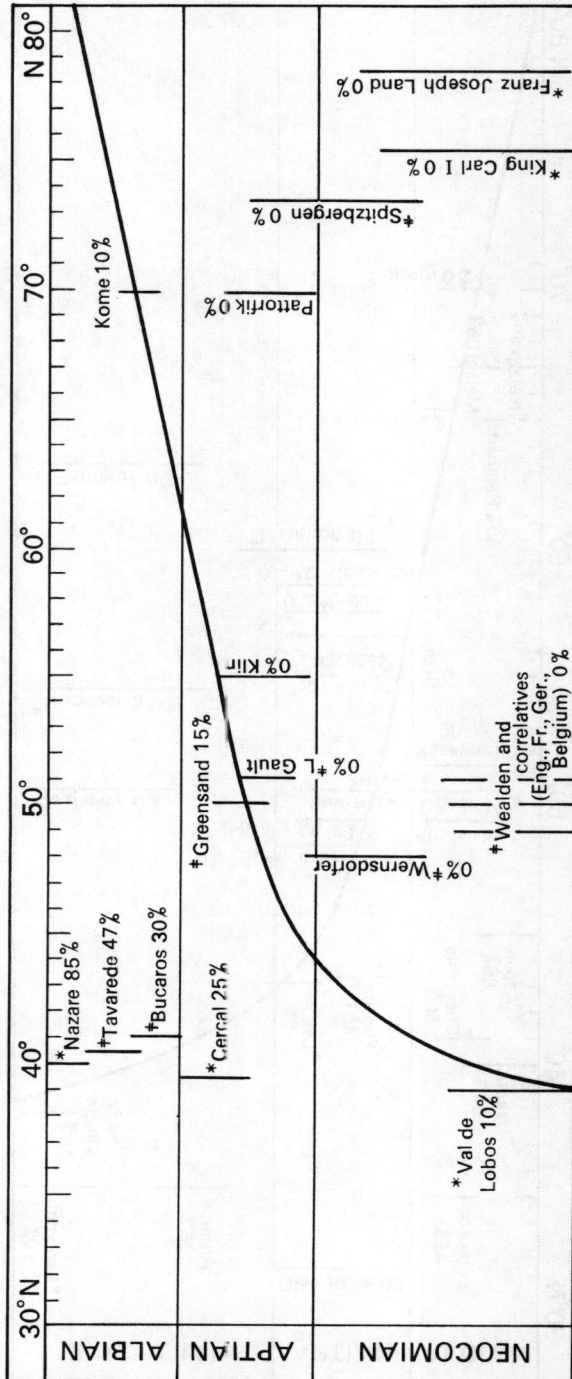

FIG. 4.67. Diagrams showing the percentages of Angiosperms in Early Cretaceous floras in North America and West Europe respectively. The curves show the progressively later arrival of the Angiosperms in higher northerly latitudes. (*After Axelrod.*)

sperms from their mountains, possibly in central Africa, to the lowlands where there were greater opportunities for wide dispersal and also better chances of preservation. What brought about this change of habit is not explained.

The supposedly sudden appearance of the Angiosperms has probably been exaggerated because of the comparatively recent character of the geological periods concerned, which we are apt to think of as short in comparison with the vast stretches of older time. The period between the first appearance of the Angiosperms and their actual achievement of dominance was, in fact, of the order of 25 million years, time enough for much normal evolution to occur. There have been other floristic revolutions in earlier times, as, for example, between the Lower and Upper Devonian or between the Middle Permian and the Trias, but these have not been called 'sudden' because we better appreciate the time factor among earlier deposits and because the changes can be associated either with stratigraphical or climatic changes. Climatic change may well have played a part in the history of the Angiosperms also. From the Jurassic through the Cretaceous periods climates seem to have been temperate and of widespread uniformity, but there was a slow warming-up going on, dating from the late Permian glaciation and progressing towards the tropical conditions of the Eocene in Europe and this amelioration of climate may well have affected the spreading of the flowering plants.

Takhtajian has also pointed out the biological advantage the Angiosperms possessed in their capacity for forming closed communities, a competitive advantage against the surviving Jurassic Gymnosperms. It also enabled them to advance and invade as communities rather than individually.

There remains the third theory, that of an Antarctic origin, the chief protagonist of which is Croizat, though Charles Darwin was convinced that radiating dispersal from the south was the only possibility that made sense of the inter-continental distribution of genera and species at the southern extremities of the continents.

Croizat has presented with an immense wealth of detail his interpretation of the tracks of global dispersal of families, genera and species of all the main Angiosperms which have attained inter-continental distribution. He insists that distributions must be considered on a worldwide basis and that local distributions taken out of their global context can only confuse. Considered in this way he maintains that at least 40 per cent of the major Angiosperm families show a well marked antarctic base. The migration tracks, however, bear no relation to modern Geography, but are the product of Cretaceous Geography with a close association of southern land-masses, especially with a continental aggregation lying in the present basin of the Indian Ocean, in other words, Gondwanaland, where Africa, India, Australia and Antarctica lay closely associated, with South America still near West Africa.

Moving northwards he postulates three 'Gates of Angiospermy', namely, the Magellanian gate in South America, the African gate, which he divides into two triangles, the Afro-antarctic triangle and an East African section, the Gondwanic triangle. Lastly there is what he calls the western Polynesian gate, which might be better called the Australasian gate. This also is divided into two centres, the Macquarian centre, which covers the present Antarctic islands lying to the south of Australia and New Zealand, and a northern centre, the Neocaledonian, including north-eastern Australia and a large area of islands to the north and east.

Croizat in 1952 still kept an open mind about continental drift. His hypothetical rearrangements of land-masses involve submergence and uplift rather than drift. They envisage larger continents and smaller seas than at present and they are different from the picture that geophysics draws today. But he insists that plant dispersal is a part of geological history and inseparable from

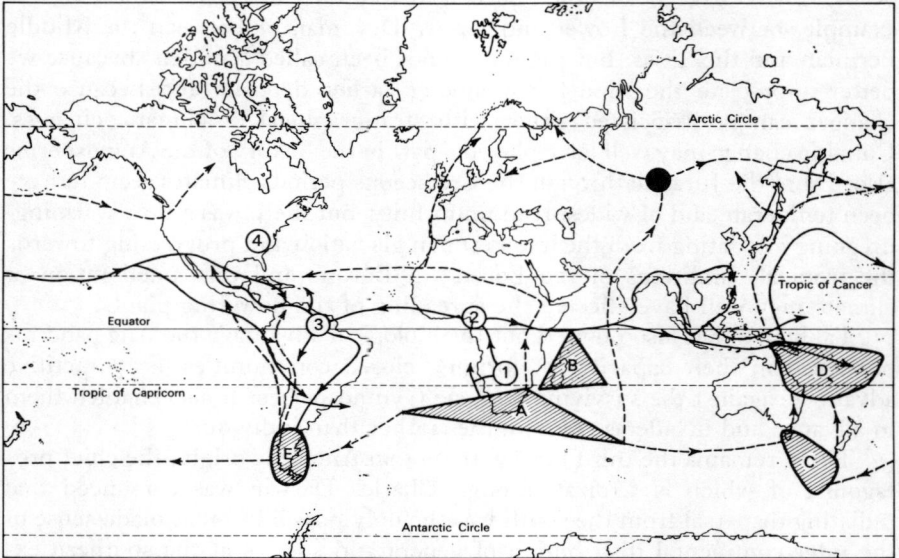

FIG 4.68. Croizat's interpretation of the 'gates' of the Angiosperms and of the main lines of migration. 'Warm' lines continuous, 'cold' lines broken. A: Afroantarctic Triangle and B: Gondwana Triangle (together making the African Gate). C: the Macquarian centre and D: the Neocaledonian centre (together making the Polynesian Gate). E: the Magellanian Gate. Important secondary centres are: 1. Kalaharian, 2. Nigerian, 3. Roraiman, 4. Ozark-Appalachian. The Altai node of redistribution shown in black. (*After Croizat*, 1952.)

it and that its evidences of geographical changes are irrefutable by any geophysical theory. He makes large claims and his argumentative style tends to grate on the reader, but his evidence is massive and deserves weighty consideration.

It is perhaps worth noticing that there is quite a number of important extra-tropical families which are confined to the southern hemisphere, for example, Proteaceae, Pittosporaceae, Cunoniaceae and Restionaceae, to mention only a selection; there are very few which are limited to the northern hemisphere,* of well-known families only Aceraceae and Juglandaceae appear

---

* This in spite of the fact that the North Temperate Zone has 1,152,500 square miles of land compared with only 49,500 in the South Temperate Zone.

to fall into this category, and both are small, together with another small family, the Diapensiaceae. The southern families do not seem to have penetrated to the north, for the leaf-fossils which have been identified with them in north temperate deposits are very dubious.

Certain dicotyledonous genera which are associated by many people with a cold or arctic distribution afford little real evidence of arctic origin. These genera, *Salix*, *Populus*, *Alnus* and *Acer* among others, are most familiar in their wide, cold-northern distribution, but they are all represented by species in the tropics or sub-tropics and some, such as *Salix*, range widely in the tropics and the southern hemisphere. Wood attributed to *Acer* has indeed been found in beds of apparently early Cretaceous age in Madagascar, and it has living species in the Philippines. Its occurrence fossil in Tertiary beds in the far north is not therefore conclusive evidence of its having originated there, any more than *Salix* or the others mentioned.

There was, as is well known, a separation of the northern and southern hemisphere floras at a much earlier period than that which we have been considering, namely, in Permo-Carboniferous times, when the Tethys Sea still formed a barrier between north and south. Indeed, it was the inter-continental distribution of these early plants, known as the *Glossopteris* Flora, which first gave rise to the concept of Gondwanaland. That land-mass must therefore have had an ancient history. The concept of it has changed somewhat with the consideration of continental drift, especially regarding the association with it of the Antarctic block. Somewhere in this enlarged land-mass it is not impossible to believe that the Angiosperms were evolved, if they did in fact evolve in any single locality. The contrary supposition that they had multiple and independent origins is very hard to believe in view of the many detailed and fundamental similarities which run through and serve to unite the class as a whole.

CHAPTER LVI

CLIMATE AND FLORA

It is a commonplace to say that climate, in the broad sense, is the most influential factor determining the growth of plants, but this is an ecologist's view, and the determination in question is chiefly effective at the species level and controls the physiognomy of the resultant vegetation.

From the distributional point of view, however, climate also has floristic effects, in as much as many of the higher taxonomic groupings, genera and families, show general preferences for one type of climate or another, and therefore we conclude that their geographical distribution is conformable to climatological control. While this is true on a large scale, it is difficult to apply it, as a principle, to details, since to every such generalization there are frequent exceptions in aberrant species which do not conform to the general pattern. It is therefore necessary to look critically at any claim that a particular distribution is determined by some specific climatological factor such as the line of a winter isotherm or the average annual rainfall. Such appearances may be genuine, but without proof of direct causality we must reserve opinion, for other, unsuspected factors may well be involved. Ecologically such factors are undoubtedly important and may be decisive for some species, but their relative importance in determining the distribution of a taxonomic group, genus or family, is another matter. From the floristic point of view edaphic factors are also chiefly important at the species level and are rarely determinative in the distribution of the larger groups.

Not only the dimensions of the chief climatic factors, but their incidence throughout the year is of the highest significance in characterizing a climate and, for a start, we should inquire how climates can best be classified in the way most directly helpful to the Plant Geographer. Are there, in fact, definable types of climate and if so what are they?

We can only deal in any detail with climates as they are now, but just as in regard to the geography of land-masses, we must emphasize that every climate has a history and that great changes have taken place during past eras, about which we have only some general ideas. The Pleistocene Ice Age must at once come to mind as the latest great revolution of this sort and our ideas about it are therefore slightly more precise than about earlier changes, but even here our arguments run rather from the plants to the climate rather than from the climate to the plants. Ice Ages are, however, rare events and the geological record shows only an extensive period in the far south in Permo-Carboniferous times and one in Pre-Cambrian times in Canada which may have reached as far as the west of Scotland.

All revolutions apart, there are certain astronomical facts which must have caused periodic changes of climate. These are, first the eccentricity of the earth's orbit round the sun, which is not constant but varies within a period of 92,000 years. This eccentricity implies that when the earth is at perihelion, the part of the orbit nearest to the sun, it receives more intense radiation than at aphelion when it is furthest from the sun. This is compensated to some extent by the fact that the perihelion sector of the orbit is shorter than the aphelion sector. The second part is the obliquity of the ecliptic, that is the angle between the plane of the earth's equator and the plane of the orbit, due to the inclination of the polar axis of rotation of the earth. This too varies, and is known to have varied between 21°39' and 24°36', being at present 23°27', with a complete period of 40,000 years. A third variable is due to the small circumnutation or 'wobble' of the axis of rotation, called the 'precession of the equinoxes' because it causes a change in the points on the orbit at which the equinoxes and the other cardinal points of the year, the solstices, occur. This has an effective period of 21,000 years, which means that if the spring equinox, for example, happens to occur at a particular point on the orbit, it will tend, with lapse of time to occur progressively earlier in the annual orbit and it will be 21,000 years before it comes round again to the same particular point.

The combined effects of these variables are exceedingly complex, but we may make a few simple deductions. First, these periods are all very short compared with geological time so that their combined effects must have operated repeatedly in geological history. Second, when the North Pole is inclined towards the sun there is a northern summer and the sun appears to rise higher in the sky. When the pole is inclined away from the sun we have a northern winter and the sun does not rise so high in our skies. The same applies, in reverse, to the South Pole and the southern hemisphere. When summer occurs in the perihelion sector of the orbit and winter in the aphelion, the difference between hot summer and cold winter will be greatest, the seasonal change will be maximal. The greater the angle of obliquity of the ecliptic the greater will be seasonal change. Similarly, if in the precession of the equinoxes the summer period falls at aphelion and the winter at perihelion, the difference between the two seasons will be minimal and this will be lessened still further when the angle of the ecliptic is lower.

The effect of the angle of obliquity of the ecliptic may be extrapolated both ways. If we imagine the angle to be zero, i.e. that the earth's axis was exactly at right-angles to the plane of the orbit, then each zone of latitude would (like the equator at present) have no seasonal changes, but the differences between the zones would be greatly exaggerated, the climate of each zone being perpetually the same. If, at the other extreme, the angle were increased to 54° the total radiation received per year would be the same at all points on the earth's surface but the seasonal differences would also be great at all points. We have no direct knowledge of such great changes of angle having happened in the past, but they may have happened, for we have palaeo-botanical evidence of the apparent absence of seasonal change at some epochs and of periods of an immensely wide uniformity of climate around a zone of latitude. It is possible that the changes of position of continental masses may have caused changes in the inclination of the earth's axis and have given rise to such profound climatic changes.

Lastly, a diminution in the eccentricity of the earth's elliptical orbit would have diminished the difference between the effects of perihelion and aphelion on the amount of solar radiation received by the earth at the respective positions, and vice versa. Although it would be theoretically possible to calculate from the astronomical data the climatic conditions in any given zone at any given date, in practice the labour would be too great. It has been done for north temperate latitudes by Milankovitch and others to cover the Pleistocene period, a maximum of one million years, but to go back beyond this involves an increasing margin of error and becomes impracticable. It is plain enough, however, that there have been large, cyclic and inevitable changes of climate throughout geological time, which must be borne in mind when the history of plant migrations is studied. Should we wish to use climatic change to elucidate any problem of migration, we must remember that the plants themselves are usually the best, if not the only, evidence for climatic changes and beware of becoming involved in a circular argument.

Climatic Zones

The term 'zone' is only strictly applicable to the classical zones of Geography, the Arctic, temperate, sub-tropical and tropical zones, for as soon as we begin to consider the vegetation of these zones we realize that they are not homogeneous but must be divided into regions with different plant coverings. The north temperate zone, for example, has forest, steppe and grassland regions, all within one zone but all quite different climatically. The sub-tropical zone has a large proportion of desert and semi-desert, but it has also extensive regions of savannah, sclerophyllous woodland and thornbush which again are climatically distinct. The tropical zone is not all rain forest, there are also the deciduous monsoon forests and tropical grasslands. Latitudinal zones, which are principally temperature zones, are therefore only partially effective as divisions in Plant Geography.

Schimper found a system satisfactory for his ecological purposes in combining the effects of temperature with rainfall. He delimited accordingly a series of climatic regions ranging from: tropical districts constantly moist; through tropical districts periodically dry, to the other extreme of arctic districts constantly cold and constantly dry (see also p. 3825). Though his purpose was primarily ecological, Schimper was alive to the floristic aspects of these divisions and he divided the families of the higher plants into Megatherms, Mesotherms and Microtherms according to their temperature demands.

Among megathermic families he lists the following:

MONOCOTYLEDONS	DICOTYLEDONS
Palmae	Piperaceae
Pandanaceae	Dipterocarpaceae
Araceae	Marcgraviaceae
Musaceae	Sterculiaceae
Zingiberaceae	Meliaceae
Cannaceae	Podostemaceae
Orchidaceae	Passifloraceae
Commelinaceae	Begoniaceae
	Melastomaceae
	Sapotaceae

The above is only a selection of the more prominent families which are almost entirely megathermic, and excludes many families which are well represented in the tropics but also in temperate latitudes. We shall have more to say about these thermal preferences later (p. 3865), but the thermal classification of families is difficult and disputable. A family like the Mimosaceae, for example, is so prominently represented in the Tropics that many would class it among megathermic families, but it is equally well represented in the sub-tropics and beyond, in warm-temperate areas.

Among mesothermic plants the Coniferales figure prominently though the Pinaceae are more extensive in microthermic regions. Taxaceae, Taxodiaceae, Cupressaceae and Araucariaceae are all distinctively mesothermic.

MONOCOTYLEDONS	DICOTYLEDONS
Gramineae	Amentiferae
Liliaceae	Ranunculaceae
Amaryllidaceae	Papaveraceae
Iridaceae	Caryophyllaceae
Cyperaceae	Papilionaceae
Juncaceae	Violaceae
	Malvaceae
	Geraniaceae
	Rhamnaceae
	Proteaceae
	Rosaceae
	Umbelliferae
	Ericaceae
	Primulaceae
	Gentianaceae
	Scrophulariaceae
	Caprifoliaceae

There is nothing floristically distinctive about microthermic families, indeed there is no family which is wholly or predominantly microthermal. All the above mesothermic families and many others are represented in the microthermic flora and whether they should be ranked in one group or the other is a question of proportionate representation, whether of more species or of more individuals, in the one region or the other. On the ground of individual numbers Pinaceae, Gramineae, Cyperaceae, Juncaceae, Salicaceae, Betulaceae and Empetraceae show a marked tendency to microthermy, especially the last named, while the families with a significantly high number of microthermic species are Cruciferae, Caryophyllaceae, Ranunculaceae, Saxifragaceae and Rosaceae. Nevertheless, the majority of species in these families are mesothermic. The microthermic flora is therefore no more than an offshoot of the mesothermic flora.

Not only latitude, but altitude controls the thermal relationships of the prevailing flora. The average fall of temperature of $1°C$ for every additional 180 m of height has floristic as well as ecological consequences and so has diminishing rainfall.

The zonation of plants observable on mountains, while generally clear, is very variable and depends on a variety of factors, for instance, whether the mountains are isolated or are part of a mountain range and on the latitude in which they stand. Zonation is most complete when the base of an isolated mountain stands in a tropical or warm temperate vegetation and is least marked on mountains at high latitudes. There are, in general, three zones, basal, montane and alpine. The high rainfall in the basal zone, with average temperatures which are little below those of the surrounding lowlands, promote a vegetation which closely resembles that of the moister parts of the lowlands both ecologically and floristically. In most cases this is an hygrophilous forest, of species belonging, in the tropics, to predominantly megathermic families. If the climate is warm and wet, forest may extend into the montane zone, but before the tree limit is reached the forest becomes dwarfed and the growth of the trees distorted, creating a boundary zone called in German *Krummholz* and in English variously as 'elfin wood', 'cloud forest' or 'moss forest' from the abundance of mosses, favoured by the prevalence of mists.

When we pass upwards from the tree limit (or tree line as it is sometimes called) we enter a zone which is chiefly herbaceous in physiognomy, with at most a few species of dwarf shrubs of chamaephytic habit. This may be true montane grassland or in other localities may be more correctly called steppe. The height at which the tree limit is reached is very variable. It is generally higher in low latitudes and descends steadily with advance towards the poles. On mountains at high latitudes there may be no forest belt at all, whereas on some tropical mountains it may rise to 1800 m or more. It is lower on isolated mountains than on those which enjoy the partial shelter of a mountain chain and it is lower also in oceanic climates than in continental climates. For example, in Britain it is about 450 m but in the Swiss Alps it rises to about 2100 m. On south slopes of the Himalayas it may reach 3900 m.

The alpine zone intervenes between the montane grassland and the line of permanent snow. It is a region of intense solar radiation, of high winds and low temperatures, inimical conditions which preclude shrubby growth and are suitable only for low-growing, creeping or cushion plants, especially such as can accommodate their growth and reproduction to a very short growing season.

The rainfall in the alpine zone is always less than in the montane zone, but in mountains of moderate height in temperate climates this may be compensated to some extent by the prevalence of cloud mists. Higher mountains, how-

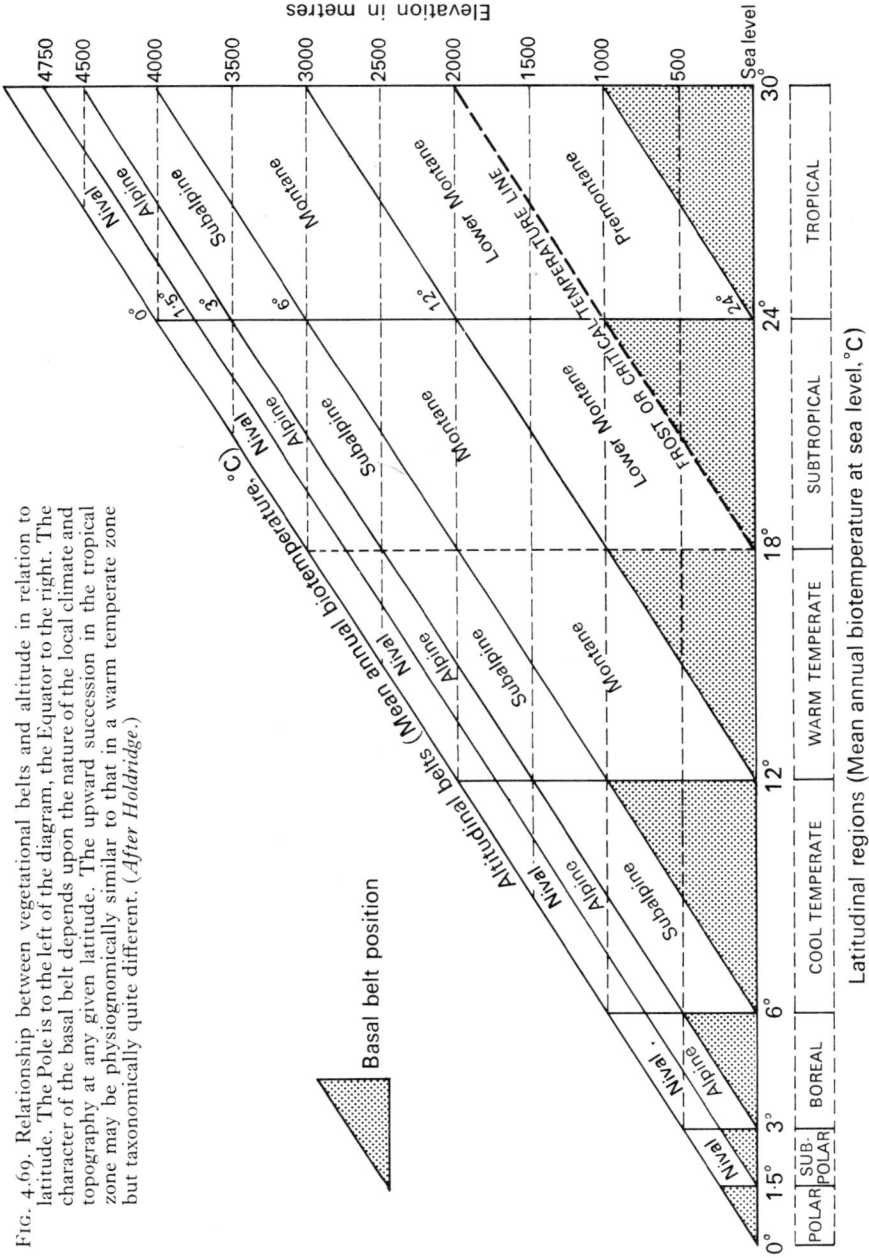

FIG. 4.69. Relationship between vegetational belts and altitude in relation to latitude. The Pole is to the left of the diagram, the Equator to the right. The character of the basal belt depends upon the nature of the local climate and topography at any given latitude. The upward succession in the tropical zone may be physiognomically similar to that in a warm temperate zone but taxonomically quite different. (*After Holdridge.*)

ever, may well rise above the height of rain clouds, especially in the stable climates of the trade-wind zones, and then the alpine zone may be veritable desert.

The earlier botanists were struck by the physiognomical resemblance of the vegetation in the montane and especially the alpine zones of tropical mountains to the vegetation of lower levels at higher latitudes and were rather too ready to couple alpine and arctic vegetation together. This is not, however, a legitimate comparison. It is true that a drop of 3°C in mean temperature due to increased height is equivalent, in terms of temperature to a horizontal distance of about 10° of latitude, but temperature is by no means the only factor involved and the differences between an arctic environment and the alpine zone of a mountain, even in temperate latitudes, in terms of seasonal changes, length of day, solar radiation and rainfall are very great.* There is often a physiognomical resemblance in the vegetation, but a community of species or even genera, except over a few degrees of latitude, is not to be expected.

Billings has recently investigated the climatic reactions of species which are truly arctic-alpine, that is, arctic species which also occur in alpine situations further south. He found that along the climatic gradient between the two areas there were local populations which were physiologically adapted to the local conditions. In other words, plants at the two extremes of climate were connected by intermediate ecotypes. We may conclude that only where such conditions existed could a single species reach from one environment to the other.

Even where temperate genera are represented in the alpine zones of tropical mountains it is often by peculiar species with growth forms sometimes very different from those of their temperate relations. Nowhere is this better seen than on the isolated mountain groups of tropical Africa: Cameroons, Ruwenzori, Elgon, Kenya, Kilimanjaro and the mountains of Ethiopia, where the giant species of *Senecio* and *Lobelia* fashion a unique and striking landscape which has often been described and illustrated. These species are highly endemic, many of them being restricted to a single mountain or mountain-group. Not that they are the only temperate genera represented in the tropical alpine zone. Good also mentions; *Anagallis, Echinops, Sonchus, Bartsia, Carduus, Alchemilla* and *Hypericum*, most of them with a similarly large proportion of endemic species. We have already mentioned (p. 3782) the presence of Ericaceae, notably of *Erica* itself, on the mountains of tropical Africa, forming a link between two great centres of the family in South Africa and the Mediterranean.

The isolation of mountains may often be a formidable barrier to migration of species from any considerable distance. J. C. Willis investigated the flora of Ritigala, an isolated mountain in the north of Ceylon. It has a moist climate but it is separated by forty miles of dry country from the exceedingly rich flora of the moist lands further south. He found that the mountain had only received 103 species across the gap and nearly all of these had light seeds or spores which might have been wind-borne. This is by no means the only

* As also is the presence or absence of pollinating insects.

instance of the effects of isolation on the alpine flora. Roraima in Guyana and Kinabalu in North Borneo have become celebrated for the number of their endemic species or even genera, which is largely owing to their being well-explored examples, though many other instances of mountain endemism on a smaller scale have been recorded. It is noteworthy that among the temperate genera found on the upper zones of tropical mountains there is generally a mixture of northern and southern hemisphere genera and the old idea that such mountains form, however imperfectly, bridges across the tropics for temperate plants is not altogether without foundation.

The incomers to the alpine zone of tropical mountains are, it will be noticed, from temperate zones, across an interval of about 20° of latitude. There is no question of any affinity with a truly arctic or antarctic flora. With high mountains in temperate climates the matter is rather different. From the European Alps or the northern Rocky Mountains, a step of 20° of latitude takes us well up into the arctic zone and consequently there are arctic affinities among their alpine plants, but only in a few cases does this involve a community of species. The climatic differences are still considerable and while common genera are not infrequent species in common are rare.

There are only a few families which can lay claim to a cosmopolitan distribution and in some cases the claim is deceptive, since it rests on the wide distribution of only one or two genera in the family, which may, moreover, be atypical in other respects. The widespread genera are also sometimes 'weeds' of broken ground, which have attained their dispersal as camp-followers of agriculture. Examples of the situation described may be drawn from Urticaceae, Euphorbiaceae and Rubiaceae.

The most truly cosmopolitan family is the Gramineae, which have pride of place not only in the extent of their distribution but in the important rank they hold in almost every sort of vegetation except tropical rain-forest. Next to the grasses must be reckoned the Compositae, which are well distributed everywhere. Although they contain nearly twice as many genera as the Gramineae (1000:600) and more than twice as many species (20,000:7500), they are more thinly dispersed and are nowhere as abundant in individuals.

Below these two families there is a considerable gap both in regard to size and in completeness of distribution. The distributional range of a family is roughly related to its size, as Willis has claimed, and this is logical since the greater the number of species involved the wider are the possibilities of ecological adaptation. Among the largest families are the Papilionaceae, Orchidaceae, Labiatae and Scrophulariaceae and all these have sub-cosmopolitan ranges in both temperate and tropical zones though their representation by species may be very different (e.g. Orchidaceae) in the two climatic regions.

Some of the most prominent families of temperate regions attain a sub-cosmopolitan range by penetrating the tropical zone, mostly at montane levels. Such are the Umbelliferae, Cruciferae, Rosaceae and Ranunculaceae, in order of size.

Certain families attain an exceptionally wide range by a species of vicarism.

Thus the Cyperaceae, although a large family in number of species, owe their wide range chiefly to *Carex* in the temperate zones and to *Cyperus* in the tropics. Among Primulaceae there are two widespread genera, *Anagallis* and *Samolus*, while the other twenty genera make up a patchwork of localized and sometimes discontinuous areas, which between them cover a large part of the world, though chiefly in temperate zones.

Lastly there is a considerable number of families which have attained very wide distribution geographically but with relatively sparse and limited occurrence because of their ecological specialization. Chief among these are the aquatic plants of families such as Alismataceae, Potamogetonaceae, Nympheaceae and Hydrocharitaceae, all freshwater aquatics, and the Zosteraceae in salt water. The predominantly halophytic Chenopodiaceae are similarly in the paradoxical position of being at the same time widespread and restricted.

When we turn to consider the taxonomic composition of tropical vegetation we are faced with several peculiar circumstances. In the first place is the immense area of land involved, greater than in any other latitudinal zone. This land, moreover, is divided between three major continents: America, Africa and Asia and in the Far East it is broken up among a multitude of islands of all sizes, including part of Australia. In America and Asia there is a more or less continuous gradation of climate beyond the tropic lines, but in Africa

TABLE 22. HORIZONTAL AND VERTICAL ARRANGEMENT OF CLIMATIC
VEGETATION TYPES IN THE TROPICAL ANDES
(From C. Troll. Presidential address, Phytogeography Section,
Stockholm Congress 1950)

Number of humid Months	Lowland	Lower Montane	Upper Montane	High Montane
12 11 10	Evergreen rain forest and semi-evergreen rain forest	Lower montane forest	Upper montane forest	Paramo
9 8 7	Moist deciduous forest and grassland	Moist forest and grassland	Moist sierra bush	Moist puna shrub and grassland
6 5	Dry deciduous forest and grassland	Dry forest and grassland	Dry sierra bush	Dry puna shrub and grassland
4 3	Thorn forest and grassland	Thorn forest and grassland	Thorn sierra bush	Thorn puna shrub and grassland
2 1	Desert shrub	Desert shrub	Desert sierra	Desert puna
0	Desert	Desert		

formidable sub-tropical deserts stand as barriers both north and south of the tropics.

The breadth of the habitat spectrum in the tropics as defined by rainfall and temperature/altitude is shown in Table 22.

The topography of this great surface offers the widest choice of habitats, with the result that it is difficult to find any family which is excluded from the tropical flora. Likewise there are very few of the large families which can be considered exclusively tropical. This category is mostly made up of relatively small and often endemic families. Families which have any claim to be called pan-tropical, that is those which are found in all sections of the zone, almost all spread beyond the tropics either north or south or both. The situation is then that the tropical flora is a blend of families which are predominantly tropical but with outliers, sometimes very extensive, beyond the tropics, and of predominantly temperate families which have inliers within the tropics, and only a minority remainder which are exclusively tropical.

While this is the situation with regard to families, it does not hold for genera and still less for species. As might be judged from the richness of the tropical floras, the number of these units which are limited to the tropics is enormous.

The family which, more than any other, is associated with the tropics in popular imagination is that of the Palmae. They have indeed a very solid pan-tropical spread, but northwards they reach the Mediterranean area and southwards, New Zealand. The palms are a large family but their rate of dispersal, with few exceptions is slow. To have reached such an extensive distribution they must have had a long time, i.e. they are an old family. There are some 250 genera of which nearly one-half are endemic, that is have a very restricted distribution. The same limitations are found among the species and the huge area occupied by the family is in reality a complex mosaic of small endemic units.

There are three other notable pan-tropical families which are so widely represented outside the tropics that they are nearly cosmopolitan. These are Euphorbiaceae, Rubiaceae and Orchidaceae. The first two are mostly trees and shrubs in tropical forests and their familiar herbaceous representatives in north temperate areas, *Euphorbia* and *Galium* respectively, are quite atypical of the families as a whole. The same is true of Orchidaceae; the soil-living species together cover the temperate zones, and because of this characteristic these three families, so prominent in the tropics, cannot be considered truly tropical, but rather sub-cosmopolitan.

Several other families are in the same position as the above, but with rather less extensive distributions; such are the Amaryllidaceae, Araceae, Cucurbitaceae and Tiliaceae.

Quite otherwise is it with families such as the Fagaceae and Magnoliaceae, both extensively represented outside the tropics but each with a heavy concentration of species in tropical South-east Asia and the East Indies.

The floristic composition of tropical rain-forest is so rich and so mixed that it has often been regarded as baffling analysis. Only locally, in edaphically exceptional areas, such as swamps, is the forest dominated by a single species. Otherwise the variety of species and genera is bewildering. Recently, however, Ashton, in Sarawak, has applied ordination analysis to a mixed Dipterocarp forest, with promising results. Coefficents of community were worked out between 100 single-acre plots including all trees over 10 cm diameter, representing 750 species. Floristic variation could be related to ecological gradients in soil texture and drainage and each species had a unique ecological range. Although the vegetation as a whole formed a continuum the plots tended to form clusters on the ordination axes, suggesting that over limited areas there was some floristic constancy, the groups being separated by broad bands of continuous variation. Such clusters appear to have the features of noda as postulated by Poore for temperate vegetation (see p. 3432).

Budowski has pointed out some of the floristic differences at various seral stages in the development of climax rain forest in tropical America. The pioneer and early secondary forest is populated by shade intolerant, fast-growing and soft wooded species among which members of Moraceae, Bombacaceae, Ulmaceae and Euphorbiaceae are prominent. Late secondary forest contains many Meliaceae, Bombacaceae and Tiliaceae but the climax forest, as in Africa and Asia, shows no general, as distinct from local, floristic dominance, though ecologically the trees are almost all shade-tolerant, slower in growth and heavy seeded, which means that their dispersal is slow. What he here describes applies almost equally, though with floristic differences, to South-east Asia, though in this region localized dominance by Diptero-carpaceae and Anacardiaceae may be more extensive. Whether the single-dominant forests are edaphically determined (e.g. by heavily leached soils) as Richards believes, or whether they represent a seral development from mixed forest, as Eggeling maintains, cannot as yet be decided.

Considering the ease with which many popular garden plants of the tropics can be cultivated in all equatorial regions, sometimes so generally that their original source is forgotten or obscured, we might conclude that the fact that only a few families have reached a pan-tropical distribution may be attributed to distance, slowness of dispersal, geographical changes in land distribution or many other possible factors, but if we reflect on the age of many, if not most, angiospermic families and the long period during which the equatorial climate has been stable, such factors sound unconvincing. Why can one genus in a family spread worldwide, when all the others remain limited? One can only say that we do not know. Absence can, in its way, be as eloquent as presence.

Croizat, in his large 1952 *Manual* warns against any easy assumption of the thermal preferences of groups, what he calls 'warm' and 'cold' groups, on the basis of local distributions. Because some genera of a family occur in a warm region this is no proof of an inherently 'warm' nature in that family. Only by tracing the migration of the family or genus to its ultimate region and there studying its distribution relative to the environment can one gain a presumption as to its warm or cold character. Nor is the distinction in any way absolute. Many groups show divided preferences or have aberrant members with preferences opposite to those of the majority, while some genera have so wide a thermal spectrum that they appear to be neutral. It is for such reasons that it becomes difficult to delimit typically tropical families. The evolutionary migration tracks of most families have passed through or entered the tropics, whether we regard the direction of their migration to have been from the north or from the south. Evolutionary offshoots of many families have remained in the tropics while other evolutionary lines have passed on to widespread temperate or 'cold' dispersal. The thermal distinction, if it is genuine, is a specific matter, it rarely applies to higher units as wholes.

Thermal preferences are certainly not the only factor in determining distribution by latitude. Equally potent is the length of day or photoperiod which for many, perhaps the majority, of the flowering plants controls development and the ability to reproduce. So closely limited are the physiological requirements in this respect that it is safe to say that it can in itself determine the range of latitude open to settlement by a given species, independently of temperature. A short-day plant may be able to grow at higher latitudes, but if it cannot flower it is quite impossible for it to become domiciled there, no matter how mild the climate. It would be too sweeping to say that photoperiodism has sifted out tropical from temperate plants, because there are many which are photo-neutral, but it has certainly been a powerful selective influence in the composition of floras. Whether this influence has been genetically selective or genetically directive, the fact remains that the proportion of long-day plants in local floras increases steadily with increasing latitude.

One remark by Croizat is deserving of emphasis as embodying a principle that lies at the root of phytogeography: 'Migration is an orderly process in time and space.' Distributions, as we find them today, are the result of this orderly process and abnormalities or discontinuities cease to appear as infractions of the rule if we can relate them to the secular changes of climate, topography and geography which have occurred during the vast period of time covered by the spreading of the flowering plants.

Climatology

We set out to look at the relationship of flora to climate. We should now consider how the various climates themselves can be systematized and classified. This is a question which has always concerned meteorologists. Apart from the broad distinctions which are obvious, there is great need for a method of analysing the essential characteristics of any climate in such a way that they can be summed up in a simple formula or diagram which can be immediately grasped by the understanding. To the Plant Geographer as well as the Ecologist a systematic method of comparing or contrasting climates is an essential need, for, as we have seen, the broad distinctions are too broad to be adequate.

One relatively simple system of classification was devised by Köppen and has been widely used. Two or three letters are used, capital letters for the substantive classes of climate and lower case letters added as required, to subdivide or supplement the main classes. The scheme, as modified by Trewartha in 1950, is given below.

Substantive Classes of Climate.

  A. Tropical Forest climates; coolest month above 18°C.
  B. Dry climates; precipitaticn variable and limit is directly correlated with temperature.
  BS. Steppe or semi-arid climate.
  BD. Desert or arid climate (BW = *Wüste* in the original scheme)
  C. Mesothermal Forest climates; coldest month above 0°C, but below 18°C; warmest month above 10°C.
  D. Microthermal (Snow-forest) climates; coldest month below 0°C; warmest month above 10°C.
  E. Polar climates; warmest month below 10°C.
  ET. Tundra climate, warmest month below 10°C but above 0°C.
  EF. Perpetual frost; all months below 0°C.

Modifying Symbols, added to the above.

  a. Warmest month above 22°C.
  b. Warmest month below 22°C.
  c. Less than four months over 10°C.
  d. As 'c' but coldest month below −38°C.
  f. Constantly moist; rainfall throughout year.
  h. Hot and dry; no month below 0°C.
  k. Cold and dry; at least one month below 0°C.
  m. Monsoon climate; short dry season but total rainfall supports rain-forest.
  n. Frequent fog.
  n'. Infrequent fog but high humidity and low rainfall.
  s. Dry season in summer.
  w. Dry season in winter.

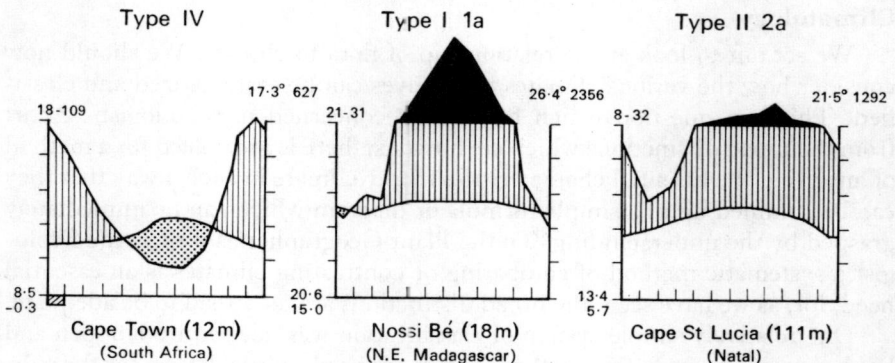

Fig. 4.70. Climatic diagrams from three stations in the southern hemisphere. The monthly bases are arranged with December in the middle. For explanation of the figures see p.3831.(*After Walter and Lieth.*)

While the above scheme is convenient it may be criticized as insufficiently flexible and particularly that it does not sufficiently indicate the different seasonal trends of climate, especially with regard to the annual distribution of rainfall. For example, it is very important to know the length of a dry season and whether this coincides with the highest temperatures or not, or whether the rainy season is the hot season and the dry season relatively cool.

Walter, in Germany, felt that only a climatic diagram could give a sufficiently comprehensive analysis of a climate. He devised a simple form of diagram which enables the annual range or trend of a climate to be appreciated at a glance and this gives a handy means of comparison. In 1960 Walter and Lieth published a *World Atlas of Climatic Diagrams*, with diagrams for all the meteorological stations in the world. The three diagrams given in Fig. 4.70 will serve to illustrate the principles of their construction.

The base line is divided into the twelve months and it is also the zero line for the vertical scale, which serves as a measure of both temperatures and rainfall. Each division equals 10°C or 20 mm of rainfall. In the northern hemisphere the months run from January to December, in the southern hemisphere from July to June. Midsummer is thus always in the middle.

The monthly averages of temperature are shown by the curve with a thin line, the corresponding rainfall averages by a thick curve. The humid season is vertically hatched. This hatching should be understood as continued to the base line, but for the sake of clarity it is stopped at the temperature curve.* The dry season is dotted. The per-humid or rainy season, where the monthly averages of rainfall exceed 100 mm, is shown in black. To avoid distortion this part of the scale is reduced to one-tenth.

Above on the left is shown the name of the station and its height above sea level (in metres).† Below this are figures for the number of years of observations. Where two such figures are given the first refers to temperatures and the second to rainfall. Above on the right are given the mean annual temperature and the mean annual rainfall in millimetres. At the bottom left-hand corner two figures appear in some diagrams. These give, when available, the mean daily minimum of the coldest month and, below this, the absolute recorded minimum.

There is a further addition to some diagrams. When the temperature curve falls below 0°C, this is shown by an oblong below the base line. Where this is black it means that the mean daily minimum for that time is below 0°C. If it is hatched it means that the absolute daily minimum falls within that period. When a figure is given below the middle of the base line it refers either to the number of days with an average temperature above 0°C (roman type) or the number of frost-free days (italic type).

Comparisons have enabled Walter and Lieth to recognize ten main types which they classify as follows:

---

* Restricting the vertical hatching also embodies the principle proposed by Gauman in 1954 that ecologically the proportion of temperature to rainfall may be taken as 1:2. Thus a period in which the rainfall is twice the average temperature may be regarded as humid. If the rainfall falls below this proportion the period is dry.

† Shown below in the diagrams reproduced.

Type I

24·7° 2250

24-49

16·2

Cairns (5 m)
(Queensland)

Type II

28·1° 1538

49-61

19·6

Darwin (32 m)
(N. Australia)

Type III

22·6° 140

15

4·0
−7·7

Baghdad (34 m)
(Iraq)

Type IV

15·9° 602

50

−1·7

Lisbon (100 m)
(Portugal)

Type V

16·1° 962

69-20

Buenos Aires (25 m)
(Argentina)

Type VI

7·6° 717

50-40

−1·8
−20·0

Kiel (47 m)
(Germany)

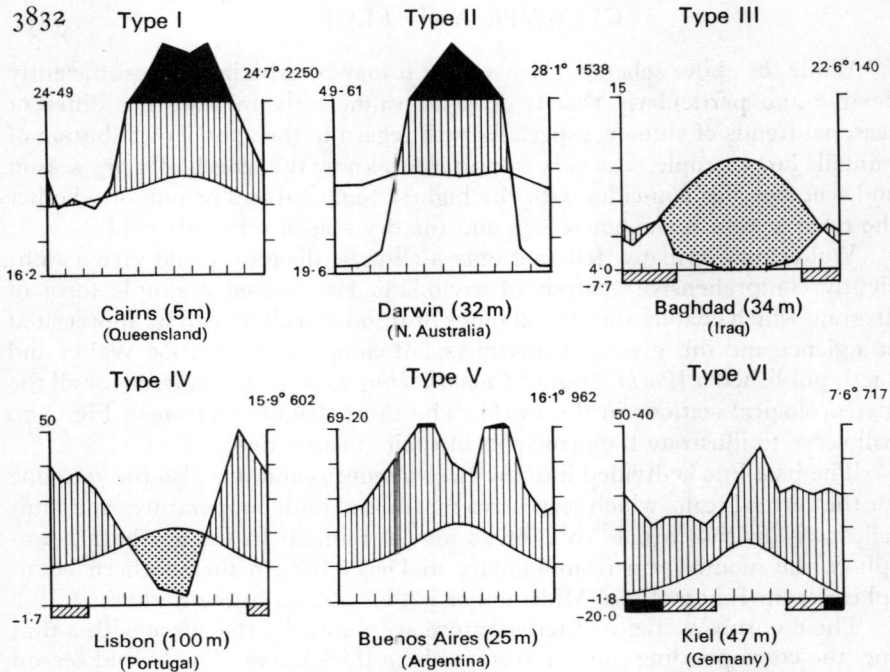

FIG. 4.71. Diagrams of the Walter–Lieth climatic types. Types I to VI. For explanation see p. 3831 in the text.

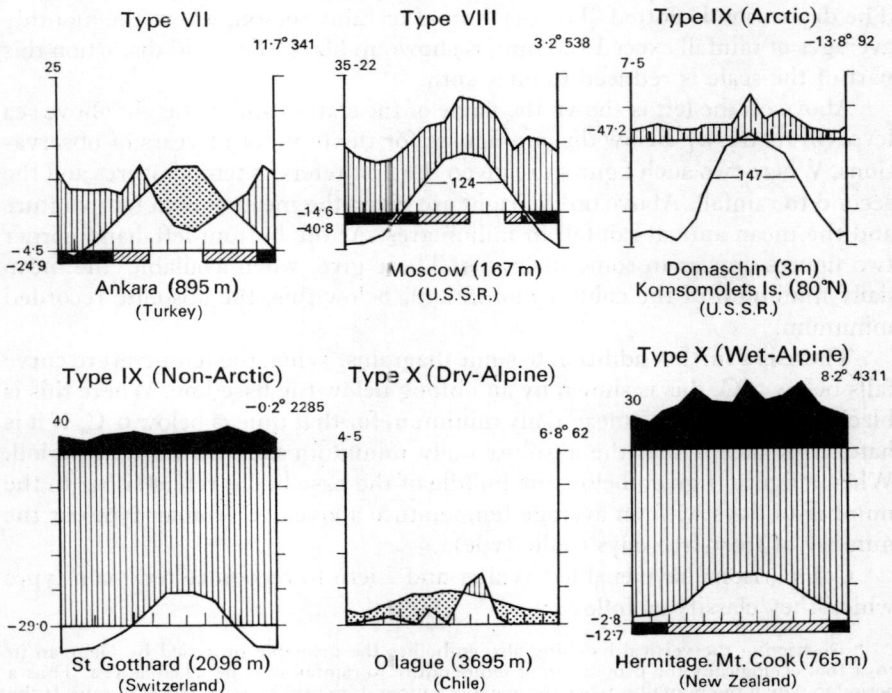

Type VII

11·7° 341

25

−4·5
−24·9

Ankara (895 m)
(Turkey)

Type VIII

3·2° 538

35-22

−14·6
−40·8

—124—

Moscow (167 m)
(U.S.S.R.)

Type IX (Arctic)

−13·8° 92

7-5

−47·2

—147—

Domaschin (3 m)
Komsomolets Is. (80°N)
(U.S.S.R.)

Type IX (Non-Arctic)

−0·2° 2285

40

−29·0

St Gotthard (2096 m)
(Switzerland)

Type X (Dry-Alpine)

6·8° 62

4-5

O lague (3695 m)
(Chile)

Type X (Wet-Alpine)

8·2° 4311

30

−2·8
−12·7

Hermitage. Mt. Cook (765 m)
(New Zealand)

FIG. 4.72. Diagrams of the Walter–Lieth climatic types. Types VII to X. For explanation see p. 3831 in text.

1. Equatorial zone, perennially humid or else with two rainy seasons. Temperatures mostly above 20°C. Small seasonal variation of temperature.

2. Tropical and sub-tropical regions of summer rain and a cool dry season. The monsoon areas.

3. Arid sub-tropical desert zone with occasional radiation-frost at night.

4. Zone of winter rainfall, not entirely frost-free but with no pronounced cold season. The Mediterranean climate type.

5. Warm temperate zone, perennially humid, with marked seasonal variations of temperature but only occasional frosts.

6. Humid temperate zone with a well marked but not prolonged cold season.

7. Arid temperate zone with hot summers and cold winters.

8. Boreal zone with prolonged cold season. Mean temperature of warmest month over 10°C.

9. Arctic region with, at most, only a short frost-free season. Mean temperature of warmest month below 10°C.

10. Alpine climate, within any of the above zones, very individual in character.

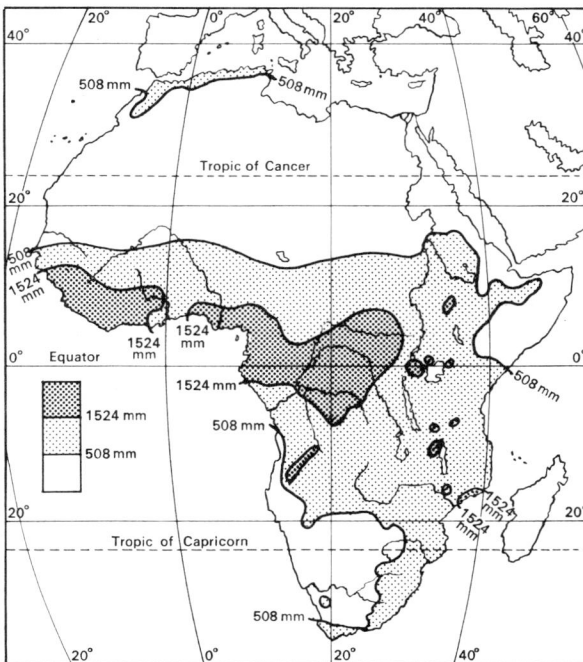

Fig. 4.73. Distribution of annual rainfall in Africa. Areas of low rainfall generally have one or more long, rainless periods each year. Note low rainfall area on the Gold Coast near Accra. (*After Milne Redhead*.)

Examples of diagrams of each of these types are given in Figs. 4.70–72. Each represents the average for the number of years indicated. The variation between individual years may be considerable quantitatively, but qualitatively the general pattern remains the same, particularly so in the relative duration of wet and dry periods, the length of which, especially of the latter, is very material to plant development.

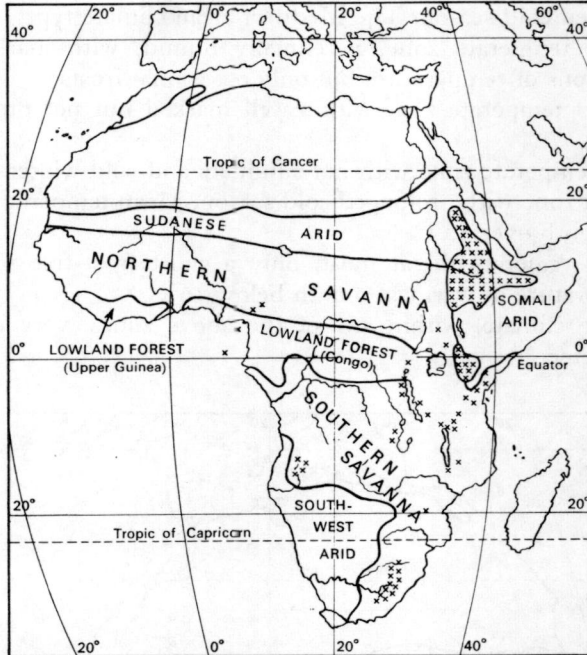

FIG. 4.74. Outline vegetation map of Africa for comparison with Fig. 4.73. Terminology after R. E. Moreau. Crosses show mountains and highlands where montane forest may occur. There are three arid regions, of which the Sudanese is the poorest floristically. (*After Milne-Redhead.*)

CHAPTER LVII

DISTRIBUTIONAL AREAS

WE now approach a subject which lies at the heart of Plant Geography, namely, the study of the actual areas of distribution of the various taxonomic entities of all grades from species upwards. In such a vast field it is necessary to be selective; even if our interests range widely we must be content to proceed step by step, to build up the area of a genus by a careful amalgamation of the areas of its constituent species and that of the family from its genera. Accuracy in such information is indispensable as the only secure foundation for wider generalizations.

Each area presents a number of significant characters to our attention: its size, its shape, the nature of its boundaries, its organic centre and its point or direction of origin, if centre and origin are distinct from one another. These are characters of the area as such, but it will usually be necessary to consider ancillary circumstances such as its topography, altitude, soil and rock types and climate, since these may often help to explain the characters of the area.

Areas are of all sizes. At one extreme there are the few species which transcend geographical and climatic limitations and are effectively worldwide. Some of these are weeds of cultivation and their ubiquity is spurious and man-made. Among natural ubiquists we may cite *Pteridium aquilinum* and *Phragmites communis*, though their prevalent biotypes are not the same everywhere. For plants like these the habitable world is their area.

At the other extreme are species which are only known from one restricted area. These are called *endemics* and they are the focus of much discussion and curiosity about their limitation.

The definition of endemism is very elastic for it all depends on the grade of unit considered, species, genus or family and on the size of the area. Without some qualification the word is almost meaningless, since practically every taxonomic unit has a limited area and few have more than one. Naturally it follows that the larger the geographical area considered the higher will be the proportion of 'endemics' in its flora. A sort of convention has therefore been adopted which makes abnormality of area the criterion of endemism. With reference to families, any which are restricted to a single continent would be called 'endemic', since they are only a minority among families. Genera usually have smaller average areas, so a genus restricted to one country or state or to any lesser area would also be reckoned endemic. Similarly with species, though at this grade the areas may become very small indeed. There are a few rarities the areas of which may be almost measured in square metres; such are *Gunnera hamiltoni*, *Metasequoia glyptostroboides*, *Cephalotus folliculiaris*, *Rhynchosinapis wrightii*, *Cupressus macrocarpa* and *Muiria hortensae*. These are endemics in anybody's book.

Endemic genera and species are more numerous in the southern hemisphere than in the northern, probably owing to the wide separation of the land masses towards the south. Probably for the same reason isolated islands or island groups are often peculiarly rich in endemics. Darwin's observations in the Galapagos Islands made them famous in this respect, besides providing him with support for his ideas of evolution. The original flora of the Hawaiian Islands was also well supplied with endemics (82 per cent of the flora), though many have disappeared under the invasion of mainland species introduced with agriculture. Skottsberg records that the endemic palm *Pritchardia* in these islands has thirty species each limited to a single island. The only island with a higher percentage of endemic species is St Helena (85) but here, too, the original flora has been devastated, in this case by goats. Some larger and less isolated islands have high percentages of endemic species the reasons for which are not obvious, though long isolation may be a factor. Such are New Zealand (79) and Corsica (58). It is notable that certain mainland areas may have unusually large numbers of endemics concentrated locally, though they are not ecologically peculiar and not climatically isolated from neighbouring areas. The outstanding example is the Cape area of South Africa, where a high degree of endemism is associated with an exceptionally rich flora, the thirty miles of the Cape Peninsula harbouring almost as many species as the whole of Great Britain. Another instance is the south-western corner of Australia.

Mountains likewise, especially isolated mountains, like Ruwenzori, Cameroons, Roraima and Kinabalu, or isolated mountain ranges, such as the Pyrenees or Itatiaia in Brazil are the homes of many endemic species.

An altogether different category of endemics includes the ecologically limited species which are peculiar to certain special soils or habitats and may rightly be said to be endemic to them (see the Serpentine soil flora, p. 3662). In that sense they may be classed here although they might be disqualified on geographical grounds.

The incidence of endemism in any taxonomic group is relative both to its size and to its grade. Thus endemic species are the most numerous and are the most limited in their area. For example, *Pandanus* is the only genus of its family to reach Madagascar, but there 75 species are known of which 74 are endemic, the sole exception being a species often cultivated.

As for families it is only on a continental scale that we can find endemics among them and the majority of these are small families, either monotypic or containing less than five genera. Out of a list of 149 families of limited distribution given by Good (1964) only 19 have 30 or more genera and 52 are monotypic, so that as far as distribution is concerned they rank as single genera and like genera may have more restricted, sub-continental areas, especially on larger islands. Thus among monotypic families, one is limited to Socotra (Dirachmaceae), one to Fiji (Degeneriaceae), three to New Caledonia (Amborellaceae, Oceanopapaveraceae, Strasburgeriaceae) and three to Madagascar (Barbeniaceae, Geosiridaceae, Humbertiaceae) and one (the Medusagynaceae) to one island in the Seychelles. Madagascar is remarkably rich in endemic families of which there are seven altogether including the small but remarkable family of the cactiform Didieraceae (Sapindales). Tropical and southern America is the richest continental region with 45 endemic families, not counting the Cactaceae which spread over the whole continent and have one representative (*Rhipsalis*) in Africa. Fifteen of these endemic American families can be reckoned 'large' in number of genera. Next in order comes Australasia in the wide sense, including Malaya and the Pacific Islands, with 30 endemic families of which six are 'large'. Europe is the poorest area in this respect. There are only five endemic families and these are all centred in the Mediterranean basin (Cneoraceae, Cynomoriaceae—monotypic root parasites—Globulariaceae, Punicaceae, Ruscaceae).

When we turn to consider genera the first difficulty we encounter is with regard to the taxonomic limits of a genus, since large genera are very liable to be split up into several smaller ones as the result of taxonomic revisions. However, the older, aggregate genera are usually sufficiently homogeneous in their general character to be retained as distributional units. A second difficulty, though a minor one from our point of view, is the variable value of species between different genera, some of which have only a relatively small number of well-separated species, while others offer a bewildering array of poorly defined, almost indistinguishable 'species' due to reproductive peculiarities.

Genera which have an almost universal distribution are very few in number and are all large. At the top of the list is *Senecio* with about 1500 species which are remarkable in assuming practically every variety of growth-form and in occupying every type of habitat. If we want to award the title of 'the highest plant' this is probably the strongest candidate, though the selection is about as arbitrary as that of a Beauty Queen. *Carex, Euphorbia* and *Solanum* are geographically almost ubiquitous, but are not evenly distributed. The same is true of several smaller genera, among which *Scirpus* (200 species) is almost as ubiquitous as *Carex*. Two others in the sub-cosmopolitan class are notable and surprising because of their biological peculiarity of insect catching, namely, *Utricularia* (210) and *Drosera* (90). Although the latter is widespread in north temperate regions its chief concentration is in Australasia and it is, oddly enough, pan-tropical, though there are large areas in Asia, Africa and western America from which it is absent.

There are a number of small genera which have attained a sub-cosmopolitan standard of distribution, notable among them being *Anagallis*, *Impatiens, Mentha, Typha, Sambucus, Vaccinium* and *Verbena*, all well known as north temperate genera, which are nevertheless widespread although deficient in some important areas. *Polygala* (475) is larger than most of these and comes nearest to cosmopolitan status because the only area in which it is absent is New Zealand and Polynesia. Widespread, but in a special category, are the plants which are ecological specialists. Among these, as we have previously mentioned, must be reckoned the ubiquitous 'weeds' or more correctly the plants of open ground, which have followed the white man's agriculture all over the world. Some of these are regular pioneer plants of the earliest stages in the succession formed on new ground, which are soon suppressed by more toughly competitive plants as succession proceeds. Their status as weeds follows naturally from the fact that agriculture provides them with new and open soils on which they can flourish without competition. Here also we should consider aquatics, such as *Callitriche*, *Lemna*, *Limosella*, *Nymphaea*, *Myriophyllum* and *Potamogeton* and semi-aquatics such as *Hydrocotyle*, *Montia*, *Nasturtium*, *Phragmites* and *Samolus*.

Lastly there are the halophytes, especially *Salicornia*, *Limonium*, *Suaeda* and *Spergularia*, widespread in one sense and restricted in another.

Endemic genera are very numerous and, doubtless for historical reasons of isolation and of migration, are much more numerous in some regions than others. Some areas have become famous for the number of their endemic genera. Among these are the islands of Madagascar and the Mascarenes, New Caledonia, and the Hawaian Islands, while among continental areas South Africa and tropical America stand out both for floristic wealth and the high proportion of endemic genera. Australia, though a much larger land area than these, has a flora of approximately the same size as that of South Africa, about 1500 genera, and a similar proportion of endemics, about 30 per cent. Tropical America also includes a very large land area with an immensely varied topography and climate. Good estimates that some 3000 genera are limited to the area, but the internal distribution of these within the area varies greatly. About 500 are limited to Brazil and only a small proportion range over the whole area. The Orchidaceae contribute largely to these very high figures.

Madagascar has to be considered as including, besides the main island, the numerous small islands in its neighbourhood, which form one isolated region with regard to Plant Geography. The main island is rich in endemic genera, some of which have reached out as far as the Comoros, the Seychelles or the Mascarene island groups. The flora of Madagascar is not sufficiently known to make numbers secure, but some 200 endemic genera are claimed. About 30 of these belong to the endemic family of the Chlaenaceae (Malvales) and they have a very limited distribution in Madagascar itself or on adjacent islands. The Mascarenes (the Mauritius group) have about 30 endemic genera among them while the Seychelles have about a dozen, including the famous 'double coconut', *Lodoicea*.

The Hawaiian islands are interesting in many ways. They are one of the most isolated groups and they are volcanic and geologically very young. Nevertheless, they have a rich flora with about 20 per cent of endemic genera* among which the Compositae and woody forms of the Lobeliaceae are prominently represented. About half of the endemic genera are monotypic. Apart from endemics some of the genera present show affinities with Polynesia and through that area with Australasia, suggesting a southern origin for much of the flora. The native flora has suffered much from the human introduction of American species which have almost replaced the natives on the cultivable lands and restricted the native plants to the mountains, at least in the more populous islands of the group.

* Grisebach gives the figure as 60 per cent. Perhaps some have died out since his time (see p. 3874).

Among larger islands New Caledonia holds a special place. Just within the tropics and in the monsoon region, New Caledonia has a drier climate than the neighbouring group of the New Hebrides and a vegetation which resembles that of tropical Australia in physiognomy, but with at least 20 per cent (100) of endemic genera. The southern half of the island bears very fine coniferous forest of mostly endemic species, especially of *Araucaria*, a link with Australia and with Norfolk Island from which comes the famous and beautiful *Araucaria excelsa*. The angiospermic endemics are also notable for their distinct character, many being aberrant or peculiar members of their respective families, a feature which is not uncommon among the endemic genera of oceanic islands.

The same is true to a large extent for the flora of Japan which has such close affinities with the asiatic mainland that the whole area—Japan, northern China, Tibet and the Tibetan Himalayas—is classed botanically as the Sino-Japanese region. Floristically it is more or less homogeneous though comparatively few genera range over the whole area. The flora of Japan is well known and in number of species is comparable to a west European country, but the flora of the mainland, despite much exploration, is still unexhausted. More than 300 endemic genera are estimated for the area and the singularity of the flora is due to the high proportion of these which are monotypic and belong to very small or monotypic families. Japan is very rich in woody Rosaceae and in Coniferae and the proportion of woody to herbaceous species is higher than in Europe. Conversely the herbaceous Compositae have about one third of the number of species characteristic of west European floras. Not only is the endemic flora rich, but so many of the genera are aesthetically attractive and have been used to stock our gardens, that a list of them reads like a nurseryman's catalogue. For this reason the endemic flora of this Asiatic region is probably more familiar to Europeans than that of any other part of the world. This might be described as one of time's revenges, for many genera now endemic in the Sino-Japanese region were part of the pre-glacial flora of Europe, but have survived only in non-glaciated Asia, which has led to it being sometimes called the world's largest plant refuge. A curious fact, which has some significance in connection with the theory of continental movement, is that a number of east-asiatic species are shared with western North America. Asa Gray has pointed out that there is an even closer linkage with the flora of eastern N America, which negatives any easy theory of migration across the Bering Straits (see further on p. 3870).

The flora of south China is distinguished from that of the northern area by the infiltration of a number of sub-tropical genera ranging from India and Malaya so that there is a considerable overlap with the Indo-Malayan region.

Europe is relatively poor in endemic genera. If temperate Siberia is included as being part of one continuous land mass, with no physiographical barriers, the total of endemic genera is still small. In eastern Siberia the intense severity of the climate is associated with a very limited flora and nearly all the endemics are in the European area. They amount to not more than one hundred, some of which extend into western Asia. Quite notable is the number of the European endemics which are alpines, often of very limited occurrence.

The Mediterranean region differs markedly from the rest of Europe in its hotter and drier climate. Spain, Italy and Greece lie within it, but these lands apart, the Mediterranean climate generally prevails over no more than a thin strip of coastal lands, limited northwards by mountains and southward and eastward by deserts, which really belong floristically to Africa or western Asia respectively. The flora of this area is rich and has a high proportion of ever-green shrubs and sub-shrubs which make up the characteristic shrublands of *maqui* and *garigue*. While there are many genera which are characteristically Mediterranean and have their greatest concentration and abundance in that area, yet relatively few of them can be called truly endemic, and they are mostly genera with very restricted distributions. Most of the well-known genera of the Mediterranean flora are not limited to the region proper but stray either northwards into western and central Europe, westwards to the Macronesian islands or eastwards into the Asiatic area. How sharp and rapid can be the change between the coastal Mediterranean climate and the semi-desert climate beyond is well shown by Boyko's observations in Israel (see p. 3507).

The natural tendency of every plant is to spread and, without physical barriers, even the slowest could travel a long way given twenty thousand years or more of life. The existence of endemism on a large scale therefore poses an important biological problem or series of problems since there can be no universal answer. Good has estimated that there are about 10,000 *genera* which are limited to one floristic region or an equivalent area. This takes no account of endemic *species*, the number of which is legion. There is no ques-tion of the great scale of the problem and it has attracted many investigators.

Two opposed views have been much stressed. One is that narrowly limited endemics, especially endemic species, are relicts, which means that they are the last representatives of species which were once widespread, but are now dying out owing to some unfavourable change in their environment. The other view is that such species are young beginners which need only time to spread normally. Such extreme views can only properly apply to endemic species. They could only apply in a very limited way to genera, not, at least, to those with several species with a combined area covering a whole floristic province, while monotypic genera are, geographically speaking, simply equivalent to species. Still less could they apply to families, whose endemism is on a continental scale.

The first, or relict, theory was a corollary of the doctrine of evolution by natural selection, and was upheld by Darwinians as necessary and practically implicit in selectionist theory. It found an articulate protagonist in Ridley. The selection theory involves the wholesale suppression of species by their 'better adapted' offspring or successors. It follows naturally that the evidences of this will be found in the large number of species which are dying out in all parts of the world, the 'relict' species. That such species were the casualties of natural selection was assumed *a priori*.

Another theory which, though unrelated to the above, led to a similar con-clusion, was that of 'Generic Cycles' (see also p. 3855). This postulated a life-cycle for species which resembled that of an individual. First, a vigorous youth, the period of active spreading during which the maximum possible range of distribution is established. Second, a 'reproductive' maturity, the period of active mutation in which new forms appear at various points in the range of the parent species and may become established as closely related species. Third, a period of declining vitality during which the species loses ground to younger and more vigorous species and in the course of which the originally continuous range may be broken up into discontinuous patches. Lastly senility in which the species has lost its competitive force and dwindles away to extinction.

This is a beautifully logical intellectual construction and our human preference for tidy theories almost forces it on our acceptance. Its chief demerit is that there is no evidence that species die out spontaneously or that they lose competitive vitality with age. Such information as there is tends to emphasize that species do not die; they are killed.

To these views Willis opposed a concept which was really an offshoot of his views on evolution. Disillusioned, as the result of his studies of the vegetation of Ceylon, with natural selection as an all-powerful agent of evolution, he adopted the mutationist standpoint and the concept advocated by Guppy of evolution by progressive differentiation from the family level downwards, which we have described on p. 3785. From this point of view species and specific distinctions are the latest comers in the evolutionary field and are naturally of different ages, according to when they came into existence, but in no case of profound antiquity. Taking as an axiom the proposition that the longer a species has existed the further it can spread (assuming equal opportunities) Willis argued that the area occupied by a species is an index of its age and that limited, endemic species were, in fact, beginners. On this he founded his theory called 'Age and Area'. He was careful to point out that this was only statistically true and could not rigidly be applied to individual cases. Also that comparisons were only admissible between groups of related species, in as much as the opportunities for dispersal vary greatly with the life-form. Thus short-lived herbs have far more opportunities for dispersal than have long-lived trees which may require many years of growth before they come to flower.

Stated thus baldly this seems no more than a statement of the obvious, as critics were quick to point out,* but Willis drew from the general principle a number of ingenious, if disputable, deductions which he backed up by the analysis of various floras, especially those of Ceylon and New Zealand. From his point of view one of the most important of these was that species multiply according to the law of compound interest, with no wholesale eliminations, which agrees with his mutationist standpoint but is incompatible with natural selection.

Another corollary was the principle that Willis called 'size and space', which postulated that larger families and genera are on the average older than smaller groups and will therefore be found to cover a wider range. From an

---

* Willis's difficult style of writing, like Croizat's, has militated against his views. This is a pity; both authors repay careful study.

analysis of the New Zealand flora he claimed another principle, namely, that in any given area (country, province, island, etc.) the range of a genus or species within the area will be proportional to the extent of its range outside that area.

Thacker has pointed out that on grounds of probability the smaller the area selected for examination the greater will be the average range of the species found in it, as they will tend to be those with the widest dispersal.

Considering an off-shore island as a separated part of the mainland, the same rule will apply. Its flora will not be a proportionately reduced flora of the mainland, but will include an excess of wide species whose average range on the mainland will be greater than the average for the mainland species taken by themselves.

Willis also showed that if genera be arranged in series according to the number of species they contain, those with only one or two species form together a majority of all genera (51·5 per cent) and that the number of genera falls regularly, proportionately to their size in number of species, up to the biggest, *Astragalus* (1600). He found this rule to hold good not only for Angiosperms as a whole, but for individual families and also for taxonomic groups of animals. He also found the same rule to apply for the floras of limited geographical areas, on which he founds a telling argument against evolution by natural selection, which, he claims, could not possibly produce such a universal regularity of numerical distribution. This could only be the result of an exponential multiplication of species in geometrical progression by dichotomous mutations.

When figures such as those Willis quotes are plotted graphically they always yield a falling or 'hollow' curve with a marked inflexion between three and five species. Curves of this sort are given by geometrical progressions but they are also given by probability distributions* of the Poisson type if the variable is never less than unity, which, of course, is the case with species. The evolutionary significance of the curves therefore remains dubious.

From all his work Willis concluded that species tend to spread at a relatively slow and uniform rate, without much regard for the means of dispersal and that those which had spread least were the youngest. This is where it touches the problem of endemism, which was not, in fact, his primary objective. In view of the criticisms levelled at this doctrine it is important to emphasize that Willis did not deny the existence of relict species. He only claimed that they were a minority of endemics and it would require a detailed analysis of the circumstances of many thousands of endemics to prove him wrong.

We may grant that the truth, as so often happens, lies between two extremes and that endemics may be either relicts or young species. Which way our beliefs incline is a matter of individual judgement. There are, however, a number of instances in which the relict status of species seems well established. There are many examples of this in the formerly glaciated areas in which certain localities, topographically favoured, provided refuges in which plants were able to survive the exceedingly rigorous conditions prevailing around

* E.g., in random samples of a mixed population.

them. Curiously enough many of these per-glacial survivors have failed to spread with the return of a mild climate and have remained endemically restricted to their refuges, which are often identifiable by the aggregation there of these 'rare' species. It has been suggested, with great probability, that this unlooked-for behaviour is due to the genetic impoverishment of the species by hardships which have eliminated all the biotypes or 'strains' in the species save one which was gifted with extra powers of resistance. Limited in this way the species can still flourish in one environment, but has lost the power of aggressive invasion or adaptation to new habitats. It is certainly true that many rare species, not necessarily endemic, show a remarkable tenacity in holding, where they occur, to one small habitat, sometimes being known for centuries, without showing any dispersal to neighbouring and apparently similar habitats. These, too, are probably single biotypes or rather ecotypes with only a narrow range of tolerance in some important respect.

It is worth while here to comment on the distinction between the terms 'endemic' and 'rare' as applied to species. The highest class of endemics, those restricted to a single locality, are also necessarily rare, but many endemics may be frequent or common within their area, and if that area is large, say, a single floristic province (p. 3891), they cannot be called rare. Truly rare species, on the other hand are not necessarily endemic. They may be widespread, but are never found in quantity in any locality. Griggs has pointed out that the majority of such species in nature belong to the early stages of ecological successions. They are plants of open ground with very limited powers of biotic competition, which limits the possibilities of their occurrence or survival. This certainly does not cover all cases. Many pioneer plants in successions are weeds, which may have limited powers of competition but whose reproductive energies are immense and nobody could call them rare. Conversely, some rare species are found in closed associations and it must be their reproductive capacity that fails.

The shapes of areas are often characteristic, so much so that the various shapes can be given a sort of classification. Vester in 1940 attempted this under twelve main groups with subdivisions. The latter need not concern us for they are simply geographical divisions, but the main groups are as follows.

1. Whole-world areas. True cosmopolitans.
2. Broad girdle areas, encircling the world outside the Arctic and Antarctic.
3. North girdle areas, lying chiefly in the northern hemisphere.
4. South girdle areas, in the southern hemisphere.
5. Narrow girdle areas, pan-tropical.
6. Girdle areas with one gap.
7. Girdle areas with two or more gaps.
8. Single region areas, with seven named regions.
9. Bi-regional areas, which cover two areas, not far removed from each other, e.g. North and South America.
10. Bi-continental areas, involving two different continents.

11. Tri-continental areas.

12. Pluri-continental areas (*Schollen Areale**). These are anomalous cases with disjunctive areas widely scattered.

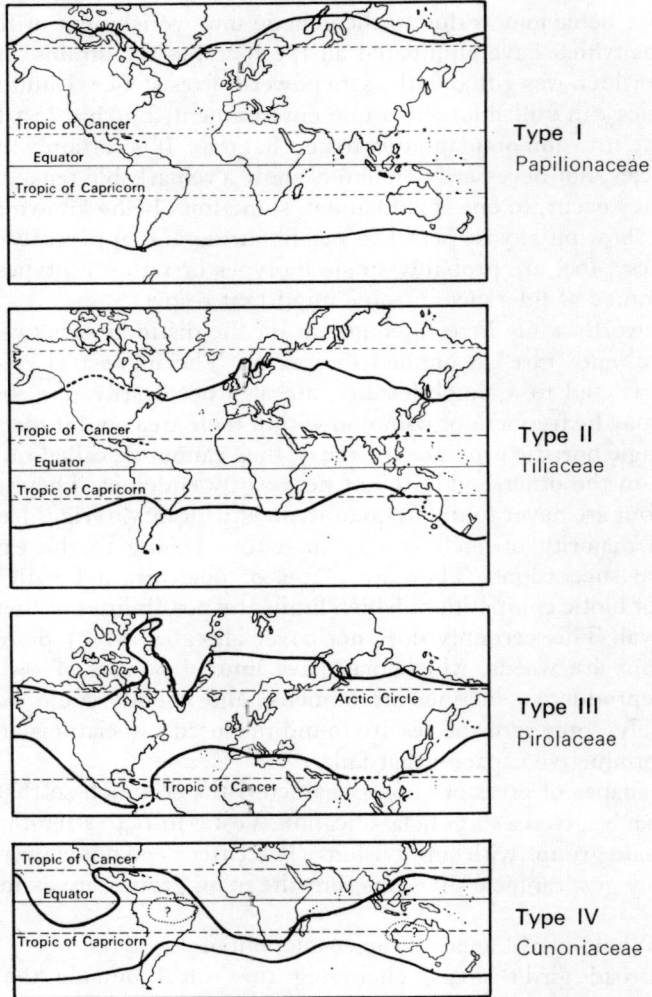

FIG. 4.75. Vester's Types of family distributions. Types I, II, III and IV. For explanation see p. 3843 in text.

From the numerous examples mapped by Vester we have chosen a few to form Figs. 4.75 to 4.78. It is interesting to note that Vester puts at the head of his list the Gramineae, as the only family which is truly worldwide, including

---

* *Scholle* is difficult to translate in this connection. Neither 'clod' nor 'flake' seem quite appropriate, though cited in dictionaries. Perhaps the word 'patch' is the best equivalent.

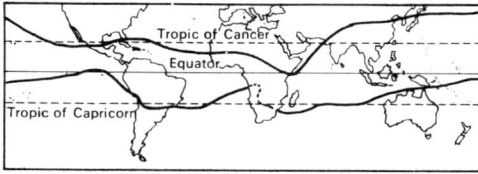

Type V
Begoniaceae

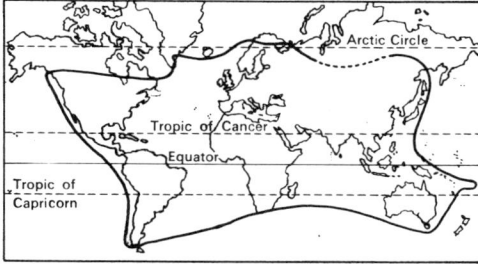

Type VI
Polygalaceae

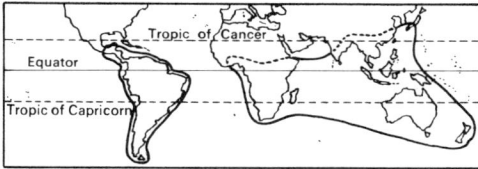

Type VII
Proteaceae

FIG 4.76. Vester's distribution types, V, VI, and VII.
For explanation see p. 3843 in text.

Punicaceae

Type VIII

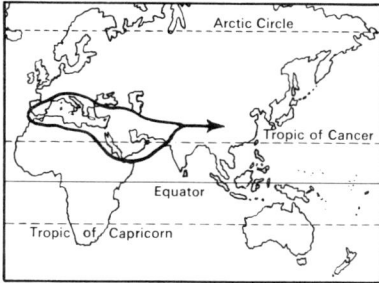

Cyclanthaceae

Type IX b

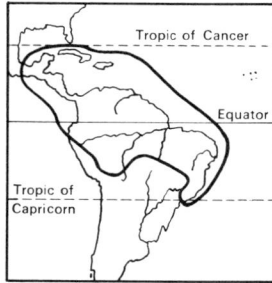

Casuarinaceae

Type IX c

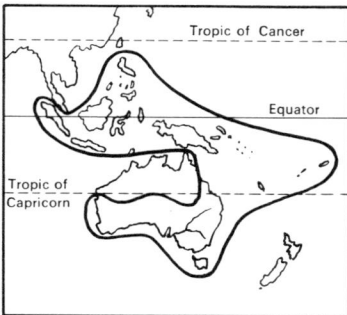

Rapateaceae

Type X b1

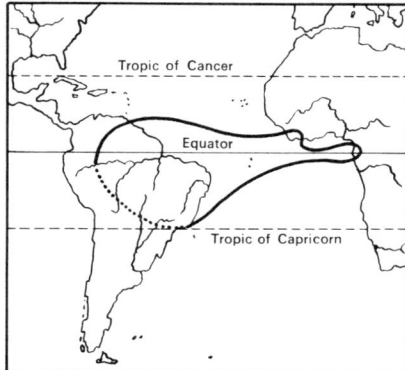

FIG. 4.77. Vester's distribution types VIII and IX. For explanation see p. 3843 in text.

both the Arctic and the Antarctic continent (from the latter of which *Des-champsia parvula* and *D. elegantula* have been described). Next in line as cosmopolitans come the following families, which are only lacking in Antarctica but otherwise are world-wide, though sometimes very unequally distributed: Cyperaceae, Ranunculaceae, Rosaceae, Papilionaceae, Scrophulariaceae and Compositae.

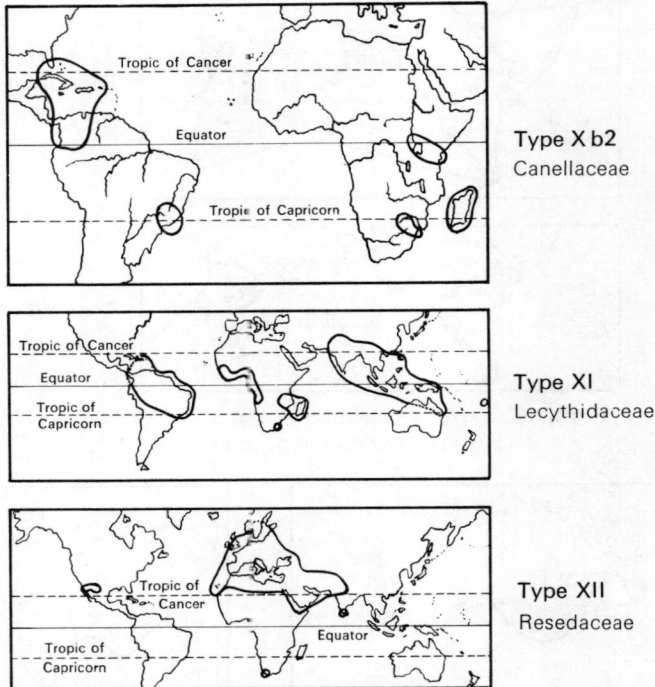

FIG. 4.78. Vester's distribution types X, XI and XII. For explanation see p. 3843 in text.

A problem which always accompanies the study of a distributional area is that of identifying its 'centre', a hypothetical entity about which very different opinions have been held. Hand-in-hand with this goes the problem of its 'origin', assuming this to be different from the centre, which would only be the case if the taxon concerned is supposed to have migrated into the area, not to have originated within it. The latter supposition is inevitable with world-wide taxa and is highly probable in the case of most uniregional taxa, so that in most cases it would seem that centre and origin are synonymous (see also polyphyly, p. 3879).

Were the doctrine of evolution by differentiation accepted, then families, or at least their prototypes, are older than the prototypes of genera. They have consequently migrated most widely, witness the number of families of cosmopolitan or almost cosmopolitan distribution, an extension reached by scarcely any genera. For many of these worldwide families, centres may be dubious or even multiple, while origins are largely speculative. In the course of a genor-heitron a family may spawn new genera at any point of its track. Generic and specific areas therefore are more likely to show a centre and a point of origin within their areas, if we consider areas as wholes, not simply the partial distribution within one country or continent. As Willis has pointed out, the status of a flora with regard to the relative proportions of immigrant and endemic taxa and the differences in this respect between different areas depend on the age of the flora. Young floras, like those of the formerly glaciated lands, are almost entirely immigrant, the species concerned having simply taken the opportunity to extend their existing areas. Older floras, like that of Ceylon, may have a considerable proportion of endemic genera and species, which are, in the true sense, aboriginal. The same is true in lands long isolated, like Australia, which only a few large, and widespread, genera have reached from outside and in which the bulk of the flora is endemic. Everywhere, Willis contends, it will be the oldest, by which he implies the largest, genera which will be the earliest immigrants, as they are the most wide-ranging. Where environments have for the longest time remained unchanged, which means, generally speaking, the tropics, there we may expect to find the greatest number of aboriginal endemics. Exceptions to this, such as temperate western Australia or the Cape Province, raise suspicions that many of their numerous endemics may be climatic refugees and in that sense relicts, but unless we learn something of their Tertiary distribution there is no proof.

The concept of centres was first promulgated by Adams (1902–09) and has been applied chiefly to generic areas, where it seems most applicable. Species, unless they are polymorphic, in which case they approach the level of a subgenus, do not lend themselves so readily to the means of analysis proposed by Adams. The concept involves two basic assumptions, one, that a genus has one geographical point of origin, and second, that its spread from the centre is proportional to the time available, i.e. is conditioned principally by time ('age and area'). These axioms define the first kind of centre, the *centre of origin*. The second kind is the *centre of dispersal*, which may coincide with the first for the oldest species of the genus. The third kind is the *centre of variation*, that is the area where the greatest number of species are found. The fourth kind is the *centre of frequency* where the greatest number of individuals occurs. Lastly, the fifth is the *centre of preservation*, which is not strictly comparable with the other kinds as it applies to what are generally called 'refuges' where a number of species have survived some unfavourable change in the general environment.

Species may arise in various ways, some abruptly some by slow genic change, by single or multiple hybridizing and by polyploidy, and the distinctions between them are of varying degrees as is also the degree of their genic isolation. Quite apart therefore from the additional possibility of parallel mutations occurring at different places, it can be said that many species had no single site of origin and that the concept of centre of origin has no validity in such cases. There are exceptions to this proposition, but it is to generic areas that the concept chiefly applies.

Adams proposed a number of criteria for forming a judgement about centres (as centres of origin) for generic areas. These were critically reviewed by Cain in 1942, with a warning against the prevalence of deductive reasoning in Plant Geography, a fault which reveals itself in the many ways in which facts do not square with Adams' criteria. Although these can be useful they can also be very misleading unless carefully checked by inductive investigation. We cannot here do more than indicate Cain's arguments without the abundant detail with which he illustrates them.*

Criterion 1. Locality of greatest differentiation, i.e. the greatest concentration of species of a genus. This may sometimes be the centre of origin, but it can be misleading. For example, a young genus may spread into an area more favourable to the formation of new specific types, which would then appear as a false origin. Moreover, it is necessary to be sure that the species concerned are of equivalent value. Szymkiewicz has categorized species in this respect and claims that the most useful species in identifying a centre are the endemic or sub-endemic species with small areas.

A valuable development was proposed by van Steenis. Having entered upon a map the numbers of species found in each part of the area of a genus, the points with an equal number of species, not necessarily the same species, are connected by contour lines which he calls *isoflors*. Ideally, if dispersal were purely centrifugal, these lines would be concentric, but this rarely happens. In practice the lines may show the existence of barriers in some directions and of favoured channels of dispersal in others. If there is more than one such channel, their convergence backwards will indicate the centre of dispersal with considerable certainty. If other genera with a similar distributional area follow the same channels of migration, it increases the certainty of the conclusions and, as Cain points out, the distribution of species or sub-species along a migratory channel may well be a useful clue to the course of evolution in the genus. Certain it is that attempts to estimate the phyletic relationships of taxa or to deduce vertical, i.e. phylogenetic relationships of taxa of whatever level, without consideration of their geographical distribution, is bound to be misleading.

A difficulty arises in dealing with the polyploid complexes of species and hybrids, often perpetuated by apomixis, which occur in such genera as *Rosa*, *Rubus*, *Crepis* and *Taraxacum*. Such complexes may arise at any point in the range of an extensive genus and give rise to a widespread swarm. Among the multitude of forms there are a small number of basic diploids, usually with restricted distributions, in older, more stable habitats. It is the centre of distribution of these diploids which represents the centre of the whole complex, though this may or may not be the true generic centre. Thus Babcock and Stebbins found such a complex in *Crepis*, centred in the Pacific north-west of America, though the genus appeared to have arisen from an Asiatic centre.

Criterion 2. Locality of greatest abundance of individuals. This is almost entirely an ecological, not an historical phenomenon and it is only in connection with species or still smaller taxa that it may have historical significance.

---

* See *Torreya*, **43**, 1943, p. 132, or CAIN (1944) *Foundations of Plant Geography*, New York, 1944, for the full treatment.

Dominant species are, moreover, usually species with a wide range outside of the association which they dominate. When a species can be shown to have expanded its range uniformly up to its physical limits, then the locality of greatest abundance may be the centre of dispersal if we presume that it is the oldest habitat of the species, though there is little assurance of this. Species differ so greatly in their range of tolerance and in their propensity to form eco-types with special tolerances that the continuity or the mosaic character of their distribution give, as a rule, little clue as to their centre. Actual analysis of specific ranges, a properly inductive test, shows that species which become progressively scarcer towards the periphery of their ranges are exceptional. Griggs examined the distribution of 123 species which reach their limits in southern Ohio and he found that 66 per cent of them were common or abundant near their limits. If we allow that reproduction follows a compound interest law then the oldest locality for a species should be the one where the most individuals are found, but this theoretical situation is subject to so many ecological limitations that it cannot be relied upon. The geological age of localities also requires consideration. The influence of physiographical changes during the Late-glacial and Post-glacial periods may well result in a local abundance of individuals in localities which happen to have been stable habitats for the longest time.

Criterion 3. Locality of the most primitive or generalized forms. This would be unexceptionable if we could be sure of what are 'primitive' forms in a genus. Otherwise it is illusory unless backed up by palaeontological evidence. The idea of primitiveness is based on deductions from comparative morphology and depends to a large extent on preconceived ideas. They are, moreover, somewhat vitiated by the prevalency of what Takhtajian has called 'heterobathmy', in other words by the co-existence in the same species of characters which are 'primitive' with others which are 'advanced' so that judgement of their status is uncertain. Some genera contain a number of species which are obviously specialized in some way, contrasted with others which are unspecialized, but unless there is good evidence from distribution we can have no assurance that the former have evolved from the latter.

Where cytological evidence is available more reliable judgements are possible. We have above referred to the contrasting distribution of polyploid and diploid forms in a complex of closely related species (see Fig. 4.38, p. 3574). Here there can be little doubt that the polyploid or tetraploid forms have originated from diploids and the distribution of the diploids may be indicative of centres, especially with aid from taxonomy and the geological history of the area.

Willis has added a further criterion which may be regarded as a modification of the above. We may quote his own words: 'The nearer one goes to the centre of distribution of a family the more do the smaller genera in world size—the relics upon the older views—come into the picture, so that the average size (of genera) becomes less and less.' He also expresses the same idea in other words, that the largest (i.e. the oldest in his view) genera are those most likely to be found near the outer edge of the area of distribution of a family, while the proportion of smaller (younger) genera will rise as we proceed towards the centre of the family's distribution. Judging by the figures he has quoted the argument appears to be well-founded and it agrees, of course, with his general age and area theorem. In this fashion Willis avoids the difficulty of identifying primitive genera by simply equating 'primitive' with 'small'.

In any area of distribution of a genus a point near the centre will have more chance of receiving species of limited range than will any given point near the periphery. If this be true, then the centre should be identifiable by the circumstance that the average range of the species in its vicinity will be lower than in any comparable region within the area as a whole.

Criterion 4. Location of biggest individuals. This is so much governed by ecological factors that it is of little value. There are cases where the biggest individuals of a species occur near the periphery of its area, while some areas are notable for the luxuriance of the growth of various unrelated species.

Allen originally based the argument on the occurrence of graded series in size away from the main centre of distribution towards the limits of the area. Such 'clines' are well known in many species and are generally regarded as due to a continuous selection of biotypes. While these series are factual it may be pointed out that they can be read both ways.

Criterion 5. Locality of greatest and most constant productivity. This again is an ecological rather than an historical concept. It may be true that productivity is most constant (not necessarily highest) in the region where the tolerances of a species are best suited, but this in no way implies that that is its centre of origin.

Criterion 6. Continuity and convergence of lines of dispersal. We have already spoken about this under Criterion 1. In some circumstances it may be valuable especially if backed up by taxonomic and cytological study of the forms found along the highways of dispersal, which may indicate the direction in which migration has gone.

Criterion 7. Location of least dependence upon a restricted habitat. Here again there is an ecological tincture. It implies a region in which the tolerances of a wide variety of species in a genus or of biotypes in a species can be satisfied, a region with no extremes of environment. Such regions may exist within the boundaries of many distributional areas, but they do not necessarily coincide with centres of origin. A concentration of nearly related species in a given region may be due to a constant and equable environment which allows mutations a better chance of survival, but the opposite may also be true, that great variability of habitats allows the segregation of ecotypes which have special habitat preferences.

On the other hand a varied range of habitats for a species may be a result either of inherently wide tolerances or of a multiplicity of biotypes rather than of uniformity of environments. That species often show greater selectivity of habitats towards their distributional limits is true (see Boyko, p. 3508) but this may be due to progressive selection of biotypes or, in other words, to the depauperization of the genetic reservoir by environmental selection and isolation (see also relict species, p. 3842). It has been claimed that primitive species have wider tolerances than more advanced species, and reference is made to the apparently wide extension of species in Tertiary times as compared with their living relations, but the argument is shaky. We know nothing of the genetic make-up of the Tertiary species and, moreover, it is uncertain whether their apparently wide areas were occupied simultaneously. If primitive species have exceptionally wide tolerances, which is not known, and if they are found in a varied choice of habitats in one region, then the argument may be valid, but it is far from conclusive.

Criterion 8. Continuity and directness of individual variations or modifications radiating from the centre of origin along highways of dispersal. This is an elaboration of criterion 6, depending on an analysis of gene-flow in a population. Mass collections analysed for changes in the frequency of characters may

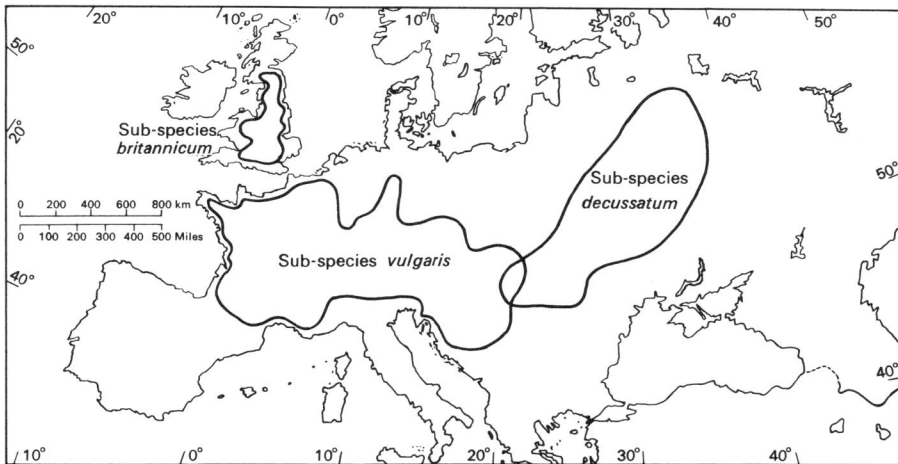

FIG. 4.79. Distribution of the three sub-species of *Cirsium eriophorum* in Europe in three contiguous areas. (*After Petrik.*)

yield evidences of gene-flow which can be reliable indications of the direction of population spread and hence of the origin of the spread. Difficulties may arise through uncertainty as to the direction in which the change of frequency in a given character should be read or in estimating the extent to which a frequency change is genotypic or due to the phenotypic or even the selective effects of a changing environment. Chromosomal changes, such as a polyploidy series, afford perhaps the least disputable evidence of the direction of migration.

Criterion 9. Direction indicated by geographical affinities. This is a criterion which can be very useful, but can also be misleading. It is chiefly applicable where a taxonomic group has a well-defined and comparatively homogeneous area of distribution from which units have penetrated into extraneous areas. Thus a genus belonging to an otherwise tropical family, which penetrates into temperate zones may be safely interpreted as having come directly from the nearest point of the family's tropical area. The opposite course of events might have occurred, it is true, but is highly unlikely. The value of this criterion depends, however, on the critical assessment of what is the 'principal area' and with regard to this appearances may be deceptive. We are liable to accept as the principal area that region from which the taxonomic group in question has been longest and most fully known which may be no more than an historical accident. The genus *Drosera* may illustrate this point. The common occurrence of a few species across all the northern continents would have signalled, until comparatively recently, the north temperate zone as its principal area, whereas it is now known that in the tropics and the southern hemisphere there is a much greater number of species of the genus, distributed over at least as great an area as in the north temperate zone. The genus *Quercus* is in a very similar position.

Another deception may arise through the mass migration of a population from one habitat to another due to a climatic or geological change in its original area. Any relic units left behind surviving in the original area might then easily be mistaken for progressive migrants coming from the new area.

These are pitfalls which are sometimes, though not always, avoidable by careful consideration of all the circumstances, especially of fossil evidence, if any.

Criterion 10. Direction of population movement indicated by bird migration routes. This is restricted to species which are disseminated by birds. Sometimes bird migration follows the same route in both directions. The question in which direction bird carriage has been effective for a given plant may be resolved by the consideration of the season in which its seeds would be available for bird carriage.

Criterion 11. Direction indicated by seasonal appearance. The suggestion behind this is that vernal flowering is associated with boreal origin or, more generally, with areas which have a short summer, which would include alpine areas, and that the vernal habit may indicate the origin of plants which have migrated into more temperate climates. Thus, it is argued, alpine plants spreading downwards retain the vernal habit, whereas lowland plants migrating upwards (or northwards) display aestival flowering. The flowering habit is generally linked to the photoperiod and once established in a species is not readily changed, hence the difficulty of cultivating many long-day (i.e. aestival) species in the tropics under relatively short-day conditions. The criterion may sometimes be useful, but is by no means absolute, since migration may well have originated in an intermediate zone and moved in both directions so that we are not dealing with a single cline but with two.

Criterion 12. An increase in the number of dominant genes towards the centre of origin. This valuable criterion originated from the classic studies of Vavilov on the origins of cultivated plants. He found that a considerable number of the plants he investigated showed centres of origin which were characterized by a great diversity of forms, most of which were distinguished by their accumulation of dominant genes. Important and perhaps decisive as

this criterion is, its application obviously poses practical difficulties as it requires extensive and prolonged genetical studies of populations not easily
carried out except within a large-scale programme.

Criterion 13. Centre indicated by concentric progressive equiformal areas.
This arises from the observations of Eric Hultén on the spread of northern
boreal species from centres which are in most cases per-glacial refugia (see
Vol. 3, p. 2592). The natural tendency of spread is centrifugal which would
theoretically proceed by concentric circles. Owing to natural barriers such
areas are, in practice, not truly circular but 'equiformal', that is areas of
approximately the same shape. As different species or biotypes of one species
do not spread at exactly uniform rates, the different taxa will form roughly
concentric areas surrounding the centre of distribution. Where such a pattern
is found the conclusion as to the central locality can be definite.

There can be no doubt that the study of distributional areas and the
analysis of the factors which have been operative in determining their respective characters can throw a great deal of light upon the history of the taxonomic groups involved and hence upon their evolution, with likewise, important
evidence upon the subject of evolution in general. If extensive geographical
studies had been possible in earlier times, the dogma of the constancy of
species could scarcely have survived and it was the beginning of such studies,
especially by Charles Darwin, which finally made it unacceptable.

Willis's proposition that the extent of a distributional area is directly correlated with the age of the group concerned may be accepted in a general way,
but it is subject to many provisos and limitations which make it unsafe to
found conclusions on it, unless they are borne out by inductive evidence. The
corollary drawn deductively from the age and area theorem, that taxa with a
limited distribution, otherwise endemics, are necessarily young, is equally unsafe as a guiding principle, as is also the older view that they are all relics.
The observed facts point to a more partial opinion, that some endemics, more
especially in the southern hemisphere, are indeed beginners, that others,
especially in the formerly glaciated regions of the northern hemisphere, are
unquestionably relics, which is frequently evidenced by the wide occurrence
of their fossil remains; while yet a third group fall into neither category, but
are limited in their dispersal by inherent characters, such as a narrow range of
tolerance in some particular, very specialized ecological preferences or simply
by an exceptionally slow, limited or hazardous means of reproduction.

It may sometimes be difficult, paradoxically, to distinguish between a
young endemic species and a relict. The breakdown of a wide area under the
influence of unfavourable climatic changes may leave a few disjoined pockets
of survivors which are essentially relics. On the other hand the scattered
remnants may consist of new ecotypes selected from the original stock by the
changed conditions, which either are, or are on their way to become, ecospecies in their own right. Hultén has pointed to an example in *Agropyron*
sect. *Elytrigia*, which consists of seven species, or perhaps sub-species as they
are all closely related. Between them they define a large area in north-eastern
Asia from which two arms stretch westwards, lying to the north and south re-

spectively, to reach eastern Eurcpe. Hultén considers this to have been the pre-glacial extension of *Elytrigia*, which was broken down during the Pleistocene changes. The seven surviving representatives are indeed relicts in one sense, but they are also new evolutionary growing points in another. Willis has pointed to a similar position in the genera *Anemone* and *Clematis*, where one widely ranging species is accompanied at different parts of its range by different endemic species. The supposition is the same, that the endemics have been segregated from a widely ranging parent species, though in the case of the two genera mentioned the parent species still occupies its extensive area.

We have previously referred (p. 3842) to the conservatism of some endemic species as being due to the impoverishment of their genetical potential by the extinction of almost all the biotypes of the species during the Pleistocene. Their narrow limitation is not in such cases attributable to age or consequent senility; relicts they may be, but they are not necessarily to be regarded as senile or as doomed to extinction. Manton's study of *Biscutella laevigata* in central Europe showed that the species contained a diploid and a tetraploid race. The diploids show the marks of relict survivors of the glaciation in their conservative limitation and uniformity of type. The tetraploids have arisen from these relict diploids and are therefore younger. They are variable and wide-ranging, but can one assert that this is due to their being younger or is it the result of polyploidy? (See also p. 3864.) Either way it shows that a conservative relict is not necessarily moribund, but that through autotetraploidy it may regain much of its lost flexibility.

The study of *Iris setosa* by Anderson provides another example. The type species and its var. *interior* are found in Alaska; the var. *canadensis* around the Gulf of St Lawrence, separated from the others by several thousand miles of formerly glaciated country. There is no doubt that they come from a common stock, surviving now only on the periphery of its former extensive area. In Alaska they enjoyed an extensive refuge from glaciation, and there the species has retained variability and capacity to spread; it has invaded the glaciated land for a considerable distance southwards. Around the St Lawrence, however, the stock experienced the full glacial rigour and survived only in nooks and crannies where the hardiest biotypes lingered. Hence it has become genetically poor and shows typically relict conservatism. Obviously it is not a question of age. The conservative race and the aggressive race are both approximately the same age. It is their history which has made the difference.

Biotic competition would seem to be perhaps the most potent factor in limiting the spread of relict species. There are numerous cases of species of very limited distribution which are conservative in their native area but which flourish easily under very diverse conditions when brought into cultivation. Such are, for example, *Cupressus macrocarpa* and *Pinus radiata* from Monterey in California, *Polygonum baldschuanicum* from Central Asia and *Ginkgo biloba* and *Metasequoia glyptostroboides* from central China, to which we might add *Delonix regia*, universal in tropical gardens, but probably extinct in its home in Madagascar.

There are, however, many cases in which conservatism is scarcely attributable to either age or genetical impoverishment by hard conditions, but seems rather to result from an inherent limitation in the species itself. There are many examples of pairs of nearly related species whose age and history are apparently closely similar, but one of which is rare and of limited occurrence, the other common and widespread. From the flora of Britain many such pairs can be drawn, of which the following are a few, taken at random.

RARE AND LIMITED	COMMON AND WIDESPREAD
Ononis reclinata	*Ononis repens*
Koeleria vallesiana	*Koeleria cristata (gracilis)*
Scleranthus perennis	*Scleranthus annuus*
Halimione pedunculata	*Halimione portulacoides*
Sonchus palustris	*Sonchus arvensis*
Artemisia campestris	*Artemisia vulgaris*

Cain has also cited the two members of the Diapensiaceae from the Appalachian mountains, *Galax aphylla* and *Shortia galacifolia*, where the extreme rarity of the latter in comparison with the former can be blamed upon its imperfect mode of seed dispersal (or rather the absence of seed dispersal) and reliance upon vegetative spread. Here again is a conservative species which is quite adaptable to varied conditions in cultivation.

There is little understanding of the meaning of the different tolerances shown by different species. These differences are basically of genetic origin, but there has been, so far, practically no information about the gene differences which determine these qualities. It seems to be the case that the amount of DNA available in cells is capable of forming far more genes than are observed to be active and it may be that widely tolerant, 'common' species produce more different or alternative gene groups than do other species.

It has been dramatically said that every species has some strength or it would not survive and some weakness or nothing else would survive.

Good has tried to summarize the phases in the life of a species or a genus in what has been called the 'theory of generic cycles' covering the history of what we had better call a taxon, to avoid too definite a concept of genus or species, from its first appearance to its final disappearance (see also p. 3840). At the beginning its areal extension is nil and the *juvenile phase* is occupied simply by the extension of this area towards the maximum possible to it, which, however, it may be long in reaching, if·it ever does. As we have already seen the earliest stages of this expansion are not easily distinguished from the last stages of disappearance. Only long observation can assure us whether a taxon is advancing or retreating, though circumstances may enable us to form an opinion about it. An extraordinary number of species have only been found once and never seen again, even in regions where the flora is fairly well known. Some of these may have existed in one individual only and their fate is unknown, they may have been extinguished by their discoverers or, as has happened, they may lurk unnoticed for a century before being rediscovered. Further observations of any of these 'missing' species could be of great interest since they are obviously in a critical stage of their existence.

The second phase of the cycle is the *mature phase* in which the taxon displays its maximum genetic activity in giving rise to new forms which may appear in limited incipient areas either within the area of the parent taxon or arising from it and extending it in some direction. The third or *declining phase* is marked by retrogression in the area occupied, which may break up into disjunctive portions, and by the cessation of genetic activity. The last phase is the *disappearance phase* in which the area contracts towards the vanishing point.

Such a scheme is useful as a skeleton which the observation of nature may clothe in many different ways. As it stands it is too formal to be accepted as a blue print of the lifespan of a genus or a species, for the range of possible variations is very wide. Environmental conditions, especially changes experienced during migration which may sometimes be more linear than centrifugal, may prolong one phase indefinitely or may turn the scheme topsyturvy. There is no proof of the senility of species or that they die out spontaneously and we have already seen in the case of *Biscutella* mentioned above (p. 3854) that even a relict species can stage a come-back through polyploidy.

CHAPTER LVIII

MIGRATION AND DISPERSAL

DISPERSAL and migration are activities which are fundamental in the life history of almost all plant species and are indeed involved in reproduction. Among higher plants the fact that a multiplicity of structures exist which are held, on grounds more or less secure, to assist dispersal, is familiar to all botanists. The dispersal units, whether vegetatively produced, such as bulbils, turions or simply detached fragments, or the sexually formed seeds and fruits, are often collectively described as 'propagules' or 'disseminules' to emphasize their dispersal function. This being so there can be no question of finality with regard to the area occupied by a species at the present time. Expansion of the area will continue until it is halted by environmental barriers and observation shows that many species have not been halted but are continuing to spread, some apparently indefinitely. Wulff mentions a number of species which seem to be in the latter category, including three Veronicas, *V. tournefortii, polita, agrestis* and *Erigeron canadensis*. These are plants of open ground, otherwise weeds, but other weed species with extensive distributions: *Agropyrum repens, Capsella bursa-pastoris, Chenopodium album* and *Taraxacum officinale*, owe their apparent flexibility to the multifarious biotypes they have produced and their very variable populations.

There are also cases known where a species, which has for long been stable, succeeds in overcoming a barrier which has halted it and flows on in a phase of rapid extension over a new area. A striking case in which the spread of a species over a new continental area has been recorded is that of *Senecio vernalis* in Europe. Its career in Europe was first noted by Ascherson and subsequently worked out by Hegi and Beger. The species is Sarmatian, that is its principal area lies in the steppes of middle and south-eastern Russia reaching a height of 2800 m on Ararat. It is a ruderal plant in Europe, a haunter of railway yards and roadsides, like several of its near relatives. The first record of its arrival in Europe, was at Angerburg in East Prussia (now Augustow in Poland) in 1726. A century later, by 1826 it had reached the Vistula, but in less than another hundred years it had reached Saarbrucken and was on the border of France, and by 1894 it was in Switzerland. The general drift has been to the west and south-west rather than south, but has been extraordinarily irregular, despite the fact that this is an annual plant readily and copiously dispersed by wind. Even allowing for the imperfections of botanical records it is difficult to see in the migration of this species anything like the continuous or uniform process which is accepted as normal in theory (Fig. 4.80).

Migration is probably never continuous or uniform. It is sporadic, spasmodic and erratic. Above all it is generally slow. The few exceptions of rapid

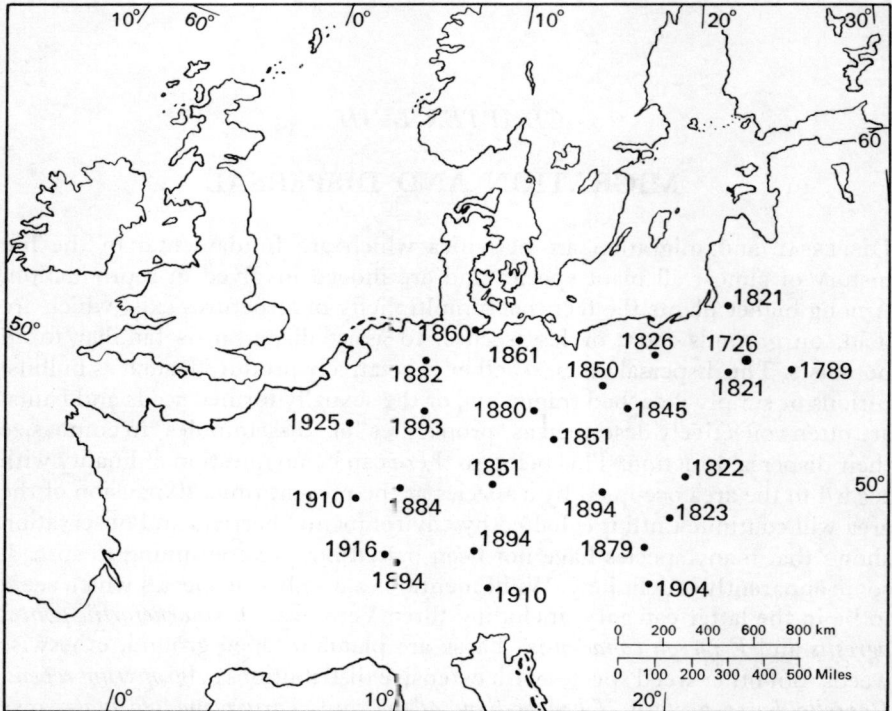

FIG. 4.80. The migration tracks of *Senecio vernalis* across Europe. See in text. (*After Beger.*)

spread by alien plants in a new country only serve to emphasize their exceptional character. Moreover, some of the most striking examples of rapid invasion are of plants with no special mechanism for rapid dispersal, e.g. in Britain, *Elodea canadensis* and *Matricaria matricarioides*.

Dispersal can be considered as 'regular' or 'irregular', the former referring to the employment of one of the 'regular' modes of dispersal, that is by wind, water or animals, all of which involve the production by the species of propagules which are suitable for one or other of these modes. 'Irregular' refers to accidental dispersal by any other means whatsoever. This does not involve any special structural advantage in the plants concerned and includes unconscious carriage by human travellers and passive transport in mud or soil on the feet of birds or transport by sea on floating wreckage of any sort. Such means of dispersal may only rarely occur, but they have the peculiarity that they have practically no limits of distance. This brings up another division of dispersal into 'short range' and 'long range' and similarly a division into 'dispersal into vacant land' and 'dispersal into occupied land' each of which involves different conditions and different considerations.

The slowness of dispersal in general, already referred to, points to short-range dispersal as the usual proceeding. The most famous inquiry about rates of dispersal was that on the island of Krakatao in the Straits of Sunda between Java and Sumatra. In 1883 the island blew up in the most tremendous eruption of modern times and inspection soon afterwards seemed to show that vegetation had been entirely destroyed by a thick layer of burning ash. The island is only thirty miles from Java which has a flora of about 5000 species and is about twenty miles from Sumatra. In 1886 an inspection by Treub found 16 species of flowering plants on the shores and 8 species inland, of which 6 were not found on the shore. Ferns were plentiful, in 11 species. The latter were almost certainly wind-borne and the shore plants water-borne. The inland plants were partly wind-borne and partly were probably bird-borne. Regular modes were thus obviously predominant over this relatively short distance. Fifty years later, in 1933, the number of species found had risen to 271, of which 30 per cent were classed as water-borne and 40 per cent as wind-borne. The remaining 30 per cent had no obvious relationship to a regular mode of dispersal, but may have been bird-borne or brought by man. A striking fact is that of the pioneer species, 115 of which were recorded before 1908, 57 had disappeared before 1928 because the vegetation was rapidly becoming closed forest.

Different conclusions may be drawn from these figures. One is that the 25 miles of open sea was no insuperable barrier to dispersal, the other inference is that in fifty years less than 300 species out of the immense reservoir on Java and Sumatra had made the crossing and part of this number may be discounted by the unsuspected survival of seeds and roots on the island itself, for which there is some evidence.

This evidence of slowness in surmounting a barrier is also brought out by the investigation of the mountain Ritigala in north Ceylon by Willis. Ritigala stands alone in an area which is too low lying to benefit from the south-west monsoon and is hence dry. The mountain, however, intercepts the monsoon wind and has a wet climate. The nearest mountains which have also a wet climate lie forty miles to the south. Willis found that apart from the dry-land plants on the mountain there were 103 species belonging to the wet zone. That is to say that in the ages since present conditions were established, only this handful of species had succeeded in crossing the intervening forty miles of dry country and the great majority of them had light seeds or spores which could have been wind-borne.

The 'long jump' is obviously not a regular feature of dispersal, rather is it a step-by-step procedure. We must not lose sight of the immense periods of time available for extensive dispersal even by the slowest modes. Ridley observed the Dipterocarpaceae in Malaya. This family of big trees produce beautifully winged fruits which seem specially suitable for wind dispersal. Yet Ridley found that the average distance at which fruits fell from the trees was no more than 40 m and that 100 m was the limit. In fact in the closed forest there is rarely any wind strong enough to carry off the fruits, which are no more than parachutes. Ridley estimated that the average age at which the trees began to bear fruit was thirty years. At this rate it would take a species 60,000 years to advance 160 km. Judged by the human time-scale this is exceedingly slow, but even so it would only take about two million years to traverse from India to the Philippines, the whole longitudinal range of the family, and this is by no means an extravagant period for the history of the family, which dates from the Oligocene.

When such lengthy periods are considered the advantage of special dispersal mechanisms is largely annulled. Consider the families which are conspicuous in the floras of all parts of the world: Compositae, Orchidaceae, Leguminosae, Rubiaceae, Gramineae, Umbelliferae, Cruciferae, Scrophulariaceae. They are all large families and may therefore be judged to be old. Among them the Compositae and Orchidaceae alone have a characteristic long-range dispersal system by wind. Even among Compositae some which have no pappus, e.g. *Bellis*, are more widely dispersed than others, e.g. *Cirsium tuberosum*, which have. Hooker, in 1853, remarked that among the Compositae common to the three isolated southern areas, Fuegia, Kerguelen and the Auckland Is. south of New Zealand, *none* have a pappus, while of those which have a pappus *none* are common to two of the three areas. Transport mechanisms may be very effective in increasing the frequency of a species locally, but in the long run they do little to affect its total range, which depends much more on its inherent flexibility and aggressive vitality. Spreading is usually only a few yards per generation, but for old species there are long periods available. Even at a metre per year a species possessing sufficient flexibility might travel linearly from the Pyrenees to England in a million years or spreading centrifugally could occupy the whole of western Europe or could indeed encircle the globe in less time than has elapsed since the Eocene period. Genetic potential, i.e. the gene reservoir, and tolerance (summed up in the old term 'adaptability') are ultimately the deciding factors controlling range, given enough time. As few families, fewer genera and hardly any species have achieved the range which geological time has offered them, it follows that barriers of one sort or another, climatic, material or biotic have been widely effective.

Herbs have an inherent advantage over trees in dispersal because of their short generations and rapid seed production. Nevertheless, trees have a biological advantage in their solid and enduring occupation of the ground they win, and in the long run they hold their own against the herbs, but as Arber has pointed out, they are phylogenetically some thousands of generations behind the herbs and cannot compare with them in flexibility or motility. If we look at the list of cosmopolitan families given on p. 3825, we see that only two, Leguminosae and Rubiaceae, contain any notable number of tree species and that it is the herbaceous sections of these two families which are most widely dispersed.

Among the various kinds of barrier there are few which are absolute or insuperable. Many climatic and material barriers can be overcome if the genetic potential of the species or genus can provide a biotype with enlarged tolerances. Heat, cold or drought among climatic factors may all be overcome by an appropriate biotype, though its appearance may often be indefinitely delayed. Material barriers such as mountain ranges or stretches of sea may both be overpassed in time, the former by alpine biotypes, the latter by accidental transport. As Heraclitus said, given time enough everything happens. Paradoxically it is the biotic barriers which can be the most formidable. A belt of dense and well-established vegetation is extremely resistant to invasion by seed-borne migrants or indeed by seedlings of its own constituents. It is for this reason that, as Salisbury has observed, free reproduction by seeds is characteristic of the plants in the early stages of an ecosere and that as succession progresses towards stabilization the plants involved come to rely more and more upon vegetative spreading. Long-distance transport therefore provides but little chance of permanent establishment except where the ground is vacant. Invasion, like infection, generally requires mass action for its success and this can best be achieved, unless resistance is too great, by slow step-by-step advances. Clements has said, 'most of the evidence available shows that effective invasion in quantity is always local', to which Willis added, 'It is clear that to think of plants in general as travelling rapidly about the world by aid of their dispersal mechanisms is to take a completely incorrect view of the situation.'

Elton and Pearsall have emphasized that the study of invasion by plants and animals conclusively shows that successful invasion is inversely related to the degree of stability in the invaded community. The most stable communities are also, in most cases, the oldest, those which have maintained themselves against the forces of change. They are also complex communities with a very diverse population of both plants and animals. Such communities are resistant to invasion. Communities which have been ravaged or decimated, whether by natural agencies or by man, are inherently unstable and it is these which are most successfully invaded by migrants.

A further advantage of short-range dispersal is in economy of seed. If we imagine a hypothetical case of an isolated plant whose seeds are dispersed uniformly in all directions, and if x represents the seed output per annum and y equals the number of available sites for germination, then ideally $x = y$ and $x/y = z$ where z is the efficiency of dispersal and when $x = y$, $z = 1$. If x remains constant and y increases the value of z will drop. With increasing radius of the area of dispersal, the increment of area per unit increment of radius increases very rapidly, with a concomitant increase in y. If therefore the efficiency of dispersal is to be maintained, there must be a corresponding rise in x, the seed output, to an extent far beyond biological possibilities. In other words the efficiency of dispersal, as measured by the utilization of available sites, is inversely proportional to the range of dispersal.

The ability of species to acclimatize themselves to differing conditions and thus to overcome many natural barriers to their dispersal depends on two factors, first, a sufficiently large population with a wide genetic potential and, second, abundant time. Neither of these conditions is usually fulfilled in long-range dissemination. This is illustrated by the copious evidence of the failure of plants introduced by man into gardens to spread beyond the limits of cultivation or to become integrated into native vegetation. In the great majority of cases the plants concerned are clones, descendants of a single imported individual or of a small number of related individuals and hence lacking in the potential of variation which is necessary, neither have they had a sufficient allowance of time. Quite opposite, but equally instructive, are the cases in which plants have been introduced by man into environments which are essentially similar to those of their native regions, but which they have failed to reach by natural means. Such species, aided by the absence of natural predators or parasites, have sometimes spread spectacularly, e.g. the famous invasion of *Opuntia* in Australia, which was only eventually checked by the introduction of a parasite. Similar, though not disastrous, has been the spread of *Matricaria matricarioides*, which, though ignored in nineteenth century floras, is now to be found at almost every farmyard gate in Britain.

The general rule, that aliens from a different environment have little chance of becoming truly indigenous in a new region, is supported by the paucity of exceptions. Wulff quotes Thellung as authority for the estimate that the species introduced into the Montpellier area over the last 300 years which have become naturalized only amount to 3·8 per cent of the flora. He does not state whether these species are truly integrated or still the casual followers of man. Most impressive is the report of de la Bâthie on Madagascar. He estimated that there were on the island some 900 species which had either been introduced or were naturally occurring aliens. They have flourished because of the very extensive destruction of the native vegetation by man, but only *one, Adenostemma viscosum*, has become integrated into natural communities.

Although the flora of Britain has been built up in post-glacial times by natural invasion from Europe, the greater part of this recruitment must have occurred during early stages when successions were still developing and vegetations changing. Later comers have not been particularly successful, as, for example, southern species like *Polypogon monspeliensis* which has failed to penetrate beyond the south and south-west coast.

These examples illustrate the difficulty we have indicated before, of overcoming a biotic barrier, success in which will depend largely on the frequency and the quantity of reproduction. Otherwise it will depend on an odd coincidence of chances which may rarely, if ever, occur.

The abruptness of a barrier may sometimes increase its effectiveness. Such conditions are usually physical, rarely climatic. A precipice, the edge of an inundated area or a sudden change of soil, may be almost impassible for there is no ecocline, no transitional zone offering a bridge across which variant biotypes of a species may make their way. On the island of Hawaii one may pass in a few yards from evergreen forest to dry semi-desert where the ground has been smothered by ashes from Kilauea and the barrier seems absolute, the change of vegetation total.

Highly effective also is the double barrier in which physical and climatic elements are combined. This is most generally found where a mountain range is big enough to cast a rain shadow on its leeward side, so that on one side the climate is wet and on the other side dry. The Cascade range in the northwestern United States shows this strikingly. On the Pacific side magnificent conifer forests rise as far as the watershed, while on the inland side is an extensive plateau covered with semi-desert sage brush.

Mountain ranges are not always barriers. A range lying along the line of migration of a species may afford a convenient highway along which an herbaceous species, for example, may travel above the limits of a barrier forest. Especially useful in this respect are ranges running north and south, which have served many species as bridges across the tropical forest zone or across formidable desert barriers.

Of material importance in dispersal is the degree of congruity between the conditions obtaining at the source of dispersal and the area reached by it. Dispersal is valueless if it leads the propagules into areas where they cannot survive. This is a factor which tells heavily against long-distance dispersal. Ocean currents can be of great service in the long-range dissemination of strand or mangrove plants, but they are almost useless to inland plants even if their fruits are carried to the sea by rivers and can survive the effects of salt water. The most famous illustration of this is *Lodoicea seychellarum*, the *Coco de mer*, whose 'double coconuts' are drifted over the Indian Ocean but have not succeeded in colonizing beyond its native Seychelles Islands, presumably because it is an inland plant of river banks. One is reminded of the Parable of the Sower and of the seed that fell, some on stony ground and some among thorns and was choked and some on good ground.

A factor in dispersal which may be overlooked is the comparative longevity of seeds. Long-lived seeds are exceptional and are mostly found among the heavier and most hard-coated seeds, which are the least adapted to long-range carriage, while the minute seeds or fruits, which are best suited to carriage by wind are, for the most part, comparatively short-lived. This is another hindrance to long-distance transport.

Wind transport is probably the most prevalent method by which plants surmount the barriers of mountains and of straits of the sea. Winged fruits have been found on the snow of high mountains as much as 1000 m above the nearest trees. It is also significant that Vogler found, in analysing the flora of Switzerland, that among true alpine species 59·5 per cent had adaptations favouring wind transport, while the non-alpine species only showed 37·9 per cent. That alpine plants living in a region of high winds tend to utilize wind transport is not surprising and it would be easy, and probably fallacious, to read this in the old-fashioned way as an instance of adaptation. More probably it is because of their reliance on wind transport that they have been preferentially enabled to reach their alpine height, a warning of how purely deductive methods may confuse cause and effect.

Across the sea, wind, drift and birds are the principal carriers. The present writer has often seen, on the coast of South Wales, the wind bringing thistledown across the Bristol Channel from Somerset thirty-seven kilometres away. As the pappus is unwettable it bounces freely on the water surface, but very few of these migrants still carried the akene fruits with which they set out and this, I think, must be a common experience so that such transport, though feasible, has probably a low efficiency.

There are instances where the appearance of trans-marine dispersal may be deceptive, in as much as the marine transgression is so recent geologically that the bulk of dispersal may have occurred while there was still a land bridge. The Straits of Dover provide a case in point, for the post-glacial period was well advanced before the sea cut through and made Britain an island. Turrill has shown that the same is true of the Aegean Sea basin where the breakdown of land is no older than the Holocene. Some 700 species have migrated across the basin from Asia Minor towards Thrace, north-westwards, or to Crete and Greece, south-westwards, but only a small number have crossed eastwards. There is, however, no evidence that these migrations took place after the sea was formed.

In the long term, evolution plays a part in dispersal. We have already several times mentioned the importance of variant biotypes within a species in determining its ability to occupy new areas. Through the long history of the Angiosperms the action goes much further than that. The genorheitron of Croizat is essentially an evolutionary track of migration and is built up by the continued differentiation of new forms from a parent prototype, with new and differing capacities for migration and for the occupation of differing habitats, so that the track of migration through space and time becomes, when seen as a whole, an evolutionary tree.

The phenomenon of *polyploidy* has been accorded an importance in regard to the distribution of plants which, although it is still a matter of discussion, deserves to be stated.

Polyploidy is the term applied to the increase of the number of chromosomes in the cells of an individual or a race by doubling, quadrupling or by some higher factor. The sporophytic number of chromosomes being $2n$ (diploid) we then have individuals with $4n$ (tetraploid), $6n$ (hexaploid) and sometimes still higher grades of polyploidy, though $8n$ (octoploid) is generally as far as the process goes in nature. Polyploidy is generally associated with morphological and physiological changes in the organism, some of which may be favourable though some are not. These changes cannot be indefinitely increased. There is a limit and though von Wettstein was able to produce 32-ploid mosses, the plants were weak and sterile.

When the series involves increases of number by cardinal multiples of n the series is called *euploid*. There are two categories of euploids, which are distinct in origin. In the first place it may arise from an autonomous doubling of the chromosome number in an individual. This is called *autopolyploidy* and it may be conditioned or favoured by a variety of factors. Second, it may arise through hybridization or by various forms of abnormal fertilization, though the results are often sterile. This is called *allopolyploidy*. The only allopolyploids of genetic importance arise by a combination of both processes, namely, hybridization followed by autonomous doubling of the diploid number. The significance of this lies in the following facts. A normal diploid nucleus contains two sets of chromosomes which are homologous, that is they are so closely similar genetically that they can pair off, each to each, in meiosis. If the two sets in a diploid hybrid are not homologous, they cannot pair, meiosis is impossible and the plant is sterile. But, if the two sets are doubled then each set has a second homologous set with which its chromosomes can pair and fertility is restored. The hybrid thus acquires the stability and permanence of a species.

Hybridization may produce unbalanced numbers of chromosomes, for example, triploid, with two sets of A and one of B, or pentaploid. This forms a series called *aneuploid* and such a condition is generally highly sterile, but here again autonomous duplication of the chromosomes restores fertility.

No full understanding of the causes of autonomous duplication of chromosomes has yet been reached, but it seems that a combination of internal and external conditions is required and that the requisite internal conditions are lacking in some families, e.g. Orchidaceae and Leguminosae (where the lack of endosperm has been blamed) which show a very low incidence of polyploidy. Artificial means of stimulating it have been found, one of the most certain of which is the alkaloid colchicine. Since 1937, when it was described by Blakeslee and Avery this method has been widely exploited for producing tetraploid races of cultivated plants which, in many cases, are larger, more robust and with larger flowers than the original diploids.

All this may seem to pertain rather to Genetics than to Plant Geography, but one fact does bring it into relationship, namely, that rapid and extreme changes of temperature have a marked effect in promoting the occurrence of polyploidy and that this is the only known means which is operative in nature.

Certain changes associated with polyploidy could be ecologically significant. Increased size and vigour, a longer vegetative period and wider tolerances in certain respects have been recorded, but it is clear that effects such as these are by no means uniform. Greater aggressiveness and competitive power may result from some of these changes and in some cases have been clearly shown, as, for example, by the allotetraploid *Spartina townsendii* in competition with one of its diploid parents, *S. stricta*, on the south coasts of England.

Although tetraploidy is frequently associated with reduced seed fertility it has also been shown that in many species it is associated with the development of vegetative propagation, either by stolons or bulbil formation or by vivipary, and this could increase both the aggressiveness in competition and the survival power of a species. It could, and this is important, also widen its ecological range and permit the occupation of areas which would otherwise have been unavailable.

Hagerup in 1928 was the first to draw attention to the significance of these considerations with regard to distribution. He compared four pairs of closely related species of Ericales and found that in each pair the polyploid member was the one with a more northern distribution. This idea was strongly supported by Tischler and by Müntzing. The former, in 1935, drawing on an immense knowledge of chromosome numbers, analysed various floras and found a significant increase of polyploidy in northern floras as compared with southern floras such as that of Sicily. This northern, i.e. cold temperate to sub-arctic, distribution of tetraploid species and of tetraploid races of single species has been widely confirmed as factual by numerous investigators. The deduction, however, that tetraploids were more resistant to cold than diploids, is theoretical and has been questioned by Bowden, who compared a large number of species in family groups, contrasting with regard to frost hardiness those of tropical or southern distribution with those (often of the same genus) having a temperate distribution. He failed to find a correlation with chromosome number.

Another point of view relates the occurrence of polyploidy with formerly glaciated areas. The forced migrations and intermingling of floras caused by the glacial advance could promote hybridization through what Anderson calls 'hybridization of the habitat'. Previously isolated diploid species are brought into contact and if doubling of chromosomes produced fertile tetraploids which were viable in such a severe environment, unusually high numbers of such polyploids might well occur.

This view lays no special stress upon hardiness but assumes only that which seems well established, the greater aggressiveness and colonizing power of polyploids, which enabled them to multiply themselves on the land vacated by the eventual recession of the ice.

Many investigators have reported the apparent association of high percentages of polyploidy with special environmental conditions apart from cold, such as those of maritime and coastal areas, true halophily, aridity and intense insolation. Results, however, seem to depend on the particular plants investigated and are in consequence somewhat inconsistent. It does not seem possible to affirm any universal rule of association between polyploidy and abnormal conditions though this may obtain in some taxonomic groups. The differences found by investigators may well arise from different historical causes and this is probably also true of the differences found in respect of the size of the areas occupied by diploids and by closely related tetraploids respectively. In several genera the polyploid species, although genetically younger, are more widely distributed than the diploids.

The centre of dispersal in general containing diploid and polyploid species tends to be the same for both, but the polyploids often spread further than the diploids and form a higher proportion near the boundaries of the area of dispersal, which may be a selective effect due to a higher rate of migration or may be related to more difficult conditions encountered at the limits of the generic area.

Apomixis is the condition in which normal sexual fertilization has been superseded or deflected by abnormal cytological changes (see Vol. 2, p. 1475) in the production of seed. It is widespread in some families of flowering plants. It is claimed that it is significantly correlated with polyploidy and similar claims have been made that it is correlated with extreme climates. For example, a high proportion of European species in Spitzbergen are apomictic, although their relatives in Europe are normally pollinated. Allowing for the fact that many pollinating insects are absent from Spitzbergen, it may well be that the apomictic strains have been climatically selected as the only ones which can survive in the local conditions.

All in all it must be concluded that in this realm of cytological plant geography, no certainty is yet possible only some degree of probability.

CHAPTER LIX

DISCONTINUITY

THE discontinuity here contemplated is geographical rather than genetic. It implies the distribution of a species, or more rarely of a genus, in two or more separate areas, sometimes widely asunder. It occurs at all levels of organization, not among Angiosperms alone, and it occurs also among animals, though less frequently.

Discontinuity is not a term which can be used with precision, because its limits, like those of endemism, are almost indefinable. No species, of course, occupies solidly its area of distribution. Edaphic factors themselves preclude this and in many cases climatic or topographical variations are also effective, so that gaps and irregularities of dispersal are everywhere to be found. These irregularities do not qualify as true discontinuity because wherever there is sufficient information it is possible to mark an area on a map and say that within this area the species is distributed and outside of the area it does not occur. Discontinuity exists where there are two or more such areas, geographically distinct, though the separation may be anything from a few miles to intercontinental distances.

Such separation of areas has always been a puzzle to botanists. In pre-evolutionary days it involved the supposition of several distinct and separate creations, a difficult idea to accept, while later workers have sometimes leaned towards a belief in the separate evolution of species in more than one locality. This is called *polyphyly* and we shall say more about it later (p. 3879).

In some instances, where very special conditions of life are concerned, a

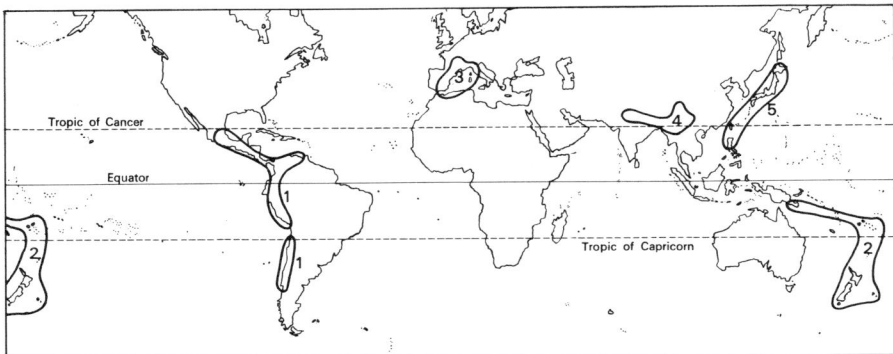

FIG. 4.81. Discontinuous intercontinental distribution of *Coriaria*. The species of the southern hemisphere are closely related, but they are quite distinct from those of the northern hemisphere. Approximate numbers of species in each area: 1·4 spp., 2·6 spp, 3·1 spp., 4·3 spp., 5·3 spp. (*After Good.*)

discontinuity has its explanation in the rarity of the appropriate conditions, but in the majority of cases an historical cause must be sought and it is in intercontinental discontinuities especially that biological support for the doctrine of the movements of continental masses is to be found.

Discontinuities are not only a phenomenon of species, they can be seen in the distributions of both genera and families. Many examples can be cited in all three categories. For a species there are two examples well known to British botanists. (1) *Eriocaulon septangulare*: north-eastern North America and west coasts of the British Isles. (2) *Spiranthes romanzoffiana*: northern North America and Ireland. Another case, typical of many, is that of *Salix herbacea*, an arctic-alpine plant, found on mountains all over temperate Europe as far as the Urals and also with a circumpolar distribution in the Arctic. This may be traced to post-glacial changes of climate.

Among discontinuous genera we may cite the following: *Liriodendron*— *L. chinense* in China and *L. tulipifera* in Atlantic North America; *Platanus*— *P. orientalis* in the eastern Mediterranean and *P. occidentalis* in eastern North America; *Torreya* (Gymnospermae)—*T. nicifera* (Japan), *T. grandis* (China), *T. taxifolia* (Florida) and *T. californica* (California). *Coriaria*, with five separate groups of species scattered across the world (Fig. 4.81).

Family discontinuities are shown by, e.g. the Magnoliaceae, which inhabit two areas, one in South-east Asia and the other in eastern North America. Two others with special connotations for continental drift, are Cactaceae; 25 genera which are pan-American, with a single genus *Rhipsalis* in tropical and southern Africa, Madagascar and Ceylon; also Vochysiaceae; 5 genera in tropical America with 1, *Erismadelphus*, closely related to the American *Erisma*, in West Africa. To these we may add the small family Rapateaceae with 8 genera in tropical America and one in West Africa.*

Discontinuities fall into several geographical types according to the areas involved. Thus there is the Arctic-alpine type of *Salix herbacea* mentioned above. There are also the North Atlantic type and North Pacific types, with corresponding types where areas are sundered by the South Atlantic and South Pacific respectively. There is a Europe-Asia type and an Asia-African type. Lastly we may mention an Indian Ocean type, which might also be called a 'Gondwanaland' type as it covers East Africa, Madagascar, with sometimes India and Australia, all lands included in the former conglomeration of continents called Gondwanaland. Actually the break-up of this continental mass began in the early days of Angiosperm dispersal in the mid-Cretaceous, and comparatively few of the very numerous Angiosperms whose present distribution indicates their Gondwanic origin have remained localized around its area. One such is *Adansonia* (Bombacaceae), the baobab, an antique and slow-

* This south Atlantic discontinuity also exists in many genera, e.g. *Saccoglottis* (Humiriaceae), which has 17 species in tropical America and 1 in West Africa, and *Paullinia* (Sapindaceae) 80 species in South America, only one of which occurs in Africa. Objection has been made to the relevance of the community of flora between America and Africa as evidence of a former continental connection, on the ground that only a small number of genera were involved, but Good lists 63 genera as distributed in both America and Africa and another 26 which also occur in Madagascar. There are also 3 which appear to be shared between America and Madagascar only.

growing tree, represented by ten species in Madagascar, one (which is wide-spread) in east Africa and one in Australia. If we take a wider geographical sweep, however, we find that many of the families and genera of Gondwanic origin have travelled further afield, some, like *Lobelia* (sect. *Rhyncopetalum*), include not only Africa, Ceylon and south India, but have progressed to the East Indies and even Hawaii. Others, like *Biophytum* (Oxalidaceae) while strongly occupying all the fringes of the Indian Ocean, have pushed both eastwards as far as the Phillipines and westward from Africa to tropical America. One section of the genus, *Dendroidea*, has one species in West Africa and fifteen in America.

It is perhaps noteworthy that in the cases of South Atlantic discontinuities the balance of species seems to be normally on the American side, even where the general plan of dispersal suggests that migration has passed from east to west. There is no recorded genus with a majority of species in Africa and an outlier in America.

Another geographical type of discontinuity is the South Pacific type, which involves such immense distances of separation between relatively small land masses, that it has for long been regarded as almost mysterious. In fact it has been recognized since Darwin's time that the only view which makes sense of these distributions is that the plants concerned are ancient radial migrants from Antarctica as a common centre. To contemplate such an explanation involves the recognition that the present terrible climate and accompanying sterility of that continent are only phases in its history. Fossil evidence shows that it has formerly borne temperate forests and the evidence for continental drift endorses the belief that it has not always occupied its present polar position (or the alternative Wegener view that the pole itself has shifted).

The number of genera involved is small but important, the most notable being the prolific forest tree *Nothofagus* in South America, New Zealand and Australia and *Fitzroya* (Coniferales), another forest tree, in South America and Tasmania. Others are: *Jovellana* (*Calceolaria*) in South America and New Zealand, *Hebe* (*Veronica*) in South America and New Zealand and *Pernettya* (Ericaceae) in South America, New Zealand and Tasmania. Alone in this group *Drimys* (Magnoliaceae), while found in South America and Australia, has intermediate stations on some Polynesian islands. To these may be added the small family of Eucryphiaceae, ranging from Chile across to Australia and New Caledonia.

Two important families exhibit a similar but even more extensive discontinuity involving South America, South Africa and Australasia. These are: Epacridaceae with an area divided between India, Australia and New Zealand and reaching out to South America and Hawaii; and Proteaceae which are massively represented in South Africa and Australia but spread over south-east Asia and New Caledonia and have 43 species in South America.

One of the most perfect examples of radiating dispersal from Antarctica is shown by a species of *Taraxacum*, *T. magellanicum*, which is distributed between South America, South Africa and New Zealand.

A transoceanic disjunction on a large scale, which has attracted much attention, is that between east Asia and North America, especially with eastern North America. Linnaeus himself noticed the floral affinities between the two regions, but it was Asa Gray who first analysed it in detail after the flora of Japan became known.

A peculiarity of this trans-Pacific disjunction is that there is a closer relationship between the Asiatic flora and that of eastern North America than there is with that of western North America. Gray reported 21 *species* as common to Japan and eastern North America and 15 in common with western North America only, while another 19 were found in Japan and on both sides of North America. These are identical *species*; the list of common *genera* is much more extensive. Gray listed 59 genera common to Japan and eastern North America but absent from the western side and Irmscher in 1922 listed 108 genera common to Japan and China and to North America generally. These are all genera absent from Europe. As the Chinese flora became better known the lists were increased and Hu, in 1935, dealing only with the woody plants of China, listed 959 genera, of which 156 were common to China and the eastern States of America.

The obvious geographical conjunction of North America and Asia in the region of the Bering Straits has led to a prevailing conception of the common elements in the two floras as resulting from a general southward migration from arctic lands during the Pleistocene, to which Hooker added the rider that it was the mountain masses of the western States which had deflected so many migrants eastwards towards the Atlantic coast. This is still regarded as orthodox in many quarters. It is correct that in Miocene times the Alaskan flora was temperate and that there was a southward drift in the Pleistocene, but most of the forest elements of the northern flora kept to the Pacific coast, as shown by their fossil remains and their present distribution. It is perhaps not without significance that about 90 per cent of the genera which apparently migrated to the eastern Atlantic seaboard are herbaceous or shrubby, mostly unsuited to forest life and with shorter generation-times and adapted for more rapid migration than the forest trees which occupied the Pacific slopes.* The wedge which divided these two divergent components of the Pleistocene migration may have been partly ecological as well as topographical. For Hooker's suggestion that the western mountains were responsible we may substitute the beginning of glaciation in the Canadian Rocky mountains, spreading eastwards to join the Keewatin glaciation and really forcing the floras apart.

This Pleistocene migration was, however, only a recent event in the long history of angiospermic migrations. We cannot conclude from these late Tertiary happenings that the migrant temperate flora *originated* in what is now arctic land, for migration goes much further back than that. Investigation of the history of a typical 'North Temperate' genus, *Euphrasia* (Scrophulariaceae) has a lesson to teach. Du Rietz in 1931 examined its distribution as an example of what has been called 'bi-polar' distribution between far north and far south, with a great discontinuity between. There are a number of other bipolar plants, and they have always provided material for speculation. Du

* But see Hu, above. His woody plants are largely shrubs.

Rietz showed that the North American species of *Euphrasia* are not closely related to those of South America, but the latter, on the contrary, are related to those of New Zealand and Australia. On the island of Juan Fernandez, off the coast of Chile there is, for example, an endemic species, *E. formosissima*, which although it belongs geographically to the South American group, is most nearly related to the New Zealand species. In New Zealand there is, interestingly enough, also a unique, monotypic plant, *Siphonidium longiflorum*, which is closely related to *Euphrasia*, almost its only close relation in the southern hemisphere. In Australia there is also a discontinuity between the area of the south coast and the disjunct areas of *Euphrasia* in the East Indies, reaching north as far as Formosa.

All this suggests a long and puzzling history, long antedating the Pleistocene. Croizat has given it a logical interpretation on his usual basis of a far-southern origin, somewhere in what is now the South Pacific, but which is presumed to have been part of the former land complex of Antarctica, Australia and New Zealand. From this centre the genorheitron runs westwards in two forks into southern Australia and New Guinea respectively. This line connects to the northern area in China, from thence passing westward into Europe and eastwards into North America. The other line from the origin went eastwards to Chile, but did not progress northwards from there and remains disjunct. Croizat points out that the conifer *Libocedrus* followed an almost identical course from the same source as *Euphrasia*, except that it died out of Europe (fossil remains) and out of most of North America (also fossil).

We have stressed this case at some length because it well illustrates the necessity of considering distribution *globally*. Conclusions based upon local distributions only, divorced from the world scene, can only lead to error.

The floral relationship between east Asia and North America has been urged by Diels as an argument against continental drift, on the ground that if it were true, then the Pacific gap was formerly much wider than at present, not narrower. Looking at the globe from a southern point of view, however, this argument loses its force since the drift theory envisages a close approach of South America, South Africa and Australasia in a complex with Antarctica, the latter being the supplier and the three former being the receivers in the primary dispersal of the Angiosperms.

It is idle to press speculation too far, unless the arguments for and against a theory can be discussed in detail, as Croizat has tried to do in his immense volume, but it can be admitted that his 'southern origin' view gives more logical and comprehensible accounts of migration than any other theory.

Van Steenis, in 1962,* published a survey of trans-oceanic plant distribution. He emphasized that botanical affinities show an essentially *latitudinal*, east–west, zonation in all climatic zones and that this zonation is crossed by two principal oceans, the Pacific and the Atlantic, with *longitudinal* axes. The important floristic links which exist transversely to these oceans, especially in the tropics and the southern hemisphere where the oceans are widest, constitute one of the most considerable problems in Plant Geography.

* J. VAN STEENIS (1962) 'The Land-Bridge Theory in Botany'. *Blumea*, **11**.

The author points to the present existence of three land-connections between continents, namely: Panama, the Bering Straits area (Beringia) and the Malaysian archipelago which links Asia to Australia. These provide what is, from the biogeographical point of view, one great land-mass.

He rejects the Wegener theory of continental drift as it then was, and argued strongly in favour of land-bridges within the tropics, across the Pacific and Atlantic Oceans and between India and south-west Africa (in place of Gondwanaland). Another southern link, between Antarctica and Australia on the one hand and South America on the other, is in line with Croizat's views and not inharmonious with recent developments of the drift theory.

Such assumptions certainly provided consistent answers to the distributional problems and the evidence of amphi-oceanic and pantropical genera now widely scattered on oceanic islands across the Pacific serves to support his views. He summarizes his belief in continental stability and the persistence of climatic conditions, at least throughout the Tertiary, as the 'steady-state' principle, while admitting the reality of geomorphological changes on a scale which seems hardly compatible with such a principle.

The emergence and submergence of such extensive land-masses, or even of archipelagoes, across the great ocean basins seems quite out of harmony with the recent developments in the study of palaeomagnetism and the history of the oceans which they have revealed, developments which had made little impact in 1962.

We have described this work of van Steenis because it is one of the latest and strongest counter-arguments to continental drift. The work is full of valuable data though the argument is not now as strong as when it was written.

Discontinuities within a single continent are of wide occurrence. Conspicuous cases are those between North and South America and between east and west Australia. In the Americas there is a parallelism between the plants of the dry sub-tropics north and south of the equator respectively, which are separated by 3000 miles, mostly of wet tropics. Some of the semi-desert shrubs are identical in the two regions, notably the well-known creosote bush, *Larrea divaricata*. Other genera, e.g. *Acacia* and *Prosopis*, the mesquite, are represented by closely related pairs of species. That this is an ancient type of vegetation is shown by the fact that some members of it have also disjunct areas in east Africa and one, *Fagonia*, has an area in the Mediterranean. We have written earlier (Vol. 3, p. 2563) on the geographical aspects of this discontinuity.

Australia has many instances of discontinuity between the south-eastern and the south-western corners of the continent. In late Cretaceous times the two areas became separate by a vast invasion of the sea over what is now the Nullarbor plain. This was not re-elevated until the Pleistocene. Wood and Bass-Becking consider that *Eucalyptus* originated in the south-west and that dispersal eastwards took place before the areas were separated. Each region has developed its own endemic species which are nevertheless closely related in pairs. Other smaller genera show a similar discontinuous separation east and west of related species pairs, e.g. *Drosera*.

Intra-continental discontinuities exist in all the continental areas. In Europe we may cite *Rhododendron ponticum*, the mauve flowered shrub which naturalizes so easily in our forests. It is native in Europe only in the region of the Black Sea and in the extreme south of Spain and Portugal (as var. *boeticum*). This is a relic of the Tertiary flora of Europe found fossil in a number of inter-glacial deposits in central Europe, but extinct before the end of the Pleisto-cene except in the extreme south-east and south-west.

On a minor scale geographically but botanically important are altitudinal discontinuities. Examples of this are frequent in mountainous areas where a species may be isolated at a high level on separate peaks. Another form of it, which is rarer is that of a single species occurring at two different levels, for example, *Armeria maritima* in Britain, which grows on seashores and on cer-tain mountain tops respectively.

Many examples quoted of this type of discontinuity are not truly such but are examples of vicarism, the replacement of one species by another. Turrill gives an example from the mountains of Greece and Crete. A genus may be represented by a number of alpine species each of which is restricted to a single mountain or a group of mountains. but which together are closely related to a neighbouring lowland species of wide range. Here the lowland form is replaced at high levels by alpine variants which may be specifically or only varietally distinct.

One cannot leave the subject of discontinuity without some mention of isolated oceanic islands. The definition of an oceanic island is that it is not and never has been part of a continental mass. This excludes islands situated on a continental shelf, which lie therefore within the range of probable tectonic elevation or submergence. Truly oceanic islands are separated from the near-est large land-area by deep ocean and apart from possible movements of the continents they have probably never been closely associated with any. Most of them derive from the mid-oceanic ridges. Their populations must therefore have reached them by migration, often by exceptional and rare migrations. Such populations must have been for long isolated, and it is common to find that they include a high percentage of endemic species or even genera. This, however, is a different thing from discontinuity as we have been treating it. The latter is only involved if we find that species of a genus are shared between an oceanic island and a remote land-area or with other islands, i.e. are non-endemic.

Special interest in this connection attaches to the islands of the Antarctic Ocean, which are all oceanic in the strict sense and are widely scattered, from South Georgia in the west, which is 1200 miles from Fuegia, and include Tristan da Cunha, the Crozets and the Kerguelens, and extend to Macquarie Island, 660 miles south of New Zealand. Their floras are all scanty, only about one hundred species for the whole group, of which half are endemic to the region but of which the other half is shared with Australasia and South America. There are two species shared with Antarctica itself (*Deschampsia antarctica* and *Colobanthus crassifolius* (Caryophyllaceae)) and a few species which are very wide-ranging, some of which, like *Luzula campestris*, have an extensive north temperate range.

Another most interesting group of islands is the Hawaiian group in mid-Pacific, 2400 miles from San Francisco. Not only are they about the most isolated of all but they are of comparatively recent volcanic origin. They have been subjected to massive invasion of foreign species from America, due to human agencies, which has been going on for so long that it is difficult now to recognize truly native species, but a careful estimate by Fosberg puts the native flora at 216 genera and 1729 species and varieties. Over 90 per cent of the species are endemic, but Fosberg estimated that these could all have originated from 272 migrants. Where these came from is questionable, but there is certainly a strong element which has close affinities with the south-west Pacific, i.e. Australasia and Polynesia. In fact the isolation of the Hawaiians is less impressive when one realizes that Oahu is only 1100 miles from Palmyra Island and that from there south-westward there are many links towards the multitudinous islands of Melanesia.

Those who are sceptical of the possibilities of exceptional long-range migrations would do well to reflect on Good's estimate that, accepting the age of the Hawaiians as 5 million years, the 272 original immigrants could have been recruited by only one new arrival every 20,000 years. It is difficult for us to grasp the importance of the prodigious stretches of time inherent in geological history. Taking 140 million years as the period since the beginning of the Cretaceous, when the Angiosperms appear to have started on their primary travels, a species would only require to have advanced on the average 126 yards every 1000 years to have spanned the globe from the Antarctic to the Arctic by now. If we take the much shorter period of 25 million years as measuring the actual advance to dominance in the world's floras, a species could have circled the globe with a speed of 1·5 km per 1000 years or about 700 m in the same period to have advanced from pole to pole. How few have done it, how sluggish must many have been and how powerful the ecological pressures which have held the rest in check!

The consideration of discontinuity is greatly widened if we include in it the numerous cases in which discontinuous areas are occupied, not by a single species but by a pair of closely related species, which is commonly called *vicarism* (see also p. 3348). The term is used in a wide sense and is applied to any pair of species or of families or genera or of sub-species or varieties, or indeed of communities, which are closely related and of which one replaces the other either geographically in a different area, or ecologically on a different soil or at different altitudes or at different seasons. It is even sometimes extended to cases where two organisms are not closely related, but which occupy the same ecological niche in separate habitats. In this latter sense, *Hippocrepis comosa* in Britain is vicarious to *Lotus corniculatus* on the older limestones. This seems to make the concept much too wide and loose and for cases such as this, where there is no close common descent shared between the paired units it is better to use Vierhapper's term, *substitution*. This is the best term to use where the larger units, families or genera, are concerned. For example, the Aizoaceae may be said to be substituted for the Cactaceae in semi-desert environments in South Africa.

There are a considerable number of vicarious species separated between

North America and Europe, between which species the differences are slight. Many of them are herbaceous, e.g. (from Fernald)

EUROPE	NORTH AMERICA
Anemone nemorosa	*A. quinquefolia*
Luzula pilosa	*L. saltuensis*
Oxalis acetosella	*O. montana*
Hepatica triloba	*H. americana*
Scrophularia nodosa	*S. marilandica*

They are associated in America with a tree-flora the ancestors of which are known to have ranged widely around the northern hemisphere (i.e. circumboreal) in pre-glacial and Tertiary times. It seems reasonable therefore to regard these disjunct species as descendants of a common ancestry in the earlier circumboreal flora which were driven southwards and separated in the Pleistocene. A similar separation occurred between species whose descendants are now found in North America and East Asia (p. 3870), nearly all of which have disappeared from Europe.

Mountain vicariants we have already referred to (p. 3873). Not uncommonly a widespread lowland species shows a close relationship to an alpine species which replaces the lowland species at high altitudes, e.g. *Phleum pratense* and *P. alpinum*. The alpine member of the pair may be itself of wide occurrence, though confined to high levels, or it may be endemic to one area. Turrill has also pointed to cases in Greece and Crete where a lowland species is related to several alpine species, each on a different mountain. One alpine species may even be vicarious to another alpine species in a different mountain area; thus *Saponaria ocymoides* of the Swiss Alps is replaced by *S. pumila* on the mountains of the Austrian Tyrol.

Ecological vicariants depend mostly on contrasting edaphic factors such as fresh or saline water, light or heavy (clay) soils, or on soils of low pH (acidic) and soils of high pH (lime or carbonate soils). Examples of these in the British flora are:

Saline soils—*Trilochin maritima*.	Fresh water soils—*T. palustris*.
Light or dry soils—*Lotus corniculatus*.	Heavy, wet soils—*L. uliginosus*.
Geum urbanum.	*G. rivale*.
Acidic soils—*Galium hercynicum*.	Lime soils—*G. sylvestre*.
Listera cordata.	*L. ovata*.

Other soil peculiarities may also be marked by the occurrence of vicariant species, such as serpentine soils (p. 3662) or gypsum soils, the latter especially in the United States.

The different habitats of vicariants, are not always to be taken as dictated by positive preferences, but rather by differing ranges of tolerance in conjunction with natural competition.

Climatic differences seldom govern vicarious association. The case of the replacement of *Ononis natrix* by *Limonium thouini* in the climatic ecocline between Jerusalem and Jericho, described by Boyko (p. 3508) is one of substitution rather than of vicarism.

The term 'relic', or 'relict', is one which is frequently used by Phytogeographers, but often with so undefined a meaning that it is virtually useless. At one time every endemic which had a restricted area was indifferently dubbed a 'relict' and written off as moribund. Such an uncritical attitude is now a thing of the past. We may well inquire in what way the term may be legitimately used, that is whether it has a field of application which is distinct from the term 'endemic', even though the two concepts may overlap. To put it another way, all endemics are not relicts and all relicts are not endemics.

To improve the situation the first requirement is to restrict the application of the term by eliminating ideas which are merely ancillary, such as the age of a relict or the palaeobotanical evidence about its past. For use in a geographical sense it should also be separated from the taxonomic use of the term, that is its application to taxonomically isolated species or genera in large families when they retain primitive or presumably ancestral characters.

The essential meaning of 'relict' is that it is a remainder of something which has gone, and this condition entails some degree of disharmony with the present situation, not necessarily with the actual habitat of the organism, but with the general condition of affairs. Some endemics qualify as relicts in this sense, particularly pre-glacial species which have survived in special refuges and have not reoccupied the territory from which the ice has disappeared, though they may continue to flourish in their own limited area. These we may call *climatic relicts*. Plants which have survived climatic changes and are indicators at the present day of the former prevalence of a different climate, are numerous and almost every flora can provide examples. They continue to survive because of their wide tolerances, though not in harmony with existing conditions. An example is the survival of small colonies of *Fagus* in northern Sweden, far to the north of the present limits of beech forest, which are relicts of the period of the climatic optimum in the post-glacial climatic sequence. Similar relicts are the steppe species found in East Anglia, which are considered to be survivors from the xerothermic post-glacial period.

Climatic relicts are not always restricted to special areas. Wulff introduces *Buxus sempervirens* as a Tertiary relict in Europe, where it now grows in a number of disjunct areas under different conditions, climatic and edaphic, but which nevertheless occurs as an integrated member of the natural vegetation in those areas. It is a relict but not an endemic.

The opposite condition may also occur, that is to say localities in which for some physiographical reason the climate has *not* changed and the area may conserve a whole relict flora.

While climatic relicts are probably the majority there is also a smaller group which are relicts of geological changes. These Wulff calls *geomorphological relicts* and they are for the most part relicts of the ancient shorelines of seas which have vanished. He instances the occurrence of isolated patches of *Quercus ilex* on the southern side of the Alps. This Mediterranean species is normally associated with maritime conditions and its occurrence in the Alps can be attributed to the former existence of a Tertiary sea which covered the basin of the Po and whose shoreline lay along the southern base of the Alps.

Stomps drew attention to the survival in East Anglia of some species which, in Europe, are characteristic of the Rhine Valley. He regards them as relics of a period before the submergence of the North Sea, when the Rhine extended north-westward to an outlet between Norway and the Shetlands and drained the whole of the North Sea basin.

Discontinuity of distribution is not a rare phenomenon. If we consider genera as units we may reckon there are 12,500 genera among Angiosperms. This figure can only be approximate, since taxonomic work is constantly splitting or combining genera and their limits are often neither very precise or universally agreed. Uncertainty abounds particularly in regard to a few families whose generic distinctions are based upon rather minute characters about which mistakes are quite possible. Such are the Compositae, Umbelliferae, Cruciferae and Acanthaceae, among which generic changes are frequent. A wrong attribution of a species to a genus may make a great difference to the map of its distribution and in this way false discontinuities may be recorded.

Correct taxonomy, or at least the most recent taxonomy, is essential to the study of plant geography and the two disciplines should run in double harness, each contributing to a common objective, a correct understanding of the relationships and history of the Angiosperms. The geographer must be satisfied as to the taxonomic verity of the units he is considering and the taxonomist should consider geographical facts as a check on his morphological conclusions.

Good endeavoured to make a list of genera with wide discontinuities based upon careful estimates of the taxonomic reliability of the genera concerned and rejecting all doubtful cases. This was first published in 1927 and has been four times revised, the latest edition being in 1964. The list may therefore be considered definitive, so far as that is possible. It now includes 765 genera, which is 16·4 per cent of the putative total of 12,500. This shows that generic discontinuities are by no means rare.

When we turn to families we find that the difficulties, which were mentioned above as applying to genera, are even more conspicuous. Opinions as to what suffices to distinguish a separate family vary greatly, and the total number recognized has risen steadily from Bentham and Hooker (1862–83) who recognized 200, to Hutchinson who, in 1959, described 411. There is thus no secure basis for estimating the frequency of discontinuity in families, though a number of families with discontinuous distribution can be counted if we confine ourselves to families well-established and recognized. Irmscher (1922–29) used the 289 families of Engler's system, and of these he claimed that 228, a very high proportion, showed one or more major disjunctions between separate continents, 32 being divided between two separate continents, 46 between three continents and 150 occurring on all four of the continental areas he employed, namely, the Americas, Europe with Africa, Asia, and Australia with Polynesia. The remainder, 61, being limited to a single continent. He only considered inter-continental discontinuities and he did not analyse discontinuities within a single continent, as, for example, between temperate North and South America, which would have increased his total.

Good, in 1964, seeking to be as inclusive as possible combined the totals of former systematists and added some other families to make a grand total of 435. In his *Geography of the Flowering Plants* he lists in detail the following categories of discontinuity.

A. Divided between north temperate and south temperate zones, 14.
B. Divided between America and Eurasia and/or Australia, 32, of which 9 are tropical and 9 in the southern hemisphere.
C. Divided between America and Africa, some also in Madagascar, 12.
D. Divided between Africa and/or Madagascar and Asia and/or Australasia, 16.
E. Divided between all three regions of the temperate southern hemisphere, 6.
F. Divided in various other ways, some scattered, e.g. Coriariaceae (*Coriaria*) (see Fig. 81, p. 3867) which is divided between alpine South America from Mexico southwards, the western Mediterranean, alpine east Asia, New Zealand and New Caledonia, 16.

The traditional view of discontinuities, that the disjunct areas represent the remnants of a formerly extensive and continuous area is certainly not applicable to the great majority of the instances we have cited. It could only apply to a discontinuity within one continental area and only there to species which belong to the relatively small class of relicts. Only a few genera and very few families can be considered to be in this class and they are either monotypic or very small. As things stand at present the old view cannot apply to a large and flourishing family like the Proteaceae which are spread over Australasia and east Asia, the whole southern half of Africa and most of south and central America, since no possibility of geographical continuity exists between these areas. To trace a possibility of former continuity we must go back to earlier times before continental separation took place, and even if the continents were conjoined it would not make the area of the family any more extensive than it is, even supposing it to have existed at that remote time. It might be considered as open to question whether purely inter-continental discontinuities such as that of the Proteaceae should not be regarded as false or pseudo-discontinuities, where the possibility of continuity does not exist.

A better conception of discontinuity, including the cases within the same land-mass, may be reached through Croizat's interpretation of the genorheitron, in which we can see complex and branched tracks of migration, sometimes converging on a region by different routes, so that the same species or genus may have been established on two separate areas which never have been united.

This view of the independent origin of the population in two disjunct areas is known as *polytopy*. It is a concept which may apply where two areas of one species are separated by more than the normal range of dispersal, or, it may apply to cases where two closely allied (apparently) species or genera are also widely sundered geographically. Willis raised the question by what conceivable genetic bond could two close genera of Agrostideae, *Lagurus* (Mediterranean) and *Pentapogon* (Victoria and Tasmania) have been connected. Where any two taxa lie far apart, the alternatives seem to be, either that, whatever appearances suggest, they are not closely related, or, if they are, that they must formerly, somewhere, have been close together where they could have sprung from a common ancestor.

The possibility of a third explanation has often been considered, namely, that species which have exceptionally disjunct areas might have *evolved* independently in more than one locality, an idea called *polyphylesis* or *polyphyly*. Where larger groups than a single species are involved there is little doubt that in a number of cases units, either species in a genus or genera in a family, which have been united taxonomically because of morphological resemblances, are in fact disparate, that is that they have not a common genetic origin, but are polyphyletic and have converged towards a common form. There is nothing controversial in such a belief, it is simply a mistake in taxonomy, due to imperfect knowledge, to be set right, as it frequently is, by fuller investigation. Whether individuals as closely related as are the members of a single species could be of independent origin is another question and the weight of probability has been generally held to be against it.

In pre-evolutionary times the realities of migration were not recognized and naturalists found no difficulty in the idea of double, or multiple, creations in different localities. Now, however, when we know something of the enormous time and spatial complexities involved in the migrational history of the flowering plants, the case for the polyphylesis of a single species would have to be very strictly proved by the elimination of every other possibility. This would be a formidable task, and it may be said at once that it has never been achieved. There is always an element of doubt and judgement is reserved. A diphyletic origin for two populations of a species must, in any case, be a very rare event. To admit its frequency would be to nullify at one blow all inquiry into migration and to put us back to the standpoint of the special creation theory.

Polyphylesis must not, however, be ruled out as an impossibility. The requirement is that two stocks of similar constitution should arise independently from one and the same parental syngameon at different places or at different times. That two parallel gene mutations could happen independently may be rare but is not impossible. It might indeed be much less than rare, but the result only preserved by the environment at certain points of space or time. Were the new taxon the result of hybridization plus polyploidy, or of an abnormal reproductive process, such as apomixis, then the probability is still more favourable. It is not necessary to suppose the production of two *identical* stocks, but only that they should have certain specific characters in common, since even accepted species may contain quite wide varietal differences.

Under the differentiation view of evolution, the polytopic origin of related species from an ancestral species or syngameon must be regarded as a regular process. The instances of isolated alpine species each of which had a limited endemic distribution but which are all related to a widespread lowland species provide a striking illustration, of which there are numerous examples. This, however, is quite distinct from polyphylesis, which presumes an independent phylogeny for two units of one taxon. This latter is probably not uncommon among larger taxonomic groups, as we have already pointed out, but it remains unproven for species.

Simple polytopy offers great problems where the present separation of the related forms is very great, as Willis, quoted above, points out. The extreme in this respect is the case of species or genera of bi-polar distribution, those, that is to say, which have a more or less extensive boreal area and one or more austral centres at the southern extremities of the continents. Most of the species have no intervening tropical stations, but some occur at alpine or high-level localities in the tropics. Species with bi-polar distribution almost all belong to the genus *Carex* (6 species), with the exception of *Armeria elongata*, *Draba magellanica*, *Deschampsia atropurpurea* and *Koenigia islandica*, but the Angiosperms are quite outnumbered by the Bryophyta and Lichens (26 species). There are also some genera the representatives of which are divided bi-polarly: *Littorella*, *Empetrum*, *Chrysosplenium* and *Primula*. Some of the outstanding families of temperate latitudes, Ranunculaceae, Cruciferae, Umbelliferae and Juncaceae, are also divided between north and south temperate zones with very sparse representation in the tropics, and there only at high levels.

These plants must, supposedly, have crossed the unsuitable tropics in some way and opinions have varied between those who believe in the sufficiency of long-distance dispersal and those who consider that the traverse was accomplished along mountain chains by comparatively short jumps. Where this view fails is in the far East, in the route from Malaya to Australia (or vice versa) where high mountains are few and far between. However, as this part of the world has certainly undergone much tectonic change in geologically recent times it is not improbable that there was an alpine path in the required direction not too long ago.

It is stirring to reflect, when we look down upon even the tiniest flowering herb, on what a history it hides of fabulous journeyings through time and space, creeping slowly across the world, over mountains which have disappeared and through vanished lands that we shall never see, holding its own among the vast changes of an hundred and twenty million years, till it arrived here at our feet.

CHAPTER LX

FLORISTIC COMPONENTS IN VEGETATION

WHAT do we mean when we speak of a 'flora'? In a narrow, traditional sense it is simply a descriptive catalogue of the plants now growing in a certain defined area, generally an area politically defined, a single national territory or a section of it, whose boundaries may have little or no significance in Plant Geography. This is no doubt convenient, but it is not scientific, unless the boundaries happen to be also natural boundaries, as is the case with islands or desert areas. The Plant Geographer must seek a wider concept with a more natural basis and his areas will be wider and fundamentally climatic, though with geographical subdivisions. Thus we may recognize a specific arctic flora, which, as it happens, does not require subdivision. Most other climatic zones do require continental division, the floras of temperate and of tropical America, Africa, Europe and Asia being all distinct. Australia, New Zealand, Japan, Britain and the East Indian islands are already natural units as are also the oceanic islands. The flora of temperate Europe, the flora of the Mediterranean basin, the flora of tropical Africa, these are examples of the kind of floristic units that Plant Geography can recognize. The concept has been further refined in the idea of Floristic Provinces as we shall see later.

A natural flora* is then an assemblage of plants which have become integrated, through their common harmony with the regional climate and by the balance of competition among themselves, into a stable whole. To call it stable does not mean that it is exempt from changes within itself, quite the contrary, but the flora as a whole is quasi-permanent except on a geological time-scale. It results from a complex of migrations in the past and it is still open to additions by further migrations. It is therefore a composite assemblage since its members have arrived from various directions and every species has its individual history. Only in a few tropical regions which have remained relatively unchanged since Tertiary times is the flora to any marked degree autochthonous, that is to say that it consists to any notable extent of species evolved on the spot, and even there the ancestors of such species also arrived by migration, the universal background.

The provenance of the members of a flora may be deduced from a study of their general distribution; their total area and the centre of dispersal, if that can be identified, give obvious clues. In this way the flora may be analysed into a number of components (sometimes called 'elements') according to their various origins.

Robyns has shown that a component analysis is possible even in the tropical

---

* The term is also applied, by an association of ideas, to the assemblage of fossil plants in a geological horizon or over a given geological period.

forest region of the Congo. He recognizes seven different components characterized by their distribution outside the Congo region. (1) Pan-african. (2) Tropical African, but of limited distribution, including ten subdivisions with different geographical affinities. (3) Guinean. (4) Oriental. (5) Zambesian. (6) Sudanese. (7) Ethiopian.

This is an example of an analysis on a purely geographical basis. Wulff has argued that this is not enough and has proposed a number of other bases of analysis, namely, migrational, genetic, historical and ecological. The first depends upon identification of the actual routes by which species have migrated, the second upon their actual source of origin as species. The last-named is based upon habitat preferences and is only marginally a part of Plant Geography. The migrational and genetic components are only idealistically realizable, but the notion of historical elements or components, classified according to the epoch of their arrival in the flora, does offer useful possibilities in countries where the Quaternary remains have been sufficiently explored. Wulff and most other writers admit, however, that the geographical basis is the most important and generally valuable. Not all the species in a flora are eligible for a geographical analysis. There is a core of species which must first be subtracted from the total. These are, first, the autochthonous species, if any, likely to be few in temperate floras, unless they cover a very large area like North America, but many more in the tropics. Second, endemic species, a list which will overlap the previous one, and, third, those species with so wide a distribution, including weeds of cultivation, that they give no useful indication of their immediate origin. These abstracted, the remainder may be assumed to be immigrants or 'penetrants', which have come from otherwhere.

The assignment of species to a particular geographical group is not always easy. Its complete area of distribution may not always be known and may require monographic research. Writers have frequently assigned species to different geographical areas, depending sometimes on their own personal experience of the species and its occurrence. Regard must be had to the frequency of the species in different regions and an endeavour made to discover its 'mass centre', that is the area of its commonest occurrence, which is statistically more probable as a centre of migration than areas where the species is rarely found. In the case of Britain there has been a tendency to accept the nearest continental area of its occurrence as its source, which by no means follows.

The consideration of 'mass centres' may avoid the type of error, pointed out by Wulff, influenced by which Hooker classified certain Himalayan species as a European component in the flora of India, whereas Wulff maintains that the Himalayas are their mass centres, and that they truly represent a Himalayan component which migrated to Europe in post-glacial times. We have previously (Vol. 3, p. 2556) also pointed out the error of classifying certain components of the Tertiary flora of Europe as 'East Asiatic' or 'North American' simply because of their present day survival in those areas, whereas in the Tertiary times they had no such territorial associations, but were just as indigenous to Europe as anywhere else and some were probably pan-temperate.

Britain is in an exceptional position as an island peripheral to Europe and its flora lends itself therefore most easily to a geographical source analysis. The comparatively small number of per-glacial survivors in the islands means that a majority of the species in the flora are immigrants. Pollen-analysis of post-glacial peat deposits has made possible the building up of a history of the immigrations in a fair amount of detail (see previously, under 'The Quaternary Era', Vol. 3, p. 2595).* The number of endemic or autochthonous species is also very small, whereas in more isolated areas, like the Hawaiian Islands, the very large proportion of endemic species makes the tracing of the sources of the flora correspondingly difficult. With such floras, or the flora of a large area like North America which naturally have a large proportion of endemic species, it is not possible to trace specific sources outside the area, except in a small minority of species, though their internal migrations may often be followed. Recourse must be had to larger units, the genera or families to which the endemic species belong, whose external distribution and courses of migration, when they can be traced, may be enlightening. Croizat's fruitful use of the genorheitron concept may be, in such cases, the clue to geographical analysis.

A number of geographically defined elements in the flora of Europe were recognized by Steffen in 1935, and have been widely accepted.

1. *The arctic-alpine element.* Species with areas both in the Arctic and on temperate Alps, now widely separated. Formerly it was held that these were arctic species which were driven southwards in the glacial period, and that when the climate moderated, one part re-migrated northwards while another part found a refuge from insupportable competition by retreating to high levels. This, no doubt, is a part of the truth, but the whole picture is probably more complex, and it has been possible to distinguish between arctic species which are genetically alpine in origin and alpine species which are genetically arctic.

2. *Boreal-montane element.* Species whose main areas are sub-arctic or north temperate, but which are found further south in mountainous but not truly alpine habitats.

3. *Pontic element.* This refers to species whose mass-centre lies in a somewhat ill-defined region of south-eastern Europe. It is now generally regarded as the area described by Steffen in 1935. It is held to begin on the northern flank of the Caucasus mountains and to include the north coast of the Black Sea, excluding the Crimea. Northwards and westwards it covers the main steppe area of Romania and Hungary and south Russia to where forests begin. The steppe species are primarily xerophytic and their eastward range towards and into Siberia makes the eastward limit of the pontic element rather indefinite. Some Russian

* For the full story, in detail, see H. GODWIN (1956) *The History of the British Flora.* Cambridge U.P.

writers have taken the Ural mountains as the boundary, others regard the species in the east which range into Siberia as westward migrants, not true pontic species. Kerner used the special term 'pannonic' for the species of the Romanian and Hungarian steppes.

4. *Atlantic element*. Species which are most developed on the western seaboard of Europe, including the British Isles and the Norwegian coast as far as the Arctic Circle. They are conditioned by the humid oceanic climate and do not extend much into central Europe except in special localities. Naturally over so great a range of latitude there are floristic differences between various zones of the seaboard and these may be distinguished as Atlantic—sub-arctic and Atlantic—sub-Mediterranean, divisions northwards and southwards respectively.

5. *Mediterranean element*. The region of the characteristic sclerophyllous Mediterranean vegetation, which includes Spain, Portugal, Italy, the French Riviera, Greece, Albania and the Adriatic coast of Yugoslavia, most of Asia Minor and the Levant with a broad coastal belt of North Africa, also including the Mediterranean islands. This area corresponds to a characteristic climatic province of hot, dry summers and winter rains. Average temperatures are high and frost or snow at low levels are rare.

Similar climates occur in other parts of the world, for example, in Chile, in parts of South Africa and Australia and in California. These areas are marked by a predominant vegetation of hard-leaved shrubs resembling that of the Mediterranean in physiognomy though quite different floristically. The characteristic trees of the Mediterranean are *Olea europaea*, *Quercus ilex* and *Pinus cembra*. Species belonging to the Mediterranean flora are rare north of the Alps but in south-west Ireland and in Cornwall there occurs a small group of species of southern origin, often called the 'Lusitanian element' in the British flora, since Portugal (Lusitania) is the nearest area within the Mediterranean region, from which they appear to have come. (See Vol. III, p. 2592.)

In Britain the first person to take a systematic interest in the distribution of the flora was H. Cottrell Watson who published in 1835 his *Remarks on the Geographical Distribution of British Plants*. In this work he divided British plants into seven geographical types according to their internal distribution in Britain. The types were:

1. *Atlantic type*. Species limited to the west, rare or absent from the east coast.
2. *Germanic type*. Species chiefly found in the east and south-east including the Chalk flora.
3. *English type*. Species only occurring in England, decreasing northwards.
4. *British type*. Widely dispersed over Britain as a whole.
5. *Scottish type*. The opposite of 3, species disappearing southwards.
6. *Highland type*. Montane species of the Scottish Highlands and of the higher mountains in the north of England.

7. *Hebridean type.* A small group of species limited to the extreme north and west, with North American affinities.

The Lusitanian element, mentioned above, belongs to the Atlantic type. While some part of these divisions plainly has a climatic basis, they also hint at differences of origin. Watson therefore considered in detail the range of British species within Britain which he expanded to a major work, the *Cybele Britannica*, Volume 1 of which appeared in 1847, the term 'Cybele' being chosen to designate a work devoted to the distribution of species rather than to their description. He pointed out that a considerable number of British species of the Highland type are also widely dispersed in the Arctic, belonging to what is now recognized as a circumpolar flora. He also points out that, while the proportion of British species in other floras is roughly proportional to the distance between them, latitudinal distance is more effective in point of separation than is longitudinal distance. Thus the proportion of British species in the flora of Sicily (as far as then known) is roughly the same as that in Japan, though the distance of the latter is about six times as great.* He also remarked that no one could understand the flora of Britain unless they had an adequate knowledge of the flora of Europe and at least some knowledge of the floras of North America and of temperate Asia.

Watson based his analysis of the distribution of British species upon 112 divisions of the country, which are either counties or subdivisions of larger counties. These are collectively referred to as the 'vice-counties', and they have been used in almost all subsequent surveys. Good estimated that only 7 per cent of species occurred in all the vice-counties, while at the other extreme 6 per cent occurred in only one.

It is interesting to notice that the category of species classed as 'very rare', which are confined to one vice-county, is more numerous than that of 'rare'. Thus Mathews gives eighty-one species as limited to one vice-county (in which, however, they may be relatively frequent) and only eighteen as occurring in ten vice-counties. What these very limited distributions mean requires individual investigation, there is no general explanation. Some of the 'very rare' species have a considerable distribution outside Britain and are here apparently at their geographical limits.

Edward Forbes, in 1846, took a broader view than Watson, in that he considered the components of the British Flora in relation to their continental affinities. The five components or sub-floras which he recognized resembled those of Watson, but were based rather on their continental than their British distribution. The expansion by British botanists of Watson's work on internal distribution, has gone on continuously and only reached a nearly final stage in the publication, by the Botanical Society of the British Isles, of the *Atlas of the British Flora* in 1962, in which the distribution of every species is mapped by National Grid squares.

---

* This is another way of saying that the areas of distribution of species are often more extensive in an east–west direction than in a north–south direction.

The questions raised by Forbes, however, attracted little attention until the beginning of the present century, when Clement Reid began the controversial discussions on the effect of the glacial period on the British Flora. Reid's doctrine of a 'clean sweep' was strenuously contested and has indeed proved to be erroneous, but it did raise the question of whence came the plants that repopulated the islands and do undoubtedly form the bulk of the present flora. Stapf in 1914 recognized and discussed the southern element in the flora, but J. R. Mathews was the first, in 1923–24, to tackle the whole problem and to analyse the continental connections of the species belonging to the different Watsonian 'types' of British distribution. This led him in 1937 to formulate a general scheme of fifteen floristic components, divided according to the main areas of the component species outside Britain, showing a surprising admixture of geographical origins. At least half of the flora consists of wide-ranging and common plants with no clear geographical affinities and these were excluded from the analysis. The remainder were classified as follows:

1. Mediterranean element, 38 species.
2. Oceanic, southern (south-western Europe), 74 spp.
3. Oceanic, west European (west European, Atlantic coasts), 76 spp. This includes the Lusitanian element previously referred to.
4. Oceanic, northern (north-western Europe), 20 spp., of which 6 have American connections.*
5. Continental, southern (central-south Europe), 127 spp.
6. Continental (central Europe), 82 spp. This includes the 'steppe' element in East Anglia.
7. Continental, northern (central or northern Europe), 91 spp.
8. Northern, montane (northern Europe and also in central European highlands), 25 spp.
9. Arctic, sub-arctic (exclusively northern Europe), 30 spp.
10. Arctic-alpine (circumpolar species reappearing on high mountains), 76 spp.
11. Alpine (European Alps, etc., absent from north and arctic), 9 spp.
12. European, general, 130 spp.
13. Eurasian, 480 spp. Some go as far as Japan.
14. Northern hemisphere general, 205 spp. Widespread, more or less cosmopolitan.
15. Endemic, 20–28 spp. This includes some micro-species and some which are little more than varieties.

In 1933 E. J. Salisbury drew from a study of the East Anglian flora conclusions about the geographical relations of the British flora as a whole, which were similar to those of Mathews, though differently arranged. He chose

* Later (1955) Mathews separated these as an additional North American element.

'components' which were subdivided into 'Elements'. The components were the following:

1. Northern.	5. West-central.
2. Southern.	6. Endemic.
3. Oceanic.	7. Widespread.
4. Continental.	

Salisbury emphasizes that these geographical groups have nothing to do with types of distribution in Britain, which he considers to be chiefly influenced by climate rather than by soils. For example, calcicole species of the Oceanic component are found on the western, oceanic, limestones, not on the chalk which lies in the south and east of the country. Some southern component species appear to be limited by hours of sunshine and consequently are only found in the dry eastern counties.

We have detailed these British studies because Britain is peculiarly well suited for them, due to its insular and peripheral position with regard to Europe and to the comparatively recent immigration of most of its flora.

Wilmott has pointed out that most of the endemic species are found in the unglaciated areas of Britain and suggests that they originated there during the severe glacial climate which prevailed. Some of them show the relict character of apparent inability to spread and have remained limited to the areas in which they were first observed two or three centuries ago.

Domin, in Czechoslovakia, classed as 'dealpines' certain naturally alpine species which took refuge during the Pleistocene on lowland limestones and have remained fixed in the warm spots they then occupied. There are two good examples in Britain, *Helianthemum polifolium* at Brean Down in Somerset and *Draba aizoides* in Gower on the other side of the Bristol Channel. Both are on limestone cliffs facing south, which are just outside the limits of the greatest of the ice-sheets. Each species is plentiful and vigorous within a small area but shows no tendency to spread elsewhere. Wilmott maintained, indeed, that the spreading of species is associated only with the early stages of ecological succession and that in stabilized vegetation spreading is very limited.

The facts of migration inevitably suggest the question, 'What is a native plant?' Especially is this so in the peculiar constitution of the British flora, with its long history of repopulation from Europe. Much discussion has resulted in Britain from the uncertainties involved. Endemic and autochthonous species may be allowed at once to be native, if their status is certain. At the other extreme species consciously and intentionally brought by man may be excluded, but in between these is a vast mass of uncertainties. Many would have unhesitatingly excluded weeds of cultivation, but Quaternary pollen analysis has shown that weeds like *Senecio vulgaris*, which, a generation ago, would have been scorned, were in fact among the earliest post-glacial invaders and are thus among the aristocracy of natives. Since those days invading species have continually arrived and still occasionally do so. Many of them settle down to occupy a permanent place among their predecessors. Is *Mimulus guttatus* which arrived in 1830 and is now completely naturalized to be accepted as 'native'? If not, how far back must the ancestry of a species in this country extend before it is accepted? There is no answer to such a question. Some would exclude all species for which human agency is suspected, but if a seed carried in on a rabbit's fur in the twelfth century may be a 'native', why not a species whose seed clung to the tail of a man's coat in the nineteenth?

We all like to think that we know what we mean by a 'native species', but every attempt to define it results in inconsistencies and incongruities. Having pointed this out we shall not add another inconsistency by attempting a definition ourselves.

When a species is dubbed in a flora, as 'doubtfully native', this implies that there is an opposite category of 'certainly native' and there argument begins, since certainty is an illusory ideal. Unless the title is restricted solely to endemics we ourselves, faced with a mob of species, none of which is truly 'native' in the strict meaning of the term, would favour the most liberal interpretation by regarding any species which maintains and propagates itself in the country by purely natural means, whatever the date and whatever the means of its entry, as being as much a 'native' as any other.

Reference to autochthonous species suggests the query whether, as species have originated within a flora in the past do they still do so. The answer is a qualified affirmative. New autochthonous species may be added to an existing flora, but the conditions governing such an event are very complex and the event itself correspondingly rare. It would be impossible here to enter into any detailed discussion of the conditions for speciation, i.e. the development of new species, a subject on which volumes have been written. One or two relevant points may be made, however.

Small, isolated populations of freely interbreeding individuals tend to become homozygous and inherently stable. Large, inter-breeding (pan-mictic) populations provide greater chances of mutations occurring and, under selection pressure, discontinuities when they arise, provide evolutionary potential. If mutations in such a population have selective value and are favoured by any degree of isolation, they may develop into new species. On the other hand, under specially equable conditions, with low selection pressure, then mutations which are not definitely prejudicial or are selectively neutral may spread rapidly through the whole population. Such may have been the conditions when Angiosperm evolution was apparently most active, in the later Cretaceous.

Even some conspicuous characters of the adult organism may have little or no selective influence, but they may be genetically linked to juvenile or even to embryonic characters which are of high selective importance and they are thus preferentially preserved. In this way quite large differences can be quickly established. Polyploidy or the adoption of apomixis can isolate a biotype at one stroke and in this way many small specific differences or microspecies have come into existence, but their evolutionary potential is, in most cases, slight.

Stability and concomitant uniformity are quite exceptional among species.

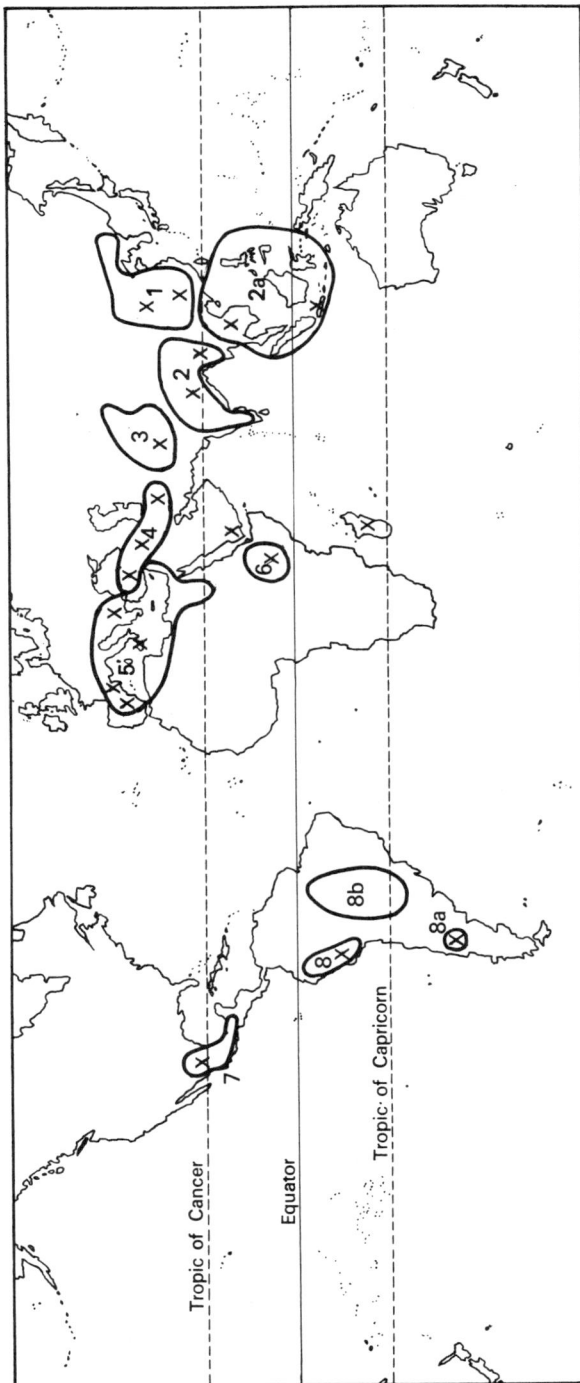

Fig. 4.82. The chief centres of origin of cultivated plants. 1. Chinese. 2. Indian. 2a. Malaysian-Indonesian. 3. Central Asian. 4. Middle Eastern. 5. Mediterranean. 6. Ethiopian. 7. Mexican-Central American. 8. Andean. 8a. Chilean. 8b. Brazilian-Paraguayan. Crosses mark the chief centres of the formation of cultivars. (*After Vavilov. Eng. trans. 1949.*)

In a large population, spread over a wide and varied area, pan-mixis is only a theoretical possibility and in fact a swarm of local populations exist more or less separated from each other. These are most favourable conditions for the appearance and perpetuation of variations, just as two human communities separated by a mountain range will, in the course of a few generations, develop differences of custom, outlook and language which make each seem 'foreign' to the other. The immense survey of economic plants carried out under Vavilov all over the world, showed that there was an almost limitless field of variation, not only among cultivated crops, but also in wild populations of plants which are utilized in some way for their products. In some cases the variations showed geographical relationships; for example, seeds of *Lathyrus* species become larger and paler in colour and the prevailing flower colours change in going from east to west. This, one might say, is the raw material of a genorheitron. Vavilov marked out a number of areas which he considered to be the world centres or origin of cultivated plants, and it is notable that these are nearly all mountainous regions of varied topography which are precisely the kind of areas in which populations tend to break up.

An interesting sideline in Plant Geography is that of statistics about the relative density of species in different regions, or, in other words, the relative richness of floras. Good in 1964 gave a number of interesting tables to illustrate the varying density of species. A coefficient of density is derived by dividing the size of the area, in square miles, into the number of species present, which is equivalent to the proportionate number of species for every square mile of area.

Very large areas tend to have very low ratios. North-western Canada has the lowest of those quoted, 0·004. In smaller areas the coefficients rise: e.g. Korea, 0·025, Austria, 0·07. The smallest areas recorded are mostly islands, where the coefficient may rise well over unity. The richest small area of continental land is probably the Cape Peninsula in South Africa, which has a coefficient of 12·5. Good points out that the higher coefficients for islands give a false impression, as the high figures are not correlated with their insular nature, but rather with their small area. There is a lack of continental floras for similarly small areas which would probably show coefficients as high, since the figures agree throughout that the coefficients increase inversely with size of area.

These area-relative figures are plainly weighted in favour of the small areas. Comparability demands that they should be calculated with reference to a standard unit area. Imagine an area of 10 k(m^2) with 10 species uniformly dispersed. The density would be 1, but the density in any single square kilometre of the area would be 10 although the flora is the same.

Obviously differences of climate have a considerable effect upon the density values. Viewing the world as a whole it appears that the density of species increases with distance from each pole to about latitude 35°. Thereafter it drops in the north and south zones of aridity until the equatorial zone is reached where the density again rises to a peak.

CHAPTER LXI

FLORISTIC PROVINCES

THE growth of knowledge of the distribution of plants soon impressed Plant Geographers with the fact that distribution was not indiscriminate, but that certain families and genera were characteristic of certain geographical regions where they played a conspicuous role in the vegetation and which might be called their kingdoms. They were not necessarily confined to these areas, but outside of them they were scarce and relatively unimportant and they gave place to other families and genera which dominated the flora of other regions. Some few families, such as Compositae and Gramineae, are almost universal, but even among their members certain sections or genera show characteristic distributions.

When a traveller moves across a continent he cannot help noticing the disappearance of some familiar species and the incoming of unfamiliar ones, and the further he goes the more strangers he meets among the plants until he finds himself surrounded by what is, all in all, a different flora. The vegetation may retain familiar aspects, but the species which compose it are not the same. Now between two areas which have substantially different floras there must somewhere be a boundary line even though it be a broad one, more of a zone than a line, unless it happens to coincide with some marked physiographical boundary.

Such reflections gave rise to the idea that it should be possible to draw a map of these floristic differences showing the approximate boundaries between substantially different floras. Thus the notion of 'floristic provinces' was born. The difficulty naturally arises from differences of opinion about what constitutes 'substantially different floras' and no settled agreement exists, nor in fact do the grounds for a settlement. It is possible to compare the floras of two areas numerically, as, for example, by Jaccard's index of similarity (p. 3455).

$$\frac{\text{Number of species (or genera) in common}}{\text{Total number of species (genera)}} \times 100$$

which is really the percentage of species in common. This, however, has been stigmatized as naïve, since it makes no allowance for the differences in the sociological importance of species in their respective areas, but only takes into account the bare fact of their occurrence.

The consequence of this uncertainty is that, while there is a certain measure of agreement respecting certain areas, like the Mediterranean, authors are at variance over the number and the extent of the floristic provinces which they recognize. A map based on floristic differences will be quite different in many respects, from a vegetation map. The latter is based almost entirely on broad physiognomical characters of the vegetation and does not show that the same type of vegetation, e.g. tropical rain forest, in South America and in central Africa, while physiognomically and ecologically similar, are of quite distinct floristic composition. Conversely, a floristic province may include very varied types of vegetation or plant communities.

The floras of many parts of the world are still imperfectly known especially as regards their boundaries. While further exploration may not change fundamentally our ideas of their composition, the present delimitation of boundaries is necessarily tentative and generally follows well-marked climatic or physiographical boundaries, such as mountain chains.

The earliest attempt to form a floristic analysis of the world was that of Schouw in 1823. He laid down twenty-five 'kingdoms', those in the better known parts of the world being subdivided into provinces. His intention was plainly to characterize his divisions by assigning to them certain significant families or genera which were either endemic to the area or were particularly prominent in it. This he was only able to do for about one half of his kingdoms, owing to absence of information at that early date. Thus Kingdom 3, the Mediterranean region, is labelled the Kingdom of Labiatae and Caryophyllaceae; Kingdom 7, India, the Kingdom of Scitamineae; while 17, the Kingdom of the lower levels of the northern Andes, was called after the endemic *Chinchona*. Some, however, like 14, the Kingdom of Tropical Africa, had to be left without any floristic expression.

Grisebach, writing about 1870 was in a period when the idea of evolution was still a novelty and the doctrine of special creations was still active though losing ground. His attitude to Plant Geography was thus somewhat ambivalent. He states his creed quite clearly:* Each natural flora is a separate creation . . . Its boundaries lie where the climate sets a limit to the majority of the endemic plants. The more sharply defined the natural floras are, the less the possibility of the admixture of their productions by migration.'

From such a starting point his conception of floristic regions was primarily one of climatic and physiographic regions and only secondarily floristic. Neither could it include an historical background, though he was aware both of Darwinism and of Lyell's uniformitarian geology. He puts aside the conception of a genetic, i.e. evolutionary, theory of Plant Geography as being possible, but beyond the frontiers of knowledge at that date.

The upshot of his considerations was the delimitation of twenty-four regions, each based on the following considerations, in order: climate, predominant life-forms of plants, types of vegetation, and lastly, vegetational centres, which implied in effect a discussion of the internal and external relationships of the flora. These regions are illustrated on his map (Fig. 4.83). Most of the divisions are continental but the last, 'Oceanic Islands', is an *omnium gatherum* of islands, in which both Madagascar and New Zealand are included, which have nothing in common beyond being islands.

---

* GRISEBACH (1872) *Die Vegetation der Erde*, p. 4.

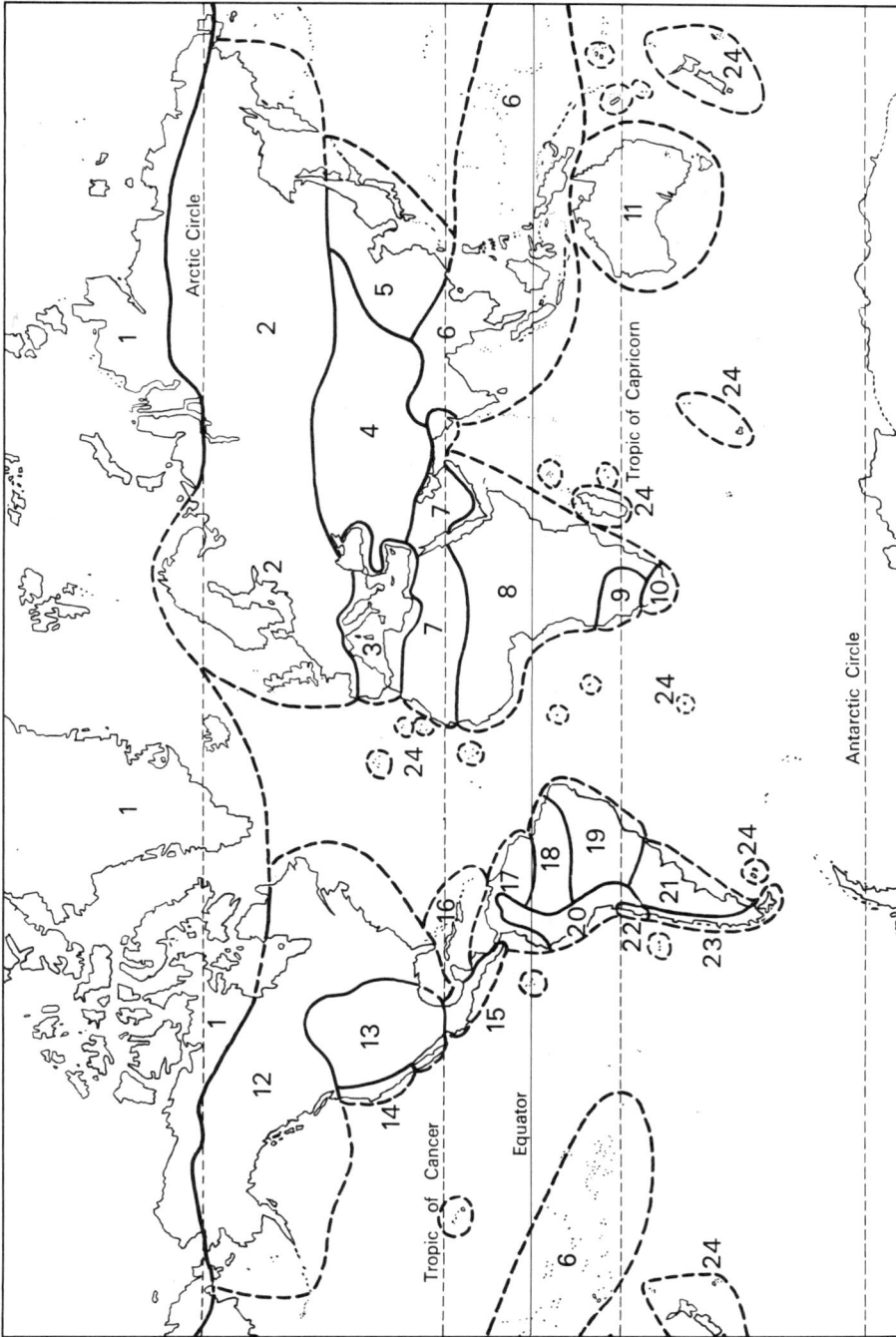

FIG. 4.83. Grisebach's Floristic Regions. 1. Arctic flora. 2. Eastern forest. 3. Mediterranean. 4. Steppe. 5. Sino-Japanese. 6. Indian monsoon region. 7. Sahara. 8. Central African. 9. Kalahari. 10. Cape. 11. Australian. 12. Western Forest. 13. Prairie. 14. Californian. 15. Mexican. 16. West Indian. 17 Equatorial South America. 18. Hylaean. 19. Brazilian. 20. Tropical Andean. 21. Pampas. 22. Chilean Transition region. 23. Antarctic forest. 24. Oceanic islands. (*Grisebach, 1872.*)

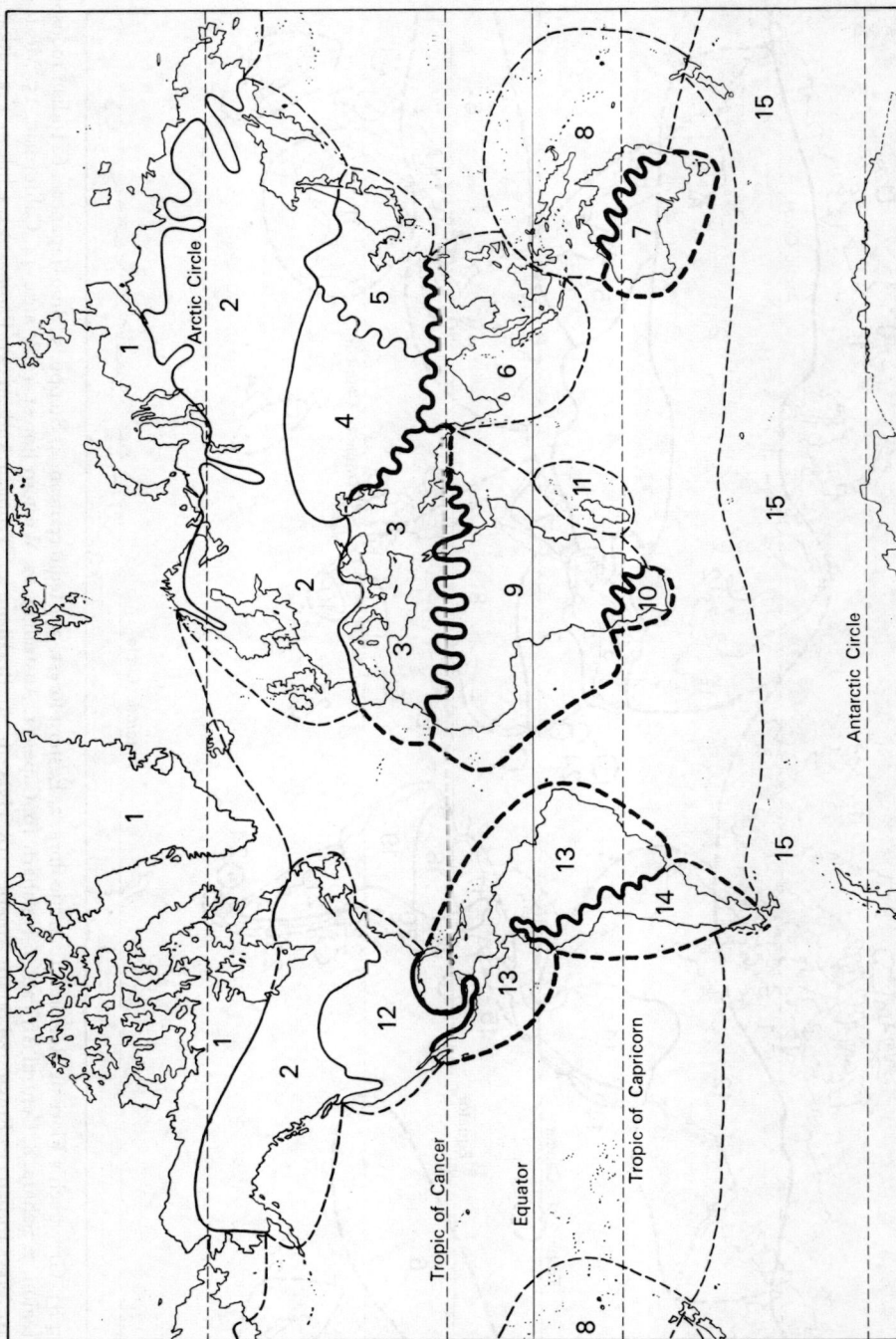

FIG. 4.84. Drude's Floristic regions. 'Those areas which are distinguished by the predominance of characteristic genera belonging to certain definitive orders.' 1. Arctic. 2. Boreal. 3. Mediterranean. 4. Central Asian. 5. Eastern Asian. 6. Indian. 7. Australian. 8. Melanesian–New Zealand. 9. Tropical African. 10. South African. 11. Malagassian. 12. North American. 13. Tropical American. 14. Andean. 15. Antarctic. The more important boundaries in heavy lines. (*Drude, 1890.*)

Drude in 1890 took a more definitely floristic viewpoint, but although his divisions have a floristic basis they are related to climate. He recognized two kinds of division: The 'floral kingdom' and the 'floral region'. He describes the floral kingdoms as areas distinguished by the principal concentrations (*Hauptmasse*) of endemic genera in certain predominant families. These are very broad divisions and though their names are printed across his map their boundaries are left largely undefined. They are as follows: Arctic, Northern, Middle North American, Mediterranean and Oriental (i.e. Pontic), Inner Asiatic, East Asiatic, Indian, Melanesian with New Zealand, Australian, Tropical African, South African, Malagassian, Andean, Tropical American, Antarctic.

So far as they go these are validly distinct floral regions, but they do not go far enough and are not all homogeneous in themselves. Drude therefore recognized twenty-one floral regions as subdivisions based, as he says, on species and the predominance of various genera, to which he assigned boundaries which he called the divisional lines of the land floras (Fig. 4.84). In some areas these divisional lines are represented by broad zones in which a considerable interpenetration of floras occurs and indeed this might apply to nearly all such boundaries.

A more recent effort to define floristic provinces is that of Engler, which appeared about 1912 as an appendix to the *Syllabus der Pflanzenfamilien*. No general statement of principles was offered, but the divisions are set out in considerable detail in a hierarchy of grades. The principal divisions are: (1) Northern extra-tropical and Boreal Floral Kingdom. (2) Palaeotropical Floral Kingdom. (3) Central and South American Floral Kingdom. (4) Australasian Floral Kingdom. (5) Oceanic Floral Kingdom, which deals with marine organisms. These kingdoms were divided into twenty-nine Regions and these again into provinces and sub-provinces, the last being listed in geographical detail in the better known parts of the world. Each region is preceded by a short analysis of its essential characteristics.

This scheme is so clear, definite and detailed that it has afforded a basis for later efforts. The most recent is that of Good,* which takes into account the extension of our knowledge of world floras since Engler's time. The number of provinces recognized has now risen to thirty-seven (see Fig. 4.85). The extent of the relationship between floras and climate can be judged by a comparison of the floristic map with the map of thirty-four climatic zones given in Good.† That there is a general resemblance is evident, but in a number of instances a single climatic zone may be occupied by two or more floristic groups. This may be due to several causes, among which the history of migrations and geological changes are especially important.

Comparison of floristic maps with vegetation maps, such as that of Linton in *The Oxford Atlas*, 1952, is difficult because vegetation maps, based on ecology, are much more detailed, in as much as they recognize the variations brought about by differences of soil, of altitude or of inequalities of rainfall by

---

* In the 3rd edition of his *Geography of the Flowering Plants*, 1964 (page 64).

† *Op. cit.*, p. 80.

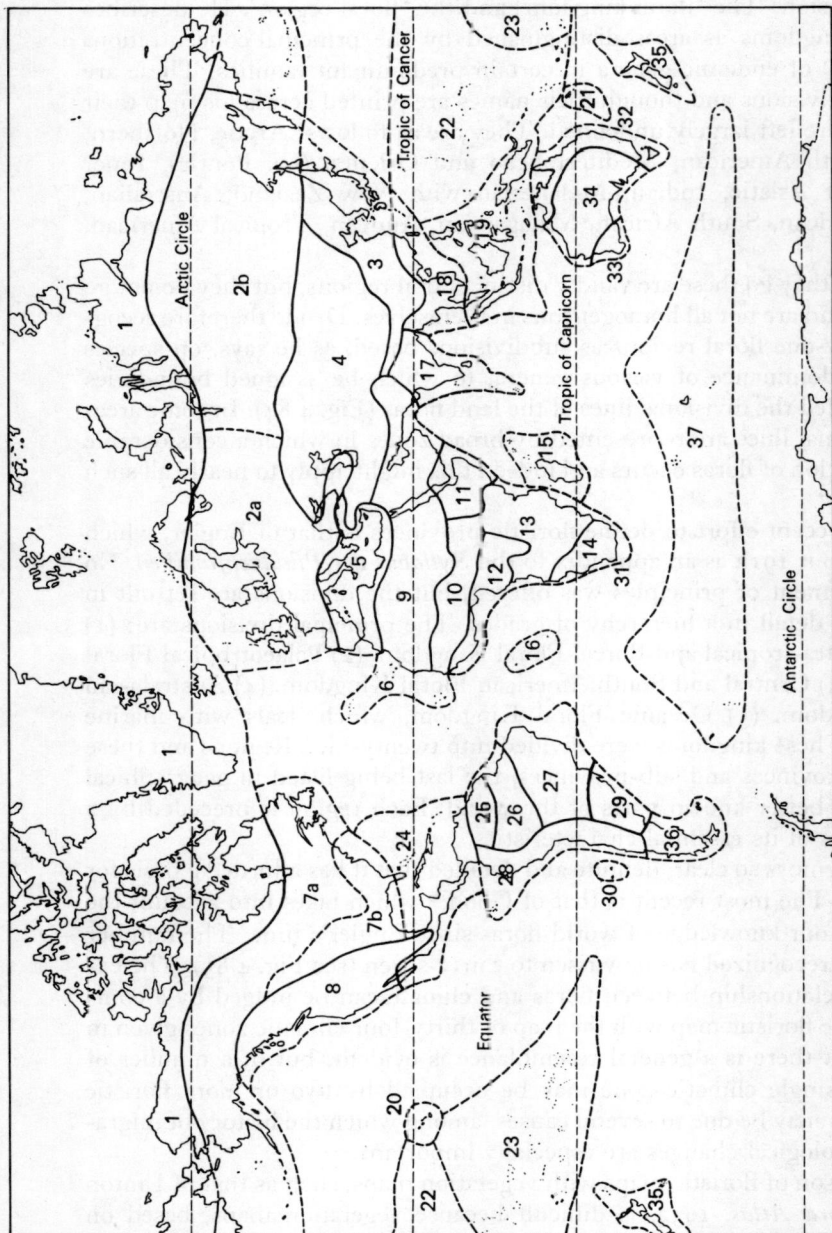

Fig. 4.85. Good's Floristic regions. Transferred to Mercator's projection from the Mollweide original. 1. Arctic. 2(a) and (b). Euro-Siberian. 3. Sino-Japanese. 4. West and Central Asia. 5. Mediterranean. 6. Macaronesia. 7(a) and (b). Atlantic North America. 8. Pacific North America. 9. Afro-Indian desert. 10. Sudanese steppe. 11. North African highlands. 12. West African rain-forest. 13. East African steppe. 14. South Africa. 15. Madagascar. 16. Ascension and St. Helena. 17. India. 18. South East Asia (continental). 19. South East Asia (islands). 20 Hawaiian Islands. 21. New Caledonia. 22. Melanesia and Micronesia. 23. Poly-nesia. 24. Caribbean. 25. Venezuela. 26. Amazon. 27. South Brazil. 28. Andes. 29. Pampas. 30. Juan Fernandez. 31. Cape. 32. North East Australia. 33. South West Australia. 34. Central Australia. 35. New Zealand. 36. Patagonia. 37. Southern Ocean islands. (Good, 1964)

which the flora of a single region is parcelled out, thus disguising the system-atic consistency impressed on the flora by the prevalence of certain families or genera over the area. Nevertheless, there is no essential contradiction between them and it is possible to see that in many areas vegetational and floristic areas have a general correspondence.

It may be asked what relationship the idea of a floristic province bears to the ecological concept of a plant association which also connotes a degree of floristic uniformity at the species level. The former is, of course, a much wider and, at the same time, less precise concept than the latter. As there is argu-ment about the nature and definition of plant associations, there is much more scope for disagreement about floristic provinces. Their areas are to a not in-considerable extent indicated by physiographical and climatic features, since these impose a formative influence upon the regional vegetation, but even vegetational uniformity does not necessarily imply floristic uniformity as evi-denced, for example, by the African rain-forest area. Again, in the enormous boreal-temperate area of Europe and northern Asia there are some character-istic species which range from end to end, but there are many others which are restricted to the eastern or to the western sides. Even the vegetationally uni-form belt of coniferous forest shows floristic diversity between west and east. In spite of an overall uniformity of vegetational types in this great area it is not floristically homogeneous. On the other hand there are many of the larger provinces which are, from an ecological point of view, highly heterogeneous but have been claimed to be floristically homogeneous. One is entitled to ask why. Information on the floristic status of some areas is available in print, but it is scattered and partial. A categorical account of a system of proposed pro-vinces, detailing the floristic considerations on which each province and its boundaries are delimited, is lacking and until such a review is available opinions on the validity, not only of particular provinces but of the whole con-cept of floristic provinces, is scarcely open to argument, for the necessary basis is wanting.

Plant Geography is not necessarily conducted on a global scale. Few of us have the extensive knowledge of Geography and of the world's floras which is requisite, but the same principles can be applied to the flora of a comparatively small area, where local knowledge is valuable. Boyko has illustrated this in his work in Judaea, described on pp. 3507–3509. On his results he formulated his geo-ecological law, which is that the topographical distribution of species, their micro-distribution, reflects their macro- or geographical distribution. This is a useful generalization, but it cannot be applied indiscriminately, since it refers chiefly to the influence of micro-climatic differences on local distribu-tion rather than to edaphic influences.

CHAPTER LXII

THE INFLUENCE OF MAN

THE impact of man on vegetation, extending back to Neolithic times when he first began to cultivate the soil, has been immense and in some long-settled countries truly natural vegetation may be difficult to find, for the whole landscape is man-made. The flora has been altered and wild nature survives precariously. Human adaptability is very great but the cosmopolitan spreading of mankind has forced him in many regions to modify the environment to the limit of his powers as a condition of survival. Much of this influence has been inevitably destructive of natural ecosystems, but it has seldom been wanton destruction. Except perhaps for the Mongol irruptions in the thirteenth century, the destruction was purposive and intended, more or less successfully, to benefit humanity.

The destruction of natural ecosystems has often had widespread effects upon vegetation far beyond the limits of the actual areas of destruction. These effects have largely been negative, either by extinguishing species or limiting their areas of distribution. There is, however, another side of the coin, namely, the positive spread of plants by human action. Crop plants and decorative or medicinal plants have, of course, been deliberately carried far and wide and it has often happened that such plants have reverted to a wild condition in new countries, either as escapes from or as relics of cultivation, merging into the native flora. Much more numerous are the species which are the camp-followers of man, carried by human movement unintentionally or else by taking advantage of man's clearances of the soil or of the special conditions in and around his settlements. This includes all the host of the 'weeds of cultivation', but, apart from weeds properly so-called, areas of waste or neglected land may become populated by a mixture of native plants and introduced weeds, quite unlike any association found in truly wild vegetation and generally dubbed 'anthropogenous'.

Every country has a large number of species which owe their introduction to human agencies and are called 'adventives'. They introduce a new element of competition into the native flora which is sometimes formidable, even disastrous. Many countries, in all parts of the world, can record the appearance of some plant which has become a pest. In some places they may be numerous and aggressive enough to supplant the native flora almost entirely. In the province of Entre Rios in Argentina it is said that introduced species make up three-quarters of the flora. The author remembers his surprise and disappointment on exploring around Punta Arenas on the Straits of Magellan at finding the ground occupied by common European weeds, almost to the exclusion of native plants. Europe has contributed some 500 introductions to

North America and the opposite traffic, though not so great, has included some species, like *Erigeron canadensis*, which have rapidly spread across Europe, as the European *Melilotus alba* has spread along railways across Canada to the Rockies.

It is not surprising that the number of wild species which have thus extended their distribution should greatly outnumber the crop plants which have been similarly successful, in spite of their deliberate introduction and wide dissemination. The plants of world-culture have been so long in cultivation that they have all, by mutation and selection, departed considerably from their wild prototypes, frequently in ways which would handicap them severely in a competitive wild life or would render it impossible, as in species in which sexual reproduction has been practically suppressed, e.g. in the Sugar-cane and the Banana. For these reasons the number of cultivated species for whom escape is possible is small. This applies not only to crop plants but to cultivated plants generally. The records of botanical gardens have repeatedly shown that only a minority of the introduced species grown in the gardens are capable of self-propagation, even when they produce viable seed and the climate is within their ranges of tolerances. Wulff quotes records from the Nikita Botanic Garden in the Crimea which showed that out of thousands of species introduced to the gardens, both woody and herbaceous, only the woody plants survived for many years and only five, all woody, had escaped to become naturalized in the neighbourhood. This limitation controls even the majority of alien weeds accidentally introduced to gardens.

The life of adventive plants which become apparently naturalized is usually precarious and circumscribed. They rarely endure for more than a few years and readily succumb to a severe season and disappear unless, as in some shipping ports, they are repeatedly introduced. The docks at Cardiff had once a rich adventive flora on the ballast heaps brought by ships coming to take back cargoes of coal, but with the cessation of coal export they have almost all gone. Along the banks of the river Tweed in Scotland there is a well-known adventive flora which originated from the washings of imported wool discharged into the river. Here the flora is maintained by frequent reintroduction. A curious case, which shows how fortuitous successful introduction may be, is that of the occurrence of the peculiar liverwort *Riella helicophylla* in the Reddish Canal in Manchester. Brought, most probably, with Egyptian cotton, it would have had no chance of survival had it not happened that the canal is warmed by the hot-water effluent from a neighbouring factory. The uncommon exceptions are the adventives with aggressive powers of spreading; the chances of becoming naturalized are slim for the generality of aliens.

Railways have been extremely effective in aiding the spread of introduced species. Longitudinally they may be carried to the remotest areas, but latitudinally there is practically no effect at all, for the adventive plants remain localized in the special habitat offered to them by the railway track, where they remain, in effect, weeds of open ground. Thus, in England, the Italian species *Senecio squalidus*, which probably escaped from the Oxford Botanic Gardens, remained local and inconspicuous until the building of the (former) Great

Western Railway which rapidly carried it throughout southern and western England, though away from the rail track it is only occasionally found on old walls or waste ground.

The chief crop plants have been in cultivation for so long a time that little trace of their origin remains except in myth and legend but research has been able to supply the missing knowledge to some extent by a combination of geographical, historical, agricultural and genetic data. De Candolle began the study in his book *The Origin of Cultivated Plants* (English translation 1884) and it was vastly extended by N. I. Vavilov between 1923 and 1933. He eventually recognized five regions which have been outstandingly important as centres from which cultivated plants have originated. These are: (1) South-west Asia. (2) South-east Asia. (3) The Mediterranean basin. (4) Ethiopia. (5) Mexico to Peru. These he regarded as primary centres, but he also recognized a larger number of secondary centres, i.e centres of development of crop plants in cultivation, where the greatest concentrations of cultivars exist.* The greater number of these centres is scattered along a sub-tropical belt from the Mediterranean to eastern Asia. It is supposed that here the earliest cultures were established, and it is from them therefore that the principal food crops have been carried northwards and southwards. (Fig. 4.82 p. 3889.)

Among the positive effects of human influence two stand out in importance. One is the transformation of forest into cultivated land, especially grassland, which is so widespread in climatically forested regions that truly primary forest has become uncommon. This requires constant effort to hold back the encroachments of the forest, which quickly reclaims abandoned ground. Indeed, in some parts of Africa, forests which had for long been regarded as primary are now known to have been cleared and cultivated in ancient time by forgotten peoples, although the existing forest showed no sign of its secondary origin.

The other effect is just the opposite of the first, namely, the bringing into cultivation of vast areas of arid land by irrigation. The potential productivity of many semi-desert soils is almost unbelievable until the effects of irrigation are seen, when thin scrub is replaced by rich crops and flourishing orchards. Some of the world's most prosperous areas, like the State of California, would be poor indeed if it were not for the money, labour and skill that have been devoted to great schemes of irrigation which have an hundredfold repaid their cost.

Man's influence upon the Geography of plant distribution is thus seen to be both positive and negative. The positive contribution is in carrying plants with him all over the world, in the cultivation and exploitation of useful plants and by experiments in acclimatization which have extended the areas in which important plants can be grown, e.g. arctic varieties of wheat in Canada and northern Russia. The range of human interests in this respect is vast but it is properly the realm of Economic Botany, which is reserved for another volume of this work.

All this activity, however, may be said to concern plants under human con-

* These secondary centres would be best named as 'centres of genic diversity'.

trol, not wild vegetation. This is not wholly true for there have been widespread repercussions on the wild plants. There have been, as we have pointed out, large numbers of 'hangers on', carried about more or less accidentally. Many of these species have spread themselves widely over new territories and have settled themselves there as new elements in the flora, sometimes even ousting to a large extent the native species. We have mentioned the invasion of South American provinces by European species. To this we may add the aggressive colonization of parts of South Africa by Australian species and the alteration of the Hawaiian flora by newcomers, to such an extent that many native species have disappeared.

Furthermore, both afforestation and disforestation have had profound effects on local climates thus changing indirectly the distribution of many species outside human control.

Man's negative effects on plant distribution have been largely due to promotion of desert conditions. This is an old story and goes back to the earliest attempts to clear forests for agriculture, sometimes successfully and sometimes disastrously, leading only to decreasing rainfall, accelerated drainage and destructive erosion. The deliberate destruction of old irrigation systems for military reasons led to widespread desiccation in the Middle East, and only in the most recent times have efforts been made to remedy the impoverishment thus caused. The destruction of stable ecosystems by the introduction of arable agriculture has often, unfortunately, resulted in dust-bowl conditions and deep gully erosion when the natural grassland has disappeared. The attempt of the early Turkish invaders to turn parts of Asia Minor into grassland for the benefit of their flocks was a move in the opposite direction, but almost equally effective in erasing the natural vegetation.

How can we balance up man's worldwide influence on plant distribution? He has created a multitude of new habitats and spread species which can utilize them. Incidentally this has created conditions for natural hybridization which never existed before. On the other hand he has restricted or destroyed the natural habitats of many other plants. Only the arctic lands have so far escaped his more fundamental efforts, and in them we see a circumpolar distribution of species that has no parallel elsewhere and recalls the wide ranges characteristic of some previous geological epochs.

The total effects, however, have been principally at species level. They have probably had much less influence on the distributional areas of families or even of many genera, which are on the whole wider and older than those of species and depend on factors long antedating the arrival of man.

INDEX

Certain names, in particular those of Braun-Blanquet, Clements and Tansley, which are seminal to the story of Ecology, recur so often in this volume that they have been omitted from this Index rather than submit a confusingly long list of page references.

Adams, 3847
Adaptation,
    direct, 3511
    latent, 3490
Adventives, 3898
Aeon, 3803
Aerenchyma, 3739
Africa,
    rainfall (map), 3833
    vegetation (map), 3834
African mountains, 3824
'Age and Area', 3363, 3841
Aggregation, types of, 3463
Aggressiveness, 3560
Agricultural surveying, 3350
Air, ecological influences, 3555–9
Air flow, laminar and turbulent, 3557
Air mass, 3543
Albedo, 3515, 3540
Allelopathy, 3369
Allen, 3850
Altitude limits of trees, 3511, 3512
Altitudinal zonation, 3822, 3823
Aluminium in soil, 3666
Amann, 3513
Anderson, E., 3491, 3854, 3866
Anderson and Sax, 3571
Andes mountains, 3807
Andosols, 3611
Animals in the biome, 3575
Angiosperms,
    maritime, 3669
    nurseries of, 3800
    origins of, 3881
Anthropogenous vegetation, 3898
Apocrenic acid, 3638
Apomixis, 3572, 3866
Aquatics,
    Carbon assimilation, 3738
    physical conditions, 3702–16
    physiological factors in, 3736–40
    pollination, 3739
Arber, A., 3860
Areas, shapes of, 3843
Aridity, index of (de Martonne), 3590
Armstrong, 3675
Ascherson, P. 3857
Ashby, E., 3436, 3457
Ashton, 3828
Aspect, 3397
    pre-vernal, 3398, 3551
Association, 3334

as super-organism, 3389, 3406
basic units (Hopkins), 3411
characteristic species, 3440
coefficient of, 3409
complexes, 3417
constants, 3439
differential species, 3441
fragments, 3410
minimal area, 3410
stand, 3404
table, 3437, 3475
theories of, 3410 *et seq.*
Associations, 3317
    between species, 3472, 3474
    Braun-Blanquet classification, 3423
    Clements' nomenclature, 3424
    complementary, 3568
    Dansereau's symbols, 3420, 3421
Associes (Clements), 3405, 3436
Asthenosphere, 3803
Atkins, W. R. E., 3757
Atmospheric pressure, variation, 3556
Autodynamism, 3461
Autoecology, 3331, 3337
Autumn turnover, aquatic, 3706
Axelrod, D. J., 3812
Azonal soils, 3593, 3612

Babcock and Stebbins, 3848
Baker, Sarah, 3760
Barriers to dispersal, 3860–3
Basal area, 3478
Base-exchange capacity in soils, 3597, 3598
Base status and pH, 3618
Basic ratio in lake waters, 3726
Baule, 3661
Bauxite, 3607
Beck, von, 3408
Beech woodlands, 3670
Beechwood soils, 3611
Beger, H., 3857
Beijerinck, W., 3784
Belussov, 3509
Bentham and Hooker, 3877
Benthon, 3499, 3700
Bentonite, 3635
Bevor and Farnsworth, 3678
Billings, W. D., 3345, 3361, 3824
Binomial series, 3447, 3465
Bi-polar distributions, 3880
Bioclimatic diagram (Dansereau), 3496
Biocoenosis, 3358, 3575